T0225298

ROUTLEDGE HANDBOOK OF HEALTH GEOGRAPHY

The places of our daily life affect our health, well-being, and receipt of health care in complex ways. The connection between health and place has been acknowledged for centuries, and the contemporary discipline of health geography sets as its core mission to uncover and explicate all facets of this connection.

The *Routledge Handbook of Health Geography* features 52 chapters from leading international thinkers that collectively characterize the breadth and depth of current thinking on the health-place connection. It will be of interest to students seeking an introduction to health geography as well as multidisciplinary health scholars looking to explore the intersection between health and place. This book provides a coherent synthesis of scholarship in health geography as well as multidisciplinary insights into cutting-edge research. It explores the key concepts central to appreciating the ways in which place influences our health, from the micro-space of the body to the macro-scale of entire world regions, in order to articulate historical and contemporary aspects of this influence.

Valorie A. Crooks is a professor in the Department of Geography at Simon Fraser University, in Canada. She currently holds the Canada Research Chair in Health Service Geographies and a Scholar Award from the Michael Smith Foundation for Health Research. She focuses on identifying ethical and equity issues associated with transnational health care mobilities.

Gavin J. Andrews is a professor in the Department of Health, Aging and Society at McMaster University, in Canada. His books include *Soundscapes of Wellbeing in Popular Music* (2014), *Health Geographies: A Critical Introduction* (2018), *Geographical Gerontology* (2018), and *Non-Representational Theory and Health* (2018).

Jamie Pearce is a professor of health geography at the University of Edinburgh, in the United Kingdom, where he is co-director of the Centre for Research on Environment Society and Health (CRESH) and director of the Scottish Graduate School for Social Science. His research seeks to understand various social, political and environmental mechanisms in health and health-related behaviour.

ROUTLEDGE HANDBOOK OF HEALTH GEOGRAPHY

Edited by Valorie A. Crooks, Gavin J. Andrews and Jamie Pearce

LONDON AND NEW YORK

First published 2018
by Routledge
2 Park Square, Milton Park, Abingdon, Oxon OX14 4RN

and by Routledge
52 Vanderbilt Avenue, New York, NY 10017

First issued in paperback 2020

Routledge is an imprint of the Taylor & Francis Group, an informa business

© 2018 selection and editorial matter, Valorie A. Crooks, Gavin J. Andrews and Jamie Pearce; individual chapters, the contributors

The right of Valorie A. Crooks, Gavin J. Andrews and Jamie Pearce to be identified as the authors of the editorial material, and of the authors for their individual chapters, has been asserted in accordance with sections 77 and 78 of the Copyright, Designs and Patents Act 1988.

All rights reserved. No part of this book may be reprinted or reproduced or utilised in any form or by any electronic, mechanical, or other means, now known or hereafter invented, including photocopying and recording, or in any information storage or retrieval system, without permission in writing from the publishers.

Trademark notice: Product or corporate names may be trademarks or registered trademarks, and are used only for identification and explanation without intent to infringe.

British Library Cataloguing-in-Publication Data
A catalogue record for this book is available from the British Library

Library of Congress Cataloging-in-Publication Data
A catalog record has been requested for this book

ISBN 13: 978−0−367−65990−5 (pbk)
ISBN 13: 978−1−138−09804−6 (hbk)

Typeset in Bembo
by Apex CoVantage, LLC

CONTENTS

Contents

Contents

FIGURES

TABLES

CONTRIBUTORS

Gavin J. Andrews is a professor in the Department of Health, Aging and Society at McMaster University, in Canada. His empirical interests include the dynamics between space/place and all the following: aging, health-care work, holistic medicine, health histories, fitness cultures and music. He is particularly interested in theory and progress in health geography, his most recent books being *Soundscapes of Wellbeing in Popular Music* (Ashgate, 2014), *Health Geographies: A Critical Introduction* (Wiley-Blackwell, 2018), *Geographical Gerontology* (Routledge, 2018), and *Non-Representational Theory and Health* (Routledge, 2018).

Tom Baker is a human-geography lecturer at the University of Auckland, in New Zealand. His research examines the making and mobility of policy. His previous and current research focuses on homelessness, housing, social security, drug treatment and other social issues.

Clare Bambra is a professor of public health at Newcastle University's Institute of Health and Society. Her research examines the political, social and economic determinants of health inequalities as well as how social policies and interventions can promote health equity. She has published extensively, including *How Politics Makes Us Sick: Neoliberal Epidemics* (Palgrave Macmillan, 2015) and *Health Divides: Where You Live Can Kill You* (Policy Press, 2016).

Caroline Barakat-Haddad (PhD) is an assistant professor of environmental and occupational health at the University of Ontario's Institute of Technology. She was trained as a health geographer. Her research interests include the links between environments and health, child and adolescent health, spatial patterns of health and health care, environmental epidemiology, and health inequities.

Jamie Baxter is a professor in the Department of Geography at the University of Western Ontario. His research interests include social science methodology, geography of health, social construction of risks from technological hazards, environmental justice and noxious facility siting. He has studied the impact of waste facilities, urban pesticides, and wind turbines on communities. He has produced more than 50 refereed publications involving qualitative and mixed methods designs and has taught qualitative methods for over two decades.

Katie Big-Canoe is Anishinaabe from Chippewas of Georgina Island First Nation. She is currently the research coordinator in the Indigenous Health Lab at Western University, where she completed a master's in health geography in 2011. Since that time, she has worked in several applied community-based research contexts.

Elijah Bisung is an assistant professor in the School of Kinesiology and Health Studies at Queen's University. His research focuses on social and environmental production of health and well-being, with a focus on issues of food and water insecurity, environmental collective action, and environmental justice.

Candice P. Boyd is an artist-geographer and Honorary Senior Fellow in the School of Geography at the University of Melbourne, in Australia. Her interests are in the geographies of mental health, cultures of sense and movement, therapeutic spaces, rurality, and contemporary museum geographies. She is the author of *Non-Representational Geographies of Therapeutic Art Making* (Palgrave Macmillan, 2017).

Tim Brown is a senior lecturer in the School of Geography at Queen Mary University of London. He is the coauthor or coeditor of four books, including *Health Geographies: A Critical Introduction* (Wiley-Blackwell, 2017), and has published widely on the geographies of global health. He is currently engaged in an interdisciplinary project examining malnutrition interventions in Zimbabwe and in research into the emergence of the WHO Healthy Cities agenda in the mid-20th century.

Vera Chouinard is a professor of geography at McMaster University. She has dealt with disability and chronic illness throughout her career and has written about her struggles for accommodation and inclusion in academia. She has written extensively on disability issues, including the legal rights of disabled persons in Canada, disabled women's activism, disabled women and housing, lived experiences of mental illness and the lives of disabled people in the Global South, specifically Guyana.

Jonathan Cinnamon has a PhD from Simon Fraser University and is currently a lecturer in human geography at the University of Exeter. His broad research interests are in geographic information science and digital health geographies. He is currently examining how novel geospatial technologies, data, and actors are being enrolled in public health surveillance and promotion, including the practical, social, and ethical implications of these sociotechnical developments.

Tara Coleman is a professional teaching fellow in the social sciences and a researcher in the School of Environment at the University of Auckland. Her research has considered aging in place, interdisciplinary approaches to health and well-being, geographies of education, phenomenology, health policy, and qualitative research methods.

Damian Collins is an associate professor of human geography at the University of Alberta. His current research focuses on human rights in homelessness policy, as well as how institutional and post-institutional spaces can promote well-being. He is one of the authors of "Parallel Worlds? French and Anglophone Perspectives on Health Geography," published in *Social Science and Medicine* (2016).

Eric Crighton (PhD) is a professor at the University of Ottawa. An award-winning educator, he has more than 15 years of experience in environmental health research, education and advocacy. His research is focused on better understanding environmental and geographic relationships to human health – in particular, children's health.

Valorie A. Crooks is a professor in the Department of Geography at Simon Fraser University, in Canada. She currently holds the Canada Research Chair in Health Service Geographies and a Scholar Award from the Michael Smith Foundation for Health Research. Much of her research is focused on identifying ethical and equity issues associated with transnational health-care mobilities (e.g., medical tourism, offshore medical schools).

Sarah Curtis (FBA, FAcSS, FRGS, DPhil) is a professor emeritus at Durham University, UK, where she was previously professor of health and risk and directed the Institute of Hazard, Risk and Resilience. She is also a visiting researcher at Edinburgh University. Her extensive publications include *Space, Place and Mental Health* (Ashgate, 2010); *Health and Inequality: Geographical Perspectives* (Sage Publications, 2004); Health and Inequality: Geographical Perspectives (Sage, 2004); and *Tipping Points: Modelling Social Problems and Health* (Wiley, 2015). Her knowledge exchange and consultancy has impacted public health agencies nationally and internationally.

Frances Darlington-Pollock is a lecturer in human geography at the University of Liverpool. Her research focuses on ethnic inequality, health inequality, migration, housing and place. She primarily uses quantitative methods applied to a mixture of cross-sectional and longitudinal data. Her work covers the UK and New Zealand.

Alec Davies, PhD Researcher, Geographic Data Science Lab, Department of Geography & Planning, University of Liverpool, Liverpool, UK. Research interests: applications of big data for health; use of consumer data for supplementing existing health data; generation of open-source health indicators. Key publications: Developing indicators for measuring health-related features of neighbourhoods. In: P. Longley, J. Cheshire and A. Singleton, eds., *Consumer data analytics* (UCL Press, 2017).

Jenna Dixon is a postdoctoral fellow in the Department of Geography and Environmental Management at the University of Waterloo. Her research interests center around four streams of health inequalities, health and development, gender and health and knowledge translation. Together, these streams intersect toward research with a critical focus on questions of social justice.

Cameron Duff is the Vice Chancellor's Senior Research Fellow at RMIT University in Melbourne, in Australia. His research explores the role of social innovation in responding to complex health and social problems in urban settings. His first book, *Assemblages of Health: Deleuze's Empiricism and the Ethology of Life*, was published in 2014 by Springer.

Michelle Duffy is an associate professor of human geography at the University of Newcastle. Through the lens of the arts, and more specifically that of sound, music and performance, her research explores how our private and public selves are articulated and/or challenged in public spaces and events.

Craig Duncan is a senior research fellow in the Department of Geography at the University of Portsmouth, in England. He is currently working on a two-year national project examining the use and effectiveness of Community Treatment Orders. His previous research has included highly cited empirical work on local and regional variations in smoking and drinking behaviors, as well as more theoretical work developing a critical understanding of the geographies of public health.

Michael Emch (PhD) is the W. R. Kenan, Jr. Distinguished Professor of Geography and Epidemiology at the University of North Carolina at Chapel Hill (UNC). He also chairs UNC's Department of Geography and is a fellow of the Carolina Population Center. He has published widely in the subfield of disease ecology, mostly of infectious diseases of the developing world. He directs the Spatial Health Research Group (spatialhealth.web.unc.edu).

Joshua Evans is an assistant professor of human geography at the University of Alberta. He studies geographies of exclusion and inclusion. His most recent research focuses on spaces of care, home, and work and their role in shaping the lived experiences of socially marginalized and vulnerable individuals, as well as spaces of policy development and implementation and their role in the creation of healthy, enabling and equitable urban environments.

Daniel J. Exeter is a quantitative health geographer at the University of Auckland. His research focuses on identifying solutions to health disparities due to socioeconomic position, area deprivation or ethnicity through the use of geographic information systems, statistical modeling and the secondary analysis of large datasets.

Jessica M. Finlay is a doctoral fellow in geography and gerontology at the University of Minnesota. As an interdisciplinary health geographer and environmental gerontologist, Jessica applies mixed methods to investigate person-place dynamics that influence health and well-being in later life. Jessica's publications include "Therapeutic Landscapes and Well-Being in Later Life: Impacts of Blue and Green Spaces for Older Adults" (2015) and "Geographies on the Move: A Practical and Theoretical Approach to the Mobile Interview" (2017).

Sebastien Fleuret is research director for the French National Center for Scientific Research (CNRS), as health geographer. He is also head of the research unit for social geography (ESO) at the University of Angers. He published various articles and books on the local organization of health-care services and developed the idea of construction locale de la santé (locally constructed health conditions). He is editor of the francophone journal *Revue Francophone sur la Santé et les Territoires*.

Ronan Foley is a senior lecturer in the Department of Geography at Maynooth University, in Ireland. He teaches a range of modules on geographies of health and GIS and has a particular interest in therapeutic landscapes and healthy blue space. He is the author of *Healing Waters: Therapeutic Landscapes in Historic and Contemporary Ireland* (Ashgate, 2010) and edited (with Thomas Kistemann) a special issue on healthy blue space for *Health & Place* in 2015.

Melissa Giesbrecht (PhD) is a research collaborator in the Department of Geography at Simon Fraser University and the Institute for Aging and Lifelong Health at the University of Victoria, in Canada. Her research interests include critically examining diversity within the context of formal and informal palliative caregiving.

Holly Gordon (BA) has an honors degree in environmental studies and geography and is a research assistant in the Department of Geography, Environment and Geomatics at the University of Ottawa. She has a strong interest in environmental issues in the Canadian context, particularly as they relate to First Nations health and well-being.

Mark A. Green, Geographic Data Science Lab, Department of Geography & Planning, Lecturer in Health Geography, University of Liverpool, Liverpool, UK. Research interests: understanding the social and environmental drivers of obesity; exploring geographical inequalities in mortality; generation of open-source health indicators. Key publications: The geography of a rapid rise in elderly mortality in England and Wales, 2014–2015 (*Health & Place*, 2017) and Population trends in the distribution of body mass index in England, 1992–2013 (*Journal of Epidemiology and Community Health*, 2016).

Neil Hanlon is a professor of geography at the University of Northern British Columbia, in Canada. His research interests include rural health-care delivery, community adaptations to social and economic change, and regionalized health governance. His work appears in journals such as *Health & Place* and *Social Science & Medicine*, and he edited (with Mark Skinner) a book about the role of voluntary sector activities in adapting resource-dependent places to population aging *Ageing Resource Communities: New Frontiers of Rural Population Change, Community Development and Voluntarism* (Routledge, 2016).

Rachel Herron is an assistant professor in the Department of Geography at Brandon University, where she specializes in health geography, rural mental health, and aging. Her current research examines the vulnerability and complexity of care relationships, social inclusion and meaningful engagement for people with dementia, and the diversity of lived experiences of rural mental health.

Jana A. Hirsch is an interdisciplinary researcher whose work examines the way neighborhoods shape health behavior and health disparities. She works at the intersections of urban planning, geography, and epidemiology with a focus on built environments and physical activity. She is an assistant research professor in the Urban Health Collaborative, Department of Epidemiology and Biostatistics, Dornsife School of Public Health, Drexel University. Her work includes longitudinal examinations of neighborhood change and subsequent impacts on health. She is an eager, active walker and a people-focused urbanophile.

Vivienne Ivory is a social scientist specializing in place, mobility and infrastructure. She has a public-health research background, specializing in the relationships between where we live and our health and well-being. Her research interests include resilience to disasters, health inequalities, child and family well-being, and transport and neighborhood mobility. Her work calls on multiple quantitative and qualitative data collection and analysis methods, population-level thinking about health and well-being, and theoretical concepts of place and health.

Robin Kearns is a professor of geography in the School of Environment at the University of Auckland. He has interests in the nature of place in health services and patient experience. His recent books include *The Afterlives of the Psychiatric Asylum* (Ashgate, 2015) and the coedited *Children's Health and Well-Being in Urban Environments* (Routledge, 2017). He has published more than 200 papers and book chapters on topics across social and cultural geography. He edited the health section of *AAG's International Encyclopedia of Geography* and is an editor of the journal *Health & Place*.

Corinna Keeler is a doctoral student in the Department of Geography at the University of North Carolina at Chapel Hill and a trainee at the Carolina Population Center. Her research focuses on the relationship between mosquito-borne diseases and spatial patterns of environmental and demographic change.

Jennifer Lea is a lecturer in human geography at the University of Exeter. She is interested in geographies of bodies and embodiment, particularly in relation to producing, consuming and experiencing health and disability. Empirically she has carried out research on yoga, meditation and therapeutic massage, giving rise to a number of publications (e.g., in the journals *Cultural Geographies* and *Body and Society*). She is currently doing work on geographies of motherhood, maternal mental health and recovery.

Daniel Lewis is a quantitative health geographer. He holds a Medical Research Council (MRC, UK) skills-development fellowship at the London School of Hygiene and Tropical Medicine (LSHTM, UK). His background is in human geography and geographic information science, and he is interested in how urban systems shape and constrain population health across scales and over time. He is one of the authors of *Health Geographies: A Critical Introduction* (Wiley-Blackwell, 2017).

Nathaniel M. Lewis is a lecturer in human geography at the University of Southampton. He is currently interested in life-course approaches to LGBTQ health, sexuality and well-being in the workplace, as well as global health mobilities related to HIV/AIDS. His work is published in *Health & Place*, *Social Science &*

Medicine, the *Annals of the Association of American Geographers*, and other journals in both geography and the health sciences.

Joan Liaschenko (PhD, RN, FAAN) introduced the significance of the geographical concepts of space, place and moral geographies to nursing 25 years ago in *The Moral Geography of Home Care*. In subsequent publications, she demonstrated the morality embedded in spatial relationships between nurses and patients and in nursing practice. Her work uses geographical concepts and is feminist in orientation. She has analyzed the role of spatiality in moral distress. She is a professor in the Center for Bioethics and the School of Nursing at the University of Minnesota.

Isaac Luginaah is a professor in the Department of Geography at Western University. His broad research interests include environment and health, population health and GIS applications in health. This work involves an integrative understanding of the broad determinants of population health and the evidence of environment and health linkages in both Canadian and African contexts.

Sarah M. Mah is a doctoral student in the Department of Geography at McGill University. Grounded by her previous front-line crisis work, her research interests center on assessing how health-care burden is shaped by the neighborhoods in which we live. Currently, she uses linked data to examine neighborhood walkability as a population health intervention. She is also trained in epidemiology and biology and has contributed to research in cancer genetics and epigenetics.

Sara McLafferty is a professor of geography and GIScience at the University of Illinois. Her research investigates place-based inequalities in health and access to health services and use of GIS and spatial analysis methods in exploring these inequalities. Her books include *GIS and Public Health* (with Ellen Cromley, Guilford Press, 2012) and *A Companion to Health and Medical Geography* (with Tim Brown and Graham Moon, Wiley-Blackwell, 2009).

Christine Milligan is a professor of health and social geography and the director of the Lancaster University Centre for Ageing Research, in the UK. Her research interests focus around innovations to support active aging and home-based care and support for older people – including new technologies to support aging in place. She has published more than 100 books, refereed journal articles and book chapters around these topics. She is associate editor of the journal *Health & Place*.

Graham Moon is a professor of spatial analysis in Human Geography at the University of Southampton. He conducts research on smoking, dietary and drinking behavior, psychiatric epidemiology, small area estimation, access to general practice and post-asylum geographies. Author of numerous publications, he collaborated on the formulation of the new health geography and the introduction of multilevel modeling to geography. He was founding editor of *Health & Place* and is an elected fellow of the UK Academy of Social Sciences.

Nina J. Morris is a lecturer in human geography in the School of GeoSciences at the University of Edinburgh with research interests in the cultural geographies of landscapes and embodiment; sensory perception and the creative arts; the links between green space, health and well-being; and research ethics. She chaired the School of GeoSciences Ethics and Research Integrity Committee from 2009 to 2016 and was a member of the University of Edinburgh Research Ethics and Integrity Review Panel from 2015 to 2017.

K. Bruce Newbold is a professor of geography at McMaster University, in Hamilton, Canada. He has held guest positions at the University of Glasgow and the University of California at San Diego. Trained as a population geographer, he is interested in migration, immigration, health, and human capital. He is the author or coauthor of more than 100 academic articles, as well as *Population Geography: Tools and Issues* (Rowman & Littlefield, 2017).

Paul Norman is an applied demographer and health geographer whose work includes the harmonization of small-area sociodemographic, morbidity and mortality data to enable time-series analysis of demographic deprivation and health change; use of area typologies to understand migration-health outcomes; and development of local-level population-projection methods.

Meghann Ormond is an assistant professor of cultural geography at Wageningen University and Research, in the Netherlands. A health geographer, she focuses on the intersections of various forms of transnational mobility, health and care. She is the author of *Neoliberal Governance and International Medical Travel in Malaysia* (Routledge, 2013).

Katie J. Oven is a research associate with leading responsibility for work packages in major research projects at Durham University, in the UK. A geographer working at the interface between physical and social science, she is interested in disasters and development in the Global South – in particular, the social production of vulnerability and resilience to natural hazards. She is engaged in applied research and works closely with a number of government and non-government partners within the UK and internationally.

Jamie Pearce is a professor of health geography at the University of Edinburgh, in the UK, where he is co-director of the Centre for Research on Environment Society and Health (CRESH) and director of the Scottish Graduate School for Social Science. His research seeks to understand various social, political and environmental mechanisms operating at a range of geographical scales that establish and perpetuate spatial inequalities in health and health-related behavior.

Amber L. Pearson is an assistant professor in the Department of Geography, Environment & Spatial Sciences at Michigan State University. She is a health geographer with a focus on social justice and understanding the unexpected tenacity of the underprivileged, while paying careful attention to the structural and social factors that led to disadvantage in the first place.

Elizabeth Peter (RN, PhD) is an associate professor at the Lawrence S. Bloomberg Faculty of Nursing and a member of the Joint Centre for Bioethics and the Centre for Critical Qualitative Health Research at the University of Toronto, Canada. Her scholarship reflects her interdisciplinary background in nursing, philosophy and bioethics. She has used geographical concepts to examine the unique ethical concerns that arise in home-care services delivery and has explored the spatial dimensions of moral distress that arise as a result of nurses' proximity to patients.

Andrew Power, an associate professor of geography at University of Southampton, has research interests in disability and caregiving, with a recent focus on community-led responses to declining statutory social care. He has published widely in this field, including the recent monograph *Active Citizenship and Disability* (Cambridge University Press, 2013).

Benet Reid is a health geographer and sometimes sociologist at St. Andrews University, currently researching international health volunteering. He is interested in research and teaching on health inequalities, social theory and the politics of health-care work.

Chantelle A. M. Richmond (Bigitigong Anishinabe, Bear clan) is an associate professor in the Department of Geography at Western University in London, Ontario (Canada), where she directs the Interdisciplinary Development Initiative in Applied Indigenous Scholarship. Her participatory research program focuses on the intersection of health, environment and Indigenous knowledge.

Mylene Riva is an assistant professor at McGill University, jointly appointed to the Institute for Health and Social Policy and Department of Geography. Her scholarly work contributes to understanding the socio-environmental determinants of health in urban, rural, Inuit and First Nations communities. Her research activities focus on housing and communities as settings for interventions to improve population health and to reduce inequities. She holds a career award from the Fonds de recherche du Québec – Santé.

Charlene Ronquillo is a PhD trainee at the University of British Columbia's School of Nursing and an associate research fellow in implementation science at PenCLAHRC South West (Collaboration for Leadership in Applied Health Research and Care). Her research interests include nursing and health informatics, mobile and connected health technologies, and the implementation of health-information technologies in health systems. She has a clinical background in nursing and is a registered nurse in Canada and the United States.

Richard C. Sadler is an assistant professor in the Department of Family Medicine and the Division of Public Health at Michigan State University in Flint, Michigan. His work is rooted in the establishment and nurturing of community partnerships and often emphasizes the novel use of GIS. The overarching aim of his work is to strengthen understanding between the built environment and health behaviors/outcomes to shape land-use policy for healthier cities.

Meryn Severson is a recent BA graduate of the University of Alberta, with a double major in human geography and sociology. With particular interests in housing and the life course, she recently completed her undergraduate thesis on the housing choices of Canadian young adults.

Nichola Shackleton is a research fellow at the University of Auckland. She is a social scientist specializing in quantitative research methods and social statistics. Her research interests include measuring and reducing inequalities in child and adolescent health, longitudinal data analysis, and latent variable modeling.

Mark Skinner is a professor of geography at Trent University, where he holds the Canada Research Chair in Rural Aging, Health and Social Care, and is founding director of the Trent Centre for Aging & Society. His research examines how rural people and places are responding to the challenges and opportunities of population aging, particularly the evolving role of the voluntary sector and volunteers in supporting older people and sustaining aging communities.

Dianna Smith is a lecturer in geographic information systems at the University of Southampton, where she conducts research on small-area estimation methods to model health outcomes (food insecurity, diabetes, obesity, common mental disorders) and behaviors (alcohol consumption, smoking, diet). Her work has been funded by the MRC and UK health agencies.

Marcie Snyder (PhD) is a health geographer interested in interdisciplinary and community-based approaches to understanding the relationship between health and mobility and the social, economic and political contexts within which this relationship occurs. Her work is primarily concerned with Indigenous health and well-being; holistic health; and equitable health care and service access, particularly among mobile and migrating populations.

Matt Sothern is a cultural geographer at St. Andrews University with interests in the politics of health, identity and governance.

Stephen Taylor is a lecturer in human geography at Queen Mary University of London. His research focuses on the geographies of global health and biomedicine. He has conducted research for the World Health Organization on the eradication of polio in Pakistan and Nigeria and has also worked at the Rockefeller and Bill & Melinda Gates Foundations on understanding the place of philanthropy in contemporary efforts to ensure global food security.

Mika Toyota is a professor at Rikkyo University's College of Tourism, in Japan. Trained in social anthropology, sociology and development studies, she focuses on ethnicity, migration and tourism in Southeast Asia. Her work has been published in *Global Networks*, *International Development Planning Review* and *International Journal of Sociology and Social Policy*.

Liz Twigg is a professor of human geography at the University of Portsmouth, in England. Over the last 25 years, her work has focused on using quantitative multilevel modeling techniques to investigate the links between individual health outcomes, health-risk behaviors (e.g., smoking and alcohol consumption) and place. She has worked on several national studies to understand geographies of common mental disorders, and her recent work focuses on more severe outcomes resulting in compulsory hospital or community treatment.

Sarah Wakefield is an associate professor in the Department of Geography and Planning and directs the Health Studies Program at the University of Toronto. Her research focuses on food policy and practice and on improving neighborhood health through participatory community development. These areas are connected by an interest in understanding how individuals and organizations work together to create just, healthy, sustainable communities. She works closely with community organizations and health-policy actors.

Michael J. Widener is an assistant professor in the Department of Geography and Planning at the University of Toronto–St. George. He is a transportation and health geographer and has published on the role of transportation in accessing food, emergency medical transport, oral health and related topics.

Janine Wiles is an associate professor of population health at the University of Auckland, in New Zealand. A geographer and gerontologist, she uses a critical positive-aging framework. Her research encompasses three disciplinary areas: social/health geographies, critical social gerontologies, and community health, linking the themes of care, place and aging.

Allison M. Williams is trained as a social scientist, specializing in caregiving and critical policy/program evaluation. She is a professor at McMaster University (in Hamilton, Ontario) and has held previous academic appointments at the University of Saskatchewan (in Saskatoon, Saskatchewan) and Brock University (in St. Catharine's, Ontario). In addition to teaching courses and supervising graduate students, she has an active program of research addressing caregiver-friendly policies and program interventions. She is on Facebook (www.facebook.com/ghwchair) and Twitter (@ghwchair), and her website is http://ghw.mcmaster.ca.

Meghan Winters is a population health researcher leading a research program on the linkages between transportation and city design. She works in close collaboration with cities and stakeholders to conduct research and create tools that address real-world challenges. She is an associate professor in the Faculty of

Health Sciences at Simon Fraser University, in Burnaby, Canada. Personally, she aims to spend as much time outside as possible and is equally happy to be wandering urban streets as she is to be immersed in nature.

Karen Witten is a psychologist and geographer interested in relationships between neighborhood places and the social relations and health-related behaviors of residents. Her current research includes investigating children's use and experiences of urban neighborhoods, enablers and barriers to community participation for disabled young people, community formation in new housing developments and the social and transport impacts of investment in walking and cycling infrastructure. She is a professor of public health at Massey University, in Auckland, New Zealand.

Sandy Wong is a recent PhD graduate from the Department of Geography & Geographic Information Science at the University of Illinois at Urbana-Champaign (in the United States), soon to begin a postdoctoral fellowship at the Icahn School of Medicine at Mount Sinai. Her research interests include health inequalities and applications of geospatial, statistical and qualitative techniques to the study of health issues. She has published in journals such as *Social Science & Medicine* and *The Professional Geographer*.

Nicole M. **Yantzi** is an associate professor in the School of the Environment at Laurentian University and the director of the Evaluating Children's Health Outcomes research center. Her research, teaching and advocacy work focuses on ensuring the accessibility and inclusiveness of children's environments. She advocates for the inclusion of all children and youth as active research participants and valuing their ideas for change. Her work has been published in *Children's Geographies*, *Social Science & Medicine*, and *Disability & Rehabilitation*.

ACKNOWLEDGMENTS

We would be remiss in not taking a moment to acknowledge those who assisted with this sizeable undertaking. Dr. Melissa Giesbrecht greatly helped us with coordinating this book project as an editorial assistant. We are delighted that she has contributed a chapter on her own work on incorporating diversity perspectives into health geography research, which is Chapter 49. Dr. Marcie Snyder also provided much needed help as an editorial assistant, which included the painstaking work of reviewing the entire volume for formatting and language consistency. She joins us as a contributing co-author in Chapter 1. Some funding to cover the editorial support for developing and assembling this book was provided through Dr. Valorie Crooks' Canada Research Chair in Health Service Geographies and Scholar Award from the Michael Smith Foundation for Health Research. Finally, but certainly not least, we are incredibly indebted to all those who contributed chapters and who recommended other potential authors. Without their labor and enthusiasm, this book would never have come to fruition.

1

INTRODUCING THE *ROUTLEDGE HANDBOOK OF HEALTH GEOGRAPHY*

Valorie A. Crooks, Gavin J. Andrews, Jamie Pearce and Marcie Snyder

Health geography is a vibrant, engaged, and methodologically diverse sub-discipline of human geography that explores all aspects of the relationship(s) between health and place. Research by health geographers is not only contributing to wider disciplinary debates in geography, but also proving increasingly influential across other social science and health disciplines and in the development of local, national, and global public-health policy. This edited volume provides the most comprehensive overview to date of contemporary research in the field of health geography. As a collection, it chronicles the diverse ideas, debates, and pressing questions that have driven this sub-discipline forward. The contributed chapters are written by those with recognized expertise in specific domains, which serves as a significant strength of this landmark edited collection. Each chapter offers an overview of the intellectual trajectory of a particular domain, identifies major claims and developments related to this domain in the context of health geography and health research more broadly, discusses the principal contributions of the domain to the sub-discipline, articulates the main criticisms or limitations of this domain of study, and offers ideas regarding future research directions or developments.

Our goal in this chapter is not to provide an exhaustive or detailed overview of health geography and its evolution. There are excellent resources to turn to for such accounts (e.g., Andrews et al., 2012; Elliott, 1999, 2017; Kearns and Gesler, 1998; Kearns and Moon, 2002). Instead, we seek to contextualize this volume by briefly providing insight into the relationship between health and place and the sub-discipline of focus and then introducing the structure and organization of the book. We build on this initial contextualization in our own short chapters that are placed at the start of each of the five sections. These short chapters provide further insight into important facets of health geography and the study of health and place, including syntheses of some of the most significant contemporary debates in the sub-discipline.

The acknowledged relationship between health and place

Space and place affect people's health, well-being, and access to and experiences of health care. The connections between health and place have been recognized as far back as Hippocrates' *Airs, Waters and Places* (written in 400 BC) showing early recognition of the causal influence of place (water, food, climate and seasons) on health and well-being. Similarly, the work of the social reformers of the 19th century in Great Britain, such as Edwin Chadwick, Friedrich Engels and Benjamin Seebohm Rowntree, was significant in making the connections between the rapid industrialization of British cities and the health of people living there (Pearce, 2014). These early reformers not only documented the stark differences in health across urban areas of the country but also identified some of the underlying living conditions that were integral to understanding the observed inequalities. The same can be said about the importance of the earliest works that described traditional medicines, and the later writings of Florence Nightingale (1859) and other public-health pioneers, in terms of articulating the importance of the local environment in understanding people's opportunities for health and well-being (Barrett, 2000). The interconnection between health and place has thus been acknowledged for centuries, and the sub-discipline of health geography sets at its core the task of uncovering and explicating all of its facets. Indeed, this interconnection is widely recognized, and the relationship between health and place has been researched not only by geographers, but also by those from public health, sociology, epidemiology, biostatistics and other disciplines (see Macintyre, Ellaway and Cummins, 2002).

In 1946, the World Health Organization defined "health" as being composed of our "physical, mental and social well-being, not merely . . . the absence of disease or infirmity" (World Health Organization, 2014, p. 1). Acceptance of this widened definition of "health" prompted a gradual shift away from an exclusively biomedical model of health toward a framework of health that also emphasized its economic, political, social and cultural antecedents not only in the health professions but also among health researchers. This shift also led to wider interest in understanding and acknowledging social influences on health; in doing so, it helped popularize consideration of the role of place (e.g., where we live, where we work, where we access health services) in shaping health. For example, by the time of the World Health Organization's Declaration of Alma Ata in 1978, which was a milestone declaration enforcing the necessity of primary health care in achieving health for all globally, the infusion of heavily geographic concepts such as community, (spatial) equity and (socio-spatial) access into public health vocabulary had become reasonably commonplace (Crooks and Andrews, 2009a, 2009b).

In 1994, Evans, Barer and Marmor spoke to the importance of using a population-based lens for understanding the social determinants of health, outlining a perspective that has become a guiding framework for health and health-promotion policy internationally. A population-health focus considers health outcomes at the group level, including its distribution, and seeks to identify why some populations have better health than others. From a policy perspective, population-level approaches have been increasingly influential, as they seek to identify the policy levers that will have the most wide-ranging and sustainable benefits. A population-health lens also positions health inequities as a central public-health concern and has resulted in a wide-ranging interdisciplinary and international literature on the ubiquitous social gradient in health that has been observed in most countries (Smith, Bambra and Hill, 2015). In recent years, the social determinants of health have come to provide a widely used framework for examining how societal factors and forces, including socioeconomic and political inequalities (e.g., social exclusion, racism, and housing security) shape health status locally and globally. Most of these determinants have explicitly, such as housing and the built environment, or implicitly, such as gender and income distribution, spatial or geographic components, further reinforced a wide acknowledgment of the connections between health and place.

This volume shows that the sub-discipline of health geography is influenced by many conceptualizations of health, particularly those informed by the social model of health proposed by the World Health Organization, as well as the "social determinants of health" framework (see also Elliott, 1999; White, 1981). The first two sections in particular do an excellent job of showcasing this breadth. Health geographers have played a significant role in articulating the role of spatial and place-based factors and forces in shaping health through a social-determinants lens. Examples from the health-geography literature include examining how geographical processes such as local labor market dynamics, urban segregation, neighborhood resources, social capital and networks, housing quality and security, and colonial legacies shape the relationships between place and health outcomes (e.g., Brown et al., 2018; Curtis, 2004). Increasingly, health geographers are giving attention to the role of these determinants in influencing health and well-being among diverse people who are exposed to different structural and social inequalities (Giesbrecht et al., 2014). In particular, work in the sub-discipline has revealed many of the key barriers to, and facilitators of, positive health experiences among some of the most marginalized groups in society. These contexts are particularly well highlighted in the third section of this book, which explores some of the people and groups that have received significant attention in health-geography research.

Health geographers are interested in understanding and exploring health as it relates to place. As we shall see in this book, place can be a specific location, fixed set of geographic coordinates or politically defined space, or it can be a social and cultural phenomenon, one that is performed, sensed, felt and imbued with meaning and identity (Andrews, 2018; Gatrell and Elliott, 2009). Places are complex in that various elements and facets converge within them to dictate people's overall experiences. Meanwhile place is susceptible to outside influences from officialdom that can facilitate the health of a population, a group, or an individual and can even contribute to exclusion and marginalization. Health geographers widely acknowledge this complexity. Indeed, we expand on the complexity of place as it relates to health in Chapter 32, our introduction to the fourth section of this book. While in this section of the current chapter we have shown that there is wide interest in and acknowledgment of the connection between health and place, it is health geographers' intellectual engagement with place in all its facets that sets this sub-discipline apart from other social-science and health disciplines that are informed by social models of health and social-determinants frameworks.

The sub-discipline of health geography

Health geography has evolved considerably since its inception as a sub-discipline of human geography in the mid-20th century. Initially referred to as *medical* geography, this sub-discipline was heavily positivist, focusing on ecological perspectives of disease (e.g., geographical patterns of disease and environmental determinants) and the spatial distribution of health services (e.g., identifying geographical gaps in health-care provision) (Meade and Earickson, 2000). Although these and related questions were important, this narrow focus resulted in a limited and incomplete understanding of how and why place is implicated in understanding health. What is often referred to as the "cultural turn" in human geography greatly informed the trajectory of the sub-discipline, directing it away from a biomedical understanding of health and disease toward a theoretically informed socio-environmental perspective on health. At the same time, developments in the sub-discipline mirrored shifts in other health-related disciplines by adopting population-level perspectives to examine wider determinants of health (see the previous section in this chapter). An outcome of these changes was a transformation in name, from *medical* to *health* geography, but more importantly a diversification in the methods, theories, and approaches that health geographers employed (Kearns, 1995).

It is widely recognized that this important shift in the sub-discipline from medical to health geography was in part inspired by Kearns, who in his landmark 1993 publication called for a *postmedical* or *reformed medical* geography. He suggested it was time for a repositioning of the sub-discipline within social geography and for the incorporation of stronger links to social theories that recognize situated experiences and qualitative research as valid ways of knowing (for a review, see Andrews, 2003; Elliott, 2018). This approach suggested that facets of social theory, together with a place-sensitive approach, were critical to the evolution of the former medical geography. Several of the theories and conceptual approaches now used in health geography are showcased in the second section of this book and have heavily influenced the work covered throughout the volume. This reformed health geography has also seen increasing attention given to other significant concerns in the wider human-geography literature, such as scale and difference (Parr, 2004) and life-course perspectives that recognize the importance of time and the durability of place (Pearce et al., in press), which are also concepts that run throughout the chapters contributed to this collection. Recently another paradigm shift has begun; a *post-human* turn whereby place is conceived as an assemblage – both it and the bodies within it are understood to be networked, material, open, active and performed phenomena. This turn has seen sub-disciplinary attention increasingly paid to such things as the sensory, atmospheric and affective qualities of health and place, all within a broader objective to think about how health emerges and *takes place* (Andrews, 2018; Duff, 2014).

Although we provide here only a snapshot of its evolution, it is evident that the sub-discipline of health geography has grown from its positivist roots to become a more broadly defined, interdisciplinary field of study informed by diverse and dynamic perspectives. Indeed, health geography is known for its pluralism, and its study has come to be informed by diverse methodological and theoretical approaches. Health geographers often draw upon methods, ideas, concepts, and even research questions from other social-science and health-science disciplines to inform understandings of the complex relationship between health and place. This interdisciplinary and collaborative approach is seen throughout this volume, though perhaps most especially in the final section, which examines the practice of health geography. From a theoretical perspective, health geographers draw upon an array of lenses to better understand the connection between health and place, including humanism, social constructionism, post-structuralism, political economy, complexity theory, relational approaches, non-representational theory, idealist theory and political ecology. The methodological diversity and innovation of health geographers is also extensive, having employed creative methods such as autoethnography, storytelling, photovoice, participatory arts-based approaches, and innovative geospatial approaches using GIScience, GPS-informed and smart technologies and spatio-temporal statistical modeling. Although this growing theoretical and methodological pluralism may have created a methodological divide between quantitative and qualitative approaches and the research questions they inform (Rosenberg, 2016), many health geographers view the methodological and theoretical diversity and innovation to be one of the greatest strengths of the sub-discipline. This view of sub-disciplinary diversity as strength is echoed throughout a number of the contributed chapters.

The practice of health geography is ever changing, and in recent years health geographers have increasingly addressed applied health problems in order to provide evidence for policy debates, stimulate media dialogue, or create interventions (Crooks and Winters, 2016). This engagement reflects a greater expectation of academics to demonstrate the broader societal value of their research through affecting public discourse and engaging with important policy needs. Health geographers are particularly well-placed to respond to this challenge and are well attuned to working alongside policy-makers in the production of research and ensuring research findings are disseminated in ways that are tailored for their intended target audience(s) (Shortt et al., 2016). Although this shift toward placing greater value on academic impact through engagement has not been unproblematic, the increased visibility of health-geography research has resulted in a growing recognition among policy-makers, practitioners and the public of the complex relationship between health and place. This can be seen, for example, in the greater attention to place in public-health strategies such as improving community network, reducing inequities and addressing the obesity epidemic. It can also be seen in the growth of nursing geographies as a dedicated area of study by some nurse researchers (e.g., Andrews, 2003; Kyle et al., 2016).

Once critiqued for their overwhelming research focus on the Global North, over the past decade health geographers have increasingly taken up development and global health frameworks that have pushed research into ever-expanding geographic territories. This has also widened the net of policy and practice relevance for research in the sub-discipline and is beginning to bring perspectives from health geography to some of the major challenges in global health, such as global governance, global inequalities, transnational movements, the control of infectious disease, and the rise of non-communicable diseases. As several contributors to this volume point out, particularly in the first section, health geography has much evolving to do to truly engage the world – in terms of both where health geographers are based and the topics and places they study – and to produce evidence that is relevant to global debates and issues concerning health for all.

Overview of the book

We have organized the *Routledge Handbook of Health Geography* into five sections, each of which has nine or ten contributed chapters. These sections are an organizational structure rather than reflecting any dominant classifications of health-geography research, and there is inevitably a certain amount of conceptual overlap between them. As noted above, each section beings with a short introductory chapter by us that provides important context about the sub-discipline of health geography and offers cross-cutting themes. The lead author of each of these chapters is the editor who took overall responsibility for the tone and content in that particular section of the book.

Section 1 provides an overview of key perspectives and debates in the sub-discipline. Taken together, the chapters in this section outline many of the most significant and enduring issues and questions that health geographers and those involved in geographical research on health focus on. Further, the chapters provide insights into the ways in which health geographers are making vital contributions to many of the most pressing public-health challenges across the globe, including immensely complex issues such as transmission of infectious diseases, global food insecurity and the obesity epidemic, and the continued burden of tobacco-related disease. Several chapters identify the latest thinking on very well-established areas of the sub-discipline, such as environmental health and justice and infectious disease. Other chapters highlight emerging perspectives and debates and emphasize where health geographers should be bringing perspectives that contribute to understanding the importance of place in these rapidly evolving concerns. As we observed above, health geography is interdisciplinary; scholars draw on concepts and ideas from a range of social-science and health disciplines, and they tend to work in highly collaborative teams of researchers with diverse disciplinary and methodological backgrounds. It is increasingly recognized that addressing the major global health challenges of the 21st century will require interdisciplinary approaches, with perspectives from health geography firmly embedded. Section 1 serves to reflect this interdisciplinary nature of health geography.

The second section focuses on theoretical traditions and concepts that inform health-geography research. As we point out in our introductory chapter to this section, health geography is not often regarded as a theoretically rich sub-discipline of human geography when compared to, for example, the theoretical contributions of cultural geography (Kearns and Moon, 2002). There are likely a number of reasons for this, including that much health geography research is tackling applied problems and that the heavily networked nature of much scholarship involves collaborations with health-professional groups in ways that makes the inclusion of theory somewhat prohibitive. However, the chapters in this section show that health geographers are successfully using theories of all types in driving forward research questions, methods and methodology selection, and that some concepts championed by health geographers have been taken up more broadly into health research.

The groups and people of focus in health geography research form the basis of the third section of this collection. Health geography has many foci in this regard, and much of it explores the health and health-service access of the most marginalized and disempowered individuals, and various health- and social-care-provider groups. Health geographers have championed the ways in which place-based and multiscalar experiences shape their experiences and ultimately their health outcomes, identifying the barriers to positive health-related experiences. We introduce this section of the volume in Chapter 22 and in doing so identify several themes that cross-cut the insightful contributed chapters. First, the chapters in this section illustrate health geographers' established interest in showing how recent socioeconomic and political changes at the global scale, such as neoliberal policy agendas (e.g., the shrinking of the welfare state) and transnationalism, are profoundly and negatively affecting the health of the most marginalized and vulnerable groups. Second, these chapters show how the social and structural processes that continue to relegate particular groups to the margins of society and exclusionary places have enduring implications for their health and everyday well-being. Finally, this section illustrates the ways in which health geographers are increasingly employing methods that enable them to work with marginalized people and groups throughout the research process in collaborative, engaged and revealing ways.

Given that this is a health-geography collection, obviously themes of place and space run throughout the volume. In the fourth section, however, we give dedicated consideration to some of the places and spaces that health geographers explore most in their work. The chapters in this section show the power of considering both the physical and social understandings of place in order to truly decipher the impact of place on health.

Some authors also offer insight into the ways that place and space contribute to health or health promotion, rather than simply detract from health. This section also sees the most explicit ongoing consideration of health-care and health-service sites and those who work there, both formally and informally. Collectively, these chapters highlight the importance and ongoing nature of the study of the sites of health-care delivery and receipt within health geography, the value of exploring everyday or taken-for-granted spaces (versus exceptional ones) for understanding the role of place in shaping health, and the uptake of the mobilities turn in social-science research into health geography.

The fifth and final section addresses how it is that geographers do what they do and, in part, the challenges that they must navigate in their research. From research-ethics processes and methods to applied collaborations and incorporating diversity approaches into studies, this section details multiple aspects of health-geography practice. As we note in our chapter at the outset of this section, there are three clusters of chapters. The first cluster is focused on the practice of qualitative research in health geography. The second cluster examines aspects of quantitative health geography, including the push toward big-data approaches. The final cluster cross-cuts the first two, exploring issues of practice relevant to all health geographers regardless of the methods they employ.

We think it is important to acknowledge that the chapters contributed to this book do not provide an exhaustive overview of health geography. There are many voices, places, theories, approaches, and practices that we sought to include in this volume but were unable to find contributors for; others we decided against due to space restrictions. Some of the key omissions that we had explicitly sought to include as chapters in this collection are concerned with race, ethnicity and exclusion; feminist theory; gender; affect; critical realism; ecological perspectives; workplaces; the body and embodiment; knowledge translation; public geographies; and health-services research. In our chapters that introduce each section, we reflect on some of these gaps in greater detail and consider some of the reasons it proved impossible to secure chapters on these themes that in some cases point toward important gaps in the literature. Therefore, what is missed in this collection might be important, and we encourage readers to look at the tables of contents of journals such as *Health & Place*, the *International Journal of Health Geographics*, and *Social Science & Medicine* along with recent programs from the biennial *International Medical Geography Symposium* as possible entry points. Meanwhile, we believe that a significant strength of our approach has been to prioritize the inclusion of emerging, cutting-edge aspects of the discipline, such as post-humanism and non-representational theory, in the hope that the contribution of this volume endures.

References

Andrews, G. J. (2003). Locating a geography of nursing: space, place and the progress of geographical thought. *Nursing Philosophy*, 4(3), pp. 231–248.

Andrews, G. J. (2018). *Non-representational theory and health: the health in life in space-time revealing.* London: Routledge.

Andrews, G. J., Evans, J., Dunn, J. R. and Masuda, J. R. (2012). Arguments in health geography: on sub-disciplinary progress, observation, translation. *Geography Compass*, 6(6), pp. 351–383.

Barrett, F. (2000). *Disease and geography: the history of an idea.* Toronto: Becker Associates.

Brown, T., Andrews, G. J., Cummins, S., Greenhough, B., Lewis, D. and Power, A. (2018). *Health geographies: a critical introduction.* Chichester: Wiley-Blackwell.

Crooks, V. A. and Andrews, G. J. (2009a). Community, equity, access: core geographic concepts in primary health care. *Primary Health Care Research & Development*, 10(3), pp. 270–273.

Crooks, V. A. and Andrews, G. J. (2009b). Thinking geographically about primary health care. In: V. A. Crooks and G. J. Andrews, eds., *Primary health care: people, practice, place.* Aldershot: Ashgate, pp. 1–20.

Crooks, V. A. and Winters, M. (2016). 16th International Medical Geography Symposium special collection: a current snapshot of health geography. *Social Science & Medicine*, 168(2016), pp. 198–199.

Curtis, S. E. (2004). *Health and inequality: geographical perspectives.* London: Sage.

Duff, C. (2014). *Assemblages of health: Deleuze's empiricism and the ethology of life.* Dordrecht: Springer.

Elliott, S. J. (1999). And the question shall determine the method. *Professional Geographer*, 51(2), pp. 240–243.

Elliott, S. J. (2018). 50 years of medical health geography(ies) of health and wellbeing. *Social Science & Medicine*, 196, pp. 206–208.

Evans, R., Barer, M. and Marmor, T. R. (1994). *Why are some people healthy and others not? The determinants of health of populations*. New York: Aldine de Gruyter.

Gatrell, A. C. and Elliott, S. J. (2009). *Geographies of health: an introduction*. Chichester: Wiley-Blackwell.

Giesbrecht, M., Cinnamon, J., Fritz, C. and Johnston, R. (2014). Themes in geographies of health and health care research: reflections from the 2012 Canadian Association of Geographers annual meeting. *Canadian Geographer*, 58(2), pp. 160–167.

Kearns, R. (1993). Place and health: towards a reformed medical geography. *The Professional Geographer*, 45(2), pp. 139–147.

Kearns, R. (1995). Medical geography: making space for difference. *Progress in Human Geography*, 19(2), pp. 251–259.

Kearns, R. and Gesler, W. M. (eds.) (1998). *Putting health into place*. Syracuse, NY: Syracuse University Press.

Kearns, R. and Moon, G. (2002). From medical to health geography: novelty, place and theory after a decade of change. *Progress in Human Geography*, 26(5), pp. 605–625.

Kyle, R. G., Atherton, I. M., Kesby, M., Sothern, M. and Andrews, G. (2016). Transfusing our lifeblood: reframing research impact through inter-disciplinary collaboration between health geography and nurse education. *Social Science & Medicine*, 168, pp. 257–264.

Macintyre, S., Ellaway, A. and Cummins, S. (2002). Place effects on health: how can we conceptualise, operationalise and measure them? *Social Science & Medicine*, 55(1), pp. 125–139.

Meade, M. S. and Earickson, R. (2000). *Medical geography*. New York: Guilford Press.

Nightingale, F. (1859). *Notes on nursing: what it is and what it is not*. London: Lippincott.

Parr, H. (2004). Medical geography: critical medical and health geography? *Progress in Human Geography*, 28(2), pp. 246–257.

Pearce, J. (2014). Geographies of health inequality. In: W. C. Cockerham, R. Dingwal and S. R. Quah, eds., *The Wiley-Blackwell encyclopedia of health, illness, behavior, and society*. Chichester: Wiley-Blackwell.

Pearce, J., Cherrie, M., Shortt, N., Deary, I. and Ward-Thompson, C. (in press). Life course of place: a longitudinal study of mental health and place. *Transactions of the Institute of British Geographers*. DOI: 10.1111/tran.12246

Rosenberg, M. (2016). Health geography II: "Dividing" health geography. *Progress in Human Geography*, 40(4), pp. 546–554.

Shortt, N. K., Pearce, J. and Smith, K. E. (2016). Taking health geography out of the academy: measuring academic impact. *Social Science & Medicine*, 168, pp. 265–272.

Smith, K. E., Bambra, C. and Hill, S. (eds.) (2015). *Health inequalities: critical perspectives*. Oxford: Oxford University Press.

White, N. (1981). Modern health concepts. In: N. White, ed., *The health conundrum*. Toronto: OECA, pp. 5–18.

World Health Organization. (2014). *Constitution of the World Health Organization: basic documents*. 48th ed. Geneva: World Health Organization.

SECTION 1

Perspectives and debates

2
INTRODUCING SECTION 1
Perspectives and debates

Jamie Pearce, Gavin J. Andrews and Valorie A. Crooks

Geographers have a long history of enriching our understanding of many of the most pressing and complex issues of the day. In the health arena, geographers are not only providing insights to some of the key global public-health challenges such as obesity, tobacco use and the transmission of infectious diseases, but also identifying the implications for health and well-being of a much wider set of social, political and environmental concerns, including transnational migration, globalization, the financial crisis and climate change. The chapters in this section consider some key perspectives and debates that have occupied health geographers in recent years and examine the contribution that work in health geography has made to addressing these international challenges. Subsequent sections in the book will consider *how* health geographers have gone about answering some of these questions, *who* has been the subject of this work, and *which* places and spaces have been integral to answering these questions.

It is widely accepted that deep understandings of the processes underpinning most key global public-health challenges, and developing strategies to address or mitigate these concerns, requires a recognition that these outcomes emerge from highly complex systems. Health geographers have revealed this complexity through interdisciplinary thinking that recognizes that the drivers of health and disease are inherently multiscalar, historically embedded and simultaneously related to the social, physical, political and environmental milieus. In developing our understanding of the key debates and perspectives identified in this section of the book, it is clear that health geographers have drawn on an extraordinary range of theoretical perspectives, conceptual frameworks and methodological approaches to consider a multitude of global health challenges and therefore brought refreshing and novel insights to these issues. Further, health geographers have developed their understanding of these concerns in a broad range of empirical contexts, showing not only that many of the drivers of health have global reach and are interconnected, but also that more localized contingencies are often highly pertinent.

The nine chapters in this section are a collection of some of the key perspectives adopted by health geographers to make better sense of how place and space are implicated in a multitude of health-related concerns. In particular, this section is concerned with the ways in which health geographers understand health and the factors shaping it. Collectively, a number of significant and interrelated themes emerge, each pointing toward some of the most pressing drivers of health and well-being. Here we examine three of these cross-cutting themes.

First, inequalities and inequities are central themes in all the chapters in this section. Geographical work recording and revealing health inequalities extends back at least as far as the pioneering work of Edwin

Chadwick, Friedrich Engels and others in England during the mid-19th century (Pearce and Dorling, 2009). In recent years, health inequalities and inequity have begun to feature more prominently in the global health agenda. A notable contribution to these debates was the World Health Organization's Commission on Social Determinants of Health 2008 report entitled *Closing the Gap in a Generation: Health Equity Through Action on the Social Determinants of Health* (CSDH, 2008). The report unequivocally emphasized the causal relationships between an array of social factors and various health outcomes. The Commission forcibly concluded that the "toxic combination of poor social policies and programmes, unfair economics, and bad politics is responsible for much of health inequity" (p. 35). The Commission also formulated a collection of ambitious policy priorities that provided an agenda for action.

Authors of the chapters in this section provide some vivid examples of the inequalities that exist at global, national and subnational levels. The chapter by Brown and the chapter by Bisung, Dixon and Luginaah both document the vast differences in mortality, morbidity and health experiences between countries in the Global North and those in the Global South. However, inequalities in health are also ubiquitous *within* most countries and, as Bambra's chapter shows using data collected in the United Kingdom, spatial inequalities in health continue to rise. Similar conclusions are drawn in Moon and Smith's chapter, which examines the socio-spatial patterning of health-related behaviors (focusing on drinking, smoking, eating and physical activity). The authors find unhealthy behaviors to be disproportionately and increasingly prevalent among lower socioeconomic groups and in more disadvantaged places.

A second and related cross-cutting theme is a focus on work concerned with understanding the structural drivers of the substantial global and local differences in health. Bisung, Dixon and Luginaah, for example, examine the post-1945 development agenda to reduce inequalities and improve living conditions and its influence on global health concerns. They chart the evolution of this program from an initial focus on economic growth, to greater attention to human conditions and need, and then with increasing concern with environmental sustainability and equity. The authors draw on perspectives from the political economy of health literature to critique the development agenda and to emphasize the role played by economic and political determinants of global health and well-being. The political-economy perspective is central to the arguments furthered in Bambra's chapter, which steps through the various theoretical perspectives adopted in health geography and elsewhere to explain why inequalities in health persist even within the wealthiest countries. The chapter draws on the notion of *fundamental causes* to emphasize that good health is driven by social, political and economic structures outside of the control of the individual. This argument is extended in Crighton, Gordon, and Barakat-Haddad's chapter, which adopts the framework of environmental justice to explain how and why poor, disadvantaged and disenfranchised groups tend to be disproportionately exposed to environmental disamenities (e.g., poor air quality, toxic sites, low-quality infrastructure). Using striking case studies, the authors demonstrate that not only do these groups suffer the greatest burden of environmental toxins and major environmental disasters, but also it has a disproportionate effect on their health and well-being.

Similar conceptualizations have been adopted by health geographers with interests in infectious disease. Keeler and Emch's chapter discusses how ideas from the field of political ecology have been adopted to explicate how social and economic factors including race, gender and livelihood are implicated in risk and transmission of disease (e.g., HIV/AIDS). These conceptual issues are picked up in Curtis and Oven's chapter, which operationalizes the concepts of *risk* and resilience and highlights the diverse ways in which health geographers have identified aspects of place that are risky for health (e.g., highly polluted areas) and increasingly how places (e.g., communities with strong social networks) can help foster health resilience. Curtis and Oven also consider interesting new possibilities, including the notion of *equigenesis*, which emphasizes how places might not only be therapeutic but also support narrower inequities in health (Mitchell et al., 2015). An alternative, and often overlooked, explanation for the growing geographical differences in health between regions of a country is selective migration flows. This is the theme of the chapter by Darlington-Pollock and colleagues, who evaluate the health consequences of migration events. They show that migration is often highly selective according to social, demographic and health characteristics, which may help explain the geographical concentration of (un)healthy people.

The final cross-cutting theme, and as noted above, is that many of the most important concerns in health geography emerge from complex, multilevel and often global systems, and health geographers have been prominent in investigating these complex arrangements (Curtis and Riva, 2010). As the chapter by Brown details, health geographers are increasingly contributing to scholarship that falls within the remit of *global health*, with more attention given to the global processes affecting health and inequalities. This complexity is apparent, for example, in the discussion in Bambra's chapter on the multiple and relational factors shaping health inequalities. It is clear that reducing health inequalities requires multisectoral action across a range of traditional domains including welfare, employment, education, health care and many other areas. As the chapters by Wakefield and by Moon and Smith demonstrate, other major global public-health challenges – including those related to human behaviors, such as the obesity epidemic, tobacco use and alcohol consumption – also benefit from the application of systems thinking. Wakefield, for example, highlights how elements of food production are driven by political and economic imperatives, with notable implications for the range of foods available, price structures and, ultimately, diet and consumption patterns. Similar arguments are noted in the chapter by Moon and Smith, particularly in relation to the pervasive and detrimental influence of large multinational companies in the tobacco and alcohol sectors. Collectively the chapters in this section emphasize that understanding and successfully responding to many contemporary public-health challenges requires health geographers to embrace new conceptual models and methods suited for examining complex systems.

Although this section covers a wide-ranging set of perspectives and debates in health geography, it is almost certain that another trio of editors would have selected a radically different set of chapters. Inevitably some important issues have been overlooked. In some cases, the reasons for omitting an important theme were prosaic, usually because we were unable to secure a health geographer with the appropriate research background who was available to contribute. For example, we intended to include a chapter that considered health geography's contribution to the field of health-services research, including work on the health implications of inequitable access to health care or people's experiences of health care in different geographical contexts. However, in other cases, themes of significant interest to health researchers and practitioners– for example, the role of urban planning in affecting community public health, or the significance of climate change as a global public-health threat – are domains where few geographers are currently engaging. Nonetheless, these and other important themes are considered at various junctures throughout the book. The writers of this section identify and conceptualize some of the key issues in contemporary health geography and offer a range of opportunities for improving global public health.

References

CSDH. (2008). *Closing the gap in a generation: health equity through action on the social determinants of health*. Final Report of Commission on Social Determinants of Health. Geneva: World Health Organization.

Curtis, S. and Riva, M. (2010). Health geographies I: complexity theory and human health. *Progress in Human Geography*, 34(2), pp. 215–223.

Mitchell, R. J., Richardson, E. A., Shortt, N. K. and Pearce, J. R. (2015). Neighborhood environments and socioeconomic inequalities in mental well-being. *American Journal of Preventive Medicine*, 49, pp. 80–84.

Pearce, J. and Dorling, D. (2009). Tackling global health inequalities: closing the health gap in a generation. *Environment and Planning A*, 41(1), pp. 1–6.

3

GLOBAL HEALTH GEOGRAPHIES

Tim Brown and Stephen Taylor

The interdisciplinary field of global health has grown exponentially in recent years (Sparke, 2009). Prior to the 1990s, the term *global health* was mentioned in the titles of only a small handful of research papers. At that moment in time, though global health research was certainly being conducted, it was not being done in the name of a specific field of inquiry. To an extent, this picture has changed. In recent times, a field has cohered around the idea of global health, and it is one that is increasingly ubiquitous in the degree programs of universities around the globe (Neely and Nading, 2017). Moreover, a diverse range of disciplines, including geography, are now much more inclined to contribute research and scholarship labeled as *global health* (Brown and Moon, 2012; Herrick and Reubi, 2017).

With this in mind, we aim in this chapter to interrogate the scope of this geographical work by focusing on three broad thematic areas: global health, inequality and justice; global health governance; and more-than-human global health. In so doing, we identify key areas where a distinctly geographical perspective has contributed to and, we argue, broadened interdisciplinary scholarship.

Health, inequality and justice

The specific question of health inequality arguably should be at the core of geographical research into global health; especially if we consider that "global health," at least with regard to one of its more prominent definitions, is described as a concern to achieve "[h]ealth equity among nations and for all people" (Koplan et al., 2009, p. 1994). This was certainly a starting point for a recent review of geography's engagement with global health, which highlighted the significance of the World Health Organization's Commission on the Social Determinants of Health (Brown and Moon, 2012).

The question of global health justice remains a pertinent one. As Simon Reid-Henry (2016) suggests, while significant resources have been ploughed into global health initiatives in recent years and have gone some way to achieving health-oriented Millennium Development Goals – citing as examples a reduction in the number of women dying in childbirth and an increase in people living with AIDS receiving anti-retroviral treatment– little has been achieved with regard to an oft-stated desire to reduce global health inequalities. As a starting point for responding to the latter, Reid-Henry posits that what is needed is not more research into why health inequalities are so persistent, but a different focus entirely. As he argues, "[w]hen it comes to the question of global health *inequality* in particular, the particular model of justice that is in play matters greatly" (2016, p. 721). Further, Reid-Henry encourages researchers to consider more carefully how competing models of justice – he identifies market justice and social justice as the dominant competing models at play in global health discourse – frame and ultimately shape global health policy. The significance of this attention to justice and its shaping of policy lies in the argument that market-justice models, which Reid-Henry identifies as the dominant model at play in mainstream global health policy today (cf. Sparke, 2009), have led to targeted, vertical interventions that cure rather than prevent a fairly narrow range of global health problems.

Where Reid-Henry calls for refocusing on the models of justice at play in global health policy, other geographers have begun to argue for a response that builds upon the contextual or place-oriented approaches that have dominated research on health inequalities. To an extent, such a call mirrors the critique of health-inequalities research offered by Susan Smith and Donna Easterlow (2005) in their account of the "strange geographies of health inequalities," which for them meant an overemphasis on commonsense understandings of the influence of place or context on health. Though responding to the conceptualization of health inequality as it stood at the time – that is, as a reflection of either individual (compositional) or area (contextual) factors, but rarely both – their paper is especially helpful because it acted to remind health geographers that health inequalities are intimately connected to social and spatial processes, such as discrimination and social inequality in all their varied forms. Moreover, it argued for a much more embodied and people-oriented approach to our understanding of how different health experiences shape a person's position in society and space. At issue here is a questioning not so much of health geographers' commitment to equity and justice as of how they ought to go about doing global health inequalities research.

One way of bridging this gap between observed health inequalities and an understanding of the social, cultural and political processes shaping global health has been the adoption of a political ecology of health perspective. Such an approach is certainly not new to health geographers (or, perhaps more appropriately, *medical* geographers; see Mayer, 1996). While some of the more recent work in this tradition has sought to outline what a political ecology of health perspective brings to understanding the uneven geographies of health – that they are "*problems*, which are relationally intertwined, produced over time, inherently political, and always simultaneously material and symbolic" (Jackson and Neely, 2015, p. 48) – other studies have more explicitly engaged with global health research agendas. Abigail Neely and Alex Nading (2017), for example, argue that a focus on embodiment and place is crucial to critiques of global health. Drawing on their study of a local nutrition intervention in Nicaragua and another study into the experiences of tuberculosis treatment in South Africa, they seek to understand how global health priorities and protocols are enacted locally and to explore how power relations that operate at multiple scales shape people's everyday experiences of global health in place. As they argue, a political-ecology perspective enables analyses that focus on the effects of global health interventions on people, places and diseases, as well as on the ways in which place produces and is produced by sickness and health.

Improving the health of others

As a multibillion-dollar enterprise with significant social, political, and economic benefits, the task of improving global health weaves together ministries of health, physicians, patients and super-rich donors in a complex web of multinational institutions, bilateral partnerships, and advocacy networks. A growing number of geographers have begun to interrogate the governance of interventions in the lives and health of others in order to better understand the motivations of those involved and how they, in turn, shape global health approaches and agendas (Laurie, 2015).

It is perhaps unsurprising that individual states and multinational institutions play a key role in global health decision-making. In the 19th and early 20th centuries, the growing circulation of people and things – made possible by developments in trade and transport – led to some of the first international agreements between states, meant to secure borders against the spread of infectious diseases (Ingram, 2009). As interconnectedness increased after the Second World War, global health was further entrenched as a geopolitical concern; it is no surprise, for instance, that one of the first specialized agencies of the United Nations (UN) was the World Health Organization (WHO). Under the WHO umbrella, Geneva-based global health institutions pioneered expert-led, disease-specific (*vertical*) interventions that succeeded, among other things, in eradicating smallpox and improving the lives of millions living with HIV/AIDS. Such governance approaches have, however, been criticized for their perceived entrainment within geopolitical agendas – particularly the prescriptions of the cozy *Washington consensus* on international development existing between the International Monetary Fund, the World Bank, and the US Treasury Department – and for their focus on anachronistic and highly technical forms of intervention that have been frequently insensitive to local priorities and concerns (Taylor, 2017).

Alongside – and frequently in competition with – this network of state-focused initiatives are a series of increasingly influential non-state global health actors. Loose alliances of non-governmental organizations and grassroots initiatives have come to play a prominent role in global health governance. Clare Herrick (2014), for instance, notes the increasing prominence of the NCD Alliance, a lobbying group of 2,000 civil-society organizations, in advocating patient-centered approaches to non-communicable disease. Private interests have also become key players in the field of global health, and the logics of the market have frequently been lifted into efforts to improve human health and well-being. These approaches have prioritized return-on-investment and have deployed metric-based forms of evaluation, often borrowed from the financial sector. Others have raised concerns about conflicts of interest that emerge when corporations – some of which market products that negatively impact health – are involved in health research and interventions. David Reubi (2013), for instance, traces the influence that tobacco companies and their proxies have had on distorting public-health messaging in the Global South. Likewise, Herrick (2016) examines corporate behavior within the alcohol industry and the efforts of the industry to sway public opinion and shape global policy agendas through extensive lobbying.

Market habits and behaviors are inculcated through global health interventions in other ways too. In their work on the influence of philanthropic donors in global health, Katharyne Mitchell and Matthew Sparke (2016) trace the contours of a *new* Washington consensus, an approach to global governance driven not by the macroeconomic development policies of Washington, DC, institutions but rather by the cluster of philanthropic actors based in Washington State. Spearheaded by the Seattle-based Bill & Melinda Gates Foundation – the world's wealthiest private foundation – this cluster of organizations (including the health innovation company PATH) promotes market-mediated innovations in global health. Big global problems, the Gateses propose, are best fixed not by dependency-inducing *charity* but by targeted micro-financial *investments* that catalyze innovation and change. This approach is typified by the Gates-backed initiative to finance immunizations for children through the issuing of vaccine bonds into the capital market, which release vaccine funding while promising investors a fixed return. The growth of this so-called *philanthrocapitalism* has not been without controversy, and a chorus of critical voices have interrogated the motives of the super-rich intervening in the lives of (distant) others (Nally and Taylor, 2015). Can creative forms of capitalism, for instance, solve health problems that many consider to be the products of market injustices in the first place? Or is this philanthropy, as the son of one major donor suggests, merely "conscience laundering" through which the wealthy assuage their guilt for past corporate practices through high-profile investments (Buffett, 2013, p. A19)?

From state interventions at the border to the geopolitics and geo-economics of global health, all attempts at improving the lives of others are political in the strongest sense. *What* is prioritized for intervention, as well as *how* such efforts are designed and deployed, are vital questions of governance to which geographers of global health must remain attentive.

More-than-human global health

A growing body of geographical research has sought to decenter established accounts of global health interventions that prioritize concerns of the human. Inspired by a *more-than-human* turn within geography more broadly (Whatmore, 2002), this research approaches global health concerns from an alternative vantage point: that of the much-maligned bugs, viruses, bacteria and parasites that live within and beyond the human body. What, we might ask, do efforts to appraise and improve global health look like if we position a mosquito, or even the microscopic malaria parasite, as the object of our attention, rather than the infected child or the weakened health system in which they struggle to get care?

Where the non-human world has been considered in past global health interventions, it has often been positioned as an unruly and adversarial space to be controlled, conquered and even destroyed through the application of human effort and technology. The Rockefeller Foundation, a leading philanthropic funder of international health efforts in the 20th century, financed ambitious campaigns to *eradicate* non-human vectors of infectious diseases, such as hookworm, malaria and yellow fever. In his work on the attempted eradication of malarial mosquitoes in 1940s Argentina through widespread spraying of the insecticide DDT, Eric Carter (2009) documents the huge financial, civilian and military mobilizations necessary to launch an ultimately futile *war* against legions of tiny insects. We can see present-day parallels of this approach in the work of the Bill & Melinda Gates Foundation, which has prioritized the eradication of polio and – once again – malaria (Taylor, 2016). Learning lessons from the spraying of toxic DDT, the Gateses' approach to malaria eradication has harnessed the power of modern genetic technology to edit the genome of malarial *Anopheles* mosquitoes to eventually make them sterile and unable to reproduce; the deployment of what we might term *extinction technology* raises real concerns about how much control one wealthy couple has over the fate of an entire species.

And yet despite these efforts at eradication, Ian Shaw and colleagues (2013, p. 266) remind us that "the mosquito, it seems, always persists; its monstrous reputation is at least in part due to its ability to escape strategies and tactics of control." While we may vilify the pesky mosquitoes – and viruses and bacteria for that matter – as mindless spreaders of infectious disease, these non-human species have a liveliness of their own that is difficult to contain. The *Anopheles* mosquitoes spreading Zika virus across South America in 2016 survived and reproduced in stagnant water collected in the detritus – empty barrels, upturned bottle caps and dog bowls – of everyday human existence (Carter, 2016). Likewise, the non-human world also has little respect for increasingly sophisticated health surveillance, control and containment policies being rolled out by state actors. As Vincent Del Casino and colleagues (2014, p. 546) point out, "polioviruses maintain themselves by seeping through the boundaries – real or imagined – we use to contain them." A more-than-human global health, then, must be attentive to the ways in which the fate of humans and non-humans is increasingly entangled, as research on the vital importance of the trillions of bacteria in the human microbiome attests (Lorimer, 2017). Rather than develop interventions that cast humans and non-humans as mortal enemies, we might reconsider how important these companion species are to our health. "What if managing mosquitoes is not about how best to eliminate them," Uli Beisel (2015, p. 153) asks in her research on malaria in Ghana, "but about asking how we might find ways to tolerate coexisting with each other?"

The question of coexistence has gained international traction through the popularity of the One World, One Health initiative. This approach to human and animal health has attempted to better integrate the traditionally distinct domains of veterinary, human, and environmental health in order to tackle the health impacts of pathogens – such as avian and swine influenzas – capable of moving between species. Now integrated into the work of the UN, WHO and Food and Agriculture Organization (FAO), this approach has led to the establishment of standardized barrier techniques and biosecurity practices in agriculture designed to halt transmission of microbes between humans and non-humans. And yet for the shared geographical imagination that the One World model of coexistence promotes, Steve Hinchliffe (2015) argues that this approach to zoonotic disease tends to reassert the primacy of human control over the non-human world. Coexistence, in this approach, is a highly managed process with internationally accepted and applied norms around the production, surveillance and regulation of non-human lives. A more-than-human critique of such an initiative would emphasize that control of the non-human world is never straightforward; coexistence is, in other words, messy and contingent. Global health approaches that champion human mastery over the non-human world – from DDT to One World, One Health – ignore the lively adaptability and slipperiness of bugs, viruses, and bacteria at their peril.

Thinking geographically about global health

Global health is not short of inspirational leaders and voices, but few articulate the centrality of geographical concerns as astutely as Paul Farmer, a Harvard-trained physician and founder of the international health charity Partners in Health. In his powerful reflections over the course of several decades of professional medical practice and HIV advocacy in rural Haiti, Farmer (1992; 2003) documents a growing conviction of the need to acknowledge and remedy the social, political and cultural determinants of ill health that are so impactful yet poorly understood in many parts of the world. "Our analysis [of global health problems] must be geographically broad," argues Farmer (2003, p. 159), because "the world that is satisfying to us is the same world that is utterly devastating to them." It is encouraging to hear an attentiveness to geographical concerns given such prominence in a call for reimagining extant approaches to global health. We can but hope that further opportunities will come into being for geographers to collaborate meaningfully with medical researchers to improve the health and well-being of individuals and communities across our increasingly interconnected world and, in so doing, respond to the many pressing challenges facing global health – whether the unevenly felt impact of global climate change (Curtis and Oven, 2012), the emergent crisis of antimicrobial resistance (WHO, 2015) or the pressing issue of toxic air affecting many of the world's major cities (WHO, 2016).

References

Beisel, U. (2015). Markets and mutations: mosquito nets and the politics of disentanglement in global health. *Geoforum*, 66, pp. 146–155.

Brown, T. and Moon, G. (2012). Geography and global health. *The Geographical Journal*, 178(1), pp. 13–17.

Buffett, P. (2013). The charitable-industrial complex. *The New York Times*, July 27, A19.

Carter, E. (2009). "God bless General Perón": DDT and the endgame of malaria eradication in Argentina in the 1940s. *Journal of the History of Medicine and Allied Sciences*, 64(1), pp. 78–122.

Carter, E. (2016). Zika anxieties and the role of geography. *Journal of Latin American Geography*, 15(1), pp. 157–161.

Curtis, S. E. and Oven, K. J. (2012). Geographies of health and climate change. *Progress in Human Geography*, 36(5), pp. 654–666.

Del Casino, V. Jr., Butterworth, M. and Davis, G. (2014). The slippery geographies of polio. *The Lancet Infectious Diseases*, 14(7), pp. 546–547.

Farmer, P. (1992). *AIDS and accusation: Haiti and the geography of blame*. Berkeley: University of California Press.

Farmer, P. (2003). *Pathologies of power: health, human rights and the new war on the poor*. Berkeley: University of California Press.

Herrick, C. (2014). (Global) health geography and the post-2015 development agenda. *The Geographical Journal*, 180(2), pp. 185–190.

Herrick, C. (2016). Alcohol, ideological schisms and a science of corporate behaviours on health. *Critical Public Health*, 26(1), pp. 14–23.

Herrick, C. and Reubi, D. (eds.) (2017). *Global health and geographical imaginaries*. London and New York: Routledge.

Hinchliffe, S. (2015). More than one world, more than one health: re-configuring interspecies health. *Social Science & Medicine*, 129, pp. 28–35.

Ingram, A. (2009). The geopolitics of disease. *Geography Compass*, 3(6), pp. 2084–2097.

Jackson, P. and Neely, A. H. (2015). Triangulating health: toward a practice of a political ecology of health. *Progress in Human Geography*, 39(1), pp. 47–64.

Koplan, J. R., Bond, T. C., Merson, M. H., Reddy, K. S., Rodriguez, M. H., Sewankambo, N. K. and Wasserheit, J. N. (2009). Towards a common definition of global health. *Lancet*, 373(9679), pp. 1993–1995.

Laurie, E. (2015). Who lives, who dies, who cares? Valuing life through the disability-adjusted life-year measurement. *Transactions of the Institute of British Geographers*, 40(1), pp. 75–87.

Lorimer, J. (2017). Parasites, ghosts and mutualists: a relational geography of microbes for global health. *Transactions of the Institute of British Geographers*, 42(4), pp. 544–558.

Mayer, J. D. (1996). The political ecology of disease as one new focus for medical geography. *Progress in Human Geography*, 20(4), pp. 441–456.

Mitchell, K. and Sparke, M. (2016). The new Washington Consensus: millennial philanthropy and the making of global market subjects. *Antipode*, 48(3), pp. 724–749.

Nally, D. and Taylor, S. (2015). The politics of self-help: the Rockefeller Foundation, philanthropy and the "long" Green Revolution. *Political Geography*, 49, pp. 51–63.

Neely, A. H. and Nading, A. M. (2017). Global health from the outside: the promise of place-based research. *Health & Place*, 45, pp. 55–63.

Reid-Henry, S. (2016). Just global health? *Development and Change*, 47(4), pp. 712–733.

Reubi, D. (2013). Health economists, tobacco control and international development: on the economisation of global health beyond neoliberal structural adjustment policies. *BioSocieties*, 8(2), pp. 205–228.

Shaw, I., Jones, J. P. III and Butterworth, M. (2013). The mosquito's umwelt, or one monster's standpoint ontology. *Geoforum*, 48, pp. 26–267.

Smith, S. J. and Easterlow, D. (2005). The strange geography of health inequalities. *Transactions of the Institute of British Geographers*, 30(2), pp. 173–190.

Sparke, M. (2009). Unpacking economism and remapping the terrain of global health. In: A. Kay and O. Williams, eds., *Global health governance: transformations, challenges and opportunities amidst globalization*. New York: Palgrave Macmillan, pp. 131–159.

Taylor, S. (2016). In pursuit of zero: polio, global health security and the politics of eradication in Peshawar, Pakistan. *Geoforum*, 69, pp. 106–116.

Taylor, S. (2017). Making space for restoration: epistemological pluralism within mental health interventions in Kinshasa, Democratic Republic of Congo. *Area*, 49(3), pp. 342–348.

Whatmore, S. (2002). *Hybrid geographies: natures, cultures, spaces*. London: Sage.

World Health Organization. (2015). *Global action plan on antimicrobial resistance*. Geneva: World Health Organization.

World Health Organization. (2016). *Ambient air pollution: a global assessment of exposure and burden of disease*. Geneva: World Health Organization.

4

DEVELOPMENT

The past, present and future contributions of health geography

Elijah Bisung, Jenna Dixon and Isaac Luginaah

We live in a world characterized by major disparities in terms of the conditions in which people live, grow and work. The daily realities for many of the world's citizens stand in stark contrast to those in high-income countries. For instance, 2.5 billion people lack access to adequate sanitation facilities, one in nine people in the world are undernourished and 2.7 billion people struggle to live on less than two dollars per day (FAO/FAD/WFP, 2015; WHO/UNICEF, 2015). These same realities determine local and global patterns and disparities in health and well-being. Even within high-income countries where living conditions are much better, income inequalities are significant. They are believed to manifest in inequalities in health outcomes through erosion of social cohesion, disinvestment in social services, and disparities in other social determinants of health. In the face of these disparities, physical environmental conditions, particularly climate change and the associated severe environmental events, are posing new and reemerging health risks to societies. This is especially so for the vulnerable and those already living in impoverishment in the Global South. The question of how to tackle these issues (i.e., how to improve living conditions, reduce inequalities, and use the environment sustainably) has been the focus of development since the end of World War II.

This chapter seeks to present various perspectives on geographies of health and development. First, accounts of development and articulation of alternative approaches to development post–World War II is presented. Post-development is then explored as a critique of, and alternative to, development. Finally, the chapter explores the links between geographies of health and developments and presents anticipated future research directions.

Defining "development"

The meaning – and debate around the meaning – of development is well documented and need not be repeated here (Willis and Kumar, 2009; Potter et al., 2012). However, it is important to set the stage for understanding how current approaches to, and agenda for, development evolved. Literature on "development" traces the origins of modern usage of the term to the late 1940s. Specifically, many researchers cite a speech by US president Harry Truman in which he referred to some areas as "undeveloped" and challenged "prosperous" nations to take responsibility for developing these areas in their own image:

[W]e must embark on a bold new program for making the benefits of our scientific advances and industrial progress available for the *improvement and growth of underdeveloped areas*. More than half of the people of the world are living in conditions approaching misery. Their food is inadequate. They are victims of disease. Their economic life is primitive and stagnant. Their poverty is a handicap and a threat both to them and to more prosperous areas. For the first time in history, humanity possesses the knowledge and the skill to relieve the suffering of these people.

(Truman, 1949, n.p., emphasis added)

The focus on development, as conceptualized through a lens of modernity, continued to dominate in Cold War–era politics. The rise of free-market or neoliberal order from the 1960s gave birth to a development agenda based on the assumption that "well-being can best be advanced by liberating individual entrepreneurial freedoms and skills within an institutional framework characterized by strong private property rights, free markets and free trade" (Harvey, 2005, p. 2). This rise has, in part, led to a dominance of the market, rising inequalities, erosion of collective values, and rolling back of the state with respect to the provision of welfare and social services. In response, alternative conceptions of development emerged to shift attention from economic growth and to put human well-being at the center of development (Payne and Philips, 2010). This wide terrain of alternative development theories coalesces around two major approaches: human-development and environmental approaches.

Human-development approaches

Initial work on basic needs spearheaded by the International Labor Organization in the 1970s helped lay the foundation for the human-development approach to development. This approach stressed the importance of creating social and economic opportunities for all persons to realize their human potential. The major idea is that basic human needs must be the first priority of development. Thus, the approach placed emphasis on two elements: (a) meeting families' minimum survival needs, including adequate food, shelter and clothing; and (b) ensuring access to essential social services such as health, education, transport, water and sanitation. The approach also underscored the importance of creating opportunities for people to participate in political and decision-making processes (International Labor Organization, 1976).

Another major contribution to the human-development approach came from Amartya Sen's work on poverty, deprivation, inequalities, and gender. Sen (1999) suggests that development should be seen as the removal of factors that leave people with little opportunity to exercise their free-agency capabilities and hinder the fulfillment of their freedoms. Sen focused on *freedoms* as the major driver of human capabilities and achievement. Thus, the object of development should be about the enlargement of choices and removal of *unfreedoms*, which create opportunity for individuals to exercise their reasoned agency. This ultimately allows people to live free from economic and social deprivation and marginalization. Sen's ideas and extensive earlier writings helped influence the landmark Human Development Report, first launched in 1990, which tried to conceptualize and measure human development based on a range of choices. These include the choice to live a long and healthy life, to be educated, and to have access to resources needed for a decent standard of living. Additional choices include political freedom, human rights and personal self-respect (United Nations Development Program, 1990).

Though the human-development approach is a major intervention in development theory and practice, bringing together a wide range of perspectives that previously received little attention, critical questions remain about how far this approach departs from concerns with economic growth. For example, two decades after the launch of the first Human Development Report, the Human Development Index (HDI) still uses Gross National Product (GNP) – the total value of all final goods and services produced and income earned by citizens within a nation in a particular year – as a key component and indicator of people's standard of living. Notwithstanding these concerns, development researchers, practitioners, and national statistics offices have used HDI as the basis for measures of well-being that reflect the aspirations, values and priorities of societies beyond economic measures. Examples of such measures include the Canadian Index of Wellbeing, the Australia National Development Index, the New Zealand Social Report, and the UK Measures of National Wellbeing.

Environmental approach

For much of the 20th century, the environment was regarded as natural capital that could be exploited for productivity and economic growth (Wellis and Kumar, 2009). There was a general optimism that technological innovation would allow humans to "harness nature on an ever-larger scale" (Woodhouse, 2002, p. 141). However, environmental critics of development objected, warning against unabated economic growth without due regard to the environment. Early contributors to these debates include Meadows et al. (1972) and Erlich (1972). When the World Commission on Environment and Development (the UN's Brundtland Commission) published its report in 1987, it presented sustainable development as a consensus aimed at redefining the trade-offs between the environment and economic growth. The Commission's report defined "sustainable" development as that which "meets the needs of current generations without compromising the ability of future generations to meet their own needs" (Brundtland, 1987, p. 16).

The Commission's concept of development supports strong economic and social development that meets the needs of populations equitably. At the same time, it is concerned with the impacts of economic growth on the environment – now and in the future. Debates around the meaning, usefulness, and practical implementation of sustainable development started almost immediately after the report was released. Particularly, critics pointed to a mismatch between the report's emphasis on economic growth, on one hand, and the environmental agenda it espoused, on the other. In another breath, the concept was seen as a loose deployment of the value of integrating environmental concerns into development, with little theoretical contribution to the field of development (Payne and Philips, 2010).

The Sustainable Development Goals (SDGs) represent, to date, the most concrete global development agenda and blueprint that attempts to achieve development objectives sustainably. In 2012, the UN Conference on Sustainable Development (Rio+20) created the High-Level Political Forum (HLPF), tasked with tracking and facilitating the implementation of Agenda 2030 and its SDGs. This process culminated in the adoption of 17 SDGs in September 2015 at the UN General Assembly. The SDGs, presented in Table 4.1, can be understood as a universal call to action to end poverty and hunger in all parts of the world while ensuring people live in a peaceful, prosperous and environmentally sustainable global society. These 17 goals build on the successes, as well as address some of the failures, of prior efforts, namely the Millennium Development Goals (MDGs). However, there are concerns that the SDGs have failed to articulate a strong global effort that is committed to eradicating severe poverty in a sustainable way. Health geographers have contributed to this conversation, noting the contradictions between urgent economic growth and combating climate change, as well as the lack of clear funding measures and regimes (see, for example, Herrick, 2014).

Post-development as a critique of development

In the past few decades, post-development has emerged as a strong critique of the whole enterprise of development, both its practice and its theory. While there are many scholars critical of development, the defining difference between critical and post-development perspectives is the call for alternatives to development rather than alternative ways of development. Key contributors to post-development theory include Ferguson (1990) and Escobar (1992, 1995).

The key contentions of post-development theory can be summarized in three major themes: First, post–World War II development policies, and other alternatives aimed at reforming it, have failed to achieve any

Table 4.1 The Sustainable Development Goals (SDGs)

SDG #	SDG
1	No poverty
2	Zero hunger
3	Good health and well-being
4	Quality education
5	Gender equality
6	Clean water and sanitation
7	Affordable and clean energy
8	Decent work and economic growth
9	Industry, innovation and infrastructure
10	Reduced inequalities
11	Sustainable cities and communities
12	Responsible consumption and production
13	Climate action
14	Life below water
15	Life on land
16	Peace, justice and strong institutions
17	Partnership for the goals

meaningful results in terms of providing a dignified standard of living for the world's population (Andreasson, 2005). Second, development is inextricably linked to modernization, which includes an extension of Western control of the Global South (Escobar, 1995). Third, a genuine interest in localized, pluralistic grassroots movements; Indigenous ways of knowing and doing; and critical examination of scientific knowledge, knowledge creation and discourse must accompany a rejection of development (Escobar, 1992). Beyond analyzing the successes or failures of development projects and interventions, post-development shifts attention toward an analysis of what development does, the actors involved, the beneficiaries and losers, and the power of development knowledge and discourse.

Not surprisingly, post-development theorists and proponents did not escape the critical lens of researchers and practitioners. Some scholars complained that to further their arguments, post-development proponents failed to recognize some of the successes of the post–World War II development agenda. Notable among these successes include the significant improvements in life expectancy in the so-called Third World and other advancements in science and technology (Pieterse, 2000). Others raised concerns with the tendency of post-developmental authors to ignore the diversity within both the *West* and the *Rest* as well as the complex landscape of development in favor of essentialized accounts (Corbridge, 1998).

Though arguments for a post-development paradigm did not gain much traction into the 21st century, significant contributions were made in terms of critical development theory, practice and policy. Ziai (2007) identifies two enduring legacies of post-development theory that are difficult to dismiss and, in many respects, profoundly contributed to development thinking and practice. First is the idea that the theory and practice of traditional development is Eurocentric and neocolonial (Ziai, 2007). For example, the enduring legacy and Eurocentric construction of labeling Western Europe and North America as developed and Africa, Asia, and Latin America as underdeveloped and needing to "catch up" with the perceived "good life" of the Western world is increasingly being questioned in the field of development. Second is the idea that traditional development practice and thinking is technocratic, authoritarian and sometimes disastrous. For example, defining what development is, and how it should be measured and achieved, has always been the job of a technocrat or development expert, who sometimes will not be affected in any way if development is achieved or not (Ziai, 2007). Central to these contributions are "provocative investigations into the full gamut of the forms, structure and politics of inequality within nations and between individuals and groups in societies" (Payne and Philips, 2010, p. 144).

Geographies of health and development

Development geography, as with other cognate disciplines, is based on the understanding that the causes and consequences of inequalities in quality of life and standards of living involve a combination of social, economic, historical, political and environmental factors and process at various scales. In the context of development-health linkages, we often think of a good (or bad) standard of living (e.g., as measured by income and employment levels) as directly influencing health. But what about other dimensions of development, such as the quality of the institutions that govern us (e.g., security), the vitality and cohesiveness of communities (e.g., social capital), the environment we live in and grow (e.g., pollution), or the fairness of political and economic systems (e.g., access to opportunities)? These dimensions of development, too, have been shown to influence different aspects of human health in profound ways, and they are among the interests of health geographers who study the nexus between health and development. For example, over the past three decades, many studies have shown the health benefits of having trusting institutions and societies, which are sometimes tied to cultural values and fairness of economic systems.

Health geography and development geography have at their core many overlapping issues and naturally converge at some points. The most obvious grounding of the two sub-disciplines is the explicit causal relationship between health and development. That is, poor health hinders development efforts, and underdevelopment works against health (Luginaah, Bezner-Kerr, and Dixon, 2015). In addition, both sub-disciplines have a keen focus on *place*. Concern with both the attributes and subjective meanings of place unifies some of the issues that health and development geographers investigate. For example, while development geographers can be interested in how the attributes of place and interactions between places affect populations' standards of living, health geographers are interested in how these same factors influence experiences and distribution of disease and health (care).

Through these themes, development approaches to health geography are wide-ranging and share many similarities with peers in other areas of health geography. Yet a few notable, though highly overlapping, clusters have emerged in the scholarship. The first has its roots in the early days of medical geography, with a focus on documenting and mapping disease prevalence, and has extended through contemporary spatial disease ecologists who draw on new techniques, such as genetic landscape mapping (Young et al., 2017). Another area finds scholars who approach health questions from a more traditional development geography perspective and focus on how ecology and environment directly impact the health of local populations. This cluster includes a diverse array of research, from health geographers who are investigating water-insecurity challenges (Bisung et al., 2015) to those looking at the health implications of global climate change (Labbé et al., 2016). Likewise, there are those who are responding to the emergence of an increasingly globalized and urbanized Global South and its implications for both health and development (Kuuire et al., 2016; Atiim and Elliott, 2016). While health geographers in high-income countries have long been interested in equitable access to health care, this is also becoming of increasing interest in lower-income contexts (Dixon, Luginaah and Mkandawire, 2014).

Theoretically, health and development geographers have found common ground in the political ecology of health (PEH) framework, which advances knowledge on how patterns of health are produced through biological expression of ecological and societal living conditions and arrangements of power (Richmond et al., 2005). PEH has deep roots in radical development geography and Third World political ecology. PEH investigations take a holistic approach to tease out the complex interactions between socio-political and ecology factors across multiple spatiotemporal scales, with a focus on arrangements of power, economic activities, living conditions, and interactions between humans and their environment. Examples are investigations on the socioecological determinants of health among First Nations in Canada (Richmond et al., 2005) and ecological determinants of access to water in rural Kenya (Bisung et al., 2015).

In addition, traditional investigations under the broad theoretical banner of social production/political economy of disease have strong linkages within development geography. At its core, political economy of disease frameworks emphasize the economic and political determinants of, and barriers to, health and well-being. The rise of political economy of health approach in the second quarter of the 21st century was linked to global changes in politics, environment and science. Just as alternative approaches to development questioned the post–World War II dominance of neoliberal ideas during this period, health researchers engaged in a radical critique of how economic systems, particularly capitalist economic systems, operated to harm health and health care (Navarro, 1986). Attesting to the continued relevance of this framework, health geographers continue to demonstrate how diseases and health – including inequalities – are produced by the structures of economic and political systems, including the unequal power relationships inherent in and produced by these systems. Examples include work on welfare restructuring and health (Artazcoz et al., 2016), war, violence, and health (Berry and Berrang-Ford, 2016), and the political economy of tobacco control and pharmaceuticals (Bump and Reich, 2013; Holloway, 2014). Also addressed are social inequalities in health, particularly with respect to the health outcomes associated with racial and ethnic discrimination (O'Campo et al., 2016), gender violence and discrimination (Infanti et al., 2015), and historical injustices, dispossession and colonization (Richmond and Ross, 2009).

Future directions

Where is research on the geographies of health and development headed? Current links between social inequalities and health will continue to receive research and maybe genuine policy attention. So too will the issue of environmental justice, particularly in relation to disproportionate risk or marginalization of specific groups (e.g., the poor, the aged, the disabled, Indigenous populations). This last point will become especially important in light of the changing demographic profile of much of the world's low-income countries. That is, researchers' understanding of vulnerability must adapt along with an aging and urbanized Global South. Further, embodiment of absolute poverty and other vulnerabilities related to food, water, and noxious industrial substances will also receive much attention.

While the rapid spread of infectious diseases across borders has been the most manifest form of global health geographies, economic and health-policy regimes, which are often tied to the development aspirations of nations, have contributed to an interconnected health world with greater mobility of people and ideas. As the world becomes increasingly globalized, health geographers with a focus on place-awareness and social theory will be in a better position to answer some of the complex questions around mobility of people and health that impact development. For example, health geographers are likely to place medical tourism – a practice enabled by trade liberalization and globalization – under a critical lens, considering, for example, the social, legal, economic and policy implications of medical tourism for both the host and destination countries in ways that erode health equity (e.g., Crush and Chikanda, 2015). Finally, development is inherently political and power-laden, with constant struggle over resources and freedoms, including freedom of opportunity and economic choices. Taking a critical approach to the political economy of health and drawing on post-development theory, health geographers are well-positioned to advance knowledge on how these struggles are embodied to produce health outcomes.

References

Andreasson, S. (2005). Orientalism and African development studies: The "reductive repetition" motif in theories of African underdevelopment. *Third World Quarterly*, 26(6): pp. 971–986.

Artazcoz, L., Cortàs, I., Benavides, F. G. and Escribà-Agüir, V. (2016). Long working hours and health in Europe: gender and welfare state differences in a context of economic crisis. *Health & Place*, 40, pp. 161–168.

Atiim, G. A. and Elliott, S. J. (2016). The global epidemiologic transition: noncommunicable diseases and emerging health risk of allergic disease in sub-Saharan Africa. *Health Education & Behavior*, 43(1 suppl), pp. 37S–55S.

Berry, I. and Berrang-Ford, L. (2016). Leishmaniasis, conflict, and political terror: a spatio-temporal analysis of global incidence. *Social Science & Medicine*, 167, pp. 140–149.

Bisung, E., Elliott, S. J., Schuster-Wallace, C. J., Karanja, D. M. and Abudho, B. (2015). "Dreaming of toilets": using photovoice to explore knowledge, attitudes and practices around water-health linkages in rural Kenya. *Health & Place*, 31, pp. 208–215.

Brundtland, G. H. (1987). *Report of the world commission on environment and development: our common future.* [online] United Nations. Available at: www.un-documents.net/our-common-future.pdf [Accessed 1 Oct. 2017].

Bump, J. B. and Reich, M. R. (2013). Political economy analysis for tobacco control in low- and middle-income countries. *Health Policy and Planning*, 28, pp. 123–133.

Corbridge, S. (1998). "Beneath the pavement only soil": the poverty of post-development. *Journal of Development Studies*, 34(6), pp. 138–148.

Crush, J. and Chikanda, A. (2015). South-South medical tourism and the quest for health in Southern Africa. *Social Science & Medicine*, 124, pp. 313–320.

Dixon, J., Luginaah, I. and Mkandawire, P. (2014). The National Health Insurance Scheme in Ghana's Upper West region: a gendered perspective of insurance acquisition in a resource-poor setting. *Social Science & Medicine*, 122, pp. 103–112.

Erlich, P. (1972). *The population bomb.* London: Ballantine.

Escobar, A. (1992). Reflections on "development": grassroots approaches and alternative development in the Third World. *Futures*, 24(5), pp. 411–436.

Escobar, A. (1995). *Encountering development: the making and unmaking of the Third World.* Princeton, NJ: Princeton University Press.

FAO, IFAD and WFP. (2015). *The state of food insecurity in the world 2015. Meeting the 2015 international hunger targets: taking stock of uneven progress.* Rome: Food and Agricultural Organization of the United Nations.

Ferguson, J. (1990). *The anti-politics machine: "Development," depoliticization and bureaucratic power in Lesotho.* Cambridge: Cambridge University Press.

Harvey, D. (2005). *A brief history of neoliberalism.* Oxford: Oxford University Press.

Herrick, C. (2014). (Global) health geography and the post-2015 development agenda. *The Geographical Journal*, 180(2), pp. 185–190.

Holloway, K. (2014). Uneasy subjects: medical students' conflicts over the pharmaceutical industry. *Social Science & Medicine*, 114, pp. 113–120.

Infanti, J. J., Lund, R., Muzrif, M. M., Schei, B. and Wijewardena, K. (2015). Addressing domestic violence through antenatal care in Sri Lanka's plantation estates: contributions of public health midwives. *Social Science & Medicine*, 145, pp. 35–43.

International Labor Organization. (1976). *Employment, growth and basic needs: a one-world problem.* Geneva: International Labour Office.

Kuuire, V. Z., Arku, G., Luginaah, I., Buzzelli, M. and Abada, T. (2016). Transnationalism-integration nexus: examining the relationship between transnational housing investment and homeownership status in Canada. *Geoforum*, 75, pp. 168–179.

Labbé, J., Ford, J. D., Berrang-Ford, L., Donnelly, B., Lwasa, S., Namanya, D. B., Twesigomwe, S., IHACC Research Team and Harper, S. L. (2016). Vulnerability to the health effects of climate variability in rural southwestern Uganda. *Mitigation and Adaptation Strategies for Global Change*, 21(6), pp. 931–953.

Luginaah, I. N., Bezner-Kerr, R. and Dixon, J. (2015). Introduction. In: I. N. Luginaah and R. Bezner-Kerr, eds., *The geographies of health and development.* London: Ashgate, pp. 1–8.

Meadows, D. H., Meadows, D. L., Randers, J. and Behrens III (1972). *The limits to growth.* New York: Universe Books.

Navarro, V. (1986). *Crisis, health, and medicine: a social critique.* New York: Tavistock.

O'Campo, P., Schetter, D. C., Guardino, M. C., Vance, R. M. et al. (2016). Explaining racial and ethnic inequalities in postpartum allostatic load: results from a multisite study of low to middle income women. *SSM Population Medicine*, 2, pp. 850–858.

Payne, A. and Phillips, N. (2010). *Development.* London: Wiley-Blackwell.

Pieterse, J. (2000). After post-development. *Third World Quarterly*, 21(2), pp. 175–191.

Potter, R. B., Conway, D., Evans, R. and Lloyd-Evans, S. (eds.) (2012). *Key concepts in development geography*. London: Sage.

Richmond, C., Elliott, S. J., Matthews, R. and Elliott, B. (2005). The political ecology of health: perceptions of environment, economy, health and well-being among 'Namgis First Nation. *Health & Place*, 11, pp. 349–365.

Richmond, C. A. M. and Ross, N. A. (2009). The determinants of First Nation and Inuit health: a critical population health approach. *Health & Place*, 15(2), pp. 403–411.

Sen, A. (1999). *Development as freedom*. Oxford: Oxford University Press.

Truman, H. S. (1949). Harry S. Truman's inaugural address – 1949. [online] Available at: www.trumanlibrary.org/whistlestop/50yr_archive/inagural20jan1949.htm [Accessed 31 Aug. 2017].

UNICEF and World Health Organization. (2015). *Progress on sanitation and drinking water – 2015 update and MDG assessment*. Geneva: WHO Press.

United Nations Development Program. (1990). *Human development report 1990*. New York: Oxford University Press.

Willis, K. D. and Kumar, M. S. (2009). Development. In: R. Kitchin and N. Thrift, eds., *International encyclopedia of human geography*. Oxford: Elsevier.

Woodhouse, P. (2002). Development policies and environmental agendas. In: U. Kothari and M. Minogue, eds., *Development theory and practice: critical perspectives*. New York: Palgrave Macmillan, pp. 163–156.

Young, S. G., Carrel, M., Kitchen, A., Malanson, G. P., Tamerius, J., Ali, M. and Kayli, G. (2017). How's the flu getting through? Landscape genetics suggests both humans and birds spread H5N1 in Egypt. *Infection, Genetics & Evolution*, 49, pp. 293–299.

Ziai, A. (2007). Development discourse and its critics: an introduction to post-development. In: A. Ziai, ed., *Exploring post-development: theory and practice, problems and perspectives*. New York: Routledge, pp. 3–17.

5

PLACING HEALTH INEQUALITIES

Where you live can kill you

Clare Bambra

Today, Americans live three years shorter than their counterparts in France or Sweden. Scottish men live more than two years shorter than English men, and there is a 25-year gap in life expectancy across the suburbs of New Orleans (Bambra, 2016). Why are there such inequalities in health across all geographical scales – between neighborhoods, cities, regions, and countries? Health geographers have traditionally explained these health divides in terms of the effects of compositional (*Who lives here?*) and contextual (*What is this place like?*) factors. The compositional explanation asserts that the health of a given area, such as a town, region or country, is a result of the characteristics of the people who live there (individual-level demographic, behavioral and socioeconomic factors); whereas the contextual explanation argues that area-level health is also in part determined by the nature of the place itself in terms of its economic, social and physical environment. More recently, it has been acknowledged that these two approaches are not mutually exclusive and that the health of places results from the interaction of people with the wider environment (Cummins et al., 2007). This chapter will therefore also examine the *relational* approach (which tries to accommodate this *interaction*) as well as the *political economy* approach (which reconceptualizes *context*, looking beyond the local to the influence of national and international political and economic factors) (Bambra, 2016).

Who lives here: the compositional approach

The compositional view argues that *who lives here* – primarily the health behaviors (smoking, alcohol use, physical activity, diet, drug use) and socioeconomic (income, education, occupation) characteristics of the people living within a particular area (neighborhood, city, region, country) determine its health outcomes: that *poor people* result in *poor places*. Smoking, alcohol, physical activity, diet, and drugs – the five so-called lifestyle factors or risky health behaviors, all influence health significantly. Smoking remains the most important preventable cause of mortality in the wealthy world and is causally linked to most major diseases, such as cancer and cardiovascular disease (Jarvis and Wardle, 2006). Likewise, excessive alcohol consumption is related to some cancers as well as other key risks, such as high blood pressure. Alcohol-related deaths and diseases are on the increase. Poor diet and low exercise rates can lead to obesity, which, as discussed in Chapters 10 and 11, is a major risk factor for poor health and reduced longevity. Drug abuse is an increasingly important determinant of death among the young (Bambra et al., 2010). People who do not smoke, have only moderate alcohol intake, consume a high amount of fruit and vegetables and engage regularly in physical activity will on average have a 14-year higher life expectancy than individuals achieving no healthy behaviors

(Khaw et al., 2008). So, on average, areas (countries, regions, cities, neighborhoods) with higher rates of these unhealthy behaviors would have worse health than others, all else being equal.

The socioeconomic status of people living in an area is also of huge health significance. "Socioeconomic status" refers to occupational class, income or educational level (Bambra, 2011). People with higher occupational status (e.g., professionals such as teachers or lawyers) have better health outcomes than non-professional workers (e.g., manual workers). By way of example, data shows that infant mortality rates were 16% higher in children of routine and manual workers as compared to professional and managerial workers (Marmot, 2010). Having a higher income or being educated to degree level can also have a protective health effect, where as having a lower income or no educational qualifications can have a negative health impact. The poorer someone is, the less likely that person is to live in good-quality housing, have time and money for leisure activities, feel secure at home or work, have good-quality work or a job at all, or afford healthy food – the social determinants of health (Marmot, 2010).

What is this place like: the contextual approach

So, while the compositional view argues that it is *who lives here* that matters for area health – and that essentially *poor people make poor health*, the contextual approach instead highlights the fact that *what is this place like* also matters for health. Health differs by place because it is also determined by the economic, social, and physical environment of a *place*: in other words, *poor places lead to poor health*. Place mediates the way in which individuals experience social, economic and physical processes on their health: places can be saluto-genic (health-promoting) or pathogenic (health-damaging) environments – place acts as a health ecosystem. These place-based effects can also be seen as the *collective* effects of the social determinants of health. There are three contextual aspects to place that have traditionally been considered important to health; these are the economic, social and physical aspects.

The compositional view takes into account the effects of individual socioeconomic position on health status. Area-level economics instead looks at the health effects of the local economic environment, independent of individual socioeconomic position. Area-economic factors that influence health are often summarized as economic deprivation. They include poverty rates, unemployment rates, wages, and types of work and employment in the area. The mechanisms whereby the economic profile of a local area affects health are multiple. For example, it affects the nature of work that an individual can access in that place (regardless of his or her socioeconomic position). It also impacts the services locally available, as more affluent areas will attract different services (such as food available locally or physical activity opportunities) than more deprived areas as businesses adapt to the different consumer demands in each area (see "access to services" in the opportunity structures section below). Area-level economic factors such as poverty are a key predictor of health, including cardiovascular disease, all-cause mortality, limiting long-term illness, and health-related behaviors (Macintyre, 2007).

Places also have social aspects that impact health. Opportunity structures are the socially constructed and patterned features of the area, which may promote health through the possibilities they provide (Macintyre, Ellaway and Cummins, 2002). These include the services provided, publicly or privately, to support people in their daily lives. Examples are child care, transport, food availability and access to a family physician or hospital, as well as the availability of health-promoting environments at home (e.g., good housing quality, access and affordability), work (good-quality work) and education (such as high-quality schools). For example, local environments can shape our access to healthy – and unhealthy – goods and services, thus enhancing or reducing our opportunities to engage in healthy or unhealthy behaviors such as smoking, alcohol consumption, fruit and vegetable consumption, and physical activity. One example is the obesogenic environment. The local food environment – such as the availability of healthy versus unhealthy foods in the neighborhood – as well as opportunities for physical activity – whether there are parks or gyms, whether the outside space is safe and walkable – are both central components of the obesogenic environment. Research has shown that in some low-income areas, a paucity of supermarkets and shops selling affordable fresh food, alongside an abundance of convenience stores and fast-food outlets selling energy-dense junk food and ready meals has created "food deserts" (Pearce et al., 2007). Low-income neighborhoods – particularly urban ones – may also inhibit opportunities for physical activity. Associations have been found between neighborhood availability of fast food and obesity rates in a number of wealthy countries, including the United Kingdom, the United States and New Zealand (Pearce et al., 2007; Burgoine, Alvanides and Lake, 2011).

A second social aspect of place is collective social functioning. Collective social functioning and practices that are beneficial to health include high levels of social cohesion and social capital within the community. Social capital – "the features of social organization such as trust, norms, and networks that can improve the efficiency of society by facilitating coordinated actions" (Putnam, 1993, p. 167) – has been put forward as a social mechanism through which place mediates the relationship between individual socioeconomic status and health outcomes (Hawe and Shiell, 2000). Some studies have found that areas with higher levels of social capital have lower mortality rates, better self-rated health, better mental health, and healthier behaviors. More negative collective effects can also come from the reputation of an area (e.g., stigmatized places can result in feelings of alienation and worthlessness) or the history of an area (e.g., if there has been a history of racial oppression). Place attachment (an emotional bond that individuals or groups have with specific places), in contrast, can have a protective health effect (Gatrell and Elliot, 2009). Certain places become marginalized by obtaining a spoiled identity and subsequently become stigmatized and discredited. This can be as a result of environment factors, such as air pollution or dirt, as well as from social stigma – such as being labeled the obesity capital of Britain, as happened with Copeland in West Cumbria (in northwest England), or economic stigma, such as low property prices (Bush, Moffatt and Dunn, 2001). Residents of stigmatized places can also be discredited by association with these place characteristics. A notable case of such placed-based stigma is Love Canal, New York – the location of a toxic waste dump. Research has shown that such place-based stigma can result in psychosocial stress and associated ill health alongside feelings of shame, on top of the physical health effects of air pollution, such as respiratory disease (Airey, 2003). Local attitudes – say, around smoking – can also influence health and health behaviors either negatively or positively (Thompson, Pearce and Barnett, 2007).

The physical environment is widely recognized as an important determinant of health and health inequalities (World Health Organization, 2008). There is a sizeable literature on the positive health effects of access to green space, as well as the negative health effects of waste facilities, brownfield or contaminated land, as well as air pollution (Bambra, 2016). An infamous example of the latter is the so-called *Cancer Alley* – the 87-mile stretch in the American state of Mississippi between Baton Rouge and New Orleans, the home of the largest petrochemicals site in the country (Markowitz and Rosner, 2002). In 2016 it was estimated that air pollution levels in London accounted for up to 10,000 unnecessary deaths per year (Walton et al., 2015). Another example of how the physical environment of areas varies is in respect to land pollution. A study found that in the American city of Baltimore, mortality rates from cancer, lung cancer, and respiratory diseases were significantly higher in neighborhoods with larger amounts of brownfield land (Litt, Tran and Burke, 2002). Similarly, an English study of differences in exposure to brownfield land found that neighborhoods with large amounts of brownfield land had higher rates of poor health and limiting long-term illness (Bambra et al., 2014).

The health-geography literature has also established the role of natural or green spaces as therapeutic or health-promoting landscapes. So, for example, studies have found that walking in natural, rather than urban, settings reduces stress levels, and people residing in green areas report less poor health than those with less green surroundings (Maas et al., 2005). Research also indicates that green space can increase people's health by attention restoration, stress reduction and/or the evocation of positive emotions (Abraham,

Sommerhalder and Abel, 2010). Awareness of how such factors differ by place has led to the development of the concept of *environmental deprivation* – the extent of exposure to key characteristics of the physical environment that are either health-promoting or health-damaging (Pearce et al., 2010). Environmental deprivation is associated with all-cause mortality: mortality was lowest in areas with the least environmental deprivation and highest in the most environmentally deprived areas. The unequal socio-spatial distribution of the environmental deprivation has also led to commentators developing the concept of environmental justice (Pearce et al., 2010). The fact that more deprived neighborhoods are more likely to have air and land pollution and less likely to have green space can be seen as an aspect of social injustice (Pearce et al., 2010).

Poor people *and* poor places: the relational approach

The contextual and compositional explanations for how place relates to health are not mutually exclusive, and to separate them is an oversimplification that ignores the interactions between these two levels (Macintyre, Ellaway and Cummins, 2002). The characteristics of individuals are influenced by the characteristics of the area. For example, occupational class can be determined by local school quality and the availability of jobs in the local labor market, or children might not play outside due to not having a private garden (a *compositional* resource), because there are no public parks or public transport to them (a *contextual* resource) or because it might not be seen as appropriate for them to do so (*contextual* social functioning) (Macintyre, Ellaway and Cummins, 2002). Similarly, areas with more successful economies (e.g., more high-paying jobs) will have lower proportions of lower-socioeconomic-status residents.

Further, the collective-resources model suggests that all residents, and particularly those on a low income, enjoy better health when they live in areas characterized by more/better social and economic collective resources. This may be especially important for those on low incomes, because they are usually more reliant on local services. Conversely, the health of poorer people may suffer more in deprived areas where collective resources and social structures are limited, a concept known as deprivation amplification: that the health effects of individual deprivation, such as lower socioeconomic status, can be *amplified* by area deprivation (Macintyre, 2007). Figure 5.1 shows an example of these interaction effects. This figure shows how a healthy lifestyle score in a study in the east of England is affected by both individual occupation (compositional) and area-level deprivation (contextual): people from all occupational backgrounds fare worse in areas of higher deprivation than when living in more affluent areas (Lakshman et al., 2011).

Composition and context should therefore be seen not as separate or competing explanations – but as entwined. Both contribute to the complex relationship between health and place – an ecosystem made up of people, systems and structures. As Cummins and colleagues argue, "there is a mutually reinforcing and reciprocal relationship between people and place" – a relational approach should therefore be taken to understanding how compositional and contextual factors interact to produce geographical inequalities in health (Cummins et al., 2007, p. 1826). Table 5.1 provides an example of the relative role of compositional and contextual factors individually and collectively in explaining health inequalities between the most and least deprived neighborhoods of a case-study town – Stockton on Tees in the northeast of England. Stockton on Tees has a 17-year gap in life expectancy for men and 11 years for women between its most and least deprived neighborhoods. This is the largest gap in life expectancy within a single local authority in England. Table 5.1 shows the results of statistical modeling of household survey data that examines how much of the health gap in Stockton is explained by compositional and contextual factors – and their interaction. The compositional factors used relate to individual-level socioeconomic factors (income, unemployment, etc.), psychosocial factors (loneliness, isolation, etc.), and behavioral factors (smoking, drinking, diet, exercise). The contextual data relates largely to the physical environment (noise, pollution, dirt, crime, safety, housing quality). Three different measures of health are used: general well-being (EQ5D and EQ-VAS scales) and

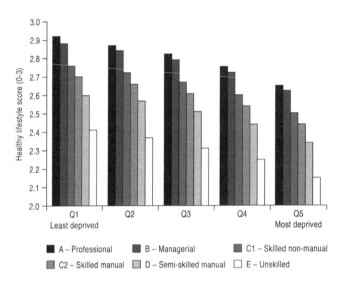

Figure 5.1 Mean healthy lifestyle score across quintiles of neighborhood deprivation and six categories of occupational social class

Data from Lakshman et al. (2011), with permission from Oxford University Press.

physical health (SF8-PCS). Compositional factors account for by far the majority of the gap, 47% across all three of the health outcomes. Contextual factors account for between 3% and 11%. Of course, the different causes of health gaps often cluster together – people who experience poor material factors often also experience poor psychosocial factors and poorer environments, and they are more likely to engage in less healthy behaviors. This interaction of compositional and contextual factors is also shown in Table 5.1, and it accounts for between 18% and 30% of the gap. As often happens with statistical models, there is a certain proportion of the gap that remains unexplained. However, this analysis shows that both contextual and compositional factors matter and that their interaction is an important cause of geographical inequalities in health – supporting a *relational* approach to health and place (Bhandari et al., 2017).

Beyond the individual and the local: the political-economy approach

The political-economy approach to explaining health divides focuses on the social, political and economic structures and relations that may be, and often are, outside the control of the individuals or the local areas they affect (Krieger, 2003). Individual and collective social and economic factors such as housing, income, and employment – indeed, many of the issues that dominate political life – are key determinants of health and well-being (Bambra, Fox and Scott-Samuel, 2005). Why some places and people are consistently privileged while others are consistently marginalized is a political choice – it is about where the power lies and in whose interests that power is exercised. Political choices can thereby be seen as the *causes of the causes of the causes* of geographical inequalities in health (Bambra, 2016).

By way of example, we can examine the causes of stroke or heart disease (Bambra, 2016). The immediate clinical *cause* could be hypertension (high blood pressure). The *proximal cause* of the hypertension itself could be compositional lifestyle factors, such as poor diet, of which the contextual cause might be living in a low-income neighborhood. The causes of the latter are political – low-income neighborhoods exist because the political and economic system allows them to exist. Wages could be regulated so that they are higher (an

Table 5.1 The relative role of compositional and contextual factors in explaining the health gap in Stockton on Tees

% gap explained	Health measure		
	Measure 1 (SF8PCS)	*Measure 2 (EQ5D)*	*Measure 3 (EQVAS)*
Compositional	47%	47%	47%
Contextual	8%	11%	3%
Interactions	18%	30%	24%
Total Explained	73%	88%	74%
Unexplained	27%	12%	26%

Data adapted from Bhandari et al. (2017)

example being the living wage), or food prices could be controlled/subsidized (e.g., in the United States it is meat and corn oil that receive government subsidies, not fruit and vegetables; likewise, in the European Union, farmers are encouraged to produce dairy) and neighborhood food provision does not have to be left to the vagaries of the market (which leads to clustering of poor food availability in poor neighborhoods).

In this sense, geographical patterns of health and disease are produced by the structures, values and priorities of political and economic systems (Krieger, 2003). Area-level health – be it local, regional or national – is determined, at least in part, by the wider political, social and economic system and the actions of the state (government) and international-level actors (supra-national government bodies, such as the European Union, interstate trade agreements such as the Transatlantic Trade and Investment Partnership [TTIP], as well as the actions of large corporations): politics can make us sick – or healthy (Schrecker and Bambra, 2015). Politics and the balance of power between key political groups – notably labor and capital – determine the role of the state and other agencies in relation to health and whether there are collective interventions to improve health and reduce health inequalities, and also whether these interventions are individually, environmentally or structurally focused. In this way, politics (broadly understood) is the fundamental determinant of our health divides, because it shapes the wider social, economic and physical environment and the social and spatial distribution of salutogenic and pathogenic factors both collectively and individually (Bambra, 2016).

An example of the influence of politics on health is demonstrated in Figure 5.2. This shows how the gap between the most and least deprived neighborhoods increased through the 1980s and 1990s. Collins and McCartney (2011) argue that this is a result of the Thatcher government's (1979–1990) neoliberal approach to the economy and society and constituted a political attack against the working class, of which Scotland (particularly Glasgow and the west of Scotland) became a particular target (Collins and McCartney, 2011). The Thatcher government radically altered the social settlement: mass unemployment became normalized via deindustrialization at the same time as trade unions and workers' rights were curtailed; there were significant reductions in welfare benefits, leading to the intensification of poverty and wage compression, as well as vast reductions in the availability of social housing (Scott-Samuel et al., 2014). While neoliberalism spread across wealthy countries in the 1980s, the United Kingdom was exposed in a way that other European nations were not – in a very rapid and intense manner that adversely affected health through unemployment, poverty, alienation and associated increases in risky health behaviors. For example, deindustrialization was implemented as a *shock doctrine* with very rapid loss of employment within a few years in the United Kingdom, while in other Western European countries it was phased in more gradually and often with more safety nets (such as employment services or inducements for new industry to come to the affected areas). A more recent example is the effects of austerity and the resulting reductions in welfare benefits and public services on social geographies and health inequalities (Pearce, 2013).

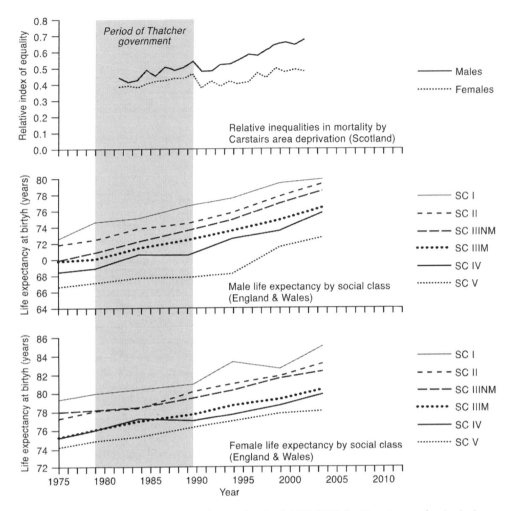

Figure 5.2 Geographical trends in health inequalities in Scotland, 1981–2001 (by Carstairs area deprivation)

Reproduced, with permission of the Sage publishing group, from Scott-Samuel et al. (2014).

Note: RII (relative index of inequality) – for further information, see Regidor (2004).

Future directions

To date, health geography has conventionally presented two main explanations – *compositional* and *contextual* – as to why there are such stark geographical inequalities in health. More recently, though, these approaches have been reconciled via the *relational* understanding of place and health. Further, a *political economy* approach has recently been taken up within the discipline so that the role of the wider macro-political, economic and societal context is beginning to be examined. The future directions and frontiers for research into geographical inequalities in health lie in taking this political-economy approach forward by examining such things as the effects of continuing austerity in southern Europe on geographical inequalities in health; the effects of major global changes, such as the migration crisis and the emergence of right-wing populist governments, on the social inequalities underpinning health inequalities; and the global economy, which is facing sudden shocks such as Brexit.

Acknowledgments

This chapter is based on Clare Bambra (2016)'s *Health Divides: Where You Live Can Kill You*, with permission of Policy Press. Clare Bambra is funded by a Leverhulme Trust Research Leadership grant to investigate health inequalities (reference RL-2012–2006).

References

Abraham, A., Sommerhalder, K. and Abel, T. (2010). Landscape and well-being: a scoping study on the health-promoting impact of outdoor environments. *International Journal of Public Health*, 55(1), pp. 59–69.

Airey, L. (2003). "Nae as nice a scheme as it used to be": lay accounts of neighbourhood incivilities and well-being. *Health & Place*, 9(2), pp. 129–137.

Bambra, C. (2011). *Work, worklessness, and the political economy of health*. Oxford: Oxford University Press.

Bambra, C. (2016). *Health divides: where you live can kill you*. Bristol: Policy Press.

Bambra, C., Fox, D. and Scott-Samuel, A. (2005). Towards a politics of health. *Health Promotion International*, 20(2), pp. 187–193.

Bambra, C., Joyce, K., Bellis, M., Greatley, S., Greengross, S., Hughes, P., Lincoln, T., Lobstein, C., Naylor, R., Salay, M., Wiseman, M. and Maryon-Davis, A. (2010). Reducing health inequalities in priority public health conditions: using rapid review to develop proposals for evidence-based policy. *Journal of Public Health*, 32(4), pp. 496–505.

Bambra, C., Robertson, S., Kasim, A., Smith, J., Cairns-Nagi, J., Copeland, A., Finlay, N. and Johnson., K. (2014). Healthy land? An examination of the area-level association between brownfield land and morbidity and mortality in England. *Environment and Planning A*, 46(2), pp. 433–454.

Bhandari, R., Akhter, N., Warren, J., Kasim, A. and Bambra, C. (2017). Geographical inequalities in general and physical health in a time of austerity: baseline findings from the Stockton-on-Tees cohort study. *Health & Place*, 48, pp. 111–122.

Burgoine, T., Alvanides, S. and Lake., A. (2011). Assessing the obesogenic environment of North East England. *Health & Place*, 17(3), pp. 738–747.

Bush, J., Moffatt, S. and Dunn, C. (2001). "Even the birds round here cough": stigma, air pollution and health in Teesside. *Health & Place*, 7(1), pp. 47–56.

Collins, C. and McCartney, G. (2011). The impact of neoliberal political attack on health: the case of the Scottish effect. *International Journal of Health Services*, 41(3), pp. 501–526.

Cummins, S., Curtis, S., Diez-Roux, A. and Macintyre, S. (2007). Understanding and representing "place" in health research: a relational approach. *Social Science & Medicine*, 65(9), pp. 1825–1838.

Gatrell, A. and Elliot, S. (2009). *Geographies of health: an introduction*. London: Wiley-Blackwell.

Hawe, P. and Shiell, A. (2000). Social capital and health promotion: a review. *Social Science & Medicine*, 51(6), pp. 871–885.

Jarvis, M. and Wardle, J. (2006). Social patterning of individual health behaviours: the case of cigarette smoking. In: M. Marmot and R. Wilkinson, eds., *The social determinants of health*. Oxford: Oxford University Press, pp. 240–255.

Khaw, K., Wareham, N., Bingham, S., Welch, A., Luben, R. and Day, N. (2008). Combined impact of health behaviours and mortality in men and women: the EPIC-Norfolk prospective population study. *PloS Medicine*, 5(3), pp. 39–47.

Krieger, N. (2003). Theories for social epidemiology in the twenty-first century: an ecosocial perspective. In: R. Hofrichter, ed., *Health and social justice: politics, ideology, and inequity in the distribution of disease – a public health reader*. San Francisco: Jossey-Bass, pp. 428–450.

Lakshman, R., McConville, A., How, S., Flowers, J., Wareham, N. and Cosford, P. (2011). Association between area-level socioeconomic deprivation and a cluster of behavioural risk factors: cross-sectional, population-based study. *Journal of Public Health*, 33(2), pp. 234–245.

Litt, J., Tran, N. and Burke, T. (2002). Examining urban brownfields through the public health "macroscope." *Environmental Health Perspectives*, 110(2), pp. 183–193.

Maas, J., Verheij, R., de Vries, S., Spreeuwenberg, P. and Groenewegen, P. (2005). Green space, urbanity, and health: how strong is the relation? *European Journal of Public Health*, 60(7), pp. 587–592.

Macintyre, S. (2007). Deprivation amplification revisited; or, is it always true that poorer places have poorer access to resources for healthy diets and physical activity? *International Journal of Behavioral Nutrition and Physical Activity*, 4(32), pp. 1–7.

Macintyre, S., Ellaway, A. and Cummins, S. (2002). Place effects on health: how can we conceptualise, operationalise and measure them? *Social Science & Medicine*, 55(1), pp. 125–139.

Markowitz, G. and Rosner, D. (2002). *Deceit and denial: the deadly politics of industrial pollution*. Berkeley: University of California Press.

Marmot, M. (2010). *Fair society, healthy lives: the Marmot review.* London: University College.

Pearce, J. (2013). Financial crisis, austerity policies, and geographical inequalities in health Introduction Commentary. *Environment and Planning A*, 45(9), pp. 2030–2045.

Pearce, J., Blakely, T., Witten, K. and Bartie, P. (2007). Neighborhood deprivation and access to fast-food retailing – A national study. *American Journal of Preventive Medicine*, 32(5), pp. 375–382.

Pearce, J., Richardson, E., Mitchell, R. and Shortt, N. (2010). Environmental justice and health: the implications of the socio-spatial distribution of multiple environmental deprivation for health inequalities in the United Kingdom. *Transactions of the Institute of British Geographers*, 35(4), pp. 522–539.

Putnam, R. (1993). *Making democracy work: civic traditions in modern Italy.* Princeton, NJ: Princeton University Press.

Regidor, E. (2004). Measures of health inequalities: part 2. *Journal of Epidemiology and Community Health*, 58(3), pp. 900–903.

Schrecker, T. and Bambra, C. (2015). *How politics makes us sick: neoliberal epidemics.* London: Palgrave Macmillan.

Scott-Samuel, A., Bambra, C., Collins, C., Hunter, D., McCartney, G. and Smith, K. (2014). The impact of Thatcherism on health and well-being in Britain. *International Journal of Health Services*, 44(1), pp. 53–71.

Thompson, L., Pearce, J. and Barnett, R. (2007). Moralising geographies: stigma, smoking islands and responsible subjects. *Area*, 39(4), pp. 508–517.

Walton, H., Dajnak, D., Beevers, S., Williams, M., Watkiss, P. and Hunt, A. (2015). *Understanding the health impacts of air pollution in London.* London: King's College.

World Health Organization. (2008). *Commission on the social determinants of health: closing the gap in a generation.* Geneva: World Health Organization.

6

ENVIRONMENTAL HEALTH INEQUITIES

From global to local contexts

Eric Crighton, Holly Gordon and Caroline Barakat-Haddad

There is general consensus that the burdens of pollution and other environmental hazards are not distributed evenly across communities and nations and that this explains an important part of the health disparities that exist between different populations around the world (WHO, 2010; Prochaska et al., 2014). Importantly, how hazards like industrial pollution, automobile emissions and agricultural pesticides are distributed, and the burden they place on health, is often determined not by chance, but rather by environmental, economic and social practices, as well as policies and laws that discriminate based on race, class, culture, age or gender. Thus, many environmental health *inequalities* also represent *inequities* in that they are unfair, unjust and avoidable (WHO, 2010).

This chapter examines the concept and theoretical underpinnings of environmental health inequity, beginning with its roots in the environmental-justice and civil-rights movements and continuing to its contribution to the discipline of health geography today, while illustrating the multiplicity, scope and scale of the phenomena through an examination of international examples.

Background

The concept of environmental health inequity is rooted in environmental justice, a social movement and research stream focused on documenting and redressing disproportionate environmental burdens faced by socioeconomically or demographically disadvantaged groups. Environmental justice was first described in terms of race (i.e., environmental racism) in studies examining community racial and socioeconomic characteristics and the location of hazardous-waste sites in the United States (Boyd, 2015). In 1979, an African-American community in Houston, Texas, filed the first lawsuit to challenge the local siting of a hazardous-waste landfill. Public interest in environmental justice grew in the 1980s, with high-profile events such as the Love Canal disaster in Niagara Falls, New York. Hundreds of homes were evacuated after a protracted battle between the government and a community built in close proximity to a toxic-waste dump, eventually leading to the evacuation of hundreds of homes (Edelstein, 2004). In 1982, more than 500 people were arrested in Warren County, North Carolina, while protesting to stop the siting of a landfill for polychlorinated biphenyl (PCB)–contaminated soil in a predominately poor, African-American community (Boyd, 2015). Prompted by the events of Warren County, the US government conducted a study comparing racial and socioeconomic characteristics and the location of hazardous-waste sites. Findings showed a disproportionate number of hazardous sites located in or near African-American communities (Lee, 2010).

These are among the many events that fed a growing awareness about environmental hazards in marginalized, low-income and racial-minority communities and a groundswell of related activism and research in the United States and internationally, on issues including air quality, toxic-waste disposal, workplace hazards and disproportionate exposure to lead (Lee, 2010).

The concept of environmental health inequity has emerged as an important theme within the disciplines of health geography and environmental health. There is a significant and growing body of evidence showing that poor and disadvantaged communities are more likely to live near sources of pollution, such as industries, refineries and highways, and as a consequence are more likely to suffer from associated adverse health effects (Masuda et al., 2008; WHO, 2010). These communities are also more likely to feature poor-quality housing, noise and odors and to have fewer environmental amenities, such as green spaces, close by (Boyd, 2015). Confirming the previous findings, Tooke, Klinkenberg and Coops (2010) have shown that in Canadian urban settings, the proportion of green space and tree canopy cover, both of which are associated with positive mental and physical health outcomes, is positively associated with neighborhood income (Tooke, Klinkenberg and Coops, 2010). Research has also shown that these inequities exist across age and gender. For example, it has been reported that, because of prolonged shortages of fuel, low-income Armenian women were commonly burning municipal waste for cooking and heating, resulting in disproportionate gender-related exposure to hazardous substances such as dioxins and heavy metals (WHO, 2010). These inequities are reproduced across geographic scales. For example, a study in Toronto, Canada, examining neighborhood levels of nitrogen dioxide, a hazardous air pollutant, showed the neighborhoods marked by low income, low education and lone-parent families were more likely to experience high ambient exposures (Buzzelli and Jerrett, 2007). At the macro scale, using World Health Organization data and a broad definition of "environmental risks" that included occupational hazards, Benetti (2007) assessed deaths attributable to environmental risk factors by nation and found that deaths attributable to environmental risk factors were substantially greater in less-developed countries (25%) as compared to more-developed countries (17%).

According to Bolte et al. (2012), there are two dimensions to environmental health inequity. The first dimension is known as distributional or geographic inequity, which refers to the spatial distribution and resulting disparities of environmental risks across socioeconomic and racial groups. This includes places where people live, work, learn and recreate. The second dimension is known as procedural inequity, which refers to opportunities, or a lack thereof, to influence decisions about pollution and development that may create environmental risks. Low socioeconomic and marginalized groups are often characterized by both geographic and procedural inequities (Kruize et al., 2014) that result in health disparities across a range of geographic contexts. How this occurs has been conceptualized by numerous authors (e.g., WHO, 2010; Morello-Frosch and Lopez, 2006; Gee and Payne-Sturges, 2004; Sexton, 2000). Morello-Frosch and Lopez (2006), for example, present an ecosocial framework outlining how the connections between spatial forms of social inequality (e.g., racial segregation), structural mechanisms of discrimination (e.g., political disenfranchisement, economic exclusion, uneven government investment patterns) and community-level conditions (e.g., land use, traffic density, civic engagement, neighborhood quality) disproportionately expose low socioeconomic and marginalized communities to environmental hazards. Individual-level stressors such as poverty, poor diet and lack of health behaviors, as well as inadequate access to health care, act to amplify *vulnerability* to the toxic effects of those environmental hazards that create health disparities. Gee and Payne-Sturges (2004), in their exposure-disease-stress framework, highlight the importance of psychosocial impacts of environmental exposures on health, as well as the role that individual-level stress may play in increasing vulnerability and environmental health inequities. This relationship has been demonstrated by health geographers in numerous studies in diverse contexts of both real and perceived environmental exposures (e.g., Crighton et al., 2003a; b; Wakefield and Elliott, 2000).

In examining place-specific health disparities, health geographers have long focused on the contribution of environmental determinants of health within the context of broad socioeconomic, health-care and individual factors. In doing so, the research has revealed evidence of significant health inequalities across a range of geographic contexts associated with the inequitable distribution of environmental hazards. In the following pages, we present a series of illustrative examples of environmental health inequity that span global to local scales in order to elucidate the nature, diversity and ubiquity of the problem. At the global scale, we examine international trade in toxic electronic waste (e-waste) and large-scale environmental disasters, namely the Aral Sea disaster, and, at the local scale, we examine urban air pollution as well as environmental exposures in Indigenous communities.

International trade in hazardous e-waste

The export of e-waste from high-income countries to low-income countries represents a significant and growing international environmental-health-inequity issue that has received considerable attention (Heacock, Kelly and Suk, 2016). E-waste, which includes obsolete and discarded computers, phones, televisions and numerous other electronic products, represents a large and fast-growing component of the global municipal solid-waste stream (Widmer et al., 2005). E-waste contains numerous valuable materials, including copper, silicon, iron and gold, but also many highly toxic substances, such as cadmium (Cd), mercury (Hg), chromium (Cr), lead (Pb), and polybrominated diphenyl ethers (PBDEs) (Xu et al., 2012). The hazardous nature of the waste, combined with strict environmental and health-and-safety regulations in high-income countries, makes it cheaper to export e-waste for recycling to low-income countries, where regulations are either weak or unenforced (Wu et al., 2015). This trade is often done in violation of international law (Widmer et al., 2005).

The environment and health impacts associated with the trade in e-waste are significant. This is demonstrated in a study by Xu et al. (2012) that examined the impacts of e-waste recycling on birth outcomes in Guiyu, China, where it is estimated that nearly 6,000 family workshops informally recycle approximately 1.6 million tons of e-waste per year using antiquated and uncontrolled methods. Comparing Guiyu to a non-recycling control site revealed significantly higher concentrations of cadmium, chromium, nickel and PBDEs in umbilical cord blood of neonates in Guiyu. Lead, a major component of e-waste and highly detrimental to neurodevelopment, was also found to be significantly higher in Guiyu. With regard to birth outcomes, the study revealed significantly higher rates of stillbirths, low birth weights and lower Apgar scores (a standard measure of the physical condition of a newborn) in Guiyu compared to the control site (Xu et al., 2012).

Unfortunately, the situation seen in Guiyu is not unique. A study examining the environmental impacts of e-waste recycling in Taizhou City, China, revealed that the improper disassembly of polychlorinated biphenyl (PCB)–containing equipment has led to PCB contamination at levels estimated to pose serious cancer risks to children and adults (Wu et al., 2015). In Ghana, the Agbogbloshie recycling site receives up to 215,000 tons of e-waste annually from Western Europe. Here, open burning is used to remove plastic insulation from copper cables, a practice that heavily pollutes the air, soil and water and exposes workers and local communities to increased health risks (Amankwaa, 2013). The international trade of e-waste and other hazardous waste from more-developed to less-developed nations is not a new phenomenon, and, given the fast pace of technological developments and speed at which electronics become obsolete, the trade in e-waste is expected to increase in the future (Widmer et al., 2005).

The Aral Sea disaster

There is overwhelming evidence of environmental health inequity in the context of both natural and unnatural environmental disasters (Bullard, 2007). Natural disasters such as Hurricane Katrina in the United States in 2005 demonstrated clearly that many low-income and black communities were particularly vulnerable to the storm and its impacts but also received the lowest priority in terms of disaster response, allocation of post-disaster assistance, and reconstruction assistance (Bullard, 2007). This inequity is also seen in the context of unnatural environmental disasters. The Aral Sea disaster is an important example of this due to its scale, the severity of the health impacts and also the lack of attention it has received internationally. The history of the disaster dates back to the 1950s, when the Soviet government in Moscow imposed unsustainable, industrial cotton-farming methods on the Soviet republics of Kazakhstan and Uzbekistan. Doing so involved diverting massive quantities of water for irrigation away from the Aral Sea, once the fourth-largest freshwater lake in the world (Micklin, 2007). The Aral Sea supported a rich natural ecosystem and a minority population of more than five million people. The region quickly became one of the world's largest cotton producers; however, the local population received few of the economic benefits, and the environmental impacts have been enormous (Small, van der Meer and Upshur, 2001). The sea steadily receded, such that today it covers less than one-fifth of its original area, and towns like Muynak, once a major fishing port and tourist destination, is more than 100 kilometers from the waters of the now biologically dead sea. Salinization of agricultural land resulting from over-irrigation, as well as the wind transport of salts from the exposed seabed, has severely impacted ground and surface water quality and reduced agricultural productivity (Micklin, 2007). Concomitantly, decades of chemical-dependent agricultural practices have resulted in high levels of toxic pesticides in the local environment and the population living there (Crighton et al., 2011). As a consequence of these environmental impacts, the people living in this area have experienced significant health declines and face some of the highest rates of anemia and certain cancers, respiratory illnesses and birth defects in the world (Crighton et al., 2011).

Although local residents and scientists have long claimed that the environmental problems have been significantly affecting their health, limited empirical evidence of the relationships, combined with a general disinterest in the disaster internationally, have been cited as barriers to action (Small, van der Meer and Upshur, 2001). Of the research that has been done, most is focused on agricultural chemicals and their impact on the most vulnerable in the population: children. Results have demonstrated high levels of organochlorine pesticides, dioxins, heavy metals and other toxic pollutants in the environment, in local food products and in the population itself (i.e., in umbilical cord blood, urine, hair, breast milk) (Crighton et al., 2011). It has also been shown that body burdens of toxic substances are significantly higher among individuals living in the most environmentally affected areas (i.e., cotton-growing areas and regions closer to the sea) and compared to reference populations in Europe (Crighton et al., 2011).

Health geographers have made an important contribution to our understanding of this disaster and the effects it has had on the population's health. For example, Bennion et al. (2007), in a study examining the relationship between airborne dust and lung function and respiratory symptoms in children in the region, demonstrated significantly higher rates of poor respiratory health closer to the former seashore. Although no association with dust deposition was found, the authors indicate that other environmental factors are likely responsible. In a study examining the psychosocial impacts associated with the disaster, Crighton et al. (2003a; b) identified high levels of emotional distress and poor self-rated health. Psychosocial impacts were found to be significantly associated with levels of environmental concern, having young children, being an ethnic minority and being female.

Urban air pollution

The aforementioned environmental health inequities occurring at global scales are similarly seen at smaller scales, with air pollution–related inequities being perhaps the best documented example of this. Health geographers have long focused on neighborhood relationships between socioeconomic status and air pollution from sources including industry and transportation networks (Buzzelli, 2008). For example, a recent study examined the relationship between ambient nitrous dioxide (NO_2) exposures and measures of material deprivation (e.g., low income, unemployment, low education) and social deprivation (e.g., living alone, linguistic isolation) in the context of Canada's three largest cities (Pinault et al., 2016). This study found that indicators of both social and material deprivation were associated with living in close proximity to a major roadway and experiencing higher concentrations of ambient NO_2 air pollution. In Europe, similar findings have been reported. For example, a Swedish study by Chaix et al. (2006) examined the relationship between local-scale data on outdoor NO_2, residential and school locations and household incomes of approximately 30,000 children living in Malmö. Results indicated that exposure to NO_2 at the school of attendance and place of residence typically increased as the socioeconomic status of the neighborhood of residence decreased. Although health outcomes were not examined in either of these studies, NO_2 exposure is a well-established risk factor for asthma and other respiratory conditions, especially among children.

Some of the strongest evidence of air pollution–related health inequity has been found in the United States. In the city of Atlanta, O'Lenick et al. (2017) examined the modifying effect of neighborhood-level socioeconomic status on the relationship between air pollution (NO_2, O_3 and $PM_{2.5}$) and pediatric asthma morbidity. Here, the authors observed significantly stronger air pollution–pediatric asthma associations in *deprivation areas* compared with *non-deprivation areas* and concluded that neighborhood level socioeconomic status is a factor contributing to children's vulnerability to air pollution–related asthma morbidity. Using an ecological level, spatial study design in New York City, Corburn, Osleeb and Porter (2006) identified neighborhood *hot spots* of childhood asthma hospitalizations and analyzed the effects of neighborhood characteristics related to air quality (i.e., presence of polluting land uses, industries and major roadways) and sociodemographic conditions (e.g., income, education, unemployment). Here, neighborhood housing characteristics were also examined, as poor quality housing is likely to increase exposures to indoor air pollutants such as rodent and cockroach allergens, molds and dust mites, which are known to trigger asthma. Results demonstrated an association between neighborhood asthma hospitalization rates, low average household income, high percentage of minorities and multiple environmental pollution burdens.

Canada's Indigenous communities

Environmental health inequities among Indigenous communities are frequently recognized as particularly significant when compared to non-Indigenous communities in contexts including Chile (Romero, Méndez and Smith, 2012), the United States (Hoover et al., 2012) and Canada (Dhillon and Young, 2010). Notably, Dhillon and Young (2010) argue that cases of environmental inequity are so prevalent in Canadian Indigenous communities that it is more appropriately labeled environmental racism. An example of this is seen in the Grassy Narrows (Ojibwe) First Nation in Ontario. Between 1962 and 1970, the Dryden paper mill dumped an estimated 10 metric tons of methylmercury into the English-Wabigoon River upstream from the community, contaminating local lakes and tributaries and all fish species (Kinghorn, Solomon and Chan, 2007). The impacts on the community included the loss of cultural practices and opportunities to transmit traditional ecological knowledge, the two most important sources of employment (commercial fishing and guiding) and confidence in the safety of the food and water (Kinghorn, Solomon and Chan, 2007). Widespread mercury exposure has also resulted in significant and long-term health impacts including a significant outbreak of Minamata disease, though the provincial government disputes this (Harada et al., 2005). While mercury levels have since decreased, river sediments and some common fish species are still contaminated above safe limits and serious health problems persist (Kinghorn, Solomon and Chan, 2007).

The Aamjiwnaang First Nation in Sarnia, Ontario, faces similar inequities. Labeled Chemical Valley, Sarnia hosts approximately 40% of Canada's chemical industries, including chemical plants, petroleum refineries, plastics recyclers and Canada's largest hazardous-waste dump. Approximately 60% of the air pollutants released by these factories occurs within five kilometers of the Aamjiwnaang Reserve, one of the closest residential areas (MacDonald and Rang, 2007). Talfourd Creek runs through the Reserve and is contaminated with substances including nickel, cadmium, arsenic and lead (Wiebe, 2016). For this community, major accidental chemical releases of benzene, toluene and other highly toxic substances and resultant evacuations are recurrent events (Wiebe, 2016). Community-based studies have identified a broad range of health issues, from headaches, asthma and skin rashes to elevated cancer rates and reproductive concerns (MacDonald and Rang, 2007). A study examining temporal changes in the sex ratio in the community concluded that male births declined continuously from 1990 to 2003, to a ratio of two females born for every one male (Mackenzie, Lockridge and Keith, 2005). The authors indicate that exposure to endocrine-disrupting industrial chemicals during the preconception and prenatal periods is a possible cause.

Even Indigenous communities living in some of the most remote areas of the Arctic are disproportionately exposed to contaminants such as mercury, pesticides, heavy metals and radionuclides. Transported thousands of kilometers by wind and ocean currents from industries in the South, these contaminants are known to accumulate in fish and marine mammals that are hunted for food (Furgal, Powell and Myers, 2005). These *country foods* are socially and culturally vital to Inuit and northern peoples and have traditionally provided an essential source of nutrients for local diets (Richmond and Ross, 2009). Problematically, country-food consumption results in northern communities commonly exceeding tolerable daily intakes and safe blood levels of numerous toxic substances including PCBs, chlordane, toxaphene and mercury (Van Oostdam et al., 2005). Poor economic circumstances coupled with remote locations means that choosing store-bought alternatives is not an option for many communities (Richmond and Ross, 2009).

Conclusion

This chapter examined the concept of environmental health inequity using a series of global and local examples to demonstrate the contextual and geographic scope of the phenomena. While these examples are highly divergent, they are similar in that each highlights how vulnerable populations are being unfairly, unjustly and avoidably affected by environmental hazards as a result of their race, class, culture, age or gender. Health geographers and environmental-health researchers have made significant contributions to identifying and enhancing public awareness and understanding many of the environmental health inequities that exist within and across communities around the world today, as well as examining the processes that create these inequities.

Despite this achievement and the considerable growth in research interest in environmental-health-inequity issues, many research gaps remain. In a systematic review of Canadian environmental-health research, Masuda et al. (2008), identify some of these gaps. The authors call for research that (a) uses broader definitions of "health" that include non-physical health outcomes (i.e., psychosocial health); (b) integrates theoretical concepts such as therapeutic landscapes, developed by geographers, to consider positive dimensions of environment and health – not just negative dimensions; and (c) takes a multiscalar approach, recognizing that contaminates often move across locales. A WHO review (2010) identified the need for environmental-health-equity research that takes age, gender and migration into account, considers the cumulative effects of multiple environmental exposures, and examines the interaction between social and economic factors and environmental exposures.

Research that addresses these gaps is critical for informing the development of effective policies and programs that do not simply reduce exposures, but also address conditions that create health inequities. While programs aimed at reducing the overall release of hazardous pollutants into the environment are important, they alone will not reduce environmental health inequities, since low socioeconomic and marginalized groups will continue to be disproportionately exposed to toxic contaminants and be more vulnerable to their effects. Thus, the development of policies targeting environmental health inequities requires highly interdisciplinary and broad-based efforts that account for the sociopolitical and economic structures that create inequities in the first place. It also requires addressing the social, economic, health-care and behavioral determinants of health that make low-socioeconomic and marginalized groups more vulnerable to environmental hazards. Health geographers have an important role to play in these efforts.

References

Amankwaa, E. (2013). Livelihoods in risk: exploring health and environmental implications of e-waste recycling as a livelihood strategy in Ghana. *The Journal of Modern African Studies*, 51(4), pp. 551–575.

Benetti, A. D. (2007). Preventing disease through healthy environments: towards an estimate of the environmental burden of disease. *Engenharia Sanitaria e Ambiental*, 12(2), pp. 115–116.

Bennion, P., Hubbard, R., O'Hara, S., Wiggs, G., Wegerdt, J., Lewis, S., Small, I., van der Meer, J. and Upshur, R. (2007). The impact of airborne dust on respiratory health in children living in the Aral Sea region. *International Journal of Epidemiology*, 36(5), pp. 1103–1110.

Bolte, G., Braubach, M., Chaudhuri, N., Deguen, S., Fairburn, J., Fast, I., Fiestas, L., Imnadze, P., Laflamme, L., Mitis, F., Morris, G. and Zmirou-Navier, D. (2012). Environmental health inequalities in Europe. *Assessment report*. Copenhagen: WHO Regional Office for Europe.

Boyd, D. R. (2015). *Cleaner, greener, healthier: a prescription for stronger Canadian environmental laws and policies*. Vancouver, Canada: UBC Press.

Bullard, R. (2007). Equity, unnatural man-made disasters, and race: why environmental justice matters. *Research in Social Problems & Public Policy*, 15, pp. 51–85.

Buzzelli, M. (2008). *Environmental justice in Canada: it matters where you live*. Ottawa, Canada: Canadian Research Policy Networks, pp. 17.

Buzzelli, M. and Jerrett, M. (2007). Geographies of susceptibility and exposure in the city: environmental inequity of traffic-related air pollution in Toronto. *Canadian Journal of Regional Science*, 30(2), pp. 195–210.

Chaix, B., Gustafsson, S., Jerrett, M., Kristersson, H., Lithman, T., Boalt, Å. and Merlo, J. (2006). Children's exposure to nitrogen dioxide in Sweden: investigating environmental injustice in an egalitarian country. *Journal of Epidemiology & Community Health*, 60(3), pp. 234–241.

Corburn, J., Osleeb, J. and Porter, M. (2006). Urban asthma and the neighbourhood environment in New York City. *Health & Place*, 12(2), pp. 167–179.

Crighton, E. J., Barwin, L., Small, I. and Upsher, R. (2011). What have we learned? A review of the literature on children's health and the environment in the Aral Sea area. *International Journal of Public Health*, 56(2), pp. 125–138.

Crighton, E. J., Elliott, S. J., Upshur, R., van der Meer, J. and Small, I. (2003b). The Aral Sea disaster and self-rated health. *Health & Place*, 9(2), pp. 73–82.

Crighton, E. J., Elliott, S. J., van Der Meer, J., Small, I. and Upshur, R. (2003a). Impacts of an environmental disaster on psychosocial health and well-being in Karakalpakstan. *Social Science & Medicine*, 56(3), pp. 551–567.

Dhillon, C. and Young, M. G. (2010). Environmental racism and First Nations: a call for socially just public policy development. *Canadian Journal of Humanities and Social Sciences*, 1(1), pp. 23–37.

Edelstein, M. R. (2004). *Contaminated communities: coping with residential toxic exposure*. 2nd ed. Boulder, CO: Westview Press.

Furgal, C. M., Powell, S. and Myers, H. (2005). Digesting the message about contaminants and country foods in the Canadian North: a review and recommendations for future research and action. *Arctic*, 58(2), pp. 103–114.

Gee, G. C. and Payne-Sturges, D. C. (2004). Environmental health disparities: a framework integrating psychosocial and environmental concepts. *Environmental Health Perspectives*, 112(17), pp. 1645–1653.

Harada, M., Fujino, T., Oorui, T., Nakachi, S., Nou, T., Kizaki, T., Hitomi, Y., Nakano, N. and Ohno, H. (2005). Follow-up study of mercury pollution in Indigenous tribe reservations in the province of Ontario, Canada, 1975–2002. *Bulletin of Environmental Contamination and Toxicology*, 74(4), pp. 689–697.

Heacock, M., Kelly, C. B. and Suk, W. A. (2016). E-waste: the growing global problem and next steps. *Reviews on Environmental Health*, 31(1), pp. 131–135.

Hoover, E., Cook, K., Plain, R., Sanchez, K., Waghiyi, V., Miller, P., Dufault, R., Sislin, C. and Carpenter, D. O. (2012). Indigenous peoples of North America: environmental exposures and reproductive justice. *Environ Health Perspectives*, 120(12), pp. 1645–1649.

Kinghorn, A., Solomon, P. and Chan, H. M. (2007). Temporal and spatial trends of mercury in fish collected in the English-Wabigoon river system in Ontario, Canada. *Science of the Total Environment*, 372(2–3), pp. 615–623.

Kruize, H., Droomers, M., van Kamp, I. and Ruijsbroek, A. (2014). What causes environmental inequalities and related health effects? An analysis of evolving concepts. *International Journal of Environmental Research and Public Health*, 11(6), pp. 5807–5827.

Lee, H. (2010). Environmental Justice. In: H. Frumkin, ed., *Environmental health: from global to local*. San Francisco: Jossey-Bass, pp. 227–256.

MacDonald, E. and Rang, S. (2007). *Exposing Canada's Chemical Valley: an investigation of cumulative air pollution emissions in the Sarnia, Ontario area*. [pdf] Ecojustice Canada. Available at www.ecojustice.ca/wp-content/uploads/2015/09/2007-Exposing-Canadas-Chemial-Valley.pdf [Accessed 1 Sept. 2017].

Mackenzie, C. A., Lockridge, A. and Keith, M. (2005). Declining sex ratio in a First Nation community. *Environmental Health Perspectives*, 113(10), pp. 1295–1298.

Masuda, J. R., Zupancic, T., Poland, B. and Cole, D. C. (2008). Environmental health and vulnerable populations in Canada: mapping an integrated equity-focused research agenda. *The Canadian Geographer/Le Géographe canadien*, 52(4), pp. 427–450.

Micklin, P. (2007). The Aral Sea disaster. *Annual Review of Earth and Planetary Sciences*, 35, pp. 47–72.

Morello-Frosch, R. and Lopez, R. (2006). The riskscape and the color line: examining the role of segregation in environmental health disparities. *Environmental Research*, 102(2), pp. 181–196.

O'Lenick, C. R., Winquist, A., Mulholland, J. A., Friberg, M. D., Chang, H. H., Kramer, M. R., Darrow, L. A. and Sarnat, S. E. (2017). Assessment of neighbourhood-level socioeconomic status as a modifier of air pollution – asthma associations among children in Atlanta. *Journal of Epidemiology and Community Health*, 71(2), pp. 129–136.

Pinault, L., Crouse, D., Jerrett, M., Brauer, M. and Tjepkema, M. (2016). Spatial associations between socioeconomic groups and NO_2 air pollution exposure within three large Canadian cities. *Environmental Research*, 147, pp. 373–382.

Prochaska, J. D., Nolan, A. B., Kelley, H., Sexton, K., Linder, S. H. and Sullivan, J. (2014). Social determinants of health in environmental justice communities: examining cumulative risk in terms of environmental exposures and social determinants of health. *Human and Ecological Risk Assessment: An International Journal*, 20(4), pp. 980–994.

Richmond, C. A. M. and Ross, N. A. (2009). The determinants of First Nation and Inuit health: a critical population health approach. *Health & Place*, 15(2), pp. 403–411.

Romero, H., Méndez, M. and Smith, P. (2012). Mining development and environmental injustice in the Atacama Desert of Northern Chile. *Environmental Justice*, 5(2), pp. 70–76.

Sexton, K. (2000). Socioeconomic and racial disparities in environmental health: is risk assessment part of the problem or part of the solution? *Human and Ecological Risk Assessment: An International Journal*, 6(4), pp. 561–574.

Small, I., van der Meer, J. and Upshur, R. E. (2001). Acting on an environmental health disaster: the case of the Aral Sea. *Environmental Health Perspectives*, 109(6), pp. 547–549.

Tooke, T. R., Klinkenberg, B. and Coops, N. C. (2010). A geographical approach to identifying vegetation-related environmental equity in Canadian cities. *Environment and Planning B: Planning and Design*, 37(6), pp. 1040–1056.

Van Oostdam, J., Donaldson, S. G., Feeley, M., Arnold, D., Ayotte, P., Bondy, G., Chan, L., Dewaily, É., Furgal, C. M., Kuhnlein, H., Loring, E., Muckle, G., Myles, E., Receveur, O., Tracy, B., Gill, U. and Kalhok, S. (2005). Human health implications of environmental contaminants in Arctic Canada: a review. *Science of the Total Environment*, 351–352, pp. 165–246.

Wakefield, S. and Elliott, S. J. (2000). Environmental risk perception and well-being: effects of the landfill siting process in two southern Ontario communities. *Social Science & Medicine*, 50(7–8), pp. 1139–1154.

Widmer, R., Oswald-Krapf, H., Sinha-Khetriwal, D., Schnellmann, M. and Böni, H. (2005). Global perspectives on e-waste. *Environmental Impact Assessment Review*, 25(5), pp. 436–458.

Wiebe, S. M. (2016). *Everyday exposure: Indigenous mobilization and environmental justice in Canada's Chemical Valley*. Vancouver: UBC Press.

World Health Organization. (2010). Social and gender inequalities in environment and health. In: *Fifth Ministerial Conference on Environment and Health. Parma, Italy 10–12 March 2010*. Copenhagen: WHO Regional Office for Europe, p. 20.

Wu, C., Zhu, H., Luo, Y., Teng, Y., Song, J. and Chen, M. (2015). Levels and potential health hazards of PCBs in shallow groundwater of an e-waste recycling area, China. *Environmental Earth Sciences*, 74(5), pp. 4431–4438.

Xu, X., Yang, H., Chen, A., Zhou, Y., Wu, K., Liu, J., Zhang, Y. and Huo, X. (2012). Birth outcomes related to informal e-waste recycling in Guiyu, China. *Reproductive Toxicology*, 33(1), pp. 94–98.

7

INFECTIOUS-DISEASE GEOGRAPHY

Disease outbreaks and outcomes through the lens of space and place

Corinna Keeler and Michael Emch

John Snow's famous map showing the spatial relationship between water-pump locations and cholera cases during an 1854 disease outbreak in London has been the subject of textbook chapters, mass-market paperbacks and documentaries. In creating a map of the relationship between the location of cholera cases and water sources, Snow demonstrated the interconnectedness of populations, environmental phenomena and infectious diseases. This map captures the synergy between studies of geography and infectious disease: space and place structure infectious disease transmission, prevalence and vulnerability, and infectious diseases, in turn, affect how people interact with and navigate their environment.

Although Snow's map was not the first example of *medical geography* – a term that has been in use since at least the late 1700s – it foregrounds the application of spatial logic to epidemiology, which has driven many of the innovations in infectious-disease geography over the past century. These innovations include the theorization of disease ecology, explorations of the spatial structure and diffusion mechanisms of epidemics, global mapping of distributions and the application of political ecology to social dimensions of disease outbreaks. Through these major developments and others, infectious-disease geography has been in conversation not only with the diverse subfields of geography, but also with disciplines such as epidemiology, ecology and demography, giving rise to novel areas of inquiry, such as landscape genetics. Infectious-disease geography combines spatial technologies and geographic methods with clinical and field data from the health sciences to further interdisciplinary research about pressing health concerns.

The contemporary field of health geography has intellectual roots in both the United Kingdom and the United States, where the postwar surge of geographic inquiry in the 1950s led to novel applications of geography to the study of infectious disease. In the United Kingdom, Peter Haggett and his student-turned-collaborator Andrew Cliff wrote extensively on disease diffusion and spatial disease processes beginning in the 1970s, applying the methodological traditions of locational analysis and model-based geography to epidemiology. Smallman-Raynor continued that tradition, with a special attention to historical and wartime epidemics (Smallman-Raynor and Cliff, 1991, 2000). Meanwhile, in the United States, Jacques May formed a Medical Geography Department in the American Geographic Society in the 1950s, establishing medical geography as a distinct subfield of geography. Notably, much of the foundational early research in the growing subfield of medical geography focused specifically on infectious diseases (May, 1954; Hunter, 1970).

Major developments in infectious-disease geography

One of the major theoretical developments in infectious-disease geography is Melinda Meade's triangle of disease ecology (1977). Meade's work expands on May's foundational writings on disease ecology that argue that disease foci in certain populations can be explained by a combination of human activity and environmental characteristics (May, 1954). The triangle of disease ecology provides a framework for considering the effects of space and place on human diseases: Meade identifies environmental factors, population-level factors and behavioral factors as three distinct yet interrelated arenas that modify health outcomes (see Figure 7.1). Meade's triangle has been influential in both geography and epidemiology, providing a theoretical framework for considering multiple contextual factors in disease processes. In her early work, Meade applied the triangle framework to study a variety of infections, from mosquito-borne diseases to respiratory illness, demonstrating that behavioral patterns such as livelihoods and activity patterns interface with environmental factors such as vegetation and the built environment in determining differential disease outcomes. Since 1977, Meade's triangle approach has been applied to infectious diseases ranging from river blindness (Hunter, 1980) to cholera (Emch, 1999) to avian flu (Carrel et al., 2012). Additionally, this framework has been extended outside of infectious-disease geography in studies of asthma, cancer and health-system inequalities (Emch, Root and Carrel, 2017).

Meade's attention to the importance of geographic factors in understanding infectious-disease dynamics is echoed in Peter Haggett and Andrew Cliff's work on spatial-diffusion models of disease epidemics. Cliff and Haggett's work represents a key contribution of infectious-disease geography both within and outside geography, as their application of quantitative spatial analysis to epidemiological processes transformed both geographic methods and public health practice. In their book *Atlas of Disease Distributions* (1988), Cliff and Haggett set forward a comprehensive set of spatial methods for epidemiological data, including autocorrelation models, space-time and lagged mapping, modeling disease-diffusion waves and bridging problems of missing data. In doing so, they provide some of the earliest spatial analyses of diseases ranging from rabies to smallpox to measles, and they set forth integrative

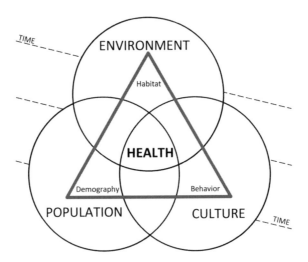

Figure 7.1 The triangle of disease ecology

Recreated from Meade (1977).

methodological approaches to measuring disease transmission and disease diffusion that are still in use today. They also theorized the geographic structure of epidemics for specific diseases of global public health concern, such as influenza and HIV/AIDS (Cliff, Haggett and Ord, 1986; Smallman-Raynor, Cliff and Haggett, 1992).

The rise of global disease mapping programs underscores the value of applying spatial methods to studies of infectious disease. At the time of writing, in the World Health Organization's online Map Gallery there are maps of the worldwide distribution of more than 20 infectious diseases, ranging from leprosy to Zika virus (WHO, 2017). While the WHO uses simple case counts to create world maps of disease incidence and prevalence at the national level, there are also sophisticated modeling teams like the Malaria Atlas Project that use spatial statistics and inference-based modeling to develop finer-scale maps of disease distribution, mosquito vector populations, and treatment programs for both *Falciparum* malaria and *Vivax* malaria (Howes et al., 2016; Kraemer et al., 2015). Both the WHO's Map Gallery and the Malaria Atlas Project mapping efforts have a shared goal: to increase geographic understanding of disease distributions in order to aid elimination and eradication programs. These maps are used by public-health practitioners, from local agencies to global policy initiatives, to plan and target interventions and to distribute resources such as diagnostic tests and drug therapies.

In response to the proliferation of quantitative spatial approaches to infectious-disease geography, in the 1990s Jonathan Mayer called for an increased application of concepts from political ecology to health geography (Mayer, 1996). He highlights the foundations of much disease-ecology work in understanding the role of economic development and environmental change in infectious diseases, but he points out that, at the time of his writing, there had been little work that applied a political ecological lens to examine how social and economic power and agency shaped relationships across spatial scales. In the ensuing two decades, scholars have begun to grapple with Mayer's call. In investigating the HIV/AIDS epidemic in southern Africa, researchers have used a political-ecology approach to explore how race, gender, and livelihood affect disease risk and transmission dynamics. For example, Butler (2005) demonstrated that the South African government's response to the HIV/AIDS epidemic used the logic of public health to enforce racial segregation, thereby reinforcing the history of colonial and apartheid racial discrimination.

Other political-ecology researchers have considered feedback processes between social and political empowerment and distributions of disease risk and burden. For example, Paul Robbins and his collaborators explore mosquito-control programs in Arizona designed to prevent further spread of West Nile virus (WNV), a mosquito-borne disease (Robbins, Farnsworth and Jones III, 2008). Through this case study, Robbins et al. show that not only do governmental bureaucracies affect the distribution of mosquito populations and WNV cases through environmental management policies, but also mosquitoes and the WNV disease agent in turn alter those very bureaucracies by prompting the creation of new government offices and programs. Robbins et al. also point out the role of political empowerment in this feedback between disease agents and government programs, noting that areas within their Arizona study site that have a larger Hispanic population may have fewer citizen reports to the mosquito-management program due to wariness in interacting with government agencies. King's (2010) review of developments in the political ecology of health synthesizes examples, such as the ones above, to highlight two key contributions of political ecologists to the study of infectious diseases. First, political ecology offers a framework for exploring reciprocal feedback between health and the environment, drawing on work from the landscapes of health tradition to understand disease transmission and distribution factors that both create and respond to environmental changes. Second, the political-ecology approach highlights the role of political economy in the formation of health protocols that produce disease vulnerabilities and alter disease patterns (King, 2010).

Connections with other disciplines

The inherent interdisciplinarity of infectious-disease geography causes distinctions between disciplinary boundaries to be a bit abstract. For example, a recent study about the temporal and spatial diffusion of Ebola virus in West Africa from 2014 to 2015 involved contributors trained in geography, computer science, molecular and cell biology, epidemiology, clinical medicine and ecology, among other disciplines (Carroll et al., 2016). In many of these fields, geographers are working on questions of infectious-disease processes, and infectious-disease experts are using approaches and theories from geography. That said, infectious-disease geography has been incorporated into the core concerns of several disciplines.

One such discipline is epidemiology, where spatial methods are a central part of both disease surveillance and analysis. Moore and Carpenter (1999) point out that disease mapping and cluster detection have revolutionized epidemiology, changing how public-health practitioners track disease and how thresholds for intervention are determined. This change can be seen in modern WHO and Centers for Disease Control (CDC) approaches to tracking disease outbreaks, where location-aware monitoring systems keep track of the spatial spread of disease in order to determine whether cases are independent of each other or connected as part of an epidemic. For example, in the United States, the CDC maintains separate spatially and temporally explicit surveillance systems for waterborne disease, food-borne disease, antimicrobial resistance and emerging and zoonotic disease, underscoring the importance of spatial record-keeping for public-health monitoring and interventions. This process of epidemiologic surveillance based on near-real-time monitoring of records from local or regional clinics is known as syndromic surveillance (Savory et al., 2010).

Relatedly, the field of landscape epidemiology has emerged in an effort to understand the affected territory for the transmission of a certain disease. The term "landscape epidemiology" was first used in the 1960s in the Soviet Union to describe the relationship between ecosystems and disease foci (Pavlovsky, 1966). In recent years, the landscape-epidemiology approach has especially been applied to vector-borne diseases such as malaria and Lyme disease, where climactic limits, ecological habitats and population characteristics interact to enforce spatial limits on certain diseases (Reisen, 2010). This field integrates contributions from ecologists and biologists in addition to public-health researchers and infectious-disease geographers to apply remote-sensing methods and landscape-ecology statistical metrics to better understand population dynamics of disease vectors and consequently the spatial patterns in disease transmission. For example, Kitron et al. (2006) used remotely sensed data of vector habitat and global positioning system (GPS) and geographic information system (GIS) data of household locations and water sources to examine the transmission dynamics of two vector-borne diseases: schistosomiasis and Chagas' disease. Their case study, which compares the relevant spatial scales for analysis of an insect-vectored disease in Chagas' disease versus a snail-vectored disease in schistosomiasis, illustrates the value of a "landscape-epidemiological approach," which combines epidemiological data on disease in humans with spatial data on vector habitat.

Another growing field that is closely linked to infectious-disease geography is landscape genetics, which combines geographic methods with molecular genetic data on pathogen evolution and host genetic factors. This field combines geographic technologies, such as advanced GPS data to track disease case locations and complex spatial statistical methods, with rapidly developing technologies in molecular biology such as whole-genome sequencing that allow for comparing specific mutations between samples to observe pathogen evolution across space and time. Landscape genetics expands on phylogenetic understandings of microbial evolution by examining when and where these specific gene changes occur, thereby allowing for integration of geographic variables such as ecological change, population migration, or climactic variation into understandings of pathogen evolution. This approach has been applied to avian flu in Southeast Asia, allowing researchers to detect that the presence of aquaculture surfaces and higher population density is associated with higher rates of virus evolution of highly pathogenic H5N1 influenza (Carrel et al., 2012).

Similarly, Patel et al. (2014) used a landscape-genetics approach to document that a drug-resistant strain of malaria detected in Guatemala could have been imported from the Democratic Republic of the Congo by United Nations peacekeepers. By leveraging genetic data and geographic methods, landscape genetics allows researchers to track diseases through space and time, which is especially valuable in studying emerging and reemerging disease processes.

Future developments and limitations

The rise of landscape genetics over the past decade illustrates the potential future of infectious-disease geography: drawing on new sources of data and integrating with diverse scholarly fields. In the current big-data revolution in health and health care, there is an increasing amount of data available to inform spatial models, including genetic data, mobile health data, GIS data and widely available high-resolution imagery. While the changing quantity and availability of data has repercussions for research in a variety of health and medical arenas, the study of infectious diseases in particular stands to be transformed by these novel data sources: communicable diseases generally have a more rapid onset and a shorter window of exposure than many chronic conditions and environmental health hazards, so the spatial and temporal richness of mobile data and other new technologies has far-reaching repercussions for the study of infectious disease.

The increasing availability of genomic-sequencing technologies has led to the creation of resources such as the International Nucleotide Sequence Database Collaboration, which allows scholars to obtain records with genotypic information for infectious-disease occurrences (Hay et al., 2013). This collaborative project includes contributions from the European Nucleotide Archive, the DNA DataBank of Japan and the Gen-Bank of the US National Institutes of Health. These data have been used in global disease monitoring as well as integrated mapping efforts to identify high-risk areas for specific infections (Hay et al., 2013). On the other hand, there is an increasing amount of data available through web activity, such as disease-reporting networks and even internet search patterns. These nontraditional sources have been incorporated into the WHO surveillance protocol to prompt a Global Outbreak Alert (Brownstein et al., 2008). A well-known application of this novel data source is Google Flu Trends, which uses search-engine entries relating to disease symptoms to track influenza rates around the world and has been shown to match closely with spatial and temporal data collected by public-health agencies (Dugas et al., 2012).

Several scholars have pointed out that GPS-enabled devices, such as smartphones, enable researchers to do space-time monitoring of infectious diseases with a faster response time and greater spatial precision than ever before. Mobile health (mHealth) programs report disease diagnoses and cases via mobile technology, allowing for faster, cheaper and more accurate data collection than paper reporting documents. This approach has been used in monitoring both malaria and HIV in sub-Saharan Africa to provide geographically explicit data about both new diagnoses and compliance with treatments (Catalani et al., 2013; Asiimwe et al., 2011). Similarly, some researchers have incorporated GPS devices into the structure of short-term research projects to provide longitudinal spatial data. For example, Vazquez-Prokopec et al. (2009) recruited participants to wear GPS devices to track people's activity space in order to examine exposure to dengue virus in Peru.

The increased availability of spatially and temporally rich data sources can facilitate more robust integration between spatial analysis of infectious disease and other modeling traditions. In particular, epidemiologic SIR/SEIR models have not often incorporated a spatially explicit component. This mathematical approach to modeling the transmission of infectious diseases between susceptible, exposed, infected and recovered members of a population is well-positioned for integration with spatial methods, but to date most work in this area has been either theoretical (Bjornstad, Finkenstadt and Grenfell, 2002) or proof-of-concept case studies using data from past epidemics (Perez and Dragicevic, 2009). On the other hand, the rich opportunities afforded by agent-based models have been underexplored with regard to infectious-disease processes. Agent-based models are a powerful tool for examining population-environment interactions, because they allow researchers to integrate spatial datasets such as land cover, climate or household-based survey data with stochastic processes to perform simulations of potential future scenarios (Miller et al., 2010). While these models have been widely applied to land-cover change and environmental management, there are only a few theoretical explorations of their use in exploring disease transmission or diffusion of epidemics (Dunham, 2005), with little applied work.

Conclusion

The history of infectious-disease geography is intertwined with the broader evolution of the field of health geography, with methodological and theoretical innovations that have shaped the trajectory of health research within and outside geography. From early work on disease ecology that demonstrated the value of considering the environmental context of disease transmission, to modern computationally intensive spatial models based on genetic sequence data, geographers have woven together space, place and infectious-disease processes to improve the collective understanding of disease transmission, surveillance and prevention.

References

Asiimwe, C., Gelvin, D., Lee, E., Amor, Y., Quinto, E., Katureebe, C., Sundaram, L., Bell, D. and Berg, M. (2011). Use of an innovative, affordable, and open-source short message service-based tool to monitor malaria in remote areas of Uganda. *American Journal of Tropical Medicine and Hygiene*, 85(1), pp. 26–33.

Bjornstad, O., Finkenstadt, B. L. and Grenfell, B. (2002). Dynamics of measles epidemics: estimating scaling of transmission rates using a time series SIR model. *Ecological Monographs*, 72(2), pp. 196–184.

Brownstein, J., Freifeld, C., Reis, B. and Mandl, K. (2008). Surveillance sans frontieres: Internet-based emerging disease intelligence and the HealthMap Project. *PLoS Medicine*, 5(7), e151.

Butler, A. (2005). South Africa's HIV/AIDS Policy, 1994–2004: how can it be explained? *African Affairs*, 104, pp. 591–614.

Carrel, M., Emch, M., Nguyen, T., Jobe, R. T. and Wan, X. (2012). Population-environment drivers of H5N1 avian influenza molecular change in Vietnam. *Health & Place*, 18(5), pp. 1122–1131.

Carroll, M. W., Matthews, D. A., Hiscox, J. A., et al. (2016). Temporal and spatial analysis of the 2014–2015 Ebola virus outbreak in West Africa. *Nature*, 524, 97–101.

Catalani, C., Philbrick, W., Fraser, H., Mechael, P., and Israelski, D. (2013). mHealth for HIV treatment & prevention: a systematic review of the literature. *The Open AIDS Journal*, 7, pp. 17–41.

Cliff, A. and Haggett, P. (1988). *Atlas of disease distributions*. Oxford: Blackwell Publishers.

Cliff, A. D., Haggett, P. and Ord, J. K. (1986). *Spatial aspects of influenza epidemics*. London: Routledge Kegan & Paul.

Dugas, A., Hsieh, Y., Levin, S., Pines, J., Mareiniss, D., Mohareb, A., Gaydos, C. A., Perl, T. M. and Rothman, R.E. (2012). Google Flu trends: correlation with emergency department influenza rates and crowding metrics. *Clinical Infectious Diseases*, 54(4): pp. 463–469.

Dunham, J. B. (2005). An agent-based spatially explicit epidemiological model in MASON. *Journal of Artificial Societies and Social Simulation*, 9(1), pp. 1–14.

Emch, M. (1999). Diarrheal disease risk in Matlab, Bangladesh. *Social Science & Medicine*, 49(4), pp. 519–530.

Emch, M., Root, E. D. and Carrel, M. (2017). *Health and medical geography*. 4th ed. New York: Guilford Press.

Hay, S., George, D., Moyes, C. and Brownstein, J. (2013). Big data opportunities for infectious disease surveillance. *PLoS Medicine*, 10(4), e1001413.

Howes, R. E., Battle, K. E., Mendis, K. N., Smith, D. L., Cibulskis, R. E., Baird, J. K. and Hay, S. I. (2016). Global epidemiology of *Plasmodium vivax*. *American Journal of Tropical Medicine and Hygiene*, 95, pp. 15–34.

Hunter, J. M. (1970). Disease and development in Africa. *Social Science & Medicine*, 3, pp. 443–493.

Hunter, J. M. (1980). Strategies for the control of river blindness. In: M. S. Meade, ed., *Conceptual and methodological issues in medical geography*. Chapel Hill: University of North Carolina Department of Geography.

King, B. (2010). Political ecologies of health. *Progress in Human Geography*, 34(1), pp. 38–55.

Kitron, U., Clennon, J. A., Cecere, M. C., Gürtler, R. E., King, C. H. and Vázquez-Prokopec, G. (2006). Upscale or down-scale: applications of fine scale remotely sensed data to Chagas disease in Argentina and schistosomiasis in Kenya. *Geospatial Health*, 1, pp. 49–58.

Kraemer, M. U. G., Hay, S. I., Pigott, D. M., Smith, D. L., Wint, W. G. R. and Golding, N. (2015). Progress and challenges in infectious disease cartography. *Trends in Parasitology*, 32(1), pp. 19–29.

May, J. M. (1954). Cultural aspects of tropical medicine. *American Journal of Tropical Medicine and Hygiene*, 3, pp. 422–430.

Mayer, J. D. (1996). The political ecology of disease as a new focus for medical geography. *Progress in Human Geography*, 20, pp. 441–456.

Meade, M. S. (1977). Medical geography as human ecology: the dimension of population movement. *Geographical Review*, 67, pp. 379–393.

Miller, B. W., Breckheimer, I., McCleary, A. L., Guzmán-Ramirez, L., Caplow, S. C., Jones-Smith, J. C. and Walsh, S. J. (2010). Using stylized agent-based models for population-environment research: a case study from the Galápagos Islands. *Population and Environment*, 75(4), pp. 279–287.

Moore, D. A. and Carpenter, T. E. (1999). Spatial analytical methods and geographic information systems: use in health research and epidemiology. *Epidemiologic Reviews*, 21(2), pp. 143–161.

Patel, J. C., Taylor, S. M., Juliao, P. C., Parobek, C. M., Janko, M., Gonzalez, L. D., Ortiz, L., Padilla, N., Tshefu, A. K., Emch, M., Udhayakumar, V., Lindblade, K. and Meshnick, S. R. (2014). Genetic evidence of importation of drug-resistant *Plasmodium falciparum* to Guatemala from the Democratic Republic of the Congo. *Emerging Infectious Diseases*, 20(6), pp. 932–940.

Pavlovsky, E. N. (1966). *Natural nidality of transmissible diseases, with special reference to the landscape epidemiology of zooanthroponoses*. Urbana: University of Illinois Press.

Perez, L. and Dragicevic, S. (2009). An agent-based approach for modeling dynamics of contagious measles spread. *International Journal of Health Geographics*, 8, pp. 50–67.

Reisen, W. K. (2010). Landscape epidemiology of vector-borne diseases. *Annual Review of Entomology*, 55, pp. 61–483.

Robbins, P., Farnsworth, R. and Jones, J. P. III. (2008). Insects and institutions: managing emergent hazards in the U.S. southwest. *Journal of Environmental Policy & Planning*, 10(1), pp. 95–112.

Savory, D. J., Cox, K. L., Emch, M., Alemi, F. and Pattie, D.C. (2010). Enhancing spatial detection accuracy for syndromic surveillance with street level incidence data. *International Journal of Health Geography*, 9(1), pp. 1–11.

Smallman-Raynor, M. and Cliff, A. (1991). Civil war and the spread of AIDS in central Africa. *Epidemiology and Infection*, 107, pp. 69–80.

Smallman-Raynor, M. and Cliff, A. (2000). The epidemiological legacy of war: the Philippine-American war and the diffusion of cholera in Batangas and La Laguna, South-West Luzon, 1902–1904. *War in History*, 7(1), pp. 29–64.

Smallman-Raynor, M., Cliff, A. D. and Haggett, P. (1992). *Atlas of AIDS*. Hoboken, NJ: Blackwell Publishers.

Vazquez-Prokopec, G. M., Stoddard, S. T., Paz-Soldan, V., Morrison, A. C., Elder, J. P., Kochel, T. J., Scott, T. W. and Kitron, U. (2009). Usefulness of commercially available GPS data-loggers for tracking human movement and risk of dengue virus infection. *International Journal of Health Geographics*, 8, e68.

World Health Organization (2017). *Global health observatory map gallery*. [online]. Available at: www.who.int/gho/map_gallery/en/ [Accessed 12 Feb. 2017].

8

RISK AND RESILIENCE

Sarah Curtis and Katie J. Oven

Research in health geography is typically concerned with variations in human health across places and spaces. We are often seeking to identify *health determinants*, the conditions in different settings that may be detrimental or beneficial to health. The nature of these systems is such that these relationships cannot be described or predicted with certainty or in simple terms. Therefore, the idea of *risk* is very prominent in our thinking. Discourses about risk incorporate diverse ideas and concepts and may be expressed in various ways. This chapter considers risk conceptualized as a combination of *hazard* and *vulnerability* as well as *opportunity* and *resilience*.

The examples discussed below also variously interpret risk as *likelihood* or *probability* and as *emergence* and *uncertainty*. Medical geography typically uses ideas of risk that are also common in fields of medicine and epidemiology and in environmental sciences. It interprets risk in terms of theories of causal pathways producing illness and disease and is focused on developing better understanding of the spatial dimensions of the factors associated with occurrence of diseases in human populations. From this perspective (as discussed in more detail below) risk, hazard and vulnerability are considered in terms of statistical evidence from samples of people, generally taken to be representative of wider populations from which they are drawn. *Risks* are often interpreted as factors that are hazardous for health and that increase the statistical chances of developing disease, whereas *resilience* is examined through analysis of population factors that seem to be *protective* against the risk of disease. The examples given in this chapter show how measures are examined for both people and places and at different geographical scales, ranging from local neighborhoods to wider regions or nations. Other work contributing to the field of health geography includes a closer articulation with human and social geography. This draws on social theories about the ways that groups of people interact with their environment, and examples discussed below illustrate how these approaches often use evidence from a range of qualitative research methods, demonstrating how we can better understand risk and resilience by studying in depth individual experience and the contingent relationships between people and their environment.

Given this diversity of theoretical approaches and methods, diverse ways of thinking about risk can lead to animated debate and also to interesting formulations that cross the boundaries between these different points of view. This chapter supports the argument that, by taking approaches that combine different theoretical and methodological perspectives, health geography has a major contribution to make to knowledge that is important for societies that are striving to make health of the population more resilient to the risks they face.

Risk as *hazard* and *vulnerability* or *opportunity* and *resilience*

In the literature on geographies of health, risk is often presented as an aspect of the physical or social environment that constitutes a challenge to human health. Risks may be considered as arising from a combination of exposure to an environmental hazard and human vulnerability to such exposure. For example, environmental hazards may comprise presence of biological pathogens, such as bacteria or viruses; pollution of air, water or land; geological or meteorological events, such as earthquakes, volcanic eruptions, storms, floods or extreme temperatures; or socio-political or socioeconomic processes operating across societies, such as structural socioeconomic inequality and exclusion, social disorder or military conflict, migration, economic recession or difficult working conditions.

Vulnerability relates to the relative impact of exposure. For example, in areas prone to earthquakes, hurricanes or floods, housing that is not structurally resilient may contribute to vulnerability to these environmental hazards among the population. Characteristics of people may also make them more or less vulnerable. Women, children and older people are commonly viewed as more vulnerable to environmental and social hazards because their physical strength is often relatively less than that of men or of young adults. Especially for women, sociocultural and political factors may limit their ability to act independently to protect themselves (Fatemi et al., 2017). Those with fewer social and economic resources may be more vulnerable to the impacts of environmental or economic crises or social disorder because they are not able to draw on social and economic capital to protect themselves during a crisis and reconstruct afterward (Rigg et al., 2016).

Resilience is a much debated term. It is often emphasized that resilience is not the antithesis of *vulnerability*, since some people and communities with characteristics that may make them more exposed or potentially vulnerable to hazards also have attributes that make them resilient (Bates and Machin, 2016). Asset-based approaches in the public-health sphere (Bartley et al., 2010; Rotegård et al., 2010) encourage us to focus more particularly on the attributes of people and places that make them stronger and promote health and well-being (Cutter et al., 2008). For example, a system of resilience indicators for localities developed in the United States (Cutter, Ash and Emrich, 2014) drew upon a review to identify different domains of capital or assets that are important at the community level, including social, economic, housing and infrastructure, institutional, community and environmental features.

Indeed, health resilience is often manifested most clearly in the face of exposure to insults to health. From a geographical perspective, interesting examples include studies that have identified resilient areas as relatively deprived settings where the health of the population is better than one might expect, given the level of disadvantage and prevailing health risks in the community (Cairns-Nagi and Bambra, 2013; Mitchell et al., 2009; Pearson, Pearce and Kingham, 2013). These studies have suggested, for example, that strong social bonds within deprived communities may be associated with greater resilience to poverty. Also, proximity and access to green space may have *equigenic* effects, which help protect the health of populations in poorer neighborhoods and narrow the health gap between rich and poor communities.

Having some degree of exposure to hazards may provide opportunities to build resilience in ways that contribute to health. Resilience often develops as a result of exposure to hazards. Inoculation programs are an important aspect of global public-health programs (WHO, 2011), which essentially build bodily immunity to infectious disease by exposure to similar types of infection (albeit in small, controlled doses). In the geographical areas where inoculation programs operate well enough to include a large proportion of the population at risk, they provide widespread immunity in the resident population. This reduces the risk of communication of infections for the whole population in those areas, so that even those individuals who are not inoculated are protected to a degree. Another example described in the psychological literature is the experience of post-traumatic growth, whereby those who survive and recover from a hazardous, traumatic event in their physical or social environment may build strength from this experience and thus be more resilient to any similar events in future (Acierno et al., 2006, Pfefferbaum et al., 2015, Villagran et al., 2014,

Yoshida et al., 2016). By extension, this also applies to whole communities, which may develop resilience to environmental, economic or social risks by sharing knowledge, gained from experiencing hazards, about ways to prepare for risks and prevent or mitigate their impacts (Curtis et al., 2018; Freitag et al., 2014; Pfefferbaum et al., 2013).

Risk and resilience as likelihood or probability

As mentioned in the introduction to this chapter, a significant part of research in health geography involves either statistical or qualitative estimation of risk in terms of the *likelihood* of a health outcome occurring in association with a given set of conditions. Often these assessments of risk are made as part of a strategy to improve and protect health in human populations by anticipating conditions in our environment that may damage health and taking action to prevent or mitigate harm. In this sense, we aim to build resilience through a better understanding of risk.

Measuring estimated risk

Risk may be measured statistically, based on theories regarding the probability of random occurrences. Much of the quantitative research in health geography involves tests of statistical significance, which indicate the theoretical likelihood that a given combination of events will occur randomly. The relationships of interest in this work are usually those that are quite unlikely to occur by chance and are more likely to reflect an association between a health outcome and a *causal* factor influencing that outcome. Evidence that an aspect of our environment is likely to act as a *health determinant* is interpreted statistically by comparing health indicators for samples of people who differ in their exposures to the environmental attribute. In this type of research, *significant* associations, are generally interpreted statistically as differences between the samples that are theoretically unlikely to happen by chance, so that it is more likely that differences in exposure to the environmental attribute is associated with differences in the health outcome of interest.

An important underlying aspect of these *probabilistic* interpretations is that they do not *prove* with absolute certainty whether or not an association exists, or the exact magnitude of the risk to health posed by a given environmental factor. Also, it is often observed that statistical *association* does not prove *causation*, so that care needs to be taken when making assumptions about causal relationships on the basis of statistical evidence. It might be argued that the interpretation of statistical research on *causal pathways* in health geography is ultimately based more fundamentally in theoretical models of causality. Statistical analysis only provides a way to critically test the likelihood that our theoretical model about causal factors that *determine* health is correct. Statistical analyses allow us to assess this likelihood within given theoretical parameters. Statistical coefficients compare what may be the *relative* risk (likely difference in probability of a given health outcome) for those who are exposed to the environmental *risk factor* and those who are not. Statistical theory also provides a way to express numerically the *confidence interval* (or likely range of values) for measures of relative risk. Statistical approaches also help us assess the relative scale of environmental impacts on health, which is often important information for those responsible for strategic planning for communities and wider national space.

Specific statistical modeling techniques such as Structural Equation Modeling enable us to examine in detail the associations between different aspects of risk and resilience operating along the *causal pathways* that help determine the health of populations (Bécares, 2014). Multilevel models help clarify how groups of people with similar individual characteristics may have different health experiences according to the kind of setting where they are located and exposed to varying environmental conditions (Duncan, Jones and Moon, 1998). Longitudinal studies (often deploying techniques shared with population geography or life-course epidemiology) also allow us to examine how the environments that we occupy at different stages of our lives may influence our health at subsequent time points. For example, data from a national sample of individuals living in England were linked to information about their place of residence in 1939, and it was found that those who had lived in areas that were worst affected by the economic depression in the 1930s, followed up into old age, tended to have worse health than others in the sample, even after controlling for conditions in their more recent place of residence (Curtis et al., 2003). Contemporary streams of statistical research have combined longitudinal information about people with information on change over time in conditions prevailing in settings where they are living. Thus, in order to understand links between places, people and health outcomes we need to consider the life course of *places* as well as *people*. For example, research in England shows that trends in economic conditions over time in one's place of residence can predict individual health more strongly than local environment at a single time point (Riva and Curtis, 2012).

Importantly for those tasked with planning how to predict, prevent or mitigate risks to health, statistical studies of risk using representative and randomly selected samples from a wider population can give some indication of the likely scale of health inequalities that may result from varying exposure to health determinants. Furthermore, Bayesian statistical methods (Congdon, 2003) allow researchers to test different hypotheses expressed in terms of prior assumptions about the characteristics of a given population and how their health may vary in response to health determinants. Related techniques can also be applied to estimate the future likelihood of given epidemiological events under different scenarios (Bissell et al., 2014, Burns et al., 2014). A growing body of research in health geography is also using agent-based modeling (Grow and VanBavel, 2017) to explore through statistical simulations the possible future health outcomes of interventions to address risk factors such as dietary behaviors (Blok et al., 2015; Giabbanelli and Crutzen, 2017; Orr, Kaplan and Galea, 2016; Zhang et al., 2014).

The context-versus-composition debate (described in other chapters of this handbook) emphasizes that while action to modify individual risk factors and health-related behaviors may have potential to address health problems that arise due to *compositional* effects, it may also be necessary to tackle *contextual* aspects of the wider environment. Geographers have therefore become very interested in the public-health literature that considers risks for health in terms of the *wider determinants* of health. Originating in a World Health Organization (WHO) report (Dahlgren and Whitehead, 1991), this perspective captures the idea that the environment that we share with others in the places where we live is important to understand if we are aiming to improve health and well-being and address the root causes of health risks. The wider determinants of health may operate at varying geographical scales, so that interventions to address determinants of health risks also need to be multiscalar, ranging from the *local* neighborhood (for example, local availability of community facilities and amenities) to the *national* or *global* scale (for example, national health systems, and actions of United Nations agencies such as the WHO).

Risk as perceived hazard from personal and individual experience

Research from an ethnographic perspective has contributed in very different ways to our understanding of how perception of risks for health varies between cultures and social groups within societies. Some research shows, for example, that some environmental factors and behaviors that are considered potential health hazards in Western medicine are seen in other cultures as socially desirable in ways that make the health risks a secondary consideration – for example, cultural dress codes may reduce sunlight exposure and increase risk of vitamin D deficiency (Quraishi and Badsha, 2016) and social protocols at funerals in Tanzania were found to conflict with use of mosquito nets to avoid malarial infection (Dunn, Le Mare and Makungu, 2016).

Also relevant to our understanding of risk are *relational* approaches to our understanding of the associations between health and place (Conradson, 2005, Cummins et al., 2007). These perspectives focus on how experience of the environment is variable from one person to another, depending on their individual characteristics and experiences over their life course, which often involve complex trajectories through many different settings, as well as wider connections to socio-spatial networks that may be geographically very extensive. Environmental conditions that may be health-enhancing or restorative for some people may, for other individuals, have little benefit or even be damaging in ways that present risks to their health. Examples may include psychosocial health impacts of sociocultural environments, as experienced in areas where particular cultural or ethnic groups are concentrated (ethnic density effects). For those of the same ethnic background, this may provide a sense of social cohesion and support (Fagg et al., 2006) that reduces risks to health or may build resilience to prevailing risks. However, for others who belong to different groups, the same social environment may seem alien or even threatening and may not be supportive for health. Other examples include variations in health-related experience of environments, associated with a person's gender, physical abilities or socioeconomic circumstances. Relational approaches in health geography often question the distinction made in some research between *contextual* and *compositional* risk factors, arguing that these are not independent aspects of health risk, but should rather be considered as closely interrelated, interactive processes.

Research of this type often bridges between geography, sociology and anthropology, drawing on shared theories and methods. Much of this research uses qualitative ethnographic and biographical approaches for small, purposive samples of people as opposed to statistical methods for larger, randomly selected population samples. These qualitative studies provide a rich picture that helps us understand in depth how the people studied interacts with places through their life course and how particular sociocultural viewpoints influence these interactions. However, it is debatable to what extent the knowledge is generalizable.

Conclusion: the potential of multidisciplinary perspectives on risk and resilience

In light of the variety of perspectives reviewed above, it is interesting to consider the relevance for health geography of the wider, interdisciplinary debate regarding the significance for policy and practice of quantitative and qualitative perspectives on risk and resilience. Some authors suggest that qualitative methods should be emphasized more widely in research and may not be compatible with research using statistical methods. For example, it has been suggested (Adams, 2016) that by focusing on *metric* aspects of health risk and resilience, for which methods of measurement are available, and applying statistical criteria to assess how these indicators vary, governmental and non-governmental agencies concerned with addressing health risks fail to understand and respond to some aspects of risk that are difficult to measure but are nevertheless important for variation in health in different settings around the world.

An alternative view may be that, in fact, globally influential agencies tackling major health risks actually understand rather well the emotive power of individual stories and accounts of conditions such as poverty and disadvantage or environmental disasters, which often have stronger influence on risk perception among the general public than quantified or technical scientific evidence (Slovic and Vastfjall, 2010). Many of the appeals by charitable organizations, for example, feature images and accounts of individuals in distress, faced with specific environmental threats. Arguably, some high-profile individual events have had an (albeit potentially beneficial) influence on policy and practice that is disproportionate to the individual experience. Examples are poignant images of infants killed or injured during attempts to flee conflict zones, and individuals evacuated from areas devastated by floods, hurricanes or earthquakes. These images' influence on social action to manage or mitigate health risks may be so strong partly because the individual cases *stand for* a more extensive problem affecting a much larger number of people but the individual cases speak more immediately to our emotions. Similarly, research on how to build resilience to environmental hazards for health and health care often draw on qualitative studies of communities in specific local settings that prove to have wide relevance. For example, a study designed to make infrastructure supporting older people's care more resilient during extreme weather conditions in parts of England (Curtis et al., 2018, Oven et al., 2012) used purposively selected local case studies that were found to be relevant for national strategic planning because local communities elsewhere recognized parallels with their own experience, making the case studies enlightening and informative in ways that usefully complemented more generalized national advice and guidance.

Thus, it can be argued that considering qualitative and quantitative evidence *in combination* is especially useful for production of knowledge related to risk and resilience, knowledge that can help us take action to promote better and more equal health for populations in different geographical settings. Arguably, health geography is one of the fields of research that has been relatively successful in developing multidisciplinary conceptual and empirical methods that bridge quantitative and qualitative methods for studying risk and resilience. Many research projects and research outputs in health geography successfully show how individual and local experience of risk and resilience interact with conditions prevailing in whole populations and wider regions. This is reflected in a significant body of the health-geography research literature, including this handbook, that combines different approaches.

References

Acierno, R., Ruggiero, K. J., Kilpatrick, D. G., Resnick, H. S. and Galea, S. (2006). Risk and protective factors for psychopathology among older versus younger adults after the 2004 Florida hurricanes. *American Journal of Geriatric Psychiatry*, 14, pp. 1051–1059.

Adams, V. (2016). *Metrics: what counts in global health*. Durham, NC: Duke University Press.

Bartley, M., Schoon, I., Mitchell, R. and Blane, D. (2010). Resilience as an asset for healthy development. In: A. Morgan, M. Davies and E. Ziglio, eds., *Health assets in a global context: theory, methods, action*. New York: Springer, pp. 101–115.

Bates, J. and Machin, A. (2016). Locality, loneliness and lifestyle: a qualitative study of factors influencing women's health perceptions. *Health & Social Care in the Community*, 24, pp. 639–648.

Bécares, L. (2014). Ethnic density effects on psychological distress among Latino ethnic groups: an examination of hypothesized pathways. *Health & Place*, 30, pp. 177–186.

Bissell, J. J., Caiado, C. C. S., Goldstein, M. and Straughan, B. (2014). Compartmental modelling of social dynamics with generalised peer incidence. *Mathematical Models & Methods in Applied Sciences*, 24, pp. 719–750.

Blok, D. J., De Vlas, S. J., Bakker, R. and Van Lenthe, F. J. (2015). Reducing income inequalities in food consumption explorations with an agent-based model. *American Journal of Preventive Medicine*, 49, pp. 605–613.

Burns, C. J., Wright, M., Pierson, J. B., Bateson, T. F., Burstyn, I., Goldstein, D. A., Klaunig, J. E., Luben, T. J., Mihlan, G., Ritter, L., Schnatter, A. R., Symons, J. M. and Yi, K. D. (2014). Evaluating uncertainty to strengthen epidemiologic data for use in human health risk assessments. *Environmental Health Perspectives*, 122, pp. 1160–1165.

Cairns-Nagi, J. M. and Bambra, C. (2013). Defying the odds: a mixed-methods study of health resilience in deprived areas of England. *Social Science & Medicine*, 91, pp. 229–237.

Congdon, P. (2003). *Applied Bayesian modelling*, Chichester: Wiley-Blackwell.

Conradson, D. (2005). Landscape, care and the relational self: therapeutic encounters in rural Englad. *Health & Place*, 11, pp. 507–525.

Cummins, S., Curtis, S., Diez-Roux, A. V. and Macintyre, S. (2007). Understanding and representing "place" in health research: a relational approach. *Social Science & Medicine*, 65, 1825–1838.

Curtis, S., Southall, H., Congdon, P. and Dodgeon, B. (2003). Area effects on health variation over the life-course: analysis of the Longitudinal Study sample in England using new data on area of residence in childhood. *Social Science & Medicine*, 58, pp. 57–74.

Curtis, S., Oven, K. J., Wistow, J., Dunn, C. E. and Dominelli, L. (2018). Adaptation to extreme weather events in complex local health care systems: the example of older people's health and care services in England. *Environment and Planning C: Politics and Space*, 36(1), pp. 67–91.

Cutter, S. L., Ash, K. D. and Emrich, C. T. (2014). The geographies of community disaster resilience. *Global Environmental Change: Human and Policy Dimensions*, 29, pp. 65–77.

Cutter, S. L., Barnes, L., Berry, M., Burton, C., Evans, E., Tate, E. and Webb, J. (2008). A place-based model for understanding community resilience to natural disasters. *Global Environmental Change: Human and Policy Dimensions*, 18, pp. 598–606.

Dahlgren, G. and Whitehead, M. (1991). *Policies and strategies to promote social equity in health: background document to WHO – strategy paper for Europe.* Arbetsrapport: Institute for Futures Studies.

Duncan, C., Jones, K. and Moon, G. (1998). Context, composition and heterogeneity: using multilevel models in health research. *Social Science & Medicine*, 46, 97–117.

Dunn, C. E., Le Mare, A. and Makungu, C. (2016). Connecting global health interventions and lived experiences: suspending "normality" at funerals in rural Tanzania. *Social & Cultural Geography*, 17, pp. 262–281.

Fagg, J., Curtis, S., Stansfeld, S. and Congdon, P. (2006). Psychological distress among adolescents, and its relationship to individual, family and area characteristics in East London. *Social Science & Medicine*, 63, pp. 636–648.

Fatemi, F., Ardalan, A., Aguirre, B., Mansouri, N. and Mohammadfam, I. (2017). Social vulnerability indicators in disasters: findings from a systematic review. *International Journal of Disaster Risk Reduction*, 22, pp. 219–227.

Freitag, R. C., Abramson, D. B., Chalana, M. and Dixon, M. (2014). Whole community resilience: an asset-based approach to enhancing adaptive capacity before a disruption. *Journal of the American Planning Association*, 80, pp. 324–335.

Giabbanelli, P. J. and Crutzen, R. (2017). Using agent-based models to develop public policy about food behaviours: future directions and recommendations. *Computational and Mathematical Methods in Medicine*, 2017, pp. 1–12.

Grow, A. and Vanbavel, J. (eds.) (2017). *Agent-based modelling in population studies: concepts, methods and applications.* Switzerland: Springer.

Mitchell, R., Gibbs, J., Tunstall, H., Platt, S. and Dorling, D. (2009). Factors which nurture geographical resilience in Britain: a mixed methods study. *Journal of Epidemiology and Community Health*, 63, pp. 18–23.

Orr, M. G., Kaplan, G. A. and Galea, S. (2016). Neighbourhood food, physical activity, and educational environments and black/white disparities in obesity: a complex systems simulation analysis. *Journal of Epidemiology and Community Health*, 70, pp. 862–867.

Oven, K. J., Curtis, S. E., Reaney, S., Riva, M., Stewart, M. G., Ohlemuller, R., Dunn, C. E., Nodwell, S., Dominelli, L. and Holden, R. (2012). Climate change and health and social care: defining future hazard, vulnerability and risk for infrastructure systems supporting older people's health care in England. *Applied Geography*, 33, pp. 16–24.

Pearson, A. L., Pearce, J. and Kingham, S. (2013). Deprived yet healthy: neighbourhood-level resilience in New Zealand. *Social Science & Medicine*, 91, pp. 238–245.

Pfefferbaum, B., Jacobs, A. K., Griffin, N. and Houston, J. B. (2015). Children's disaster reactions: the influence of exposure and personal characteristics. *Current Psychiatry Reports*, 17, pp. 1–6.

Pfefferbaum, R. L., Pfefferbaum, B., Van Horn, R. L., Klomp, R. W., Norris, F. H. and Reissman, D. B. (2013). The Communities Advancing Resilience Toolkit (CART): an intervention to build community resilience to disasters. *Journal of Public Health Management and Practice*, 19, pp. 250–258.

Quraishi, M. K. and Badsha, H. (2016). Rheumatoid arthritis disease activity and vitamin D deficiency in an Asian resident population. *International Journal of Rheumatic Diseases*, 19, pp. 348–354.

Rigg, J., Oven, K. J., Basyal, G. K. and Lamichhane, R. (2016). Between a rock and a hard place: vulnerability and precarity in rural Nepal. *Geoforum*, 76, pp. 63–74.

Riva, M. and Curtis, S. (2012). Long term local area employment rates as predictors of individual mortality and morbidity: a prospective study in England spanning more than two decades. *Journal of Epidemiology and Community Health*, 66, pp. 919–926.

Rotegard, A. K., Moore, S. M., Fagermoen, M. S. and Ruland, C. M. (2010). Health assets: a concept analysis. *International Journal of Nursing Studies*, 47, pp. 513–525.

Slovic, P. and Vastfjall, D. (2010). Affect, moral intuition, and risk. *Psychological Inquiry*, 21, pp. 387–398.

Villagrán, L., Reyes, C., Wlodarczyk, A. and Páez, D. (2014). Coping community, collective posttraumatic growth and social well-being in context february 27 earthquake in Chile, 2010. *Terapia Psicologica*, 32, pp. 243–254.

WHO (2011). *Global vaccination action plan.* [online] Available at: www.who.int/immunization/documents/general/ISBN_978_92_4_150498_0/en/ [Accessed 6 Oct. 2017].

Yoshida, H., Kobayashi, N., Honda, N., Matsuoka, H., Yamaguchi, T., Homma, H. and Tomita, H. (2016). Post-traumatic growth of children affected by the Great East Japan Earthquake and their attitudes to memorial services and media coverage. *Psychiatry and Clinical Neurosciences*, 70, pp. 193–201.

Zhang, D. L., Giabbanelli, P. J., Arah, O. A. and Zimmerman, F. J. (2014). Impact of different policies on unhealthy dietary behaviors in an urban adult population: an agent-based simulation model. *American Journal of Public Health*, 104, pp. 1217–1222.

9

RESEARCHING MIGRATION AND HEALTH

Perspectives and debates

Frances Darlington-Pollock, Paul Norman, Daniel J. Exeter and Nichola Shackleton

Population movements have long been associated with the spread of disease. As early as 1690, maps delineating the cordoned-off boundaries of Bari, a province of Italy, are revealing as to the fear that the plague was portable, spread through a mobile population (Koch, 2011). This fear of diseased migrant bodies persists today, with a wide body of research looking for the etiological clues revealed by international migration flows or the broader impacts on the spread of disease (Boyle and Norman, 2009; Gushulak and MacPherson, 2006). However, of particular interest to health geographers is the impact of population movement on spatial inequalities in health (Darlington, Norman and Gould, 2015). This dynamic field of research, a critical discussion of the definitions employed, frameworks applied and methods selected, is overdue. Understanding how such decisions shape perspectives around the relationship between health and migration is important when evaluating debates as to the influence of migration on health inequalities. These debates have a long history, which resonates today in terms of both the conclusions drawn and the methods adopted. This chapter will first look to the history of this field of research, thereby introducing key contemporary studies examining the relationship between (subnational) migration and health. Next, we review the methodological parameters of this body of work before concluding with speculation as to future areas of inquiry within and beyond existing parameters of research.

The origins of the debate

In 1864, William Farr reported that migrants moving from urban to rural areas within England had poorer health than those moving from rural to urban areas. Migration thus appeared to be selective on health, a finding replicated by one of Farr's contemporaries (Welton, 1872) and many others in the following years.

Failing to account for the selectivity of migration complicates efforts to understand the causes and mechanisms behind geographic variations in health. If the population are not static, exposure to physical or socioeconomic conditions adversely affecting, or even benefiting, health may manifest in a different location to their origin. Evaluating the success of policy interventions designed to improve health, identifying the etiological pathways of particular morbidities, and understanding the causal mechanisms driving health inequalities in a population may therefore be undermined if the movement of differently healthy groups is ignored (Bentham, 1988; Learmonth, 1988; Prothero, 1977). The act of migrating itself can also impact health, especially when migration is enacted under duress. This is particularly important for international flows of migration precipitated by war, political persecution and civil unrest (though these are not the focus of this chapter) but equally applicable to subnational migration flows prompted by insecure tenures, unstable labor markets, and declining household circumstances. Research into the health-selectivity of migration, geographic patterns of mobility and their relationship to health and mortality has come a long way since Farr's early observation.

Health-selective migration

The process of changing address is inherently selective. One's propensity to migrate (across any geographic scale) varies systematically by age according to a range of different sociodemographic attributes, including socioeconomic status, ethnicity, employment status, educational attainment, housing tenure and marital status (Champion, 2005; Cooke, 2008; Norman, Boyle and Rees, 2005). Through the selectivity of migration, mobile groups are sorted through space, often heralding a change in socioeconomic and/or physical environment. In the context of subnational moves, this is perhaps most apparent in the systematic movement of age-cohorts between differently deprived areas (Norman and Boyle, 2014). However, following Farr's early observations, it has since been widely demonstrated that migration is also selective on health (Brimblecombe, Dorling and Shaw, 1999, 2000; Norman, Boyle and Rees, 2005; Verheij et al., 1998). As the population are sorted through space, the sociodemographic composition of mobile groups and their health status will vary depending on the nature or (socioeconomic) direction of a move.

Reasons for moving shorter distances are qualitatively different from motivations for a longer-distance move. Moves across longer distances often reflect moves to new opportunity structures, particularly new labor markets. Moves across shorter distances may be necessitated by a decline in material circumstances or the diktats of an insecure housing tenure. Therefore, short-distance migrants will be compositionally different from long-distance migrants, selected according to contrasting sociodemographic attributes and health status. Key life events such as family formation or union dissolution are also triggers of migration, themselves associated with different distances of moves and particular sociodemographic attributes. Understanding the interplay between the health and sociodemographic characteristics of mobile groups, in conjunction with the reasons for, and direction of, moves is critical for disentangling the relationship between health and place.

Most migration events are made, often for employment and/or educational opportunities, by young adults who tend to be in better health than their immobile peers (Bentham, 1988). Given the widely documented social gradients of health, the notion that those in better health are more likely to move across greater distances and be of a higher socioeconomic status is perhaps unsurprising. Their motivations for moving are more likely to be positive, reflecting advances in labor-market position and maintaining the exposure of these mobile groups to salutogenic contexts. Conversely, older migrants are more likely to be in poor health, prompted by the need to seek informal or formal care (Norman, Boyle and Rees, 2005). Short-distance, frequent moves are also associated with poorer health, often linked through the relationship between insecurity of tenure and social disadvantage (Morris, Manley and Sabel, 2016; Larson, Bell and Young, 2004; Boyle, Norman and Rees, 2002). The elevated risk of poor health associated with lower socioeconomic status is therefore compounded by a heightened risk of unfavorable short-distance mobility, which likely maintains existing exposure(s) to pathogenic environments.

Better health is therefore associated with moves toward, within or between less-deprived areas and across greater distances, whereas those in poorer health are constrained in both distance and trajectory, more likely to move shorter distances toward, within or between more deprived areas (Boyle, Norman and Rees, 2002). The likelihood of sustaining, or changing, one's existing experience of deprivation is bound by the selectivity of migration, tied to compositional characteristics of the population, and varies across the life course (Norman and Boyle, 2014). These theoretical considerations underpin research into the non-linear complexities of the health-migration relationship.

The specific perspective adopted within selective migration research is largely determined by the nature of available data: how migrants are identified, how health is measured, what relevant compositional data exist and, crucially, what level of geographic and/or temporal detail are available. Existing reviews concerning the complexities of the health-migration relationship tend to neglect the methodological parameters of the research in question. While competing conclusions surrounding the influence of migration on perpetuating spatial health inequalities are defined by the methodological parameters, neglecting these comes at the expense of a deeper understanding of the relationship between health and migration. A better understanding of both the parameters and content of the debate will inform future research by health geographers. The following section will explore existing debates in this field of research, before considering what methodological parameters may be shaping the conclusions.

Existing debates

A number of studies find little or no evidence of the influence of selective migration on health, with conclusions suggesting the numbers involved are too small to be of importance (Martikainen et al., 2008; Popham et al., 2011; van Lenthe, Martikainen and Mackenbach, 2007; Verheij et al., 1998). However, a sizeable body of research conversely suggests that selective migration can contribute to changing health gradients in a variety of contexts with different health outcomes (e.g., Curtis, Maninder and Quesnel-vallee, 2009; DeVerteuil et al., 2007; Larson, Bell and Young, 2004). Sorting populations according to health status into different spaces, defined either by boundaries or by contextual attributes, with migrants selected based on their sociodemographic attributes including health, at least maintains and can exaggerate existing health gradients. However, scale matters: stronger evidence exists in support of the contribution of selective migration to health inequalities at more local scales (Connolly and O'Reilly, 2007; Norman, Boyle and Rees, 2005; O'Reilly and Stevenson, 2003) than observed for larger-area interregional moves (Popham et al., 2011). Brimblecombe, Dorling and Shaw (1999, 2000) concluded that whereas regional inequalities in mortality were not attributable to selective migration, district-level inequalities were. Noting the importance of lifetime migration histories, the authors observed that the social composition of groups with more favorable lifetime migration patterns was markedly different from those moving in the opposite direction. Thus, as selective migration redistributes differently healthy groups of the population, it contributes to the polarization of (ill) health (O'Reilly and Stevenson, 2003), thereby exacerbating health gradients.

Effective explanation of these contrasting patterns is hampered by the complexities of the relationships examined. Factors that explain propensity to migrate, the very selectivity of migration, are also closely related to health. This is evident in the well-known literature on the social determinants of health. Delineating the relationship between health and migration when framed in a binary way is problematic: the relationships are multidimensional, non-linear and complex, stretching far beyond health selectivity. Consider the observed differences in the age-structure of ethnic internal migration patterns in the United Kingdom (Finney, 2011). Ethnic minority groups that are often disadvantaged in socioeconomic status and health (Nazroo, 2003) may have different age-profiles to the majority population and may also be subject to different psychosocial pressures. Thus, existing analytical frameworks may be unable to capture these complex interactions, let alone their ability to depict likely interactions between subnational migration patterns, health status and previous international migrations by different ethnic groups across the world. Consider also the interaction between housing tenure, health and migration propensity: short-distance moves are more likely for those with insecure tenures and associated with poorer health than long-distance moves, or moves by owner-occupiers. Despite the apparent interaction between these variables, existing research has been unable to fully disentangle the nature of these complexities (Boyle, Norman and Rees, 2002). To better situate directions for future research, reflecting upon the methodological approaches used in the existing research in this field is useful.

Methodological parameters

Definitions, contexts and scale

Discussions of population movement are typically framed around the concepts of *migration, migrants, residential mobility* and *movers*. The root of purported conceptual differences relates to questions of scale, whether geographic or temporal. Over what distance, and for what length of time, does a move need to be before it is considered a migration event? Champion (1992) suggests that a move across a statistical boundary to a new abode for at least a year is an adequate definition of a (subnational) migration event. Following this definition, moves within a statistical boundary or moves that are more short-term in nature constitute residential mobility rather than migration. White and Mueser (1988) base the distinction between residential mobility and migration on differences in moves *within* or *between* communities or local labor and housing markets. Residential mobility is therefore distinct from migration, as it may not involve a sufficient change in the social or economic situation of an individual to constitute a migration event.

For health geographers interested in general population health, of interest is whether a change of address results in exposure to significantly different socioeconomic or physical conditions and/or whether there is sufficient (temporal) exposure to substantively influence health (Bentham, 1988; Norman, Boyle and Rees, 2005). For those with a more targeted focus on a particular health outcome, such as the geographic patterning to parasite-related morbidities (Prothero, 1977) or risk of cardiovascular disease (Darlington-Pollock et al., 2016; Exeter et al., 2015), short-term or frequent moves gain importance. Determining whether people are subnational *migrants* or residentially mobile *movers* may therefore depend either (a) on the ability of data to identify specific geographic, administrative or statistical boundaries that people move within or cross or (b) on the particular health focus of the research in question. As results may vary temporally, across different geographic scales, by distance and according to different health outcomes, these types of conceptual distinctions are important to consider when interpreting results of existing research.

Popham et al. (2011) counted changes of address reflecting intra-regional moves between administrative areas of Scotland as migration events, concluding that selective migration did not contribute to inequalities in mortality between those administrative areas identified in the study. This contrasts with Brimblecombe, Dorling and Shaw's (1999) study that also defined migration events as interregional moves, focusing on England and Wales rather than Scotland. Larson, Bell and Young (2004) found evidence of health-selective migration in older ages, defining moves within and between Australian postcodes, or between rural and urban areas. The authors simultaneously consider what may be defined as migration and residential mobility events. Scale, and indeed context, is important, as the implication of moves (and the health attributes of individuals involved) within and between localities may be very different from the implication of moves between regions.

Comparing rates of limiting long-term illness between migrants with non-migrants, Norman, Boyle and Rees (2005) found that health-selective migration accounted for the majority of changes in inequalities in health by deprivation in England and Wales over a 20-year period. Here, migrants are identified as those with a different address at a given time point compared to 10 years previously, and then tracked over a 20-year period. Migrants moving from more- to less-deprived areas were in better health than migrants moving in the opposite direction. Conversely, comparing risk of cardiovascular disease between mobile groups in New Zealand found less evidence of a clear difference in health between migrants moving toward or away from deprivation (Exeter et al., 2015). Here, moves are identified where addresses vary quarterly over a year rather than decennially and a particular health outcome is targeted. These studies examine mobile groups under different macro-political, economic and social contexts, identify migrants at different temporal scales and focus on different health outcomes.

Health has been measured in a variety of ways, including objective measures such as general mortality (Exeter, Boyle and Norman, 2011; Popham et al., 2011; Robards et al., 2014) or for specific outcomes such as psychological distress (Curtis, Maninder and Quesnel-vallee, 2009). A substantial health-migration literature is also based on subjective measures of health such as limiting long-term illness (Verheij et al., 1998). Insight into the complexities of the health-migration relationship evidently varies according to the health outcome used, though interpretation of these collective differences must sit alongside consideration of the wider parameters of the research in question. DeVerteuil et al. (2007) found intra-urban moves from the suburbs to the inner city were more likely among those with schizophrenia than those with no mental illness, whereas moves in the opposite direction were less likely. Jongeneel-Grimen et al. (2011) found no evidence that health-selective migration widens health inequalities between areas of the Netherlands, defining "health" through (a) self-assessments of general health, (b) the presence of absence of long-term health problems, or (c) the presence or absence of physical disabilities limiting physical functioning. The authors suggested this may contrast with wider studies (Brown and Leyland, 2010; Norman, Boyle and Rees, 2005; Martikainen et al., 2008) due to differences in the health outcomes used, also reflecting on the implications of country context and the identification of migrants, both in scale and time.

Scale and context are particularly important, exemplified by a comparison of the relevance of health-related behaviors selecting migrants internally or internationally. Contrast van Lenthe Martikainen and Mackenbach's (2007) Dutch study of internal migration between neighborhoods with Silventoinen et al.'s (2007) Finnish study of international migration: the former found only weak associations between health-related behaviors and favorable or unfavorable migration, contrasting with the stronger associations at an international scale in the latter. The defining contexts for a migration event will markedly influence the socioeconomic composition of groups moving between those contexts, thereby impacting their chances of good health and uptake of different health-related behaviors. Despite the breadth of observed differences, there is a relative paucity of research explicitly considering whether and how differences in the operationalization of *health* matters for efforts to disentangle the complexities of the health-migration relationship.

Given the selectivity of migration, efforts to define and distinguish between different types of migrants may be better served by recognition of their contrasting compositional attributes and the implications for particular health outcomes than by questions of distance, scale or time. A more nuanced understanding of the nature of a migration event, across any geographic scale and without any *a priori* assumptions, would follow from the availability of relevant compositional markers of migrants. This might include data on educational attainment, housing tenure, and indeed prior health status (however defined). Although discussion of *migrants* within studies of migration and health often speaks to some of these issues, such discussion should be more systematic.

Analytical frameworks

The studies cited thus far variously employ cross-sectional data, capturing the experiences of migrant populations at a single point in time, or link data to enable more revealing comparisons at two or more points in time through longitudinal and cohort data. The inability to retrospectively capture relevant socioeconomic, residential and health data in cross-sectional data impedes conclusions drawn from this type of research (Verheij et al., 1998). By contrast, longitudinal or cohort data more informatively allow researchers to track individuals over time and through space. Linked data can determine whether mobile groups who begin in less deprived areas and move to more deprived areas are in poorer health than those who either remain in or move to less-deprived areas (e.g., Norman, Boyle and Rees, 2005). Where the health of those moving into more deprived circumstances is poorer than the health of those moving out, and the health of those moving into less-deprived circumstances is better than the health of those moving out, health differences may widen (Boyle, Norman and Popham, 2009). Cross-sectional data that identifies migrants without retrospective contextual characteristics simply reveals associations between the current sociodemographic profile of a group and its migrant status.

While longitudinal data offer greater insight into the complexities of the health-migration relationship, they may be hampered by a lack of temporal detail and undermined by the absence of rich migration histories (though see Brimblecombe, Dorling and Shaw, 1999, 2000). Migration histories, rather than simple comparisons of address at two time points, may be more revealing as to the health-migration relationship. Particularly within the context of life-course perspectives on health and the accumulated influence of (dis) advantage over time. The frequency of moves are associated with poorer health (Geronimus, Bound and Ro, 2014): simplifying migration to a comparison of address at two points in time is not sufficient to capture this complexity, nor perhaps is a binary variable distinguishing between low- and high-frequency moves (see Morris, Manley and Sabel, 2016).

Future research

Existing research is constrained by the types of available data, whether cross-sectional or longitudinal, the availability of rich compositional and contextual data, choice of health outcome, and ability to appropriately define the nature of a migration event. Cohort studies, wherein data are collected for the sample at regular intervals, offer more potential for a richer examination of migration trajectories and the implications of frequency of moves. This may strengthen conclusions as to the specific implications of accumulated (dis)advantage and residential mobility or migration for health. However, meaningful interpretation of these patterns will depend on adequate linkage to both contextual area-level and compositional individual-level data. The availability of one is often at the expense of the other; nevertheless, the increasing availability routinely collected population registers (e.g., to Scandinavia, or New Zealand) provides may respond to these limitations.

Future research must make better use of robust analytical methods to compare *like for like* between migrant groups. Explanatory models may be employed to compare the odds or relative risks of a specific health outcome between differently mobile groups, or to compare the odds of migrating given different sociodemographic attributes (e.g., Darlington-Pollock et al., 2016; DeVerteuil et al., 2007; Exeter et al., 2015; Larson, Bell and Young, 2004; van Lenthe, Martikainen and Mackenbach, 2007; Verheij et al., 1998). However, these approaches tend to simply reveal (even when longitudinal) the association between migrant status and a health event, rather than establishing whether the association arises because of the migration event itself or the composition of the migrant groups. Green et al.'s (2015) innovative use of matching methods addresses the inherent bias in existing studies of selective migration and health, in which differences between migrants and non-migrants may simply be an artifact of their contrasting compositional profiles. Extending Green et al.'s method to different contexts, datasets and health outcomes may be revealing. Similarly, ongoing research by the authors of this chapter makes better use of the temporal detail in cohort studies to understand how mobile groups move within and across differently deprived spaces, creating complex migration trajectories, which have equally complex relationships with health (Darlington-Pollock et al., forthcoming; Exeter et al., 2017). Notwithstanding the value of methodological developments in explanatory modeling, calculating demographic rates such as standardized mortality ratios for different mobile groups may be particularly revealing as to the influence of migration on population health (Brown and Leyland, 2010; Martikainen et al., 2008; Norman, Boyle and Rees, 2005).

Efforts to fully address the complexities of the interrelationships, which in turn define the health-migration nexus, may be better served in the future through methodological developments and advances in data capture. The ability to adequately account for more temporary or seasonal moves may be enhanced with the growing availability of *big data*. However, where these data exist, reasons for a move are not captured, undermining the utility of big data for health geographers interested in migration. Nevertheless, future research need not be limited to quantitative approaches. Qualitative work on health and subnational migration typically focuses on the experience and consequences of migrating for marginalized vulnerable groups (e.g., Elliott and Gillie, 1998). This could be extended to give more insight into the multi-dimensional nature of the health-migration relationship(s), which are often simplified or constrained by available quantitative data. The insights of qualitative research may then inform big data: what should be sought, how frequently, and how should it be analyzed?

Future research must strive to explore the multidimensional qualities of the health-migration relationship(s) and look beyond simple health selectivity. The selectivity of migration in the context of health is convoluted. Research should critically examine the implications of selective migration for population health and whether relationships vary over time, between places and by health outcome. A stronger theoretical underpinning to these issues may follow from richer qualitative studies that are better placed to disentangle the multidimensional interrelationships between health and migration. Perhaps only then can quantitative perspectives provide an understanding of the multiple and varied moves made by individuals in a given time period, measuring the effect of lifetime migration histories on health or maximizing the potential for big data to fully capture the complexities of migration, mobility and health.

References

Bentham, G. (1988). Migration and morbidity: implications for geographical studies of disease. *Social Science & Medicine*, 26(1), pp. 49–54.

Boyle, P. and Norman, P. (2009). Migration and health. In: T. Brown, S. McLafferty and G. Moon, eds., *The companion to health and medical geography*. Chichester: Wiley-Blackwell, pp. 346–374.

Boyle, P., Norman, P. and Popham, F. (2009). Social mobility: evidence that it can widen health inequalities. *Social Science & Medicine*, 68(10), pp. 1835–1842.

Boyle, P., Norman, P. and Rees, P. (2002). Does migration exaggerate the relationship between deprivation and limiting long-term illness? A Scottish analysis. *Social Science & Medicine*, 55, pp. 21–31.

Brimblecombe, N., Dorling, D. and Shaw, M. (1999). Mortality and migration in Britain, first results from the British Household Panel Survey. *Social Science & Medicine*, 49, pp. 981–988.

Brimblecombe, N., Dorling, D. and Shaw, M. (2000). Migration and geographical inequalities in health in Britain. *Social Science & Medicine*, 50, pp. 861–878.

Brown, D. and Leyland, A. (2010). Scottish mortality rates 2000–2002 by deprivation and small area population mobility. *Social Science & Medicine*, 71, pp. 1951–1957.

Champion, A. (1992). Migration in Britain: research challenges and prospects. In: A. Champion and A. Fielding, eds., *Migration processes and patterns volume 1, research progress and prospects*. London: Belhaven Press, pp. 215–226.

Champion, T. (2005). Population movement within the UK. In: R. Chappell, ed., *Focus on people and migration*. Basingstoke: Palgrave Macmillan, pp. 92–114.

Connolly, S. and O'Reilly, D. (2007). The contribution of migration to changes in the distribution of health over time: Five-year follow-up study in Northern Ireland. *Social Science & Medicine*, 65(5), pp. 1004–1011.

Cooke, T. J. (2008). Migration in a family way. *Population, Space and Place*, 14(4), pp. 255–265.

Curtis, S., Maninder, S. S. and Quesnel-vallee, A. (2009). Socio-geographic mobility and health status: a longitudinal analysis using the National Population Health Survey of Canada. *Social Science & Medicine*, 69, 1845–1853.

Darlington, F., Norman, P. and Gould, M. (2015). Migration and health. In: N. Finney, D. Smith, K. Halfacree and N. Walford, eds., *Internal migration: geographic perspectives and processes*. Farnham: Ashgate, pp. 113–128.

Darlington-Pollock, F., Norman, P., Lee, A. C., Grey, C., Mehta, S. and Exeter, D. (2016). To move or not to move? Exploring the relationship between residential mobility, risk of cardiovascular disease and ethnicity in New Zealand. *Social Science & Medicine*, 165, pp. 128–140.

Darlington-Pollock, F., Shackleton, N., Norman, P., Lee, A. C. and Exeter, D. (2017). Differences in the risk of cardiovascular disease for movers and stayers in New Zealand: a survival analysis. *International Journal of Public Health*, 63 (2): 173–179.

DeVerteuil, G., Hinds, A., Lix, L., Walker, J., Robinson, R. and Roos, L. L. (2007). Mental health and the city: intra-urban mobility among individuals with schizophrenia. *Health & Place*, 13, pp. 31–323.

Elliott, S. J. and Gillie, J. (1998). Moving experiences: a qualitative analysis of health and migration. *Health & Place*, 4(4), pp. 327–339.

Exeter, D. J., Boyle, P. and Norman, P. (2011). Deprivation (im)mobility and cause-specific premature mortality in Scotland. *Social Science & Medicine*, 72(3), pp. 389–397.

Exeter, D. J., Sabel, C. E., Hanham, G., Lee, A. C. and Wells, S. (2015). Movers and stayers: the geography of residential mobility and CVD hospitalisations in Auckland, New Zealand. *Social Science & Medicine*, 133, pp. 331–339.

Exeter, D., Shackleton, N., Darlington-Pollock, F., Norman, P., Jackson, R. and Lee, A. (2017). Using trajectory analysis to explain deprivation mobility patterns in New Zealand. 17th International Medical Geography Symposium, Angers, France, 2017.

Farr, W. (1864). *Supplement to the 25th annual report on the registrar general*. London: HMSO.

Finney, N. (2011). Understanding ethnic differences in the migration of young adults within Britain from a lifecourse perspective. *Transactions of the Institute of British Geographers*, 36(3), pp. 455–470.

Geronimus, A. T., Bound, J. and Ro, A. (2014). Residential mobility across local areas in the United States and the geographic distribution of the healthy population. *Demography*, 51(3), pp. 777–809.

Green, M. A., Subramanian, S. V., Vickers, D. and Dorling, D. (2015). Internal migration, area effects and health: does where you move to impact upon your health? *Social Science & Medicine*, 136–137, pp. 27–34.

Gushulak, B. D. and MacPherson, D. W. (2006). The basic principles of migration health: population mobility and gaps in disease prevalence. *Emerging Themes in Epidemiology*, 3, 3.

Jongeneel-Grimen, B., Droomers, M., Stronks, L. and Kunst, A. E. (2011). Migration does not enlarge inequalities in health between rich and poor neighbourhoods in The Netherlands. *Health & Place*, 17, pp. 988–995.

Koch, T. (2011). *Disease maps: epidemics on the ground*. Chicago: University of Chicago Press.

Larson, A., Bell, M. and Young, A. (2004). Clarifying the relationships between health and residential mobility. *Social Science & Medicine*, 59(10), pp. 2149–2160.

Learmonth, A. (1988). *Disease ecology: an introduction*. Oxford: Basil Blackwell.

Martikainen, P., Sipilä, P., Blomgren, J. and van Lenthe, F. J. (2008). The effects of migration on the relationship between area socioeconomic structure and mortality. *Health & Place*, 14, pp. 361–366.

Morris, T., Manley, D. and Sabel, C. E. (2016). Residential mobility: towards progress in mobility health research. *Progress in Human Geography*, pp. 1–22.

Nazroo, J. Y. (2003). The structuring of ethnic inequalities in health: economic position, racial discrimination, and racism. *American Journal of Public Health*, 93(2), pp. 277–284.

Norman, P. and Boyle, P. (2014). Are health inequalities between differently deprived areas evident at different ages? A longitudinal study of census records in England and Wales, 1991–2001. *Health & Place*, 26, pp. 88–93.

Norman, P., Boyle, P. and Rees, P. (2005). Selective migration, health and deprivation: a longitudinal analysis. *Social Science & Medicine*, 60, pp. 2755–2771.

O'Reilly, D. and Stevenson, M. (2003). Selective migration from deprived areas in Northern Ireland and the spatial distribution of inequalities: implications for monitoring health and inequalities in health. *Social Science & Medicine*, 58(7), pp. 1455–1462.

Popham, F., Boyle, P., O'Reilly, D. and Leyland, A. (2011). Selective internal migration: does it explain Glasgow's worsening mortality record. *Health & Place*, 17(6), pp. 1212–1217.

Prothero, R. M. (1977). Disease and mobility: a neglected factor in epidemiology. *International Journal of Epidemiology*, 6(3), pp. 259–267.

Robards, J., Evandrou, M., Falkingham, J. and Vlachantoni, A. (2014). Mortality at older ages and moves in residential and sheltered housing: evidence from the UK. *Journal of Epidemiology and Community Health*, 68, pp. 524–529.

Silventoinen, K., Hammar, N., Hedlund, E., Koskenvuo, M., Rönnemaa, T. and Kaprio, J. (2007). Selective international migration by social position, health behaviour and personality. *European Journal of Public Health*, 18(2), pp. 150–155.

van Lenthe, F. J., Martikainen, P. and Mackenbach, J. P. (2007). Neighbourhood inequalities in health and health-related behaviour: results of selective migration? *Health & Place*, 13, pp. 123–137.

Verheij, R. A., van de Mheen, H. D., de Bakker, D. H., Groenewegen, P. P. and Mackenbach, J. P. (1998). Urban-rural variations in health in the Netherlands: does selective migration play a part? *Journal of Epidemiology and Community Health*, 52, pp. 487–493.

Welton, T. (1872). On the effects of migrations in disturbing local rates of mortality, as exemplified in the statistics of London and the surrounding country, for the years 1851–1860. *Journal of the Institute of Actuaries*, 16, pp. 153–186.

White, M. T. and Mueser, P. R. (1988). Implications of boundary choice for the measurement of residential mobility. *Demography*, 25(3), pp. 443–459.

10

FOOD IN HEALTH GEOGRAPHY

Sarah Wakefield

Food is fundamental to health, so it is no surprise that food and food-related issues, such as nutrition and obesity, have been important topics of inquiry for health geographers. While food adequacy (and the geographical and logistical challenges associated with the production and distribution of food) have been preoccupations of governments since the beginnings of human society, the modern era has exacerbated these challenges (e.g., through increasingly globalized supply chains, a human population that is likely to reach 11 billion, and a changing climate). It has also created new challenges, such as the widespread concerns about overnutrition and obesity and food contamination.

This chapter examines the engagement of health geographers with food issues, tracing the historical and intellectual trajectory of the topic within health geography but also in the broader related disciplines of public health and food studies.

Food environments: access in the deserts and the swamps

Perhaps the most obvious application of health geographers' skills in the study of food is the issue of food environments. Not surprisingly, geographic access – for example, the distance a person has to travel to access a supermarket or other food retail outlet – has been a central feature of health geographers' work in this area. The term *food desert* was coined to describe areas where geographic access to food was limited. Importantly, food deserts were envisioned as urban, low-income neighborhoods – while distance from food retail locations in other areas, particularly rural and some suburban areas, could be significantly longer than those observed in many food deserts, the low-income character of these neighborhoods was seen as prohibiting car ownership and, thus, limiting residents' options for accessing food. Other researchers observed that food deserts were often inner-city neighborhoods and that the food landscape there was a result of changes in the retail environment: the closing of many smaller groceries and niche sellers (e.g., bakeries, butchers, and greengrocers) in the urban core, paired with the consolidation of food retail in superstores in suburban and periphery locations served by major road (but not generally transit) infrastructure. More sinister was the apparently disproportionate relocation of food outlets out of racialized neighborhoods. The utility of geographic methods in this context has provided an important opening for health geographers to engage with public-health practice, as food deserts have been the subject of government interest and intervention in a variety of contexts.

The food-desert literature has been beset with measurement controversies and methodological challenges. Key questions have included not only where (and if) food deserts exist, but also how far is too far to have to travel for food, and how we should measure that distance. Can any kind of food be sold, or does only healthy food count? And above all, does simple geographic access to healthy food mean that people are actually buying it? Significant efforts have been made to characterize and refine the methods and measurements used in studies of food deserts along these lines (see Charreire et al., 2010, and Shaw, 2014, for reviews of the various techniques used). For example, some studies have begun to map daily activity spaces to create a more accurate picture of how and where people shop for food (e.g., Cetateanu and Jones, 2016; Widener et al., 2013). Researchers have also used both quantitative and qualitative methods to attempt to trace the connection between food availability and purchasing choices (e.g., Thompson et al., 2016; Vogel et al., 2016).

Research has suggested that food deserts are not observable across all urban areas or in all high-income countries. Indeed, a systematic review by Beaulac, Kristjansson and Cummins (2009) found little evidence of food deserts outside of the United States. This raises questions about the utility of the food-desert concept in explaining patterns of hunger and obesity in urban areas. At the same time, the framework is being adopted, albeit with significant caveats and reimaginings, in other contexts: for example, in Africa (Battersby and Crush, 2014), Central Europe (Cerovečki and Grünhagen, 2016), South America (Gartin, 2012) and the Middle East (Mosammam et al., 2017). A growing recognition that many low-income areas had access to considerable numbers of food outlets, particularly fast-food restaurants and convenience stores, led to the coining of the term *food swamp* to describe these areas (Ver-Ploeg et al., 2009). Ultimately, however, many researchers have tried to move away from the evocative but somewhat inaccurate language of "desert" and "swamp" to more generally characterize the food environments of communities.

Much of this work has suggested – not surprisingly – that the connections between food environments and healthy diet decisions are complex (e.g., Caspi, Sorensen and Subramanian, 2012; Sadler, Gilliland and Arku, 2016). Improved access to food (for example, through the introduction of new supermarkets or healthy corner store initiatives) appears to have a limited – albeit generally positive – impact on the diet decisions and health outcomes of most residents.

The explosion of work on food environments (as well as more generally connecting food, physical activity and obesity within geography – Rosenberg, 2014) is not surprising, given the clear relevance of geographic methods and preoccupations (e.g., with place) to the issue. The emphasis on environment in understanding diet has been seen by some as a welcome change from public-health discourses that emphasize personal choices without recognizing the importance of structural and contextual factors. Still, the importance of having sufficient income to purchase nutritious food is often neglected in the literature on food deserts, despite considerable research demonstrating that lack of money, not geographic access, is the most significant cause of food insecurity. In addition, the focus on food environments – particularly when the behaviors of low-income and racialized communities are scrutinized without reference to the structural factors that have led to the reshaped food retail environments we see today – has the potential to reembed (and re-stigmatize) inner-city, low-income and racialized communities as deprived and in need of external intervention. For example, Joassart-Marcelli, Rossiter and Bosco (2017, p. 1642) highlight the food-desert concept, which implies an absence of food, as:

> misleading and potentially detrimental to the health of poor and racially diverse communities because it ignores the contribution of smaller stores, particularly that of so-called ethnic markets. Current applications of the food desert concept in this setting reflect classed and racialized understandings of the food environment that ignore the everyday geographies of food provision in immigrant communities while favoring external interventions.

Shannon (2014) goes further, suggesting that "work on food deserts is a spatialized form of neoliberal paternalism that bounds health problems within low-income communities" (p. 248). This reembedding of marginality can also occur in discussions of obesity, a concern that has been identified by health geographers, among others. Health geographers have made many interesting contributions to discussions of diet and obesity, and it is to this issue that we now turn.

Overnutrition and obesity

Health geographers have increasingly engaged with food issues through the lens of obesity and obesity prevention. Studies by health geographers have characterized obesity patterns in a variety of contexts, and connecting these patterns with measures of physical environment and built form – in particular, access to healthy or unhealthy food and physical activity spaces such as recreation centers, parks and cycling trails. Utter (2011), after reviewing a key edited collection on the topic (Pearce and Witten, 2009), concluded that, despite lingering caveats about methodological adequacy, "[t]he overwhelming conclusion on the state of the current research is that the contribution of the environment to obesity is modest" (p. 79).

Although this work overlaps with the food-environments literature, the focus shifts away from food insecurity in low-income neighborhoods, instead centering obesity – and by extension the obese – as the object of inquiry. Much of this work is contextualized by discussions – in public health and elsewhere – about the "obesity epidemic" (for a discussion of the use of the term in this context, see Flegal, 2006). Much of this work builds on understandings of obesity that emphasize the relation between food consumed (calorie intake) and exercise (calories burned), assuming that "obesity results from an energy imbalance that occurs when energy consumption exceeds energy expenditure" (Pearce and Whitten, 2009, p. x). It also – often implicitly, but sometimes explicitly – connects obesity with a range of negative health consequences and concomitant costs to the taxpayer in terms of health-care expenditures. However, some geographers have suggested a more complex relationship between diet, exercise and body size, recognizing the importance of metabolic processes that are not currently well understood (e.g., Guthman, 2011; Shaw, 2014). Indeed, in one of several articles critiquing dominant approaches to obesity in public health, Colls and Evans (2014) suggest that many of the points that are taken for granted in studies of obesity – that body mass index (BMI) is a measure of fatness, and that being overweight is an indicator of poor health and/or lowered life expectancy – are not well supported in the literature. Instead, they suggest that health geographers should "challenge the medical-structural certainties of the obesity debate in order to expose its limitations and moral proclivities and to acknowledge alternative discursive regimes through which the fat body can be understood" (Colls and Evans, 2009, p. 1017), a theme that is also taken up in the emerging field of fat studies.

In unpacking the discourses of obesity, scholars draw on a tradition of critical health geography that emphasizes the use of social theory in understanding the complex relationships between social, biological and environmental factors in shaping people's health outcomes and their experiences of health. Del Casino (2015) additionally suggests the importance of connecting to the growing interest in bodies and embodiment in other areas of geography, given the fundamentally sensory and embodied character of food itself. There have also been calls to bring in the voices and perspectives of fat people themselves (e.g., Lloyd and Hopkins, 2015).

Importantly, Colls and Evans (2009) suggest that while the shift in research and policy away from individual behavioral interventions toward obesogenic environments might be seen as a move away from victim-blaming, this work can still stigmatize particular communities; when residents do not achieve arguably arbitrary standards of thinness, their fatness is seen as "a signifier of moral and physical decay" (Bell and Valentine, 1997, p. 36, quoted in Brown, 2014). Critical obesity scholars in geography and elsewhere have emphasized the importance of geographic work to expose the shaming discourse that underlies most obesity research and policy. For example, Guthman (2009) unpacks how her students' demand for the obese to show self-control is connected to neoliberal subjectivities, and Brown (2014) explores how ideas of contagion in relation to the obesity epidemic link to conceptions of risk and security. These strong themes emerge in other geographic work on food and health, particularly around food safety.

Food, risk and (bio)security

Considerable work in health geography explores how risk and rationality are constructed at a variety of scales, from global to local. This work draws on a range of scholars, including Ulrick Beck and Michel Foucault, as well as geographers Bruce Braun, Peter Adey and Ben Anderson. While much of this literature is not about food per se, it connects with food – and particularly issues and perceptions of food security and food safety – in the context of exploring broad patterns of risk and marginality.

A focus on risk is evident in some geographers' recent work on food security (Fredriksen, 2016; Simatele and Simatele, 2015), food allergies (Dean et al., 2016; Harrington et al., 2012) and food safety (Demeritt et al., 2015; Devaney, 2013; Ilbery, 2012). This work ranges from heavily theoretical to primarily empirical and is concerned with risk in a variety of ways. Some geographers are concerned primarily with risk management, focusing on identifying and mitigating risks from food (for example, from contamination). Others explore risk from a social-constructionist perspective, focusing on how risk perception is embedded in specific cultural contexts that shape, enhance or truncate people's fears about food – for example, Jackson (2015) uses the significant negative reaction to the discovery of horsemeat in prepared food products in the United Kingdom as a lens for understanding cultural anxieties about food. Still others investigate the connections between discourses of risk and economic, social, and environmental marginality using Foucauldian or political-economic frameworks. In addition, some of this work is beginning to engage with broader discussions of biopolitics, securitization and governmentality occurring in health geography and elsewhere (e.g., Brown, Craddock and Ingram, 2012; O'Connor et al., 2017). Much of this discussion emphasizes how our understandings of risk can build on pre-existing ways that particular populations are marginalized; for example, Brown, Craddock and Ingram (2012) explore how problems identified as global in nature are those that threaten the (food) security of Western nations, while problems that lack this element are not prioritized (Brown, Craddock and Ingram, 2012). This scholarship as a whole highlights how understandings of risks related to food (e.g., whether it is safe) are shaped by broader understandings of what good food is (e.g., what is healthy or even edible), as well as what makes us (feel) safe and secure more broadly. It also connects food not only to risk, but also to analyses of power.

Health and the food system

A final thread worth noting in health geographers' engagement with food is the large body of scholarship that examines the food system – and, increasingly, efforts to create alternative food systems. Much of this work is undertaken by scholars who may not consider themselves health geographers, but these broader interrogations of both the mainstream food system and the alternative food movements have relevance to health geographers interested in food. For example, Julie Guthman's work on both mainstream and alternative food systems has highlighted how (healthy and unhealthy) food production is embedded within the political-economic context that drives food production (e.g., the pressure to increase yields at the expense of environmental or social values), while at the same time emphasizing how discursive representations of good food (e.g., romanticized images of white male farmers) can link to narrowly imagined pasts – and futures – that exclude women, people of color, and Indigenous people (Guthman, 2008, 2011, 2014; Guthman and Brown, 2016; Guthman and DuPuis, 2006). Critiques of the recent emphasis on localizing food systems have suggested that this can too easily align with xenophobia and can romanticize agrarian landscapes.

Other relevant work has focused on emerging ideas of food justice and food sovereignty. The food-justice literature mirrors scholarship and activism around *environmental* justice: as Alkon and Agyeman (2011) explain, "[w]hile the environmental justice movement is primarily concerned with preventing disproportionate exposure to toxic environmental burdens, the food justice movement works to ensure equal access to the environmental benefit of healthy food" (p. 8). This work has tackled many of the same issues as the food environments literature, but with a more consciously critical and activist bias (Dixon, 2014). Food sovereignty, a concept emerging out of the global peasant movement La Via Campesina, is oriented around a community's "right to define their own food and agriculture systems" (La Via Campesina, quoted in Food Secure Canada, n.d.). Academic literature on food sovereignty has explored barriers to, and possibilities for, food sovereignty in a variety of national contexts – for example, by highlighting the role of peasant agriculture in food security and exploring ways of developing and promoting local control of food systems. As Del Casino (2015, p. 805) reflected in a recent review:

> [T]here is much opportunity to pick up on variegated themes that constitute the social geographies of food and expand their conversations in relation to other subdisciplinary areas, such as political ecology, and disciplines, such as health and nutritional sciences. Maintaining a focus on inequality and difference . . . geographers of food can continue to ask the questions that have long animated a subdiscipline with strong radical roots.

Health geographers interested in food could gain a great deal by enhancing their connections with food studies more broadly.

Conclusion

This chapter summarizes major threads within health geography in terms of its connection to food. Specifically, it explores the literature on food environments, obesity, food risk, and food systems more generally. Based on this review, it appears that there is still significant scope for continued methodological refinement in the food-environments literature; however, this work would benefit from the more consistent application of a critical and holistic lens. There is also plenty of room for health geographers to continue to explore food security and safety, as well as to connect our expertise in health and food environments to more embodied accounts (e.g., of eating).

Overall, this chapter also highlights how geographers have both contributed to and challenged common discourses in medicine, public health and health policy. In particular, it shows how certain conceptions of food within health geography could serve to reinforce neoliberal subjectivities and make particular individuals and populations the (ir)responsible pathologized subjects of public-health inquiry. In this context, it is important that health geographers be reflexive about how their work connects to these overarching discourses.

References

Alkon, A. H. and Agyeman, J. (2011). *Cultivating food justice: race, class, and sustainability*. Cambridge: MIT Press.

Battersby, J. and Crush, J. (2014). Africa's urban food deserts. *Urban Forum*, 25(2), pp. 43–151.

Beaulac, J., Kristjansson, E. and Cummins, S. (2009). A systematic review of food deserts, 1966–2007. *Preventing Chronic Disease*, 6(3), p. A105.

Bell, D. and Valentine, G. (1997). *Consuming geographies: We are where we eat*. London: Routledge.

Brown, T. (2014). Differences by degree: fatness, contagion and pre-emption. *Health*, 18(2), pp. 117–129.

Brown, T., Craddock, S. and Ingram, A. (2012). Critical interventions in global health: governmentality, risk, and assemblage. *Annals of the Association of American Geographers*, 102(5), pp. 1182–1189.

Caspi, C. E., Sorensen, G. and Subramanian, S. V. (2012). The local food environment and diet: a systematic review. *Health & Place*, 18(5), pp. 1172–1187.

Cerovečki, I. G. and Grünhagen, M. (2016). "Food Deserts" in urban districts: evidence from a transitional market and implications for macromarketing. *Journal of Macromarketing*, 36(3), pp. 337–353.

Cetateanu, A. and Jones, A. (2016). How can GPS technology help us better understand exposure to the food environment? A systematic review. *Social Science and Medicine: Population Health*, 2, pp. 196–205.

Charreire, H., Casey, R., Salze, P., Simon, C., Chaix, B., Banos, A., Badariotti, D., Weber, C. and Oppert, J. M. (2010). Measuring the food environment using geographical information systems: a methodological review. *Public Health Nutrition*, 13(11), pp. 1773–1785.

Colls, R. and Evans, B. (2009). Introduction: questioning obesity politics. *Antipode*, 41(5), pp. 1011–1020.

Colls, R. and Evans, B. (2014). Making space for fat bodies? A critical account of "the obesogenic environment." *Progress in Human Geography*, 38(6), pp. 733–753.

Dean, J., Fenton, N. E., Shannon, S., Elliott, S. J. and A. Clarke. (2016). Disclosing food allergy status in schools: health-related stigma among school children in Ontario. *Health & Social Care in the Community*, 24(5), pp. e43–52.

Del Casino, V. J. (2015). Social geography I: food. *Progress in Human Geography*, 39(6), pp. 800–808.

Demeritt, D., Rothstein, H., Beaussier, A-L. and Howard, M. (2015). Mobilizing risk: explaining policy transfer in food and occupational safety regulation in the UK. *Environment and Planning A*, 47(2), pp. 373–339.

Devaney, L. (2013). Spaces of security, surveillance and food safety: interrogating perceptions of the Food Safety Authority of Ireland's governing technologies, power and performance. *The Geographical Journal*, 179(4), pp. 320–330.

Dixon, B. (2014). Learning to see food justice. *Agriculture and Human Values*, 31(2), pp. 175–184.

Flegal, K. M. (2006). Commentary: the epidemic of obesity – what's in a name? *International Journal of Epidemiology*, 35(1), pp. 72–74.

Food Secure Canada. (n.d.). *What is food sovereignty*. [online] Available at: https://foodsecurecanada.org/who-we-are/what-food-sovereignty [Accessed 1 Sept. 2017].

Fredriksen, A. (2016). Crisis in "a normal bad year": spaces of humanitarian emergency, the Integrated Food Security Phase Classification scale and the Somali famine of 2011. *Environment and Planning A*, 48(1), pp. 40–57.

Gartin, M. (2012). Food deserts and nutritional risk in Paraguay. *American Journal of Human Biology*, 24(3), pp. 296–301.

Guthman, J. (2008). Bringing good food to others: investigating the subjects of alternative food practice. *Cultural Geographies*, 15, pp. 431–447.

Guthman, J. (2009). Teaching the politics of obesity: insights into neoliberal embodiment and contemporary biopolitics. *Antipode*, 41(5), pp. 1110–1133.

Guthman, J. (2011). *Weighing in: obesity, food justice, and the limits of capitalism*. Berkeley: University of California Press.

Guthman, J. (2012). Doing justice to bodies? Reflections on food justice, race, and biology. *Antipode*, 46(5), pp. 1153–1171.

Guthman, J. (2014). *Agrarian dreams: the paradox of organic farming in California*. 2nd ed. Berkeley: University of California Press.

Guthman, J. and Brown, S. (2016). I will never eat another strawberry again: the biopolitics of consumer-citizenship in the fight against methyl iodide in California. *Agriculture and Human Values*, 33, pp. 575–585.

Guthman, J. and Dupuis, E. M. (2006). Embodying neoliberalism: economy, culture, and the politics of fat. *Environment and Planning D: Society and Space*, 24(3), pp. 427–448.

Harrington, D. W., Elliott, S. J., Clarke, A. E., Ben-Shoshan, M. and Godefroy, S. (2012). Exploring the determinants of the perceived risk of food allergies in Canada. *Human and Ecological Risk Assessment*, 18(6), pp. 1338–1358.

Ilbery, B. (2012). Interrogating food security and infectious animal and plant diseases: a critical introduction. *The Geographical Journal*, 178(4), pp. 308–312.

Jackson, P. (2015). *Anxious appetites: food and consumer culture*. London: Bloomsbury Academic Publishing.

Joassart-Marcelli, P., Rossiter, J. S. and Bosco, F. J. (2017). Ethnic markets and community food security in an urban "food desert." *Environment and Planning A*, 49(7), pp. 1642–1663.

Lloyd, J. and Hopkins, P. (2015). Using interviews to research body size: methodological and ethical considerations. *Area*, 47(3), pp. 305–310.

Mosammam, Mohammadian H., Sarrafi, M., Tavakoli Nia, J. and Mosammam, A. M. (2017). Measuring food deserts via GIS-based multicriteria decision making: the case of Tehran. *The Professional Geographer*, 69(3), pp. 455–471.

O'Connor, D., Boyle, P., Ilcan, S. and Oliver, M. (2017). Living with insecurity: food security, resilience, and the World Food Programme (WFP). *Global Social Policy*, 17(1), pp. 3–20.

Pearce, J. and Whitten, K. (eds.) (2009). *Geographies of obesity: environmental understandings of the obesity epidemic*. Farnham, Surrey: Ashgate.

Rosenberg, M. (2014). Health geography I: social justice, idealist theory, health and health care. *Progress in Human Geography*, 38(3), pp. 466–475.

Sadler, R., Gilliland, J. and Arku, G. (2016). Theoretical issues in the "food desert" debate and ways forward. *Geojournal*, 81(3), pp. 443–455.

Shannon, J. (2014). Food deserts: governing obesity in the neoliberal city. *Progress in Human Geography*, 38(2), pp. 248–266.

Shaw, H. J. (2014). *The consuming geographies of food: diet, food deserts and obesity*. London: Routledge, Taylor & Francis Group.

Simatele, D. and Simatele, M. (2015). Climate variability and urban food security in sub-Saharan Africa: lessons from Zambia using an asset-based adaptation framework. *South African Geographical Journal*, 9(3), pp. 243–263.

Thompson, C., Cummins, S., Brown, T. and Kyle, R. (2016). Contrasting approaches to "doing" family meals: a qualitative study of how parents frame children's food preferences. *Critical Public Health*, 26(3), pp. 322–332.

Utter, J. (2011). Geographies of obesity: environmental understandings of the obesity epidemic. *New Zealand Geographer*, 7(1), pp. 70–71.

Ver Ploeg, M., Breneman, V., Farrigan, T., Hamrick, K., Hopkins, D., Kaufman, P., Lin, B.-H., Nord, M., Smith, T., Williams, R., Kinnison, K., Olander, C., Singh, A. and Tuckermanty, E. (2009). *Access to affordable and nutritious food – measuring and understanding food deserts and their consequences: report to Congress.* Washington, DC: U.S. Department of Agriculture.

Vogel, C., Ntani, G., Inskip, H., Barker, M., Cummins, S., Cooper, C., Moon, G. and Baird, J. (2016). Education and the relationship between supermarket environment and diet. *American Journal of Preventive Medicine*, 51(2), pp. e27–e34.

Widener, M. J., Farber, S., Neutens, T. and Horner, M. W. (2013). Using urban commuting data to calculate a spatiotemporal accessibility measure for food environment studies. *Health & Place*, 21, pp. 1–9.

11

UN/HEALTHY BEHAVIOR

A bibliometric assessment of geographers' contributions to understanding the association between environment and health-related behavior

Graham Moon and Dianna Smith

This chapter examines the contribution of geographers to the study of health-related behavior. We aim to complement recent geographical reviews of the same subject matter (e.g., Twigg and Cooper, 2010) by providing an updated assessment, identifying contemporary themes and emerging agendas. We define "health-related behavior" as encompassing smoking, drinking, diet and physical activity, the commonly listed key health behaviors implicated collectively in 18.5% of the global mortality burden (World Health Organization, 2009).

Our chosen behaviors are significant for their relevance to themes considered elsewhere in this volume. As we shall see in the next section, variations in the prevalence of health-related behavior are one disputed element in understanding geographical health inequalities; that is, why health outcomes differ from place to place. Our behaviors are also significantly implicated in global health. The perpetuation of (un)healthy behavior on a global scale is in no small part associated with the actions of global corporations seeking to maintain market position. Additionally, ideas about risk and resilience pose questions about why certain groups (and places) are more susceptible to (un)healthy behavior, and health-related behavior is increasingly recognized as playing a role in urban planning, most notably in relation to physical activity and the notion of the walkable city.

In the next section, we examine the key theoretical concerns that have motivated the study of health-related behavior by geographers, developing a critical perspective. We then go on to examine long-standing concerns and more recent themes for each of our chosen behaviors. For each of the four selected health behaviors, we searched Scopus using relevant synonyms, limited to papers reporting geography as an author affiliation and to known geographers (see Table 11.1). We then focused on the work since 2012 and the most-cited authors before highlighting emerging themes for future research. We conclude the chapter with reflections on other health-related behaviors, thoughts on co-behavior (groupings of two or more behaviors), and a brief consideration of policy implications.

Theory and critique

Health-related behavior has long been recognized as a key explanation for health inequality, often being contrasted with explanations focusing on the impact of socioeconomic structure. In reality, both have a role to play. Since at least the publication of the influential UK *Black Report* (Whitehead, Townsend and Davidson, 1992), behavioral and structural theories about health inequality have been strongly politicized. Health-related behavior has been seen as an outcome of autonomous choice by individuals, who are personally responsible for their resultant health outcomes. In contrast, structural explanations emphasize the constraints

Table 11.1 Current research on health-related behavior involving geographers

Topic	Scopus search terms	Total papers	Post-2012 papers
Smoking	(TITLE (**smok★** OR **tobacco** OR **cigarette**) AND AFFIL (**geogr★**)) AND (LIMIT-TO (SUBJAREA , **"MEDI"**) OR LIMIT-TO (SUBJAREA , **"SOCI"**))	187	92
Diet	(TITLE (**food★** OR **diet★** OR **eat★**) AND TITLE-ABS-KEY (**health**) AND AFFIL (**geogr★**) AND NOT TITLE-ABS-KEY (**safety** OR **security** OR **farm★** OR **famine**)) AND (LIMIT-TO (SUBJAREA , **"MEDI"**) OR LIMIT-TO (SUBJAREA , **"SOCI"**)	171	87
Drinking	(TITLE (**alcohol** OR **drink★**) AND AFFIL (**geogr★**)) AND NOT TITLE (**water**) AND (LIMIT-TO (SUBJAREA , **"MEDI"**) OR LIMIT-TO (SUBJAREA , **"SOCI"**))	162	95
Physical Activity	(TITLE (**"physical activit★"**) AND AFFIL (**geograph★** OR **"east anglia"**)) AND (LIMIT-TO (SUBJAREA , **"MEDI"**) OR LIMIT-TO (SUBJAREA , **"SOCI"**))	178	121

that flow from wealth and access to power. Simplistically, the contrast is between a right-wing (behavioral) and a left-wing (structural) perspective, and this has tended to be reflected in differing governmental approaches to addressing (un)healthy behavior.

If we focus on the behavior rather than the health outcome, it quickly becomes clear that choice and constraint are both important. While individuals may choose to smoke, that choice and the way it is exercised is strongly influenced by disposable income, social attitude, group norms and a range of other factors. These factors come together to provide a geography for each behavior, a geography that reflects individual characteristics, the aggregation of people possessing similar individual characteristics, broader area-level determinants such as programs to reduce (un)healthy behavior, and the interaction of all these compositional, aggregative and contextual factors (Cummins et al., 2007).

A recognition of the complex, relational and multilayered nature of health-related behavior has led much quantitative research to multilevel modeling (Duncan, Jones and Moon, 1996). This approach is all but essential in large-scale national studies and can accommodate longitudinal studies of change and complex geographical considerations such as cross-classification where individual behavior is affected by non-nested contexts such as home and work locations. At the same time, it is important not to lose sight of more experiential, sociological and cultural perspectives (Cummins et al., 2007; Thompson, Pearce and Barnett, 2007). Unhealthy behavior can be entirely normal and understandable in some contexts, with the challenges being to effect change without demonizing people or communities where the behavior is manifest and to understand why unhealthy behavior can be a rational response to disadvantage, with, moreover, the definitions of precisely what constitutes (un)healthy behavior being much debated.

Current concerns and future agendas

We begin this section by considering tobacco smoking. An early contribution was Heenan (1983), exploiting the unusual presence of smoking in the New Zealand census. The most prolific geographers currently working on this behavior are Moon (Southampton, UK), Pearce (Edinburgh, UK), and Barnett (Canterbury,

New Zealand), each involved in more than ten papers, often collaborative but also involving others. Duncan, Jones and Moon (1999) is the most cited paper led by a geographer, garnering over 200 citations. Also highly cited (70+ cites) is a collaborative paper involving two geographers on national variations in European tobacco-control policy and its association with smoking cessation (Schaap et al., 2008).

Papers produced over the past five years point to a range of shared concerns. Pearce, Barnett and Moon (2012) exemplify work on inequalities in smoking prevalence and develop an important distinction between place-based practices and place-based regulation. In the absence of routine data on small-area variations in smoking prevalence, a second group of studies has been developing estimation methods to fill this knowledge gap. Szatkowski et al. (2015) employ multilevel small-area estimation to predict the prevalence of smoking during pregnancy at delivery for hospital areas. A third theme is work on tobacco retail. Shareck et al. (2016) note that past studies have focused on tobacco retail in residential or school neighborhoods. Their innovation is to consider the wider activity spaces where individuals also spend time. The final current theme is rather different and is exemplified by Tan (2013), who takes a qualitative and critical lens to the experience of smoking in Singapore, portraying it as a transgressive practice that brings embodied sensations, helping us understand why some people continue to smoke. Qualitative work of this kind provides an alternative to what remains a dominantly quantitative field.

All these themes are brought together in Barnett et al. (2016), a monograph that serves as a statement of current geographical perspectives on smoking and tobacco. The authors begin with an examination of the longitudinal international evolution of the tobacco epidemic and then offer an assessment of the global development of Big Tobacco, setting out an economic geography of tobacco. These two initial chapters point to a need to consider smoking in its wider context, moving beyond the health silo to recognize entanglements with development, empire, politics, and economics. The central portion of the monograph then turns to inequality, taking both quantitative and qualitative approaches and extending the discussion to include research on smoking-attributable mortality and morbidity, as well as coverage of the notion of smoking as a gateway to other behaviors. Common themes in these chapters point to the need to focus on understanding why people smoke, how smoking communities can emerge, how smoking can be a tactic for resistance and how it also has consequences. Barnett et al. (2016) then turn to the various tactics that have been employed to bring about smoking cessation or reduce smoking initiation. Though there are exceptions, this is a theme that has been relatively neglected in health-geography work on smoking and tobacco. The authors highlight the uneven development and impact of these policies and their unintended consequences, as well as the need for innovative area-based policies.

Future directions for geographical work on smoking as a health behavior could arguably focus on extending the big picture and health-promotion themes identified by Barnett et al. (2016). These are certainly themes that have received little attention to date. In particular, though there have been some outputs, the economic geography of tobacco remains under-researched by geographers. Links between this economic geography and smoking behavior might include consideration of company tactics targeting less-developed countries and emerging economies; they might also chart the substantial consumption of smuggled tobacco and the impact on consumption of emerging internet markets. This said, there remains much to be done on smoking and inequality, particularly with regard to population sub-groups and, linking back to the health promotion theme, the differential uptake and impact of environmentally focused tobacco-control initiatives. There is also a clear case for enhanced attention to qualitative and critical perspectives on smoking, drawing out the experiences of the dwindling number of tobacco smokers. As Barnett et al. (2016) conclude, however, perhaps the most significant direction for research concerns the emergent geographies of vaping and e-cigarettes.

Turning next to research on diet, we focus explicitly on diet as a behavior: on geographical variations in eating behavior and access to healthy food. We acknowledge associations but chose not to cover broader issues such as food security and famine. Many papers cover diet as an adjunct to a main focus on obesity, and many cover diet alongside physical activity, another social determinant of obesity; we did not exclude these papers but focused our analysis on what was being said specifically about geography and dietary behavior. With these caveats in mind, dietary behavior is of a similar size to smoking and tobacco as a research theme among geographers. The most prolific geographer is currently Cummins (London School of Hygiene and Tropical Medicine), one of the authors of the most cited paper on the topic: Cummins and Macintyre (2006). Thirteen papers have accrued more than 100 citations, a more extensive level of citation than that for smoking.

Food deserts, (frequently urban) areas where there is limited or no access to affordable good-quality nutrition, have provided an enduring theme among papers by geographers working on diet as a health-related behavior. Alongside Cummins' work, seminal early contributions were by Wrigley and colleagues (e.g., Wrigley, 2002) working on food consumption in a Leeds (UK) food desert following the opening of a new food retail outlet. The arguments (and debates) inherent in these early studies were summarized by Cummins et al. (2005), who described natural experiments in which supermarket-scale food retailing was introduced to food deserts, with diet patterns compared between intervention and comparison communities. Whereas Cummins and colleagues tended to find against significant positive effects, Wrigley and colleagues were more positive, also identifying wider community benefits associated with urban regeneration.

Replication research has seen subsequent studies of food deserts in communities worldwide (Sadler, Gilliland and Arku, 2016). These have tended to confirm the known mixed evidence, but some have sparked new directions. Three themes can be identified. First to emerge, and still popular, were studies focusing on access to food, noting how communities differ, usually on the basis of socioeconomic status, in the distance traveled to reach food-consumption opportunities. These studies have differentiated access to large supermarkets, convenience stores, full-service restaurants, fast-food outlets, farmers' markets, allotment gardens and many other forms of provision, with convenience stores and fast-food outlets generally characterized as providing poorer nutrition. Pearce et al. (2007), for example, found that residents of more affluent areas in New Zealand traveled twice as far to get to fast food compared to people from deprived areas. The impact of such findings on youth populations has been a second theme. He et al. (2012) showed how attending a school with more than three fast-food outlets within one kilometer has a negative effect on diet quality. The third recent theme has been methodological innovation using geospatial technologies to identify food outlets and develop novel access measures. In this frame, Widener and Li (2014) used Twitter data and sentiment analysis to reveal that people living in food deserts discuss healthy food less often.

Geographers working on diet have also been active on topics other than food deserts. Thompson et al. (2016) examined the performance of eating, focusing on family meals using photo elicitation and other qualitative methods. They contrasted consultative and non-consultative strategies regarding the content of meals in the everyday setting of the home. Soller et al. (2014) considered food allergy among vulnerable populations in Canada, finding no association with income or Indigenous identity but fewer allergies among immigrants and people with low education. Although future research could usefully continue this diversification, review papers have seen a productive future in continued applied work on food access, with better measurements of both diet and geographies enhancing understanding of their association. Thus, Black, Moon and Baird (2014: 229) note that "better and more nuanced measures of the food environment, including multidimensional and individualized approaches, would enhance the state of the evidence and help inform future interventions," and Pettygrove and Ghose (2016) call for the integration of critical geographic information systems (GIS) methodologies, with a stronger theorization of the urban spaces impacted by poor diet.

For our third behavior, we examine work by geographers on drinking behavior. Qualitative studies of alcohol consumption and attitudes have been led by Valentine (Sheffield), Jayne (Manchester) and Holloway (Loughborough), each authors of more than ten papers in Scopus, usually together (e.g., Jayne, Valentine and Holloway, 2011). The focus of their collective research is the spaces and gendered behaviors of drinking and drunkenness in the United Kingdom, resulting in three of the most highly cited (>50 citations) papers, published between 2006 and 2010. Herrick (King's College London) offers a different approach in a collection of eight papers, exploring the geographies of alcohol consumption in less-developed settings, in particular

South Africa. A significant body of other work addresses the spatial patterning of alcohol outlets and local deprivation (Pearce, Day and Witten, 2008) and related inequalities (Berke et al., 2010).

A crucial area of developing work is the correct assessment of drinking levels, as described in Boniface, Kneale and Shelton (2013). At present, most research depends upon self-reported alcohol consumption, and there is acknowledgment that the measures of alcohol by units can be challenging to understand and report accurately. Boniface, Kneale and Shelton asked participants to pour glasses of alcoholic drink and estimate the number of units, with results showing both under- and over-estimation; the key recommendation was to advocate for smaller glasses, to reduce underestimation. A focus on modifying individual behavior is very much in line with the move toward individual responsibility in consumption, whether the consumption is of unhealthy food, alcohol or tobacco, with less emphasis on more upstream measures to promote healthier behaviors such as taxation or reducing availability of items (e.g., regulation of outlets selling alcohol). Herrick (2016) illustrates how the role of government intervention in alcohol regulation is less explored than that in other harmful behaviors in both the Global North and the Global South.

The work on the spatial modeling of alcohol consumption has extended to small-area estimation of population-level drinking behavior. This methodological approach may allow researchers to identify neighborhoods where binge drinking, based on data from large surveys, is more common, with a view to targeting behavior-change interventions. Twigg and Moon (2013) and Riva and Smith (2012) each modeled the estimated prevalence of binge drinking across England, in an effort to identify areas where binge drinking is more common and interrogate the potential causes for these patterns. Twigg and Moon used a multilevel approach to model drinking between 2001 and 2009, finding a clear north-south divide in England with strong differences by gender and over time, with women more likely to binge drink on one day a week only. Crucially they identify the effects of measurement/definition on the results; area deprivation is less relevant over time.

Emerging work is focused on the systematic assessment of the retail environments that supply alcohol and the impacts on local populations in terms of ill health and social disorder. In the United Kingdom, researchers have drawn on routinely collected administrative data to explore these relationships, with a large study in Wales linking the alcohol-retail environment with hospital admissions (Fone et al., 2016) and other work in Scotland (Richardson et al., 2015). The next stage for this work will demand longitudinal analysis. Other challenges are to work on services for problem drinkers, where useful Canadian work has been emerging (Evans, 2012), and to improve the accuracy in reporting of drinking in the survey data so often relied upon for the research. The regulatory role of government also demands ongoing study.

Finally, we turn our attention to physical activity. Here, by far the most prolific geographer is Jones (East Anglia). He has contributed to 42 papers on the topic; three have more than 100 citations. In contrast to our other three areas, physical activity shows stronger evidence of collaboration between public-health and geography academics. Indeed, Jones has moved to Medicine at the University of East Anglia, a pattern also seen with Cummins (LSHTM) noted above for his strong contribution in diet research. The most-cited paper involving a geographer working on physical activity was Frank, Andresen and Schmid (2004), exploring the association between obesity, community design, physical activity and car use.

This interplay of physical activity, diet and obesity is a frequent focus of recent research. The relationship was outlined in the British government's 2007 report on obesity, for which Jones was the author of the evidence review on obesogenic environments (Jones et al., 2007). He described the difficulties in singling out one element of an individual's physical environment as significantly influential – a theme we will return to in our conclusion. Also worth noting from this comprehensive review is the need to define the target population for place-based interventions to promote physical activity. Is the intention to encourage more people to activity or to enhance standards for those already active?

In a recent paper (Sallis et al., 2016), three geographic aspects of a neighborhood that promote physical activity – walkability, access to parks and density of public transport – were assessed across 14 cities worldwide. Each factor has been explored by health geographers, albeit not always solely in terms of influence on physical activity. One frequent focus is access to green space, with the assumption that good access will encourage more physical activity, though evidence is mixed. Walkability has, however, attracted the most attention. Andrews et al. (2012) call for health geographers to take a more critical perspective of the assumptions often embodied in walkability research. In particular, they point to a need to acknowledge the mobility requirements of people with chronic illness or disability. These needs are seldom addressed well in urban planning. Sallis et al. (2016) also highlight a more recent development in physical activity research: active travel. Active travel is often associated with a commute to work or school and is defined as making part of the journey without using private motorized transport. Longitudinal studies show active travel to have a positive effect on the health of English adults (Flint, Cummins and Sacker, 2014), supporting the call by Sallis et al. to consider urban design in promoting easier, safer means of active transport.

As research on physical activity expands, we are seeing greater use of technology such as Global Positioning Systems (GPS) and accelerometers to monitor physical activity and movement within environments. This will provide valuable data that will not only inform health geography's study of activity environments, but also improve our measurement of behavior in the environment more generally, moving beyond studies that focus on areas near to home or school. The scope for more informed descriptions of health-promoting environments will lead, hopefully, to a sustained level of academically rigorous input from health geography into public-policy recommendations.

Conclusion

We have concentrated in this chapter on the role that geographers have played in the study of health-related behavior. The geographical contribution has been a significant one. It has highlighted the importance of place, location, distance and separation in understanding health-related behavior, and it has shown significant methodological innovation. In a field in which there is a strong interdisciplinary tradition, extensive collaboration with other disciplines has ensured that the geographical perspective has been a strong voice with a real-world impact. Less positively, and partly an artifact of our decision to base our assessment on bibliometric analysis, we note that the focus of the contributions reviewed above has been very largely on the Global North; there is clear potential for greater engagement with global health concerns. We conclude with three observations.

First, we have been selective in our coverage of health behavior. Geographers have also been active in other behaviors. Sexual health and sexual behavior is one such area. Lewis (2014) offers one direction in his study of migration by gay men, but many more exist. Another area is the use of illicit drugs. McCann and Temenos (2015), for example, examine the role of drug-consumption rooms as a technology for harm minimization. Less studied, but hugely important and a potential area for future research, are the behaviors evidenced when seeking and consuming health care. Here there is potential for overlap with the now vigorous research taking place under the broad heading of the geography of care: asking how people and communities go about seeking health care, how care-seeking behavior varies by category of care and by social group, and how it is impacted by new technologies of care, policy change and austerity.

Second, we have looked at each of our key behaviors in isolation. In reality, of course, they are interrelated. We have noted as much in our passing comments on obesity. Pearce and Witten (2010), in their edited collection on environmental perspectives on obesity, bring together essays on physical activity and diet, showing how both contribute to the generation of obesogenic environments through the concept of the energy balance equation. Other papers have considered access to outlets selling tobacco, alcohol and indeed fast food (Schneider and Gruber, 2013). The notion of co-behavior, the associations between behaviors, is a major area for future research – going beyond the association of diet and physical activity to examine, for example, the characteristics of communities where there is high alcohol consumption and significant residual smoking, the potential for targeting areas with multiple unhealthy behaviors, and more qualitative assessments of the experience of living in such areas.

Finally, we stress the need for geographical research on health-related behaviors to consider real-world impact. A central tenet for much policy aimed at reducing unhealthy behavior is the idea of denormalization. This can be a powerful catalyst to behavioral change, yet it can also be stigmatizing and raise new challenges. Bans on smoking in indoor public places have been widely effective in reducing exposure to harm, yet they have arguably displaced smoking to outdoor settings and to the private home. In the field of alcohol control, zones where street drinking is prohibited or sales of highly alcoholic drinks are restricted can lead to similar displacement. As geographers, we should be alert to the spatial manifestations, differential experiences, and unforeseen consequences of measures that seek to reduce unhealthy behavior.

References

Andrews, G. J., Hall, E., Evans, B. and Colls, R. (2012). Moving beyond walkability: on the potential of health geography. *Social Science & Medicine*, 75(11), pp. 1925–1932.

Barnett, R., Moon, G., Pearce, J., Thompson, L. and Twigg, L. (2016). *Smoking geographies: space, place and tobacco*. London: Wiley-Blackwell.

Berke, E. M., Tanski, S. E., Demidenko, E., Alford-Teaster, J., Shi, X. and Sargent, J. D. (2010). Alcohol retail density and demographic predictors of health disparities: a geographic analysis. *American Journal of Public Health*, 100(10), pp. 1967–1971.

Black, C., Moon, G. and Baird, J. (2014). Dietary inequalities: what is the evidence for the effect of the neighbourhood food environment? *Health & Place*, 27, pp. 229–242.

Boniface, S., Kneale, J. and Shelton, N. (2013). Actual and perceived units of alcohol in a self-defined "Usual Glass" of alcoholic drinks in England. *Alcoholism: Clinical and Experimental Research*, 37(6), pp. 978–983.

Cummins, S., Curtis, S., Diez-Roux, A. V. and Macintyre, S. (2007). Understanding and representing "place" in health research: a relational approach. *Social Science & Medicine*, 65(9), pp. 1825–1838.

Cummins, S. and Macintyre, S. (2006). Food environments and obesity – neighbourhood or nation? *International Journal of Epidemiology*, 35(1), pp. 100–104.

Cummins, S., Petticrew, M., Higgins, C., Findlay, A. and Sparks, L. (2005). Large scale food retailing as an intervention for diet and health: Quasi-experimental evaluation of a natural experiment. *Journal of Epidemiology and Community Health*, 59(12), pp. 1035–1040.

Duncan, C., Jones, K. and Moon, G. (1996). Health-related behaviour in context: a multilevel modelling approach. *Social Science & Medicine*, 42(6), pp. 817–830.

Duncan, C., Jones, K. and Moon, G. (1999). Smoking and deprivation: are there neighbourhood effects? *Social Science & Medicine*, 48(4), pp. 497–505.

Evans, J. (2012). Supportive measures, enabling restraint: governing homeless "street drinkers" in Hamilton, Canada. *Social & Cultural Geography*, 13(2), pp. 185–200.

Flint, E., Cummins, S. and Sacker, A. (2014). Associations between active commuting, body fat, and body mass index: population based, cross sectional study in the United Kingdom. *British Medical Journal*, 349, g4887.

Fone, D., Morgan, J., Fry, R., Rodgers, S., Orford, S., Farewell, D., Dunstan, F., White, J., Sivarajasingam, V., Trefan, L., Brennan, I., Lee, S., Shiode, N., Weightman, A., Webster, C. and Lyons, R. (2016). *Change in alcohol outlet density and alcohol-related harm to population health (CHALICE): a comprehensive record-linked database study in Wales*. Southampton: HMSO.

Frank, L. D., Andresen, M. A. and Schmid, T. L. (2004). Obesity relationships with community design, physical activity, and time spent in cars. *American Journal of Preventive Medicine*, 27(2), pp. 87–96.

He, M., Tucker, P., Irwin, J. D., Gilliland, J., Larsen, K. and Hess, P. (2012). Obesogenic neighbourhoods: the impact of neighbourhood restaurants and convenience stores on adolescents' food consumption behaviours. *Public Health Nutrition*, 15(12), pp. 2331–2339.

Heenan, L. (1983). Cigarette smoking among New Zealanders: evidence from the 1976 census. In: *Geographical aspects of health: essays in Honour of Andrew Learmonth*. London: Academic Press, pp. 239–253.

Herrick, C. (2016). Alcohol, ideological schisms and a science of corporate behaviours on health. *Critical Public Health*, 26(1), pp. 14–23.

Jayne, M., Valentine, G. and Holloway, S. L. (2011). *Alcohol, drinking, drunkeness: (dis)orderly spaces*. London: Ashgate.

Jones, A., Bentham, G., Foster, C., Hilsdon, M. and Panter, J. (2007). *Foresight tackling obesities: future choices – obesogenic environments – evidence review*. United Kingdom: Government Office for Science.

Lewis, N. M. (2014). Moving "out," moving on: gay men's migrations through the life course. *Annals of the Association of American Geographers*, 104(2), pp. 225–233.

McCann, E. and Temenos, C. (2015). Mobilizing drug consumption rooms: inter-place networks and harm reduction drug policy. *Health & Place*, 31, pp. 216–223.

Pearce, J., Barnett, R. and Moon, G. (2012). Sociospatial inequalities in health-related behaviours: pathways linking place and smoking. *Progress in Human Geography*, 36(1), pp. 3–24.

Pearce, J., Blakely, T., Witten, K. and Bartie, P. (2007). Neighborhood deprivation and access to fast-food retailing: a national study. *American Journal of Preventive Medicine*, 32(5), pp. 375–382.

Pearce, J., Day, P. and Witten, K. (2008). Neighbourhood provision of food and alcohol retailing and social deprivation in urban New Zealand. *Urban Policy and Research*, 26(2), pp. 213–227.

Pearce, J. and Witten, K. (2010). *Geographies of obesity: environmental understandings of the obesity epidemic*. Thousand Oaks, CA: Ashgate.

Pettygrove, M. and Ghose, R. (2016). Mapping urban geographies of food and dietary health: a synthesized framework. *Geography Compass*, 10(6), pp. 268–281.

Richardson, E. A., Hill, S. E., Mitchell, R., Pearce, J. and Shortt, N. K. (2015). Is local alcohol outlet density related to alcohol-related morbidity and mortality in Scottish cities? *Health & Place*, 33, pp. 172–180.

Riva, M. and Smith, D. (2012). Generating small-area prevalence of psychological distress and alcohol consumption: validation of a spatial microsimulation method. *Social Psychiatry and Psychiatric Epidemiology*, 47(5), pp. 745–755.

Sadler, R. C., Gilliland, J. A. and Arku, G. (2016). Theoretical issues in the "food desert" debate and ways forward. *GeoJournal*, 81(3), pp. 443–455.

Sallis, J. F., Cerin, E., Conway, T. L., Adams, M. A., Frank, L. D., Pratt, M., Salvo, D., Schipperijn, J., Smith, G., Cain, K. L., Davey, R., Kerr, J., Lai, P. C., Mitáš, J., Reis, R., Sarmiento, O. L., Schofield, G., Troelsen, J., Van Dyck, D., De Bourdeaudhuij, I. and Owen, N. (2016). Physical activity in relation to urban environments in 14 cities worldwide: a cross-sectional study. *The Lancet*, 387(10034), pp. 2207–2217.

Schaap, M. M., Kunst, A. E., Leinsalu, M., Regidor, E., Ekholm, O., Dzurova, D., Helmert, U., Klumbiene, J., Santana, P. and Mackenbach, J. P. (2008). Effect of nationwide tobacco control policies on smoking cessation in high and low educated groups in 18 European countries. *Tobacco Control*, 17(4), pp. 248–255.

Schneider, S. and Gruber, J. (2013). Neighbourhood deprivation and outlet density for tobacco, alcohol and fast food: first hints of obesogenic and addictive environments in Germany. *Public Health Nutrition*, 16(07), pp. 1168–1177.

Shareck, M., Kestens, Y., Vallée, J., Datta, G. and Frohlich, K. L. (2016). The added value of accounting for activity space when examining the association between tobacco retailer availability and smoking among young adults. *Tobacco Control*, 25(4), pp. 406–412.

Soller, L., Ben-Shoshan, M., Harrington, D. W., Knoll, M., Fragapane, J., Joseph, L., St. Pierre, Y., La Vieille, S., Wilson, K., Elliott, S. J. and Clarke, A. E. (2014). Prevalence and predictors of food allergy in Canada: a focus on vulnerable populations. *Journal of Allergy and Clinical Immunology: In Practice*, 3(1), pp. 42–49.

Szatkowski, L., Fahy, S. J., Coleman, T., Taylor, J., Twigg, L., Moon, G. and Leonardi-Bee, J. (2015). Small area synthetic estimates of smoking prevalence during pregnancy in England. *Population Health Metrics*, 13(1).

Tan, Q. H. (2013). Smell in the city: smoking and olfactory politics. *Urban Studies*, 50(1), pp. 55–71.

Thompson, C., Cummins, S., Brown, T. and Kyle, R. (2016). Contrasting approaches to "doing" family meals: a qualitative study of how parents frame children's food preferences. *Critical Public Health*, 26(3), pp. 322–332.

Thompson, L., Pearce, J. and Barnett, J. R. (2007). Moralising geographies: stigma, smoking islands and responsible subjects. *Area*, 39(4), pp. 508–517.

Twigg, L. and Cooper, L. (2010). Healthy behavior. In: T. Brown, S. McLafferty and G. Moon, G., eds., *A companion to health and medical geography*. London: Wiley-Blackwell, pp. 460–476.

Twigg, L. and Moon, G. (2013). The spatial and temporal development of binge drinking in England 2001–2009: an observational study. *Social Science & Medicine*, 91, pp. 162–167.

Whitehead, M., Townsend, P. and Davidson, N. (1992). *Inequalities in health: the Black report, the health divide*. London: Penguin.

Widener, M. J. and Li, W. (2014). Using geolocated Twitter data to monitor the prevalence of healthy and unhealthy food references across the US. *Applied Geography*, 54, pp. 189–197.

World Health Organization. (2009). *Global health risks: mortality and burden of disease attributable to selected major risks*. Geneva: World Health Organization.

Wrigley, N. (2002). "Food deserts" in British cities: policy context and research priorities. *Urban Studies*, 39(11), pp. 2029–2040.

SECTION 2

Theories and concepts

12

INTRODUCING SECTION 2

Theories and concepts

Gavin J. Andrews, Valorie A. Crooks and Jamie Pearce

This section is concerned with the theoretical traditions and concepts that inform and frame research in health geography. Given this focus, it is worth thinking briefly first about the nature of theory. Certainly, theory can seem intimidating for scholars, particularly those just starting out in their careers; much of it appearing complex, often being articulated through dense, technical and sometimes abstract writings. However, it is worth remembering that, fundamentally, theory is nothing more than sets of propositions on, and interpretations of, the empirical realities of the world. These propositions and interpretations constitute a lens: a common way in which researchers look at and understand the world. Indeed, it is a lens through which all inquiry flows, helping researchers determine their empirical priorities, their research questions and the types of methods they employ.

Importantly theory adds value to any discipline or field of study within which it is employed, including health geography. Through building underlying knowledge – or what might be thought of as *surplus information* beyond empirical observations – theory has the potential to facilitate conversations between scholars about the world. As a language, then, theory provides academic direction, informing common ways of practicing, and ultimately building academic identity and cohesion. Having said this, however, no matter how broadly applicable, resilient and open to debate and modification a particular theory might be, there will always be those who disagree fundamentally with it and, in disciplinary contexts where a lot is at stake, ultimately frictions and arguments over which theory should occupy the center ground. Indeed, health geography has had its fair share of these; for three key occasions see these debates: (1) Kearns (1993, 1994a, 1994b) with Mayer and Meade (1994) and Paul (1994); (2) Litva and Eyles (1995) and Eyles and Litva (1996) with Philo (1996) and (3) Andrews, Chen and Myers (2014) and Andrews (2015) with Kearns (2014) and Hanlon (2014).

As is well documented, there are three broad types or levels of theory in any academic discipline; grand or meta theory (i.e., ways of seeing and explaining substantive aspects of the world), mid-range theory (i.e., applicable to more focused, yet still general, situations and scenarios), and specific theory (i.e., focused ideas on quite particular empirical happenings). Concepts, meanwhile, should not be thought of as inferior to theory (i.e., as simplistic-theory, part-theory or sub-theory). Rather, they are important ideas; pieces of a jigsaw that together constitute whole theoretical traditions. Indeed, concepts bring theory to life and, speaking directly to the nature of things, are often used to frame studies. While all three levels of theory are evident in health geography, the chapters in this section tend to focus on the first two, their being more influential to the sub-discipline as a whole and speaking to the broad directions it has taken.

A general observation that seems to hold true is that, over the years, health geography has not been a particularly theoretical sub-discipline, in that it has not produced a particularly large volume of dedicated theory papers. Nor has it led to *cutting-edge* theoretical trends in the parent discipline of human geography (Kearns and Moon, 2002; Parr, 2004). This observation is consistent with a more general narrative that health geography – and particularly the long-standing medical-geography component of it – is rather applied and practical, focused on servicing the needs of health sectors through providing the mapping component of broader health-services research, public-health research and epidemiology. Having said this, however, the idea of health geography as data-rich/theory-poor might well be overstated. Indeed, the chapters in this section illustrate two things: on one hand, that health geography has led to certain theoretical developments in the health sciences more broadly, both in terms of the overall social-science contribution to understanding health and in terms of geographical ideas (space, place, landscape, urbanicity, rurality, settings, etc.); and on the other hand, that health geographers have joined emerging theoretical traditions in human geography quite promptly, the lag time between theoretical developments in the parent discipline and their adoption in health geography being somewhat shorter than it once was.

What this section aims to achieve is an understanding among readers as to how theory motivates and shapes research in health geography and more broadly the nature of the sub-discipline. Moreover, it aims to develop an understanding of some of the main theoretical perspectives; their constituent ideas, interests, range of applications, and internal developments and progress. This sets readers up for a better understanding of many of the empirical topics covered in other sections of the book. The chapters tend to focus on theoretical traditions that have informed and molded the sub-discipline particularly in the last 30 years, many of which have extended histories of use and engagement outside health geography. The section starts with three chapters focused on traditions that have had powerful influences on scholarship in health geography: Neil Hanlon writing on political economy; Benet Reid and Matt Sothern, on humanism; and Sebastian Fleuret, on social constructionism. Indeed, these theoretical traditions in many ways helped kick-start the qualitative, place-sensitive turn in the sub-discipline from the early 1990s onward. The next three chapters focus on some important concepts that have emerged from these traditions: Amber Pearson and Richard Sadler writing on social capital; Jessica Finlay, on therapeutic landscapes; and Meryn Severson and Damian Collins, on well-being. Indeed, these concepts are used to frame and explain much empirical work in health geography since the early 1990s. The final three chapters in this section are more contemporary, in that they map theoretical progress since the early 2000s, providing new perspectives and opportunities for research. Here Josh Evans writes on post-structuralism; Cameron Duff, on post-humanism; and Jennifer Lea, on one style of research emerging from these, non-representational theory. Certainly, there are other ways of dividing the theoretical perspectives that inform health geography, and other theories and concepts could have been prioritized (e.g., theoretical traditions such as positivism/spatial science, ecology, feminism, sense of place and relationality). These have not been missed, as they are discussed throughout the book, both in this and in other sections.

In terms of the future, what the final three chapters in this section clearly show is the emergence of a range of variously interconnected theoretical traditions in health geography that include new materialism, post-humanism, affect theory, complexity theory, relationality, assemblage theory and actor-network theory. Collectively these emerging traditions inform a move away from the traditional twin streams of health geography (mapping health and health care across space versus digging for the meanings of health in place), instead telling us something about the networked, physical, sensory, atmospheric, energetic, performed and moving nature of health and place (Andrews, 2018). Moreover, as health geographer Josh Evans observed at a session at the 2017 International Symposium in Medical Geography, together they also constitute an onto-logical turn in health geography, this being a move in the sub-discipline away from theory as epistemology (the way we know things) to theory as ontology (what things are). Indeed, this ontological turn involves fun-damentally rethinking old ideas in health geography (such as health and place), including what they are and how they emerge. Despite these exciting developments, there is plenty of theoretical work to be done in the sub-discipline. We need to think more about these emerging traditions and what else they can tell us about the health of/in individuals, groups, communities and places of various forms and scales. Moreover, we need to think more about what they lend to both established and new areas of empirical inquiry and debate and how they can be used in conjunction with other, more established theory. Moreover, relating to other sections in this book, we need to consider what existing methods need to be enhanced, and what new methods need to be developed, so that we might maximize their potential and gain the insights we want from them.

References

Andrews, G. J. (2015). The lively challenges and opportunities of non-representational theory: a reply to Hanlon and Kearns. *Social Science & Medicine*, 128, pp. 338–341.

Andrews, G. J. (2018). *Non-representational theory and health: the health in life in space-time revealing.* London: Routledge.

Andrews, G. J., Chen, S. and Myers, S. (2014). The "taking place" of health and wellbeing: towards non-representational theory. *Social Science & Medicine*, 108, pp. 210–222.

Eyles, J. and Litva, A. (1996). Theory calming: you can only get there from here. *Health & Place*, 2(1), pp. 41–43.

Hanlon, N. (2014). Doing health geography with feeling. *Social Science & Medicine*, 115, pp. 144–146.

Kearns, R. A. (1993). Place and health: towards a reformed medical geography. *The Professional Geographer*, 45(2), pp. 139–147.

Kearns, R. A. (1994a). Putting health and health care into place: an invitation accepted and declined. *The Professional Geographer*, 46, pp. 111–115.

Kearns, R. A. (1994b). To reform is not to discard: a reply to Paul. *The Professional Geographer*, 46, pp. 505–507.

Kearns, R. A. (2014). The health in "life's infinite doings": a response to Andrews et al. *Social Science & Medicine*, 115, pp. 147–149.

Kearns, R. and Moon, G. (2002). From medical to health geography: novelty, place and theory after a decade of change. *Progress in Human Geography*, 26(5), 605–625.

Litva, A. and Eyles, J. (1995). Coming out: exposing social theory in medical geography. *Health & Place*, 1(1), pp. 5–14.

Mayer, J. D. and Meade, M. S. (1994). A reformed medical geography reconsidered. *The Professional Geographer*, 46, pp. 103–106.

Parr, H. (2004). Medical geography: critical medical and health geography? *Progress in Human Geography*, 28(2), pp. 246–257.

Paul, B. K. (1994). Commentary on Kearns's "Place and health: toward a reformed medical geography." *The Professional Geographer*, 46, pp. 504–505.

Philo, C. (1996). Staying in? Invited comments on "Coming out: exposing social theory in medical geography." *Health & Place*, 2(1), pp. 35–40.

13

ENVIRONMENTS OF HEALTH AND CARE

The contributions of political economy

Neil Hanlon

"Political economy" is concerned broadly with outlining the prevailing social relations of production and its implications for the creation, distribution, maintenance and transformation of wealth and power. The term first appeared in classic liberal treatises, and it was later reworked in various branches of neoliberal thought, promoting policies of limited government and unfettered market exchange. As it appears in geographic scholarship, however, political economy is generally critical of liberal principles, characterizing capitalist relations as inherently contradictory, unstable and inequitable (Sheppard, 2011). Health geographers have embraced various strands of critical political economy to help explain the persistence of uneven health and health-care risks and outcomes and how relations of power are intimately (i.e., socially and spatially) embedded in processes of wellness and care.

This chapter offers a brief overview of health geography's engagements with political economy. The account is organized in four substantive sections, employing an analogy of waves to highlight key differences in emphasis, approach and application, while also recognizing an underlying connectivity and continuity. The first wave describes the gradual drift in preoccupations from spatial analytics (e.g., location, distance decay, regional science, spatial interaction) to those of social welfare and territorial justice. A second wave saw a growing interest in embodied and situated processes of medical subjectivity and biopolitics. The third wave concerns the role of place in the social negotiation of care obligations and responsibilities. The fourth and final wave is characterized by an interest in political ecology as a means to integrate interests in biosocial relations and the emancipatory impulses of post-structural political economy. The chapter concludes with a call for health geographers to take a more assertive role in contributing to political-economic theory.

The drift toward territorial justice

Health geography was greatly influenced by spatial analytical developments and positivist modes of inquiry popular in the 1950s and 1960s. Geographers adapted a variety of quantitative and cartographic techniques – for instance, to map disease outbreaks, develop and test location-allocation models and measure the effect of distance decay on the utilization of various medical services. Much of this work was premised on the notion that research should strive to uncover regularities in spatial phenomena, in the expectation that such findings would inform rational policy choices (e.g., Thomas, 1992). In other words, under conditions of resource limitations and geographical complexity, the role of health geography was to produce objective scientific evidence that held the greatest likelihood of improving overall levels of social well-being.

Over time, critiques of the validity and relevance of positivist approaches that had inspired a *turn* to issues of social justice in other areas of the discipline began to take hold in health geography (Jones and Moon, 1987; Parr, 2004). Calls for more progressive and socially engaged research grew louder, and health geographers began to consider that unequal geographies of risk, access and health outcomes were not simply a matter of distributional inefficiencies, gaps in knowledge or uninformed policy-making, but rather were products of inequitable social structures built on relations of dominance and discrimination. This drift occurred with varying degrees of intensity in different areas of study, but its general trajectory is illustrated in debates about the development and refinement of theories of public facility location (Jones and Moon, 1987). Early for-mulations were based strictly on liberal concepts of efficiency and system optimization under conditions of idealized abstraction. By the early 1980s, however, call was growing for a theory that accounted for social, political, and welfare dimensions of location and accessibility (Jones and Kirby, 1982). As Michael Dear, a leading figure in the push for more socially relevant theory, opined, "public facility location theory must also be a theory of society, and it can only be understood as a manifestation of a given social order" (1978, p. 97).

Within health geography, the notion began to take hold that inequalities in health and health care were *necessarily* the products of uneven social distributions of wealth and power. Much effort has since been expended to demonstrate the magnitude and persistence of territorially expressed social injustices (e.g., neighborhood social deprivation) and its association with (poor) health outcomes and health-care (in)acces-sibility. This body of work has been criticized for tending to reinforce rather than challenge conventional understandings of place and space as a mere backdrop to the operation of unequal social relations (Cummins et al., 2007; Smith and Easterlow, 2005). This criticism notwithstanding, health geographers remain active contributors to interdisciplinary literatures on the social determinants of health, engaging in topics that span many forms of social discrimination (e.g., those based on income, race, ethnicity, gender, sexuality, age, dis-ability) and their intersections in places and spaces of everyday living.

Place and biopolitics

A second wave of political economy thought in health geography traces its intellectual lineage to post-structural interests in embodied relations of social dominance (e.g., heteronormativity, patriarchy, xeno-phobia). Much of this work draws on the notion of *biopolitics* first introduced in the later work of Michel Foucault (Foucault, 1991). Biopolitics refers here to the more general sense in which issues of health and care have become a focal point of efforts by modern states to subjectify and regulate entire populations (Ruther-ford and Rutherford, 2013). Rose (2001, p. 1) famously characterized this as "the politics of life itself" and offers the following description of our biopolitical age:

> Political authorities, in alliance with many others, have taken on the task of the management of life in the name of the wellbeing of the population. . . . Politics now addresses the vital processes of human existence: the size and quality of the population; reproduction and human sexuality; conju-gal, parental and familial relations; health and disease; birth and death.

A biopolitical lens, therefore, takes a critical and reflexive approach to familiar health-geography topics such as disease, disability and mental health. The aim is not to reinforce and reproduce the *alliance* Rose alludes to above, but rather to bring critical attention to the ways in which health authorities (e.g., health sciences, epi-demiology, public health, medical geography) identify and characterize public-health issues as problems to be classified, diagnosed and solved (Brown and Duncan, 2002) and the implications of this for individual *subjects*.

An important feature of biopolitics is the labeling of difference as deviance and threat to a *heteronormal* body politic. Health geographers have, for instance, looked at the ways in which infectious-disease-control measures (an area in which traditional medical geography sought to play a contributing role) reveal an underlying moral panic and political will on the part of authorities to sanction and control certain populations deemed risks to the wider public (e.g., Brown, 2009; Craddock, 2000). Others have drawn on aspects of critical disabilities studies and social models of mental health to shed light on the ways in which spaces (e.g., everyday urban and rural landscapes) and places (e.g., mental-health institutions, medical clinics) in capitalist society are imbued with liberal notions of heteronormativity that exclude people who do not measure up to *normal* standards of physical and mental capacity (Butler and Bowlby, 1997; Parr and Philo, 2003).

The work of Michel Foucault is again prominent in efforts to explain how powerful interests attempt to exert control over the most intimate and *vital* of activities (e.g., sexuality, family relations). Foucault's notion of a governmental rationality, or governmentality, asserts that modern states (i.e., those with capitalist economies, representative democracies, centralized political authority) have become all-encompassing sources of authority by accumulating knowledge about how to regulate their subjects at a distance; that is, by convincing citizens to regulate themselves in ways that serve state interests. Governmentality involves material and discursive practices that identify particular facets of social life (e.g., biopolitics) as problems to be managed and around which different apparatuses of state authority employ coercion and stigmatization as means to achieve social control (Ingram, 2010) and to sanction or constrain the activities of certain groups identified as risks to the body politic (e.g., gay men, drug users, refugees, the homeless). Various forms of state apparatus (e.g., census taking, disease surveillance, urban planning, border control, policing) and spheres of activity closely aligned to the operations of the state (e.g., health care, insurance, education) are implicated in governmentality through such means as influencing behavior and constructing particular forms of dependence that serve to strengthen dominant interests rather than improve overall welfare (Legg, 2005; Rutherford and Rutherford, 2013).

Interest in post-structural theory has heightened over the past two decades (see Evans, this volume), and health geographers now commonly draw on various strands of feminist, queer and cultural theories of the body to extend the work of critical biopolitics (Hall, 2000; Moss and Dyck, 2002). By way of illustration, numerous studies illuminate how public-health campaigns (e.g., weight loss, smoking cessation) invoke discourses of shaming and stigmatization aimed at individuals who do not conform to social norms of appearance and behavior (e.g., Evans, Crookes and Coaffee, 2012; Thompson, Pearce and Barnett, 2007). This literature turns the spotlight on the ways in which health authorities construct public-health problems and solutions in ways that conceal the contradictions and inequities of capitalist relations (e.g., Brown, Craddock and Ingram, 2012). Guthman and DuPuis (2006), for instance, argue that neoliberalism is responsible for both the conditions that *promote* obesity (e.g., an overabundance of -calorie-rich, nutrient-poor food, wage insecurity) and its vilification as a public-health *problem* (e.g., blaming *obese* subjects for straining health-care systems and undermining social productivity).

The place-embeddedness of care

A third wave of political-economic thought in health geography focuses on situated relations of care. As late as a decade ago, Andrews and Evans (2008) noted that health geographers remained fixated on issues of facility location and spatial distribution, while neglecting the inherent spatiality of health-care work. Indeed, the extended social and spatial processes by which health-care work is (re)produced remain a largely untapped source of potential health political-economic scholarship. Notable exceptions are studies that unpack the biopolitics of health-care work, including uneven contestations over the restructuring of formal health-care roles, responsibilities and hierarchies, and processes by which certain care roles are medicalized while others are consigned to private, gendered and largely devalued realms of individual household responsibility (e.g., Milligan, 2000; Wiles, 2011).

More recently, health geographers have taken up the challenge of thinking about care in a relational sense, rejecting the binary impulses of neoliberal thought (e.g., public/private, formal/informal, male/female) in favor of considerations of care as an ethical and political undertaking (Greenhough, 2011). These approaches effectively redirect political economy to tackle questions of what constitutes the boundaries between formal and informal care, as well as the relations of power and authority within and between these categories of care. Dominant liberal discourses and practices tend to enforce a binary identification of care as either formal (i.e., professional, medicalized, scarce) or informal (i.e., voluntary, intimate, common), whereas relational-ethical approaches highlight the degree to which the terms "formal" and "informal" are fluid, negotiated and mutually constituted (e.g., Wiles, 2011).

Geographers have been especially concerned with the tendency toward increasing constrictions on welfare entitlement as states seek to offload obligations for care to private households. Much of this scholarship illuminates the policy enactment of favored neoliberal notions of individual resilience, austerity and *independence*, on one hand, and the stigmatization of *vulnerability* and *dependence* on the other. These discursive and material valuations (and devaluations) have been shown to infuse the practice of care across the spectrum (e.g., professionals, volunteers) into agents of neoliberal austerity and governmentality (Herron, Rosenberg and Skinner, 2016; Thompson, 2008).

Finally, place and space are highlighted as important resources in the operation of neoliberal policy, invoking a more traditional political-economy conceptualization of health care as both a feature of capitalist relations and an important means by which these relations are reproduced. Health- and social-care systems are deeply embedded in the social geography of livelihood (Hanlon et al., 2007; Henry, 2015; Smith and Easterlow, 2005). Questions about changing sites of care are thus inseparable from the biopolitics of neoliberalism, as states seek to offload care obligations in favor of private economies of care in community- and home-based settings (Hanlon et al., 2007; Herron, Rosenberg and Skinner, 2016; Milligan, 2000; Skinner, Joseph and Herron, 2016).

Political ecologies of wellness and care

The fourth wave of political-economic health geography is characterized by a small but growing interest in adopting political ecology as a framework to illuminate the role of socio-spatial processes in shaping experiences of health and care. For political ecologists, decision-making must take account of environmental as well as political forces acting on individual households and care providers in order to frame the full extent of constraints and opportunities available. That is, localized articulations of wider political economic systems (e.g., livelihood routines, environmental practices, cultural practices, social relations) interact with certain bio-physical contingencies (e.g., disease agents, genetics, physiology, biogeography) to frame individual health behavior and choice in important ways. Likewise, health risks and vulnerabilities must be understood as embedded in prevailing biosocial ecologies (Hunter, 2007; King, 2010).

Much of this literature credits Mansfield (2008) for reasserting the notion that health (and care) are enactments of nature-society interrelationships, or what she calls "the biosocial." In a similar vein to those promoting an ethics of care, political ecology eschews prevailing dualisms of liberal thought (e.g., nature/society, local/global) and embraces relationality as a way to confront key conceptual challenges. Health and care are therefore products or expressions of biosocial interrelationships; that is, social relations (e.g., care, stigmatization) are critical features in the enactment of the biological body, just as the biology of human health has undeniable effects on social relations and interactions.

In a recent study, Neely (2015) asserts the importance of a nested place-based approach that connects processes at work at the smallest scales (i.e., what she terms the internal ecosystem of individual bodies) to the most expansive of biophysical and political systems (i.e., external ecosystems) operating well beyond the immediate confines of daily living. The key points she stresses are (a) the interconnection of all scales and (b) the potential for any scale to be activated in addressing a given health or care problem. The challenge for political ecology, then, is to determine which of the contingent biosocial processes are most critical to take account of, and respond to, the problem at hand.

The work of Sultana (2012) offers a useful and accessible introduction to health political ecology. Sultana sketches the intersections of environmental, cultural and political relations implicated in the exposure of millions of people in rural Bangladesh to contaminated drinking water, exploring the complexity of how people respond to, and cope with, the risk of these unintended exposures. The contaminant in question, arsenic, is naturally present in groundwater aquifers in many parts of rural Bangladesh, and exposure to this environmental hazard was an unintentional (and ironic) consequence of national and international development initiatives to promote freshwater security. Issues of uneven relations of power and dependence played a key role here, but Sultana's political-ecology framework integrates these relations of power with multiscaled, nature-society interrelationships that are also critical in understanding this public-health disaster.

Concluding remarks

Political-economy approaches are now fairly commonplace in health geography, but, as this review suggests, they have taken on very disparate forms over the past three decades. Scholars employing political economy are apt to disagree about the factors most in need of articulation and the means by which these factors operate. For some, capitalist relations of exploitation are paramount, while others stress that patriarchy, racism, heteronormativity, or the intersections of these and other forces, are paramount. What these efforts share in common, however, is an impulse that can be traced to the early days of geographical political economy; that is, the sense that individual experience cannot be understood in social, spatial and temporal isolation. Environments of health and care in which individual experiences take place are necessarily connected to wider socio-spatial networks and relationships, and the choices and opportunities available to individuals are thus framed and constrained in important ways that demand articulation.

Concerns about prevailing social relations and power imbalances have kept health geographers connected to wider developments in human geography and the social sciences. To date, there is a strong case to be made that health geography has gained much more from its engagements with geographical political economy than it has given back. That said, the central concerns of health geography (i.e., the spatiality of wellness, care and risk) have much to offer political-economic theory, both within human geography and beyond it. That is, by means of multiscaled and place-embedded articulations of biopower, health geographers can and should be more assertive in both critically applying and re-articulating political economy scholarship in its many varieties.

References

Andrews, G. J. and Evans, J. (2008). Understanding the reproduction of health care: towards geographies in health care work. *Progress in Human Geography*, 32, pp. 759–780.

Brown, M. (2009). Public health as urban politics, urban geography: venereal biopower in Seattle, 1943–1983. *Urban Geography*, 30(1), pp. 1–29.

Brown, T., Craddock., S. and Ingram, A. (2012). Critical interventions in global health: governmentality, risk, and assemblage. *Annals of the Association of American Geographers*, 102(5), pp. 1182–1189.

Brown, T. and Duncan, C. (2002). Placing geographies of public health. *Area*, 34, pp. 361–369.

Butler, R. and Bowlby, S. (1997). Bodies and spaces: an exploration of disabled people's experiences of public space. *Environment and Planning D: Society and Space*, 15, pp. 411–433.

Craddock, S. (2000). *City of plagues: disease, poverty and deviance in San Francisco*. Minneapolis: University of Minnesota Press.

Cummins, S., Curtis, S., Diez-Roux, A. V. and MacIntyre, S. (2007). Understanding and representing "place" in health research: a relational approach. *Social Science and Medicine*, 65(9), pp. 1825–1838.

Dear, M. J. (1978). Planning for mental health care: a reconsideration of public facility location theory. *International Regional Science Review*, 3(2), pp. 93–111.

Evans, B., Crookes, L. and Coaffee, J. (2012). Obesity/fatness and the city: critical urban geographies. *Geography Compass*, 6(2), pp. 100–110.

Foucault, M. (1991). Governmentality. In: G. Burchell, C. Gordon and P. Miller, eds., *The Foucault effect: studies in governmentality*. Chicago: University of Chicago Press, pp. 87–104.

Greenhough, B. (2011). Citizenship, care and companionship: approaching geographies of health and bioscience. *Progress in Human Geography*, 35(2), pp. 153–171.

Guthman, J. and DuPuis, M. (2006). Embodying neoliberalism: economy, culture and the politics of fat. *Environment and Planning D: Society and Space*, 24(3), pp. 427–448.

Hall, E. (2000). "Blood, brain and bones": taking the body seriously in the geography of health and impairment. *Area*, 32(1), pp. 21–29.

Hanlon, N., Halseth, G., Clasby, R. and Pow, V. (2007). The place embeddedness of social care: restructuring work and welfare in Mackenzie, BC. *Health & Place*, 13, pp. 466–481.

Henry, C. (2015). Hospital closures: the sociospatial restructuring of labor and health care. *Annals of the Association of American Geographers*, 105(5), pp. 1094–1110.

Herron, R., Rosenberg, M. W. and Skinner, M. W. (2016). The dynamics of voluntarism in rural dementia care. *Health & Place*, 41, pp. 34–41.

Hunter, M. (2007). The changing political economy of sex in South Africa: the significance of unemployment and inequalities to the scale of the AIDS pandemic. *Social Science and Medicine*, 64, pp. 689–700.

Ingram, A. (2010). Biosecurity and the international response to HIV/AIDS: governmentality, globalization and security. *Area*, 42, pp. 293–301.

Jones, K. and Kirby, A. (1982). Provision and wellbeing: an agenda for public resources research. *Environment and Planning A*, 14, pp. 297–310.

Jones, K. and Moon, G. (1987). *Health, disease and society: an introduction to medical geography*. London and New York: Routledge.

King, B. (2010). Political ecologies of health. *Progress in Human Geography*, 34(1), pp. 38–55.

Legg, S. (2005). Foucault's population geographies: classification, biopolitics and governmental spaces. *Population, Space and Place*, 11(3), pp. 137–156.

Mansfield, B. (2008). Health as a nature-society question. *Environment and Planning A*, 40, pp. 1015–1019.

Milligan, C. (2000). Bearing the burden. *Area*, 32(1), pp. 49–58.

Moss, P. and Dyck, I. (2002). *Women, body, illness: space and identity in the everyday lives of women with chronic illness*. Lanham, MD: Rowman and Littlefield.

Neely, A. H. (2015). Internal ecologies and the limits of local biologies: a political ecology of tuberculosis in the time of AIDS. *Annals of the Association of American Geographers*, 105(4), pp. 791–805.

Parr, H. (2004). Medical geography: critical medical and health geography? *Progress in Human Geography*, 28(2), pp. 246–257.

Parr, H. and Philo, C. (2003). Rural mental health and social geographies of caring. *Social and Cultural Geography*, 4(4), pp. 471–488.

Rose, N. (2001). The politics of life itself. *Theory, Culture and Society*, 18(6), pp. 1–30.

Rutherford, P. and Rutherford, S. (2013). The confusions and exuberances of biopolitics. *Geography Compass*, 7(6), pp. 412–422.

Sheppard, E. (2011). Geographical political economy. *Journal of Economic Geography*, 11, pp. 319–331.

Skinner, M. W., Joseph, A. E. and Herron, R. V. (2016). Voluntarism, defensive localism and spaces of resistance to health care restructuring. *Geoforum*, 72, pp. 67–75.

Smith, S. J. and Easterlow, D. (2005). The strange geography of health inequalities. *Transactions of the Institute of British Geographers*, NS 30(2), pp. 173–190.

Sultana, F. (2012). Producing contaminated citizens: toward a nature-society geography of health and well-being. *Annals of the Association of American Geographers*, 102(5), pp. 1165–1172.

Thomas, R. (1992). *Geomedical systems: intervention and control*. London and New York: Routledge.

Thompson, L. (2008). The role of nursing in governmentality, biopower and population health: family health nursing. *Health & Place*, 14(1), pp. 76–84.

Thompson, L., Pearce, J. and Barnett, J. R. (2007). Moralising geographies: stigma, smoking islands and responsible subjects. *Area*, 39(4), pp. 508–517.

Wiles, J. (2011). Reflections on being a recipient of care: vexing the concept of vulnerability. *Social and Cultural Geography*, 12(6), pp. 573–588.

14

HUMANISM AND HEALTH GEOGRAPHY

Placing the *human* in health geography

Matthew Sothern and Benet Reid

> In general, people – in the sense of acknowledged, autonomous, sentient beings – remain generally absent from the narratives of health geography. . . . A tendency to see the individual not as a person but as an observation has been largely retained.
>
> *(Kearns and Moon, 2002, p. 613)*

Humanism has something of a fraught and sometimes hidden history in geography. Livingstone's (1992) classic text *The Geographical Tradition*, for example, does not explicitly index humanism until the penultimate chapter, tellingly titled "Statistics Don't Bleed." Many histories of geography place humanism as a reaction to the spatial-science turn of the 1960s and 1970s – an attempt to re-engage with the complex, rich, experiential and creative ways in which we make our worlds through how we live our lives. Humanism was less about causal explanation rooted in epistemological positivism than it was about engaging with consciousness, imagination, meaning and understanding. Humanism encouraged a focus on the richly textured ways in which we actively interpret, engage and construct the world around us. While humanism as an explicit epistemological and theoretical tradition has largely been eclipsed by the now not-so-new critical geography, it remains an important and pivotal moment in the history of geography in general, and health geography in particular. Too often, however, humanism is marginal to the programmatic statements about the trajectory of our sub-discipline.

This chapter is divided into three parts. The first outlines the history and philosophical thrust of humanism. The second introduces how these ideas found expression in the move from medical to health geography and especially through the adoption of qualitative methodology. In the final section, we argue that critical health geography remains indebted to the legacy of humanism in terms of both empirical diversity and methodological heterogeneity and that the shift toward post-humanism represents a renewed attempt to understand what it means to be human.

Humanism in history, philosophy and geography

Humanism is a diverse philosophical tradition that places human consciousness and reason as the defining attributes of the human condition. Although its origins can be traced back to ancient Greece, it rose to prominence during the Renaissance, during which its proponents argued for ideas of the rationality and autonomy of man to act – this made it thinkable for man to supplant God in control of nature. As such, humanism informed Enlightenment thinking against religious and social dogma and became the "philosophical champion of human freedom and dignity," yet insofar as it was concerned with a very narrow spectrum of humanity (male elites) it also served as an "ideological smokescreen" for marginalization and oppression by failing to address inequalities between different types of people (Cosgrove, 1984; Davies, 2008, p. 5). By the early 20th century, philosophies of humanism framed a nostalgic fascination with the classical period and an aesthetic of human purity linked to eugenics on which rested the ideological foundations of much 20th-century state-making (Ordover, 2003). These earlier humanisms promulgated an elitist and exclusionary vision of the human: post-theological for sure, but resolutely teleological and linked to the growing influence of science as a social regulator.

The humanism that emerged in the 1960s was both an extension and a rejection of these earlier moments. On one hand, social change and the growing prominence of feminism, lesbian, gay, bisexual and transgender (LGBT) activism, and the struggle against colonialism sought to extend the breadth of human dignity, often drawing on humanist traditions challenging religious and cultural norms. These were explicitly political projects borne by, and born from, those whose humanity was all too often denied. On the other hand, postwar disillusionment with industrial military expansion and with the tarnished promise of scientific progress progressively eroded the Enlightenment belief in objectivity, rationality and the uncritical application of scientific methodologies. This led to a rejection of the rationalities on which earlier humanism had depended.

In geography, these streams of thought were most evident in two ways: first, the expansion of work that sought to uncover geographies of social difference (gender, sexuality, race, disability and the like) and, secondly, a growing critique (and sometimes outright rejection) of the legacy of a postwar quantitative revolution and of insistence on geography as a spatial science. Whether used descriptively or normatively, the numerical approaches of spatial science were viewed as too mechanistic, too reliant on a vision of the human as a rational, mappable and predictable subject; an abstracted and aggregated representation of an idealized person who did not actually exist as anything more than mere dots on a map (Cloke, Philo and Sadler, 1991). As feminist geographers would later argue, this idealized person was usually male, white, able-bodied and healthy – and claims to objective science rested on the insistence of the universality of very particular subject positions (Rose, 1993). Equally, the growing prominence of structuralist thinking, particularly Marxism, that sought to subtend cultural expression and agency to deep socioeconomic structures left little room for sentient, thinking, and active subjects. For all their methodological sophistication and political utility, the worry for some was that we were producing a human geography without man, where things were done *to* but not *by* man (Buttimer, 1993; Ley and Samuels, 1978). Humanistic approaches, meanwhile, focused attention on contingency, variability of interpretation, and contextualized human agency. Rather than theorizing away or seeking to eliminate subjectivity and contextual difference, they made *meaning*, *values* and *intention* central to geography (Entrikin, 1976).

Humanistic geographers drew inspiration from the philosophies of phenomenology and existentialism (see Pickles, 1985). Phenomenological approaches (after Husserl, Heidegger and Merleau-Ponty) focus on how phenomena are experienced by people, ask how meaning is constructed from these experiences, and seek to identify the essential elements which comprise shared categories of experience. Phenomenology relied on rich qualitative description – a methodological corrective to the overly abstracted tendencies of spatial science and structuralism. Existentialism, after philosophers including Nietzsche, Kierkegaard and Sartre, trained attention on the relationship of human consciousness to itself and emphasized human capacity to consciously act – an ethical corrective to overemphasis on the determinative forces of biology, culture and structure that seemed so devoid of agency. Both philosophical traditions called attention to the ontological difficulty of conceptualizing human consciousness as separate from an external world. They opened up ontological, epistemological and above all political questions about materiality, language and representation – questions that continue to animate contemporary critical geography.

In practice, however, much humanist geography had limited technical interest in the philosophical intricacies of phenomenology or existentialism (Curry, 1996). Instead, the explosion of humanist ideas in geography in the 1970s was motivated by a desire to allow the "individual to take pride of place" (Cox, 2014, p. 96); this was as an antidote to the dominant view of spatial science that understood space as a passive container in which processes occur or as planes over which phenomena are distributed. Humanist geography sought to understand "geographical experiences as they are 'actually' experienced – as meaningful, value laden" (Entrikin, 1976, p. 629) and as lived through bodies, emotions, senses and the performance of everyday life. In so doing, humanism sought to displace a flattened, narrow and functional vision of the human with one in which there were intimate connections between people and their environment. Key texts included Tuan's (1974) *Topophilia*, which explored the important bonds between people and places, Buttimer's (1974) *Values in Geography*, which argued for the importance of values in shaping experiences of place, and Relph's (1976) *Place and Placelessness*, which provided an ambitious framework for understanding place and landscape as an essential component of being-in-the-world. Collectively these works reimagined place as socially produced. They raised questions about how we know, from where we know and how we as geographers come to know. Contemporary critical geography's concern with positionality, reflexivity, power and the interrogation of the mundane, the taken for granted and the everyday all have their antecedents (as well as much of their language and political salience) in these early humanist geographies (Cloke, Philo and Sadler, 1991).

Humanism and health geography

In their account of the dualism between medical and health geography, Dorn, Keirns and Del Casino (2010) lament that a broad and fulsome vision of medical geography developed in the 18th and 19th centuries "as an enlightenment project to explore the physical, social, and political factors which both damaged and sustained population health" (p. 58) had morphed into a narrower concern with disease ecology and the provision of health services (see Mayer, 1982). These two traditions often relied, as did much work elsewhere in geography, on positivism and quantitative methods that had come to dominate the post-quantitative-revolution landscape. This work has had considerable influence on the policy arena and has meant that core geographic concerns with diffusion and environment as determinants of health had become firmly entrenched in public-health and epidemiology-research agendas. By the 1990s there was growing concern that medical geography was becoming isolated from broader debates elsewhere in human geography. It is telling that Kearns' call for a "reformed medical geography" – often cited as the touchstone for the birth of health geography – was explicitly framed by arguments for the "marriage of humanistic geography and contemporary models of health" (Kearns, 1993, p. 141). The focus on *place* provided a key entry point for the emergent health geography's contributions to a social (that is, non-biomedical) model of health. While a variety of epistemological perspectives characterized the move toward health geography, the "eclectic deployments of the humanist tradition and its legacy of methodological *perestroika* have allowed the most nuanced and effective contributions" (Kearns and Moon, 2002, p. 610).

It is hard to understate the impact humanism had on the dominant methodological paradigm. While we do not suggest an easy equivalence between epistemology and method, it is nonetheless clear that humanistic geography relied upon significant methodological innovation arising from and contributing to the epistemological rejection of positivism. Emphasizing the complexity (and often opaqueness) of experience, humanistic geographers turned to what Seamon and Lundberg (2017) call *methodologies of engagement* that privileged intimacy and connection to the lifeworlds of research subjects. This called for deep qualitative commitment to understand the lifeworlds and meanings participants themselves attached to places, institutions, their own bodies and experiences of health and health care. The focus of this work was emphatically on language (see Gesler and Kearns, 2002) – it deployed interviews, focus groups, participant observation, and ethnography but also engaged literature, art, photography, and film as ways to uncover the meanings people draw from place and landscape. This health geography valorized the particular and the local so as to allow for sensitivity to context and meaning. While generalization was sometimes an important element, many health geographers focused on the exceptions, the outliers and those mundane, unremarkable and taken-for-granted aspects that comprise the stuff of everyday life.

Gesler's (1992) influential work on therapeutic landscapes, for example, not only drew attention to the complex nature of places but expanded health-geographical interest in the everyday spaces people inhabit and the ways in which people experience and make sense of these places – all concerns that occupied humanistic geography through the 1970s. More recent examples in this genre make connections between the agency of healing landscapes and the agency of people who use them intentionally to pursue health. Sampson and Gifford (2010) worked with young refugees to understand their active processes of place-making for well-being in resettlement. English, Wilson and Keller-Olaman (2008) interviewed breast-cancer survivors to discover how they created and accessed places of healing through negotiating personal processes of emotional experience. Wood et al. (2015) extend humanism's concerns with experience and meaning-making to examine the emotional responses to the changing spaces of psychiatric care to argue for a more humane design of therapeutic spaces. In each of these studies, the focus was on how diverse groups experienced and shaped the spaces around them and how, in turn, these spaces helped shape an understanding of self and their experiences of health. These landscapes are not mere containers for people's health outcomes but are produced through the mundane and not-so-mundane experiences and willful action of those who make them.

Critical and post-human health geography

While humanism had a significant impact on the practice of human geography, it could never have been described as a dominant paradigm. Indeed, humanism was only one response to the perceived limitations of spatial science. Throughout the 1970s, an increasingly confident Marxist critique argued that humanism was too apolitical and ignored the productive power of capitalism, that it over-valorized human agency and focused on the local at the expense of deeper structural connections (Gregory, 1981). Critiques from feminism, disability, postcolonial and critical race geographies also charged that humanism did not engage enough with the differing and highly uneven experiences of some humans, relying too heavily on assumptions of autonomous individual agency and coherent subjectivity (Rose, 1993). Critiques of humanism for being overly descriptive, for blindness to the nature of power and for a failure to theorize the construction of subjects was a major driver of the cultural turn and of the growing influence of post-structuralism (Gregory, 1994). Humanism came to be viewed by some as a theoretical and political dead-end that had been eclipsed by critical theory.

However, the foundations of critical theory arguably lie in the same central questions that concern humanism: to ask how the world is experienced, to understand how subjects manifest active agency, and to explore heterogeneity and to reject overly universalizing laws that aspire to a scientific ideal. This continuity is well-illustrated in health geography. The skepticism of supposedly detached perspectives and placelessness of medical geography, and the call for a "renewed recognition of diversity and difference" in active and emplaced experiences of health (Kearns, 1993, p. 145), mark a humanist sentiment but are not inconsistent with post-structural thought, which embraces multiple maps of meaning. The intensified theoretical scrutiny upon *place* in health geography, meanwhile, prefigured the relational ontologies proffered by more recent post-structural geographies. Moreover, from its inception, health geography has always been concerned with questions of inequality, power and resistance to oppression. These have been most obvious in the insistence of health geography to speak *to* and *with* patients. An emphasis on broader notions of well-being, on voices that are otherwise silenced or ignored, and on the important knowledge derived from embodied experiences of ill health have always been understood as a political critique of biomedicine and spatial epidemiology (Brown, 1995; Dyck and Moss, 2002; Kesby and Sothern, 2014).

More recently, there have been calls to reconsider the legacies and limitations of humanism, not just in terms of an antihumanist form of rejection but also to acknowledge the potentials of new humanism (Simonsen, 2012) and post-humanism (Barridotti, 2013; Castree and Nash, 2006). Post-humanist work focuses on both a decentering and a reimagining of the human – not as a contained and coherent actor but as embodying a process of *becoming*, an assemblage that involves more-than-human elements. Whereas humanism focused on consciousness, intention and experiences of human subjects, post-humanism deconstructs the boundaries between nature/culture, self/other, human/non-human, flesh/technology (see Hayles, 1999). Post-humanism does not reject agency of human beings in favor of natural or structural forces per se; rather, it focuses on the complexity and deep mutual permeability of relations between fleshy, embodied people and the continually changing materiality of places. When Cummins et al. (2007) argued for *relational* understandings of health variation situated in place, for example, they expressed a growing recognition that commonsense distinctions between natural and social, active and passive, human and non-human could not unproblematically be retained by geographers seeking a thorough understanding of health.

These sentiments have been explored in a range of new geographies of health that draw on the post-humanisms developed in science and technology studies, actor-network theory and non-representational theory. While not always positioned within the sub-disciplinary boundary of health geography, much of the innovative work in this area nonetheless continues to develop the health-geography tradition of critically examining contemporary biomedicine and its complicity with regimes of neoliberalism. Greenhough (2006), for example, shows how medical records and gene mapping rest on the attempt to produce a boundary between the fully human and those parts of the body (including information) that can circulate as commodities. Similarly, Parry's (2012) work on biosecurity and swine flu demonstrates the intimacy between biomedicine, public health and new forms of biocitizenship. Fannin's oeuvre of work on the geographies of maternity traces the complex and networked journeys of the placenta as they move through institutions, repositories and knowledges to demonstrate how biomedical interventions produce new narratives and potentials for understanding the production of subjectivity (see Fannin and Kent, 2015). More recent work has drawn on the growing influence of non-representational theory to reframe core health-geography concerns with experience and well-being as emergent effects (e.g., Andrews, Chen and Myers, 2014; Boyd, 2017).

These new directions in health geography do not reject the humanist concern with bringing the human back into geography. Rather, they offer productive reformulations of the concept of the human, in which "humans, natures and technologies are co-emergent" (Murdoch, 2004, p. 1359). Moreover, post-human geographies reframe, but do not move us away from, the concerns with experience, inequality and power relations that are so fundamental to health geography (Panelli, 2010). Although predicated on a critique of original humanism, post-humanism urges us to sharpen our ideas about conditions of life and what it means to be human.

Conclusion

Humanism has a long history in geography. As an important critique of the spatial-science tradition, it argued for a geography focused on more than just spatial laws or determinative structural forces. Humanistic geography developed sensitive ways of understanding lived realities and the meanings made from collective and individual perceptions of place. These arguments were crucial in the move from medical to health geography and the recognition that health is not simply the absence of disease but is something that is lived, experienced and made. Humanism reinvigorated the study of place as one of the organizing ideas of health geography. Humanism championed qualitative methodologies and insisted upon the importance of subjectivity, including the subjectivity of researchers. More recent developments, inspired by feminism and other critical theories, have offered substantive critiques of humanism – in particular, they have been critical of humanism's all too often under-examined universalist suppositions. Yet, in important ways the post-humanist critique serves to extend humanist interest in how different groups of people are made unequal in the opportunities they have to engage with health-giving resources. Moreover, in examining how contemporary technological, scientific and biomedical practice challenge the boundaries between human and non-human, they serve to deepen a fundamentally humanistic concern with what it means to be human in particular times and in particular places.

References

Andrews, G., Chen, S. and Myers, S. (2014). The "taking place" of health and wellbeing: towards non-representational theory. *Social Science & Medicine*, 108, pp. 210–222.

Barridotti, R. (2013). *Posthumanism*. Cambridge: Polity Press.

Boyd, C. P. (2017). *Non-representational geographies of therapeutic art making*. London: Palgrave Macmillan.

Brown, M. (1995). Ironies of distance: an ongoing critique of the geographies of AIDS. *Environment and Planning D*, 13(2), pp. 159–183.

Buttimer, A. (1974). *Values in geography*. Washington, DC: Association of American Geographers.

Buttimer, A. (1993). *Geography and the human spirit*. Baltimore: Johns Hopkins University Press.

Castree, N. and Nash, C. (2006). Posthuman geographies. *Social and Cultural Geography*, 7(4), pp. 501–504.

Cloke, P., Philo, C. and Sadler, D. (1991). *Approaching human geography*. London: Sage.

Cosgrove, D. (1984). Prospect, perspective and the evolution of the landscape idea. *Transactions of the Institute of British Geographers*, 10(1), pp. 45–62.

Cox, R. (2014). *Making human geography*. New York: Guilford Press.

Cummins, S., Curtis, S. Diez-Roux, A. and MacIntyre, S. (2007). Understanding and representing "place" in health research: a relational approach. *Social Science & Medicine,* 65(9), pp. 1825–1838.

Curry, M. (1996). Commentary 1. *Progress in Human Geography*, 20(4), pp. 513–519.

Davies, T. (2008). *Humanism*. London: Routledge.

Dorn, M., Keirns, C. and Del Casino, V. (2010). Doubting dualisms. In: T. Brown, S. McLafferty and G. Moon, eds., *A companion to health and medical geography*. Oxford: Wiley-Blackwell, pp. 55–78.

English, J., Wilson, K. and Keller-Olaman, S. (2008). Health, healing and recovery: therapeutic landscapes and the everyday lives of breast cancer survivors. *Social Science & Medicine*, 67(1), pp. 68–78.

Entrikin, J. (1976). Contemporary humanism in geography. *Annals of the Association of American Geographers*, 66(4), pp. 615–632.

Fannin, M. and Kent, J. (2015). Origin stories from a regional placenta tissue collection. *New Genetics and Society*, 34, pp. 25–51.

Gesler, W. (1992). Therapeutic landscapes: medical geographic research in the light of new cultural geography. *Social Science & Medicine*, 34(7), pp. 735–746.

Gesler, W. and Kearns, R. (2002). *Culture/place/health*. London: Routledge.

Greenhough, B. (2006). Decontextualised? Dissociated? Detached? Mapping the networks of bio-informatic exchange. *Environment and Planning A*, 38(3), pp. 445–463.

Gregory, D. (1981). Human agency and human geography. *Transactions of the Institute of British Geographers*, 6(1), pp. 1–18.

Gregory, D. (1994). *Geographical imaginations*. Oxford: Blackwell.

Hayles, K. (1999). *How we became posthuman*. Chicago: University of Chicago Press.

Kearns, R. (1993). Place and health: towards a reformed medical geography. *The Professional Geographer*, 45(2), pp. 139–147.

Kearns, R. and Moon, G. (2002). From medical to health geography: novelty, place and theory after a decade of change. *Progress in Human Geography*, 26(5), pp. 605–625.

Kesby, M. and Sothern, M. (2014). Blood, sex and trust: the limits of the population-based risk management paradigm. *Health & Place*, 26(1), pp. 21–30.

Ley, D. and Samuels, M. (eds.) (1978). *Humanistic geography: prospects and problems*. Chicago: Maaroufa Press.

Livingstone, D. (1992). *The geographical tradition*. Oxford: Wiley-Blackwell.

Mayer, J. (1982). Medical geography: some unsolved problems. *The Professional Geographer*, 34(3), pp. 261–269.

Moss, P. and Dyck, I. (2002). *Women, body, illness: space and identity in the everyday lives of women with chronic illness*. Lanham, MD: Rowman and Littlefield.

Murdoch, J. (2004). Humanising posthumanism. *Environment and Planning A*, 36, pp. 1356–1359.

Ordover, N. (2003). *American eugenics: race, queer anatomy and the science of nationalism*. Minneapolis: University of Minnesota Press.

Panelli, R. (2010). More-than-human social geographies: posthuman and other possibilities. *Progress in Human Geography*, 34(1), pp. 79–87.

Parry, B. (2012). Domesticating biosurveillance: "containment" and the politics of bioinformation. *Health & Place*, 8(4), pp. 718–725.

Pickles, J. (1985). *Phenomenology, science and geography: spatiality and the human sciences.* Cambridge: Cambridge University Press.

Relph, E. (1976). *Place and placelessness.* London: Pion.

Rose, G. (1993). *Feminism and geography: the limits of geographical knowledge.* Cambridge: Polity.

Sampson, R. and Gifford, S. (2010). Place-making, settlement and well-being: the therapeutic landscapes of recently arrived youth with refugee backgrounds. *Health & Place*, 16(1), pp. 116–131.

Seamon, D. and Lundberg, A. (2017). Humanistic geography. In: D. Richardson, N. Castree, M. F. Goodchild, A. Kobayashi, W. Liu and R. A. Marston, eds., *The international encyclopedia of geography: People, the earth, environment, and technology.* Chichester: Wiley-Blackwell.

Simonsen, K. (2012). In quest of a new humanism: embodiment, experience and phenomenology as critical geography. *Progress in Human Geography*, 37(1), pp. 10–26.

Tuan, Y. (1974). *Topophilia: a study of environmental perception, attitudes and values.* Upper Saddle River, NJ: Prentice Hall.

Wood, V., Gesler, W., Curtis, S., Spencer, I., Close, H., Mason, J. and Riley, J. (2015). "Therapeutic landscapes" and the importance of nostalgia, solastalgia, salvage and abandonment for psychiatric hospital design. *Health & Place*, 33, pp. 83–89.

15

SOCIAL CONSTRUCTIVISM (IN A SOCIALLY CONSTRUCTED HEALTH GEOGRAPHY?)

Sebastien Fleuret

All knowledge is derived from looking at the world from some perspective or other, and is in the service of some interests rather than others.

(Burr, 2003, p. 6)

According to the principles of constructivism, the world is structured on the basis of systems of representation or action. Human beings interact with both their surroundings (environment) and others (society). These systems constitute perspectives for understanding this interaction in human and social sciences. Constructivism, developed in the 1920s, refers to classical philosophical principles that question the everyday concepts of truth, knowledge and objectivity (Kant is frequently cited; see Krasnoff, 1999). Born at the emergence of critical thinking in human and social sciences, constructivism contributed to a broadening and enrichment of reflection in this field and, more recently, in health geography. At the same time, however, constructivism has met with much criticism and controversy.

This chapter is structured into three parts. The first part retraces the history of constructivist thought, whereby contributions made by those from diverse disciplines (mainly sociology, philosophy and psychology) are discussed in relation to the global evolution of research in health and medical sciences in general, and more specifically in the field of health geography. The limits and criticisms of constructivism are also presented. The second part examines the various mobilizations of constructivism in health-geography research – for example, studies on diseases, behaviors, individual and collective issues, health-care systems and their organization. Consideration is given to certain disciplinary and sub-disciplinary fields and to specific research approaches (e.g., mental health, gender studies). The third part comprises a discussion on the repercussions of this research on public action in the health sector.

The development of constructivist thought

Constructivist thought developed as a rejection of naturalism, which adheres to an immutable state of things that are governed by natural order or determinism (natural or divine). Constructivist thought represents a break from epistemologies common in human and social sciences, such as functionalism (actions are driven by a social function in response to a need) or structuralism (social reality is derived from fundamental structures that are often unconscious). Inspired in part by phenomenology (what we observe is inseparable from the subject who becomes aware of it, this awareness having no value without the subject), constructivism refers to a philosophical posture for which all reality is conceivable only through categories of filters, the interpretation of perspectives and approaches, and systems of representation (Orain, 2014). In constructivist theory, reality is not what we observe; rather, it is an object constructed by science, norms and rules. In this sense, all reality is subjective in nature. In order to understand this reality, a social and normative construct, it is necessary to develop a scientific system of construction to study the structure of things hidden behind appearances, according to what Derrida (1992) called a deconstruction process. This process is subject to different understandings, which are presented hereafter.

Pierre Bourdieu (1987) wanted to go beyond structuralist theory by means of a constructivist structuralism, which stipulates that in the social world itself, objective structures are independent of the consciousness and will of the agents, capable of directing or constraining their practices or their representations. Thus, Bourdieu believes that individuals construct and reconstruct social reality from existing structures – in other words, that there are objective structures that individuals find in society and adopt depending on the leeway they allow themselves to deconstruct and appropriate them.

Two authors, Peter Berger and Thomas Luckmann (1967), apply a phenomenological constructivism perspective. In contrast to Bourdieu's approach, which uses social structures to specify the constructivist dimension of social reality, Berger and Luckmann favor an approach based on individuals and their interactions. This *phenomenological* approach seeks to understand the reality of everyday life based on the fact that the latter contains typification principles that determine the way in which social actors behave and interact. Thus, individuals exist in relation to their social position and to the representations they give rise to, on one hand, and those they make on the other. And this varies from place to place.

Constructivist theory leads us to think of space as a sum of realities constructed in relation to each other. Its tendency to question any universal truth (something can be true in one place, in one society, but false elsewhere) has resulted in much criticism being leveled at constructivism, including:

(1) Criticism with respect to relativism. Can everything be relativized? If all reality can vary depending on representations, perceptions and social and/or cultural constructions, then there is no truth, only representations. And in this case, relativism would take precedence over everything else

(2) Criticism with respect to solid scientific foundations. Constructivism is incompatible with evidence-based medicine, for example. "Whether or not a particular behavior or experience is viewed by members of a society as a sign or symptom of illness depends on cultural values, social norms and culturally shared rules of interpretation" (Mishler, 1982, p. 141). This assertion is in direct contradiction with the biomedical model, which is defined in reference to universalist criteria, rules and data. Making a diagnosis consequently becomes an interpretive act rather than a technical procedure. This leads to the fact that the diagnosis is no longer seen as a measure of deviation with respect to a biological norm, but more broadly as the action of attempting to understand the illness. This position supports criticism of the traditional biomedical model. It is also very present in certain medical fields (e.g., mental health). But this is not without a pitfall, namely the tendency to doubt everything and therefore the risk of challenging acquired knowledge (e.g., the current questioning of the need for vaccination by a significant percentage of the population).

In spite of these criticisms (or taking them into account), social constructivism has resulted in many developments in the area of (a critical) health geography.

Contributions and mobilizations in health geography

Hester Parr (2004), referring to Kearns and Moon (2002), noticed the growing place of cultural geographies in geographies of illness and impairment. This is accompanied by the questioning of outdated stereotypes of medicine and a claim for robust basic and applied research. This encourages geographers to further utilize existing theoretical resources on health and subjectivity (Parr evokes Foucault) to interpret new public-health initiatives, with a particular focus on how aspects of social life are increasingly subject to the medical gaze. Moreover, Parr's point could be reversed, since an increasing number of medical issues are now analyzed with regard to their social determinants. Globally, constructivism plays an important role in the development of critical health geography through several notions, which are discussed hereafter.

Social and situational construction of illness

One of the contributions of constructivism is to consider the cultural meaning of illness, arguing that illnesses have both biomedical and experiential dimensions (Conrad and Barker, 2010). Negative representations associated with diseases such as HIV/AIDS, cancer or mental illness impact those afflicted and can limit their access to treatment or affect the way in which they live with the illness (e.g., impairing social relationships). Another issue is the contestation of certain illnesses, which can vary from one place to another. Attention-deficit/hyperactivity disorder (ADHD) is, for instance, a subject of controversy. The rate of people diagnosed with ADHD is higher in North America than in Europe (Conrad and Bergey, 2014), particularly in southern countries, although it is not clear whether this difference is related to the populations themselves or to the way in which the disorder is diagnosed. Cultural differences, such as perceptions of what constitutes normality, might explain why a hyperactive child is more generally considered normal in southern Europe than in the United States. Europe is also more reticent with respect to medicalization than the United States, and the place of medication in society is another form of social construction in the health sector.

Moreover, illness, disability or any incident affecting health can be seen as being socially and geographically constructed. Behaviors, everyday mobility and relationships can be modified by an illness or a change affecting the body. Fournand (2009), for example, has shown a change in how pregnant women relate to their environment and their different uses of urban amenities. Such a change varies depending on the cultural, social, political and economic context. The place of pregnant women in public spaces is subject to variations. For instance, Fournand points out that as soon as the pregnancy is announced, sport, nocturnal parties, and evening dinners at the restaurant are abandoned, for various reasons (not only medical). The public sphere of pregnant women gradually diminishes, becoming almost non-existent just before birth. On the other hand, when breastfeeding in a public space, a young mother creates around her body, and that of her child, a bubble of intimacy that blurs the boundaries between body and space (Mahon Daly and Andrews, 2002). The body can thus be considered as having a geographical dimension.

Approaches through the interactions of actors, organizations and authorities

Foucault is one of the authors in constructivism theory whose works have had considerable impact, notably through the concepts of biopower and biopolitics that he developed (1974). In the construction of his field of knowledge, Foucault, consciously or unconsciously, resorts to a constructivist hypothesis when attempting to read between the lines of the prevailing discourse of 17th- and 18th-century Western society on the subject of insanity and whether it was considered a disease and/or a social reality. Moreover, Foucault considers that the biological processes affecting populations need to be ruled by a regulatory or assurance authority that he calls *biopolitical*. This authority, depending on the representations, norms, meanings and values given to the illness, would be responsible for taking measures diversely oriented by these influences (such as the choice to lock the mentally ill away). Schematically, Foucault considers that the disease can be relegated within institutions (internment) or confined within the bodies that must be *monitored and punished* (to paraphrase the title of one of his works) when they deviate too much from the norm. From then on, society's control over individuals is accomplished not only through consciousness or ideology, but also in the body and with the body. For capitalist society, it is the biopolitical that matters more than everything else – the biological, the somatic, the corporeal. The body is a biopolitical reality; medicine is a biopolitical strategy.

Geographers have examined in different ways types of connection, variously captured by the terms "biopower," "biopolitics" and "biosociality." At-risk behaviors, for example, are possibly reinforced by one being ousted or living in spaces of relegation. Craddock (1999), cited in Atkinson et al. (2015), explores the co-production of race, place and pathology in relation to smallpox epidemics in 19th-century San Francisco, illustrating how a political anatomy of the Chinese body read disease and depravity into its fundamental structure and simultaneously stigmatized both Chinatown and Chinese bodies. Thompson, Pierce and Barnett (2007), on the basis of a study on smoking in New Zealand, have pointed out that the tendency (in the prevention campaign conducted by authorities) to stigmatize smokers results in already stigmatized environments and on deprived or stigmatized groups to produce either resistance (toward a change of behavior) or a sense of helplessness on the part of disadvantaged smokers. This illustrates the complex interactions involved in biopolitics as it relates to health in particular contexts. A general prevention message may then have counterproductive effects in particular places where it could appear as a constraint or a moral injunction emanating from a distant (bio)power. Fleuret and Tusseau (2010) provide another example in their study on HIV prevention in Mali. Prevention discourses, when seen as a direct translation of a white-occidental model, contribute to reinforcing the belief in a significant part of the population that HIV/AIDS is a disease invented by the *whites* to sell their drugs. This appears to be particularly true in the most deprived townships of Bamako.

The notion of gender is also an illustration of a socially constructed reality, and Simone de Beauvoir's famous words, "[o]ne is not born, but rather becomes, a woman" (1949, p. 13), are very frequently quoted in support of the idea that inequality between men and women is culturally constructed, not biological. As it is men who produce ideology and who are dominant, women are considered as their otherness. This anti-essentialist posture has resulted in the binary categorization of humanity (we are born male or female) and of social divisions based on behavior, social roles, norms and representations, which as a whole are encapsulated by the word *gender*. Gender, homosexuality and transgender themes are relatively recent in health geography and can be linked to the contributions of constructivist thought.

It is interesting to discuss the case of homosexuality here. Homosexuality was long regarded as a mental illness or degeneration, degeneracy and perverse behavior. Naturalist arguments are still mobilized today by reactionary movements (which speak of unnatural sexuality). The definitive withdrawal of homosexuality from the American Psychiatric Association's *Diagnostic and Statistical Manual of Mental Disorders* dates from 1987, while its withdrawal from the World Health Organization's international classification of diseases is as recent as 1992. Examples of evolution in the social construction of diseases are numerous (i.e., mental illness was formerly seen as a sign of satanic possession or witchcraft; pregnant women were discouraged from driving for fear of harming the unborn child, etc.), but for the sake of conciseness, it is not possible to mention them all here.

A few words about the fuzzy repercussions of constructivist research on public action in the health sector

According to Loriol (2012), the social sciences often refer to the concept of *social constructions* without any real clarification or explanation of the notions that these constructions cover. This poses the problem of their operationalization in public action. For example, in the field of mental health in the workplace, excessive fatigue, stress and burnout are all issues whereby their definitions are not sufficiently precise or stable. Consequently, according to Loriol, much confrontation exists between the various actors in the field of psychosocial risks, each imposing their own understanding of a category, such as stress or fatigue, and each approach proposing a different definition, targets and methods. "The result of the actions and positions of actors who agree or disagree with each other is therefore variable: the putting on the agenda or scrapping of a public issue, the emergence of a notion, the media coverage or lack of recognition of an illness" (Vézinat, 2013). This raises the question of methods and of certain difficulties in clarifying matters. Methods and concepts, such as grounded theory and interpretive models, have started to provide some answers. Qualitative methods have gradually begun to receive acceptance alongside quantitative methods; the latter nevertheless is still considered more objective.

Ultimately, does adopting a constructivist posture really result in calling the dominant biomedical model into question? Do we really arrive at oriented public action? The key issue raised by medical/health geographers, then, is the field's orientation to what might be meant by *the medical* (Atkinson et al., 2015). Despite claiming to challenge the dominance of *the medical* in medical humanities, does constructivism really contest the authority associated with *the medical*? For Atkinson et al., those claiming to speak from a *medical* position continue to claim an authority that remains unchallenged.

An example of this is the domination of the curative over the preventive in the health field; the former stemming from biomedical intervention on the human body, the latter from a set of actions that are every bit as concerned with the social body and environmental social determinants as with the biology of the body itself. The second is therefore much more complex to elaborate, to operationalize and to assess. For these reasons, in a context of tightened agendas, policy-makers (as well as researchers pushed by the need to publish) will tend to move toward the first.

Conclusion

Introducing relativism in social-sciences approaches of health and medical issues, constructivism can be seen as one of the pillars of critical health geography. To paraphrase Straus (1957), constructivism can be seen as a highlighting of differences and a form of competition between *in* and *of*. There is a geography in medicine, involved in medical questions in search of solutions (e.g., epidemiological), and a geography (and sociology) of medicine dealing with health, medical, healing and care issues as shedding light on the functioning of societies or on the organization of regions. A challenge for the future is to operate the junction between a geography seen as objective (evidence-based) and a complementary one seen as subjective, mixing their methodologies to enrich each other.

References

Atkinson, S., Evans, B., Woods, A. and Kearns, R. (2015). *The Medical* and *Health* in a critical medical humanities. *Journal of Medical Humanities*, 36, pp. 71–81.

Berger, P. and Luckmann, T. (1967). *The social construction of reality: a treatise in the sociology of knowledge.* Great Britain: The Penguin Press.

Bourdieu, P. (1987). Espace social et pouvoir symbolique. In: *Choses dites.* Paris: Editions de Minuit.

Burr, V. (2003). *An introduction to social constructionism.* London and New York: Routledge.

Conrad, P. and Barker, K. (2010). The social construction of illness: key insights and policy implications. *Journal of Health and Social Behavior*, 51(S), pp. S67–S79.

Conrad, P. and Bergey, M. R. (2014). The impending globalization of ADHD: notes on the expansion and growth of a medicalized disorder, *Social Sciences and Medicine*, 122, pp. 31–43.

De Beauvoir, S. (1949). (folio 1986). *Le deuxième sexe.* Collection Folio essais no 37. Paris: Gallimard.

Derrida, J. (1992). Force of law. In: D. Cornell, M. Rosenfeld and D. Gray Carlson, eds. (translated by M. Quaintance), *Deconstruction and the possibility of justice.* New York: Routledge, pp. 3–67.

Fleuret, S. and Tusseau, S. (2010). La prévention du SIDA à Bamako, une mise en contexte. *Villes en Parallèles*, numéro 44–45, "Territoire et santé", pp. 49–74.

Foucault, M. (1974). La naissance de la médecine sociale. *Dits et écrits*, T. III. Paris: Gallimard, p. 210.

Fournand, A. (2009). La femme enceinte, la jeune mère et son bébé dans l'espace public. *Géographie et cultures*, 70, pp. 79–98.

Kearns, R. and Moon, G. (2002). From medical to health geography: novelty, place and theory after a decade of change. *Progress in Human Geography*, 26, pp. 605–625.

Krasnoff, L. (1999). How Kantian is constructivism? *Kant Studies*, 90(4), pp. 385–409.

Loriol, M. (2012). *La Construction du social. Souffrance, travail et catégorisation des usagers dans l'action publique*. Rennes: Presses Universitaires de Rennes.

Mahon-Daly, P. and Andrews, G. (2002). Liminality and breastfeeding: women negotiating space and two bodies. *Health & Place*, 8, pp. 61–76.

Mishler, E. G. (1982). *Social contexts of health, illness, and patient care*. CUP Archive.

Orain, O. (2014). *Constructivism*. [online]. Hypergeo. Available at: www.hypergeo.eu/spip.php?article407 [Accessed 1 June 2017].

Parr, H. (2004). Medical geography: critical medical and health geography? *Progress in Human Geography*, 28(2), pp. 246–257.

Straus, R. (1957). The nature and status of medical sociology. *American Sociological Review*, 22(2), p. 200.

Vézinat, N. (2013). *La construction du social*. [online]. Sociologie. Available at: http://sociologie.revues.org/1576 [Accessed 25 Oct. 2016].

16

HEALTH GEOGRAPHY'S ROLE IN UNDERSTANDING SOCIAL CAPITAL AND ITS INFLUENCE ON HEALTH

Amber L. Pearson and Richard C. Sadler

In general, social capital involves the resources obtained through social relationships and nonmonetary power and influence, with the central idea being that membership and participation in groups may have positive consequences for the individual and/or the community. The positive consequence of particular interest here is good health. Over time, definitions and usage of "social capital" have become convoluted and overlapping, as we will discuss in this chapter. The evolution of the concept highlights important dimensions of social capital as they relate to health, including pathways and measurement challenges. In this chapter, we focus on contemporary notions of social capital, tracing its origins in sociology to applications and debates in the health sciences, with particular attention to contributions in health geography. We then provide a review of the pathways to health, measurement challenges, issues of scale and the positive and negative consequences of social capital in recent research. In closing, we present emerging areas of social capital research among health geographers.

The contemporary origins of social capital in sociology

Historical origins of the idea of social capital were described in political science and economics texts in the late 19th century, but the contemporary notion of social capital (as it is widely used in health research) was not taken up by sociologists until the 1970s and 1980s. Seeded through earlier work by Durkheim and Marx, the origins of contemporary notions of social capital emerged from work by Bourdieu, Loury, Coleman, Portes and also the political scientist Putnam (Bourdieu, 1980; Coleman, 1988; Loury, 1977; Portes, 1998; Putnam, 2007). Marx developed theories of emergent solidarity due to a common fate, or bounded solidarity, while Durkheim developed the theory of social integration (Durkheim, 1893; Marx, 1894), both paving the way for social capital.

Defining "social capital" has been neither straightforward nor consistent. At times, social capital has been seen as a mechanism to obtain a resource by one individual or group from a more privileged other. Social capital thus emerged as a form of capital in which social networks are central, transactions are marked by reciprocity, trust and cooperation, and market agents produce goods and services not mainly for themselves, but for a common good. Others refer to social capital as the resource itself, one that "actors derive from specific social structures" (Baker, 1990, p. 619). Still others define "social capital" as the social network through which individuals receive opportunities to access or use other forms of capital, including financial (Schiff, 1992). Putnam defines it as both the social networks themselves (Putnam, 2007) and the norms and trust in those

107

networks that facilitate action and cooperation for mutual benefit (Putnam, 1993). Putnam's popular definition has formed the basis for much of the contemporary work in the social sciences, including geography.

This line of work in sociology outlines three important components related to social capital that perhaps ought to be, but in practice are rarely, clearly distinguished: (1) the possessors of social capital, or those making claims; (2) the sources of social capital, or those agreeing to those demands (also called donors); and (3) the resources themselves, or the objects, services or other goods that are being demanded (Portes, 1998). The resources obtained through social capital may be, for example, information about employment opportunities, mutual aid and reciprocity, or collective action for change in a community. Inherent in some of these distinctions is the nascent idea that those making claims do not have social capital unless the demands are met by another person or group. Such transactions are also seen to be less transparent than economic exchanges with unspecified obligations or time horizons, possible violation of reciprocity and high reliance on trust (Portes, 1998). However, the concept of social capital was taken up with vigor by many health scientists (and policy groups) in the late 1990s and early 2000s, sparking debate about its utility and the possible public-health implications of its uptake.

Pathways from social capital to health

The parallel movements in both epidemiology and health geography to examine social determinants of health emerged in the 1980s and 1990s, partly in response to growing neoliberal policies exacerbating socioeconomic inequalities in much of the Global North. Social determinants of health, or social conditions, became a catch-all for the ways in which the (unequal) social conditions of our lives (e.g., income inequality) influence our health. For example, socioeconomic status, the support of friends and family, neighborhood crime and fear, racism and discrimination all may influence health. Social capital may then be a response to these social processes, whereby marginalized groups adhere in order to bring about change.

The research published in this era focused more broadly on social determinants of health rather than social capital per se. Many overlapping or subsumed concepts were explored during these two decades, stemming from social support to social cohesion (Carrasco and Bilal, 2016), control over destiny and feelings of belonging or sense of community (Marmot, 2004). For example, in research on income inequality and health, the central debate hinged on whether relative income inequality exerts more of an influence over health than does absolute income (Lynch et al., 2000; Pearce and Davey Smith, 2003; Wilkinson, 2005), with each having very different policy implications. The connection to social capital here is that some social processes play a role in the way people experience poverty and its effects. To explore the pathways through which social conditions affect health, two primary theories emerged: the psychosocial and the neomaterial perspectives.

The psychosocial perspective suggests that relative deprivation and social comparison between groups have negative health consequences, including stress, neuroendocrine problems and smoking, which is considered a direct effect on health (Veenstra et al., 2005). The neomaterial perspective posits that social inequalities manifest in material differences via the opportunities and amenities available to us to protect our health. This pathway from social capital to health is considered indirect (Veenstra et al., 2005), whereby community social capital may be leveraged to ensure access to green spaces (i.e., parks), which, in turn, leads to increased physical activity and lower body mass index (BMI) of residents. Ostensibly, each of these pathways has close connections to the origins of social capital, because the development and maintenance of public amenities and equity in material investment rely on the creation and nurturing of close group ties. We could therefore ask: does social capital involve getting a *thing*, or does it more accurately entail feelings of belonging to a group? Put another way, does social capital involve *having* something or *being* cohesive (Carrasco and Bilal, 2016), or perhaps both?

If we take the psychosocial perspective, social capital is a property of a group or community, which protects its health. If we take the neomaterial perspective, however, social capital is a feature that allows the community to make demands (which may, in turn, affect health). Therein lie two dilemmas. First, it is unclear whether a community possesses social capital if demands are not met; second, the motivations of donors may be pivotal in whether the group making demands *gets* something. These two tensions in the notion of social capital are exemplified in the recent Flint Water Crisis, which took place in Michigan, in the United States.

The Flint Water Crisis (FWC)

The FWC arose as a consequence of short-sighted governance decisions made by the state of Michigan in the city of Flint. Having changed the public's water source to save money, naturally more corrosive river water began leaching lead off of pipes, sending it into the drinking-water system and causing a range of public-health problems, including elevated blood lead and water-quality issues related to bacteria. With the state government turning a blind eye to Flint residents' concerns, residents relied on their cohesive community networks for collective action to impel independent researchers to explore evidence in support of their concerns. One could take the perspective that water-quality problems persisted because residents lacked the requisite social capital required for making demands. This perspective, however, downplays the role that donors play in ignoring even loud claims made by citizens (motivations for which can include shame and fear of litigation, taking responsibility and showing negligence). Such motivations are visible to varying degrees, as several state-level officials have now been criminally charged for their responsibility in the FWC. A longer temporal view of Flint and the state of Michigan's history of race and class issues, however, suggests that many entrenched macro-level social, political and economic processes were at play in creating the FWC, particularly under neoliberal state governance (discussed more fully in Sadler and Highsmith, 2016), which may have lowered psychosocial forms of social capital among the marginalized. The same citizens who brought the FWC to light prior to national attention had been ignored by the state government for many years. Neomaterial forms of social capital were weakened through the enactment of an anti-democratic emergency manager law that stripped cities of their elected officials *and* prevented citizens statewide from voting it out via referendum.

With Flint effectively functioning as an arm of the state of Michigan – and given what we know of social capital being constrained in autocratic societies (Uslaner, 1999) – it would not have been surprising if social capital (in the sense of community cohesiveness) diminished. Yet, Flint residents – in the face of having lost their democratic voice – persisted in their efforts until researchers and physicians lent expertise and amplified the voices of residents. In this way, the cohesiveness and the willingness to persist in collective action can be seen as social capital, despite state (donor) inaction. The FWC therefore epitomizes the *being* rather than the *having* dimension of social capital (Carrasco and Bilal, 2016). As seen here and elsewhere, it is unclear whether separating psychosocial from material processes is feasible or beneficial for reducing very real inequalities in people's lives: "social capital has come to be privileged over material inequalities (between people and places) in a way which may be . . . disabling" (Mohan and Mohan, 2002, p. 205).

Measurement challenges

Determining even a singular indicator or proxy for social capital is fraught with faults because it encapsulates such a broad range of conditions and outcomes. Some researchers have operationalized social capital in terms of *structural* or *cognitive* proxies. Structural measures include the quantity or quality of social links or participation. Cognitive measures include perceptions of support, reciprocity and trust. Structural proxies for social capital include participation in voting (Lindstrom, Merlo and Ostergren, 2003), volunteerism (Blakely et al., 2006) and blood donation (MacIntyre and Ellaway, 2003), to name only a few. Cognitive social-capital proxies with evidence of a relationship to health include fear of crime (Foster et al., 2016) and generalized trust (Kim, Subramanian and Kawachi, 2008). Trust may be distinct from social capital in that it does not indicate actual social networks (Carpiano and Fitterer, 2014). But the degree to which these proxies indicate membership in a network with shared values – and not simply a network of politically engaged individuals with the interest, time and/or resources to volunteer, vote or donate blood – is unknown. The notion that participation in voluntary activities indicates social capital is perhaps a very American one: government investment in social programs is very low, so volunteerism is considered an important and perhaps essential way of getting resources to high-needs group members. This approach has been criticized for placing the decline of social capital on the leisure-time activities of individuals, rather than the lack of provision of government services or the macro-economic conditions that led to the need for government-assistance programs in the first place. In practice, these measures of social capital are likely used primarily due to data availability (Cummins et al., 2005). Equally, the measurement of health outcomes, theoretically influenced by social capital, has varied widely.

Because of its connection to perceived social conditions, social capital is protective against a range of mental-health outcomes, including low self-rated health, depression and suicide (e.g., Aida et al., 2011; Giordano and Lindstrom, 2010). One review discovered consistent, positive associations between better social capital and a range of improved health conditions including lower rates of alcohol-related mortality, psychosis, coronary heart disease and smoking (Murayama, Fujiwara and Kawachi, 2012). Most such studies were conducted in wealthy cities and countries and evaluated the positive aspects of social capital, yielding (at times weak or null) associations with improved health (e.g., Blakely et al., 2006). Much of this work concludes that social comparisons likely operate on smaller scales, whereby interactions between neighbors, across neighborhoods, or within a country are associated with negative mental-health outcomes, rather than cross-country comparisons (Pearce and Davey Smith, 2003). Due to the culturally specific nature of norms, values and beliefs, and – particularly relevant to geographers – context, it may be unrealistic or inappropriate to attempt to determine a universal indicator of social capital.

Social capital in health geography

Geographers have used social capital as a possible explanation for geographic inequalities in health, with conflicting findings. This inconsistent evidence likely relates to multiple causes including study design, scale of measurement and analytical issues. Still, the inherent aggregate scale of inquiry, the relevance of neighborhoods as important groupings and the attention to context in health-geographic thinking fit neatly with some of the tenets of social capital.

Social capital (and other social determinants of health) has been conceptualized as involving three scales: the individual, the meso or community, and the macro scales (Whitehead et al., 2016). This means that health-related benefits may be obtained or demanded by individuals (micro) who are part of a group, community or neighborhood (meso), but the outcome and the ability to exert this power depends on the macro social, political and economic contexts within which it takes place (Pearce and Davey Smith, 2003; Wakefield and Poland, 2005). Similarly, the extent to which social capital influences health may vary between places. In other words, there may be an interaction between context and social capital (Veenstra et al., 2005). Processes of social capital are not isolated experiences of individuals or neighborhoods but rather nested in larger spatial contexts. Below, we outline health geography's primary contributions to understanding social capital – the meso and macro scales.

At the meso scale, sociologists assert that the groups form in order to obtain benefits or resources. The group may be Native Americans calling for collective action in opposition of an oil pipeline "to protect land and water from environmental degradation and uphold Indigenous responsibilities to all of Creation" (Wilkins and Hkiiwetinepinesiik Stark, 2018, p. 237), joined by environmentalists and others who share similar beliefs. In this example, the norms, values, attitudes and beliefs are the shared identity that forms the group. Conversely, geographers conceptualize or measure social capital in geographical grouping, such as census units or neighborhoods. The geographical space forms the group from which these norms, attitudes, values, culture and beliefs emerge. It is in these spaces, rather than among purely voluntary networks or through groups affiliated with cultural identity, where social capital exists, diminishes or returns. The neighborhood or the community is the meso scale, where collective action transpires, where people can lobby for their own health (Wakefield and Poland, 2005) and where interactions take place between neighborhood residents. For example, Bisung and Elliott (2014) developed a framework for the role of social capital to facilitate collective action toward improving access to water and sanitation in communities in rural Kenya. Ivory et al. (2011) evaluated the lack of trust, or social fragmentation, in New Zealand neighborhoods, to understand its negative mental-health impacts. Fagg et al. (2008) explicitly investigated multiple scales at which social capital may operate, finding that (1) increased risk of psychological distress was associated with both neighborhood social fragmentation and lower individual-level social support through immediate social networks and (2) there was no evidence that a person's immediate social network was protective in highly fragmented neighborhoods.

At the macro scale, political, economic and social structures can serve to either promote or inhibit the creation of social capital. In some cases, macroeconomics and geopolitics may serve as a cause of, for example, the underdevelopment of poor countries (Mohan and Mohan, 2002) or uneven development within countries. In this circumstance, the macro context is the structural cause of decline, and the indicators of civic disorder (e.g., lack of active voters) are a symptom. In Flint's case, the macro-scale economic (neoliberalism) and political (state of Michigan) drivers inhibited social capital by ignoring community collective action and perpetuating a system that marginalized community voices. The argument that social capital alone could be used to rebuild communities is weak, because the loss of community cohesiveness itself was a symptom of decline, not the cause of it.

Negative aspects of social capital

Most early work on social capital focused on its benefits, possibly a partial response to or subversion of the neoliberal policies and rhetoric simultaneously being espoused throughout much of the world. But such sentiments can also be taken up by government and ultimately become an excuse for inaction (Mohan and Mohan, 2002). As a hypothetical example, the government may promote *communities of participation* rather than provide the resources themselves that communities are demanding, even though such communities of participation do not guarantee access to resources in the way that direct government provision can. Whatever the reason, the positive and negative consequences of social capital must be examined in tandem (Wakefield and Poland, 2005). We note several negative dimensions here.

First, within sociology, the notion of embodied cultural capital notes that individuals or groups may gain access to resources through contact with experts or prestigious individuals or groups. In health geography and epidemiology, this is referred to as "bridging social capital" (Szreter and Woolcock, 2004). If social capital involves access to opportunities or resources through social connections, those without connections may fall behind, creating or exacerbating inequalities between the connected and the marginalized. Second, the formation of groups based on shared identity, norms, culture, and beliefs implies that others are on the outside. This may lead to exclusion or restriction of outsiders from accessing resources and services, or it may lead to racism and discrimination of those who are not within the social group; both of these outcomes have been identified as harmful to health (Chatty, Mansour and Yassin, 2013; Harris et al., 2006). Similarly, the exclusion of marginalized groups engaged in unhealthy behaviors may unintentionally strengthen and reinforce the behaviors. For example, increasing stigmatization of (particularly poor, minority) smokers has been shown to actually reinforce rather than discourage smoking (Thompson, Pearce and Barnett, 2007).

Second, the norms, values, and beliefs that form the group identity may be either (1) hurtful, hateful or misogynistic or (2) downward-leveling norms with various health effects on members of the group or those outside the group. In hurtful or misogynistic groups, the negative effects may primarily be internal.

For example, members may share norms that women are of lower status and therefore women in this group may be discriminated against or not have the same life opportunities as men. In hateful groups, the negative health effects may primarily be external, whereby members may share the belief that they are better or more righteous than outsiders. A poignant example of this form of social capital can be found in violent Muslim extremist groups, such as the Islamic State of Iraq and Syria (ISIS). Many immigrants have been drawn to the group due to lack of belonging in their home country or their heritage and through marginalization and discrimination (Lyons-Padilla et al., 2015). Groups such as ISIS, therefore, provide an alternative (often virtual) community for connection and sense of purpose but use the group to inflict violence on outside groups. Downward-leveling norms begin within a group that was historically blocked from social mobility by another group, which leads to a norm of nonconformity, which ironically leads to attachment to aspects of a lack of social mobility (Portes, 1998). A variation on this is the tall poppy syndrome, whereby lower-status groups bully more successful individuals and groups as a way to *normalize themselves*, with negative mental-health outcomes reported among high-performance athlete adolescents, particularly females (O'Neill, Calder and Allen, 2014). As mentioned previously, social comparison with other (more successful) groups can also be harmful for the lower-status groups. At the neighborhood level, this may operate through direct contact with more advantaged neighbors (racism and discrimination), community fragmentation and lowered trust, or the direct stress of living in a disadvantaged area (Pearson et al., 2013).

Last, we must grapple with the possible negative effects of neighborhood ethnic diversity. Putnam (2007) found that higher ethnic diversity in a community was associated with lower trust both between and within ethnic groups, and lower trust was associated with lower political participation, collective action and perceived quality of life. This finding appears to be supported by evidence that ethnic density may be protective for mental health (Das-Munshi et al., 2010) but fundamentally at odds with the evidence that segregation leads to poorer health for lower-socioeconomic-status or minority groups (Acevedo-Garcia and Lochner, 2003) and to disinvestment in the infrastructure in segregated, minority neighborhoods (Sadler and Highsmith, 2016). Particularly at this moment in geopolitical history, the grave health consequences of social comparison, group formation, exclusion and discrimination cannot be ignored.

Emerging areas of inquiry

A number of current thrusts in research may add to our understanding of the relationship between social capital and health, while challenging geographic thinking. Here, we briefly outline two. First, we may in part demystify the psychosocial embodiment of inequality (Pearce and Davey Smith, 2003) by conducting interdisciplinary research on interactions between genes and the environment. For instance, social conditions and physical environments (i.e., chemical exposures) can affect gene expression and subsequent health outcomes, although with varying probability. This challenges the human-environment dichotomy, as bodies become somehow permeable to environments (Guthman and Mansfield, 2012). Because changes in genetic expression can be inherited by future generations, conventional health-geography approaches – whereby exposure occurs in spatial and temporal proximity to an outcome – are insufficient. Racism and discrimination may, in fact, alter gene expression. As such, we must not divert attention from the root causes of social inequality experienced now. Our efforts today could influence the health of future generations.

Second, we may interrogate both positive and negative dimensions of social capital in virtual communities. This may be increasingly important, as virtual social comparison is on the rise, and the effects of it are already under investigation. Some research has shown that negative self-perceived competence and unattractiveness was associated with negative social comparison on Facebook, particularly among those reporting low levels of happiness (de Vries and Kühne, 2015). As social networks take shape in aspatial ways, such changes challenge geographic thinking. These networks established through social media force us to grapple with possibly competing processes of social capital development and dilution. Establishing social networks online involves very little effort or investment, and maintaining the relationships is primarily passive. In contrast, some claim that social capital is generated through face-to-face interactions rather than passive (or virtual) participation. Although this claim was originally made in relation to financial donations to charities (checkbook participation) (Mohan et al., 2005), it can now be extended to online social participation. On the other hand, social media may create a platform for isolated individuals with little locational access to others who share their values or norms.

Conclusion

We have learned about conflicting conceptualizations of social capital, varying indicators and measures, underdeveloped biological pathways to health and the complexities of harnessing social capital to reduce inequalities. Although we now understand a great deal more about social capital, these problems suggest that our work as social scientists is not finished. Health geography plays an important role in continued theorizing about how social capital is created and maintained at multiple scales, particularly within neighborhoods and in the macro context.

Generally, understanding how social conditions shape our lives and the complex ways in which humans interact continues to be alluring, because it helps us uncover the implications of development patterns and economic systems that unduly disadvantage some residents by exposing them to more challenging situations with respect to the creation of social capital. By the same token, social capital may empower community groups to respond to oppression. Particularly as the possibilities of geographic information science continue to expand, the role of health geography in the study of social capital will remain an important perspective.

References

Acevedo-Garcia, D. and Lochner, K. A. (2003). Residential segregation and health. In: I. Kawachi and L. F. Berkman, eds., *Neighborhoods and health*. New York: Oxford University Press.

Aida, J., Kondo, K., Hirai, H., Subramanian, S. V., Murata, C., Kondo, N., Ichida, Y., Shirai, K. and Osaka, K. (2011). Assessing the association between all-cause mortality and multiple aspects of individual social capital among the older Japanese. *BMC Public Health*, 11, pp. 499–499.

Baker, W. E. (1990). Market networks and corporate behavior. *American Journal of Sociology*, 96, pp. 589–625.

Bisung, E. and Elliott, S. (2014). Toward a social capital based framework for understanding the water-health nexus. *Social Science & Medicine*, 108, pp. 194–200.

Blakely, T., Atkinson, J., Ivory, V., Collings, S., Wilton, J. and Howden-Chapman, P. (2006). No association of neighbourhood volunteerism with mortality in New Zealand: a national multilevel cohort study. *International Journal of Epidemiology*, 35, pp. 981–989.

Bourdieu, P. (1980). Le capital social: notes provisoires. *Actes De La Recherche En Sciences Sociale*, 31, pp. 2–3.

Carpiano, R. M. and Fitterer, L. M. (2014). Questions of trust in health research on social capital: what aspects of personal network social capital do they measure? *Social Science & Medicine*, 116, pp. 225–234.

Carrasco, M. A. and Bilal, U. (2016). A sign of the times: to have or to be? Social capital or social cohesion? *Social Science & Medicine*, 159, pp. 127–131.

Chatty, D., Mansour, N. and Yassin, N. (2013). Bedouin in Lebanon: social discrimination, political exclusion, and compromised health care. *Social Science & Medicine*, 82, pp. 43–50.

Coleman, J. S. (1988). Social capital in the creation of human capital. *American Journal of Sociology*, 94, pp. S95–121.

Cummins, S., Macintyre, S., Davidson, S. and Ellaway, A. (2005). Measuring neighbourhood social and material context: generation and interpretation of ecological data from routine and non-routine sources. *Health & Place*, 11, pp. 249–260.

Das-Munshi, J., Becares, L., Dewey, M. E., Stansfeld, S. A. and Prince, M. J. (2010). Understanding the effect of ethnic density on mental health: multi-level investigation of survey data from England. *British Medical Journal*, 341, c5367.

De Vries, D. A. and Kühne, R. (2015). Facebook and self-perception: individual susceptibility to negative social comparison on Facebook. *Personality and Individual Differences*, 86, pp. 217–221.

Durkheim, E. (1893). *The division of labor in society*. New York: Free Press.

Fagg, J., Curtis, S., Stansfeld, S. A., Cattell, V., Tupuola, A. M. and Arephin, M. (2008). Area social fragmentation, social support for individuals and psychosocial health in young adults: evidence from a national survey in England. *Social Science & Medicine*, 66, pp. 242–254.

Foster, S., Hooper, P., Knuiman, M. and Giles-Corti, B. (2016). Does heightened fear of crime lead to poorer mental health in new suburbs, or vice versa? *Social Science & Medicine*, 168, pp. 30–34.

Giordano, G. N. and Lindstrom, M. (2010). The impact of changes in different aspects of social capital and material conditions on self-rated health over time: a longitudinal cohort study. *Social Science & Medicine*, 70, pp. 700–710.

Guthman, J. and Mansfield, B. (2012). The implications of environmental epigenetics: a new direction for geographic inquiry on health, space, and nature-society relations. *Progress in Human Geography*, 37, pp. 486–504.

Harris, R., Tobias, M., Jeffreys, M., Waldegrave, K., Karlsen, S. and Nazroo, J. (2006). Racism and health: the relationship between experience of racial discrimination and health in New Zealand. *Social Science & Medicine*, 63, pp. 1428–1441.

Ivory, V., Collings, S. C., Blakely, T. and Dew, K. (2011). When does neighbourhood matter? Multilevel relationships between neighbourhood social fragmentation and mental health. *Social Science & Medicine*, 72, pp. 1993–2002.

Kim, D., Subramanian, S. and Kawachi, I. (2008). Social capital and physical health: a systematic review of the literature. In: I. Kawachi, S.V. Subramanian and D. Kim, eds., *Social capital and health*. New York: Springer, pp. 139–190.

Lindstrom, M., Merlo, J. and Ostergren, P. O. (2003). Social capital and sense of insecurity in the neighbourhood: a population-based multilevel analysis in Malmö, Sweden. *Social Science & Medicine*, 56, pp. 1111–1120.

Loury, G. C. (1977). A dynamic theory of racial income differences. In: P. A. Wallace and A. M. La Mond, eds., *Women minorities, and employment discrimination*. Lexington, MA: Heath, pp. 153–188.

Lynch, J., Due, P., Muntaner, C. and Davey Smith, G. (2000). Social capital – is it a good investment strategy for public health? *Journal of Epidemiology & Community Health*, 54, pp. 404–408.

Lyons-Padilla, S., Gelfand, M. J., Mirahmadi, H., Farooq, M. and Van Egmond, M. (2015). Belonging nowhere: marginalization & radicalization risk among Muslim immigrants. *Behavioral Science & Policy*, 1, pp. 1–12.

Macintyre, S. and Ellaway, A. (2003). Neighborhoods and health: an overview. In: I. Kawachi and L. F. Berkman, eds., *Neighborhoods and health*. New York: Oxford University Press, pp. 20–42.

Marmot, M. (2004). *The status syndrome: how social standing affects our health and longevity*. New York: Henry Holt and Company.

Marx, K. (1894). *Capital*. New York: International.

Mohan, G. and Mohan, J. (2002). Placing social capital. *Progress in human geography*, 26, pp. 191–210.

Mohan, J., Twigg, L., Barnard, S. and Jones, K. (2005). Social capital, geography and health: a small-area analysis for England. *Social Science & Medicine*, 60, pp. 1267–1283.

Murayama, H., Fujiwara, Y. and Kawachi, I. (2012). Social capital and health: a review of prospective multilevel studies. *Journal of Epidemiology*, 22, pp. 179–187.

O'Neill, M., Calder, A. and Allen, B. (2014). Tall poppies: bullying behaviors faced by Australian high-performance school-age athletes. *Journal of School Violence*, 13, pp. 210–227.

Pearce, N. and Davey Smith, G. (2003). Is social capital the key to inequalities in health? *American Journal of Public Health*, 93, pp. 122–129.

Pearson, A. L., Griffin, E., Davies, A. and Kingham, S. (2013). An ecological study of the relationship between socio-economic isolation and mental health in the most deprived areas in Auckland, New Zealand. *Health & Place*, 19, pp. 159–166.

Portes, A. (1998). Social capital: its origins and applications in modern sociology. *Annual Review of Sociology*, 24, pp. 1–24.

Putnam, R. (2007). E Pluribus Unum: diversity and community in the twenty-first century. *Scandinavian Political Studies*, 30, pp. 137–174.

Putnam, R. D. (1993). The prosperous community: social capital and public life. *The American Prospect*, 13, pp. 35–42.

Sadler, R. C. and Highsmith, A. R. (2016). Rethinking Tiebout: the contribution of political fragmentation and racial/economic segregation to the Flint Water Crisis. *Environmental Justice*, 9, pp. 143–151.

Schiff, M. (1992). Social capital, labor mobility, and welfare. *Rationality and Society*, 4, pp. 157–175.

Szreter, S. and Woolcock, M. (2004). Health by association? Social capital, social theory, and the political economy of public health. *International Journal of Epidemiology*, 33, pp. 650–667.

Thompson, L., Pearce, J. and Barnett, R. (2007). Moralising geographies: stigma, smoking islands and responsible subjects. *Area*, 39, pp. 508–517.

Uslaner, E. M. (1999). Democracy and social capital. In: M. E. Warren, ed., *Democracy and trust*. Cambridge: Cambridge University Press, pp. 121–150.

Veenstra, G., Luginaah, I., Wakefield, S., Birch, S., Eyles, J. and Elliott, S. (2005). Who you know, where you live: social capital, neighbourhood and health. *Social Science & Medicine*, 60, pp. 2799–2818.

Wakefield, S. E. and Poland, B. (2005). Family, friend or foe? Critical reflections on the relevance and role of social capital in health promotion and community development. *Social Science & Medicine*, 60, pp. 2819–2832.

Whitehead, M., Pennington, A., Orton, L., Nayak, S., Petticrew, M., Sowden, A. and White, M. (2016). How could differences in "control over destiny" lead to socio-economic inequalities in health? A synthesis of theories and pathways in the living environment. *Health & Place*, 39, pp. 51–61.

Wilkins, D. E. and Hkiiwetinepinesiik Stark, H. (2018). *American Indian politics and the American political system*. Lanham, MD: Rowman & Littlefield.

Wilkinson, R. (2005). *The impact of inequality: how to make sick societies healthier*. London: The New Press.

17

THERAPEUTIC LANDSCAPES

From exceptional sites of healing to everyday assemblages of well-being

Jessica M. Finlay

The concept of therapeutic landscapes represents a valuable application to assess the relationship between health and place in a wide variety of spaces. Gesler (1993) first introduced the concept to geographers as places with "an enduring reputation for achieving physical, mental, and spiritual healing" (p. 171). By incorporating theory from cultural and human geography (e.g., sense of place, symbolic landscapes), Gesler argued that scholars could examine sites of healing in health geography. He built upon humanistic geography's attention to individuals' and groups' attitudes and feelings toward the environments they inhabit, suggesting that intimate personal and emotional relationships between self and place could impact one's health and well-being. Initial applications of therapeutic landscapes tended to focus on extraordinary destinations, such as pilgrimage sites (Gesler, 1996); parks and wildernesses (Palka, 1999); and hot springs, baths and spas (Gesler, 1998). Subsequent scholarship critiqued that, while important, these exceptional spaces are generally encountered over short time periods and are not intended to foster long-term healing (English, Wilson and Keller-Olaman, 2008). Williams' (1999) volume expanded the concept from extraordinary locations to everyday places associated with the maintenance of health and wellness. Indeed, scholars broadened exploration to nontraditional and mundane pursuits of well-being, such as home settings (Williams, 2002), children's summer camps (Kearns and Collins, 2000), collective gardening programs (Milligan, Gatrell and Bingley, 2004) and yoga studios (Hoyez, 2007). In the last 15 years, the concept has evolved considerably to critically study complex relationships between health and place.

Sites of study

Scholars tend to focus on three main types of therapeutic landscapes: physical environments, symbolic environments and social environments.

Physical environments

This area of scholarship constitutes classic therapeutic environments that many people associate with healing and health. These sites literally promote well-being through natural landscape (e.g., scenic beauty, hot springs) and constructed design (e.g., spas, hospitals). Gesler (1993) first investigated the healing reputation of the Asclepian sanctuary at Epidauros, Greece. The site was famous throughout the ancient world for its dream healings; it was known for its appealing temple buildings, gentle waters and beautiful landscape. The climate, water quality and scenic environment produced a strong sense of place and reputation for healing efficacy. It provided visitors with a sense of refuge and security. Stemming from Gesler's original attention to sites of healing, a robust area of focus for therapeutic landscapes is modern institutional settings, such as health centers, hospitals and clinics. Scholars investigate the role of location, architecture, design and organization (e.g., room comfort, available technology) in local healing processes (Smyth, 2005). Curtis and colleagues (2007), for example, assessed the hospital design of a mental-health inpatient unit. Staff and service users described which aspects of the hospital – including cleanliness, the location of nursing stations, food quality, air conditioning, windows and natural light and gardens – were beneficial or detrimental to patient well-being. The role of nature constitutes a second primary focus area for therapeutic landscapes, which stems from Western ideology that people can attain physical, mental and social healing by spending time outdoors in natural spaces. Finlay and colleagues (2015), for example, investigated how green and blue spaces affect older adult health and well-being. They found that landscape elements including parks, gardens, street greenery, lakes and the ocean influenced participants' physical, mental and social health. Therapeutic effects resulted from direct physical engagement (i.e., being physically present in the landscape), as well as mental engagement (i.e., through retreat and restoration). Natural and built physical environments are thus relatively easy to grasp, because they are tangible, comprising what we can see, hear, smell, taste and feel.

Symbolic environments

When connecting physical environments to healing and well-being, we must consider how they are perceived and what they might mean. A courtyard fountain on hospital grounds, for example, may symbolize tranquility and rejuvenation and have implications for healing in addition to treatment received inside hospital walls. Many objects around us have meaning because they symbolize something important (Gesler, 2003). A doctor's white coat, for example, traditionally signifies medical knowledge and authority. High-tech equipment in modern hospitals symbolizes scientific advancement and the power of biomedicine. Particular localities express cultural values, social behavior and individual actions embedded in that place. Western hospitals and clinics are generally understood as powerful sites of healing, treatment and biomedical care. Symbolic environments thus arise from individual meanings and collective interpretations of place. This type of therapeutic landscape often provides support, care and healing outside of, or in parallel to, traditional biomedicine. Examples are spas, yoga studios, tai chi locations and gardens. Buzinde and Yarnal (2012) applied the framework of therapeutic landscapes to understand medical-tourism sites as curative spaces combining modern and alternative forms of medicine with travel and leisure. Under this phenomenon, wealthy Westerners travel to nations such as Thailand, India and Mexico to receive health care and enjoy touristic activities. These sites link symbolic expectations of leisure travel (e.g., fun, relaxation, a reprieve from life's everyday strains) to medical interventions (e.g., surgery, organ transplants, plastic surgery, dental care). Physical and abstract symbols, myths, stories, spiritualism and language are also important to this area of work. Healing sessions often involve symbolic language or actions to transform an individual from sickness to health. For example: the language of patient-healer interactions, the room color in a hospital or a shaman's stories recounted during a therapeutic session all affect healing (Gesler, 2003). This work emphasizes how individuals attribute personal meanings, values and emotions to spaces to perceive them as therapeutic. These perceptions may differ widely.

Social environments

Healing is a social activity that involves interactions with many different people in varying roles (Gesler, 2003). Physicians and healers, for example, with the power to diagnose ills and prescribe treatments, play a pivotal role in our well-being. Interpersonal therapeutic environments generally include kinship groups and networks of care supplied by family, friends, health professionals, community members and other agents of support (Smyth, 2005). This body of literature therefore concentrates on therapeutic landscapes in relation to others: institutional, home and community-based spaces that facilitate interpersonal interaction. Everyday public spaces, and their implications for well-being, feature across this literature. Milligan and colleagues (2004), for example, investigated older people's experiences of communal gardens. They found that communal gardening created inclusionary spaces to combat social isolation and promote the development of social networks. By enhancing quality of life and emotional well-being through socialization, the communal-garden allotments constituted a therapeutic landscape. Social relationships are an essential component of many therapeutic landscapes.

Fuzzy boundaries

The three environments described above are convenient categorizations; in practice, therapeutic landscapes transcend boundaries. In the previous example (Milligan, Gatrell and Bingley, 2004), participants derived health benefits from gardening activities and physically engaging with the space. The gardens symbolized the relaxation, peace and tranquility of nature. They also facilitated social interaction and positive connections with neighbors and passing members of the local community. Extraordinary locations also combine multiple therapeutic experiences and types of environments, such as in Palka's (1999) analysis of visitors to Denali National Park. Tourist experiences included direct engagement with nature (the physical environment), as well as individual expectations of self-exploration, restoration and renewal (the symbolic environment). The next section will consider how the substance and relative importance of these three categories vary greatly by time, place and individual (Baer and Gesler, 2004). We will see that no therapeutic landscape is perfectly healthy.

Potentially therapeutic: a contested and relational approach

Spaces are not intrinsically therapeutic, but they can promote healing and well-being depending on how people view and interact with them. In other words, no space can guarantee a positive health experience for everyone. What may be therapeutic for one person may even be harmful to another. Baer and Gesler (2004) advocated for applying the therapeutic-landscape concept to difficult and contestable examples with potentially "less positive shades of meaning" (p. 406). They used the fictional example of J.D. Salinger's *The Catcher in the Rye* to call attention to ambivalent spaces that are not completely positive and may even be destructive. A real-life example of this is the beach, which many consider a therapeutic landscape for relaxation, reflection, socialization and exercise. However, others may find the beach unhealthy given sun-exposure risk (Collins and Kearns, 2007) or a past negative experience. Public parks can also be perceived as pleasant and inviting, or conversely threatening and dangerous, depending on the individual. Perception, circumstance and embodied identity (including gender, race/ethnicity, mobility level and self-confidence) filter people's experiences of potentially therapeutic landscapes (Finlay et al., 2015).

Therapeutic landscapes are thus disputed, rather than a site of mutual consensus. In an effort to recognize diverse and complex spatial relationships, Conradson (2005) observed that "individuals clearly experience even scenic environments in quite different ways, in terms ranging from enjoyment through to ambivalence and even anxiety" (p. 338). He conceptualized therapeutic landscape *experiences* in a manner that is attentive to the relational dimensions of encounters between self and landscape: a complex set of transactions between people and their broader socio-environmental settings. What may be therapeutic must be filtered through individual context and broader socioeconomic conditions.

Foley (2011, 2014) built on the notion of therapeutic landscapes as fluid relational outcomes to understand these spaces as negotiable, conditional and undetermined. Foley examined holy wells and baths as therapeutic assemblages encompassing material, metaphoric and inhabited dimensions. The *material* of Irish holy wells incorporated tangible bricks, water, visitors' physical bodies and documented and experienced cures (an embodied therapeutic experience). *Metaphors* encapsulated the wells' cultural reputations as long-standing sites of spiritual healing and beliefs. *Inhabitation* referred to rituals, prayers and daily enactments performed at holy wells: the lived, experiential and performative dimensions of health in place (Foley, 2011). Foley applied notions of embodiment, performance and non-representational theory to critically think through material and immaterial relationships at play between people and objects in spaces of health.

Diverse pursuits of health

Geographic applications of therapeutic landscapes are applied to diverse populations and locations. Distinct populations and experiences under consideration include pregnancy and childbirth (Oster et al., 2011), women (MacKian, 2008), older people (Milligan, Gatrell and Bingley, 2004; Finlay et al., 2015), immigrants and refugees (Sampson and Gifford, 2010) and medical tourists (Buzinde and Yarnal, 2012). Literature focuses on the mentally ill, self-help groups (Laws, 2009) and drug-treatment programs (Love, Wilton and Deverteuil, 2012). Studies highlight inequalities, gendered assumptions and uneven power relationships in the production and navigation of therapeutic landscapes. They demonstrate that health is negotiated at multiple overlapping scales and involves much more than the absence or presence of illness and injury. This holistic approach to health involves broad notions such as safety, social inclusion, financial security, intellectual stimulation, empowerment and cultural self-determination.

A small but growing body of literature attends to the intertwining of land, family and community in Indigenous peoples' experiences of therapeutic landscape. Wilson (2003), for example, explored culturally specific dimensions in the context of everyday lives of Anishinabek. Alaazi and colleagues (2015) investigated therapeutic landscapes of the home through Indigenous peoples' experiences of a housing intervention. This research broadens understanding of First Nations peoples and the cultural significance of linking health and place in non-Eurocentric populations. Further non-Western scholarship includes traditional medicine in Africa. Bignante (2015), for example, explored the patient–healer relationship and its role in producing well-being in northern Senegal. The study demonstrated specificities and embeddedness within local belief and value systems of patients and healers outside of the Western biomedical model.

Bodies are increasingly recognized by the literature as central terrains of health and illness. Two studies of breast-cancer survivors (English, Wilson and Keller-Olaman, 2008; Liamputtong and Suwankhong, 2015), for example, depicted the body as a landscape that requires adjustment and recovery. Women worked on their bodies (e.g., through diet and exercise) to heal themselves both physically and mentally. The findings demonstrate complexities of healing bound up in physical, psychological and emotional aspects at the intimate scale of the body. Elsewhere, Doughty (2013) worked on the embodied production of therapeutic landscapes by walking with participants to explore the affective potency of shared movement. In this example, shared movement produced supportive social spaces and therapeutic bodywork in the pursuit of well-being.

Extending beyond physical sites, therapeutic landscapes also encompass spaces of the mind. Gastaldo and colleagues (2004), for example, narrated their experiences as four Canadian immigrants. They interpreted migration through individualized mental strategies that each author uses to enhance his or her mental health and further his or her well-being in times of change. Rose (2012) applied a psychoanalytic approach to work out psychological therapeutic landscapes. People conceptualize landscape encounters to mirror their own feelings and emotions. For example: we know that dark storm clouds are not really angry, and we know that browned leaves and withered flowers are not sad. But we interpret them in this way in order to represent and manage our own affective states. Some individuals use visualization strategies to cope with change and stress, thus creating therapeutic landscapes in their minds.

We can see that therapeutic landscapes represent a constructive metaphor to investigate complex relationships between health and physical, psychological and social spaces. The concept broadens the scope of health geography and meaningfully connects to broader fields of social, cultural and emotional geography.

Nuanced understandings of health as embodied and emplaced inform scholarship in health geography and beyond. Therapeutic landscapes are beginning to be recognized in outside disciplines such as nursing (Gilmour, 2006), landscape and urban planning (Jiang, 2014) and kinesiology (Van Ingen, 2004). The concept connects well to other health-related disciplines, including environmental psychology, sociology, medical anthropology and public health.

Future progress

Scholarship addresses increasingly broad relational experiences of health and well-being, including gendered, aged, cultural, emotional and multiscalar navigations of therapeutic landscapes. Despite growing appreciation for interpersonal approaches, our understanding of culture and race/ethnicity remains under-theorized. Research remains closely tied to white, Eurocentric contexts and experiences (notable exceptions include the Indigenous and African scholarship mentioned previously). Arguably, then, the concept of therapeutic landscapes could focus more on sociocultural intersections of identity, such as overlapping experiences of gender, age, race, income and socioeconomic status. Scholars could also do more to interrogate power relations and stereotypes that influence therapeutic experiences. This may lead to compelling areas of inquiry, such as how the clinical gaze and hierarchical power relations filter doctor-patient health-care experiences in marginalized communities. Scholarship to date has not sufficiently explored relational dynamics and individual agency contributing to the construction, navigation and maintenance of therapeutic landscapes.

Regionally, therapeutic-landscape scholarship is dominated by scholars from Commonwealth countries (e.g., the United Kingdom, Canada, Australia, New Zealand). American geographers have yet to consistently grapple with this topic, and US contexts remain understudied. Given unique sociopolitical complexities and health-care contexts, this area is ripe for investigation. Further geographic limitations include limited existing perspectives from the Global South. Research to date focuses primarily on Western sites. Patient-healer negotiations outside of the Western biomedical context, for example, remain under-theorized. This area of inquiry has potential to demonstrate how physical and emotional wellness are mediated by not only other people and place, but also sociocultural vectors (e.g., herbs, sacred objects, holy scriptures, massages, purifying baths, incense, music, sounds) (Bignante, 2015). Scholarship needs to attend to mediating non-Western elements and incorporate a broader understanding of well-being that simultaneously embraces the practical and the spiritual, cosmic and religious. A postcolonial lens could also critically unpack core/periphery relations through health access and provision, unveiling structural relations of power (Buzinde and Yarnal, 2012). For example, we can consider Hoyez's (2007) example of yogic therapeutic landscapes through the filters of global production and uneven reproduction. A minority of yogis have visited Indian sites in person, yet the majority utilize a mental practice of such sites and attempt to practice Indian reproductions in Western settings. Yoga is thus an example of re-appropriating place through therapeutic practice, with complicated implications for neocolonialism of the Global South.

Therapeutic landscapes still tend to focus on clear sites of healing and relatively straightforward health maintenance. Scholars need to pay more attention to ambivalent and stigmatized landscapes that contribute to wellness in less obvious ways. Medical anthropologist Mokos (2017), for example, studied beneficial dimensions of river-bottom homeless encampments. Camps tend to be grouped with other unmanaged sites of decline and despair that homeless people inhabit. These sites were never intended for health maintenance or human occupation, yet homeless people inhabit these untherapeutic landscapes. Mokos (2017) grapples with complexities of exclusion, systematic violence, and stigma, helping readers understand river-bottom camps as therapeutic landscapes for inhabitants. They can provide care through belonging, access to nature, privacy and social benefits. Embracing the intricacies of therapeutic landscapes generates opportunities for more just ways of promoting health.

Literature on therapeutic landscapes continues to advance from intrinsic values of place to incorporate complexities of power, individuality, agency and intersection. One promising area of future scholarship encompasses performative aspects and embodied encounters. Evans (2014), for example, attended to the therapeutic dimensions of ambient music. As people listen to and feel with therapeutic soundscapes, their embodied encounter can transform the listener's awareness and emotions: "this shift in self-awareness, disposition, or liveliness ('therapeutic') emerges as the body becomes *affected* within a particular network of relations with other objects ('landscape')" (p. 175). The body thus acquires new capacities, empowering an individual to act and feel well. This example of ambient music dissolves boundaries between the individual and environment, relating to Foley's (2011) fluid and performative notion of assemblage. We can extend consideration of the body in space and movement ranging from the mundane (e.g., cooking, caregiving, bathing) to the extraordinary (e.g., pilgrimage, refugee movement). Therapeutic experiences are always relational, embodied, multisensory and fluid.

Broader investigation of therapeutic landscapes will pose methodological challenges and opportunities for advancement. The concept is one of the few frameworks developed exclusively by qualitative health geographers. We can expect integration with quantitative approaches, such as spatial analysis, virtual technology and more precise measurement of therapeutic sites. Emerging methods applied across geography and beyond, such as the mobile interview (Finlay and Bowman, 2017), can broaden therapeutic inquiry *in place*. Longitudinal studies can capture shifting relationships and life-course circumstances that influence the relative importance of therapeutic landscapes over time. Temporality, personal history and remembrance are important elements of future consideration. For example: how does one's past exposure to particular landscapes and cultural practices influence current use? Comparative practices across more diverse settings could further investigate the individualized and often ambiguous role of therapeutic landscapes.

Gorman (2016) called for multispecies ethnographies of therapeutic spaces to highlight the way in which different human and non-human beings engage with place. He stated a need for expanded more-than-human actants, artifacts and technologies to infiltrate therapeutic-landscapes scholarship. This line of inquiry could extend the notion of intersubjectivity to how we share therapeutic spaces and practices with other beings and objects. Given the rapid rise in technology, we need to integrate the role of mechanized tools and robots in care spaces. Emerging technologies, such as voice-activated cellphones, robotic dolls for Alzheimer's patients and self-driving cars provide new opportunities for care. With widespread technological changes, our health and wellness demands also shift. Multidimensional understandings of wellness in place need to include the role of technology filtering through our everyday lives.

Armed with a better understanding of the factors influencing person–place health relations, academic research on therapeutic landscapes can ideally facilitate broader well-being. Williams (2017) called for using the concept as a tool to mobilize positive change. As inequalities in access to health and wellness grow, therapeutic-landscapes scholarship can vocalize and draw critical attention to vulnerable populations and marginalized spaces. Individuals and contexts could include impoverished neighborhoods, public-health epidemics, incarcerated individuals, displaced refugees and victims of violence. Therapeutic landscapes are always in part a political project (MacKian, 2008). Only by further exploring the intricacies of everyday marginalized peoples and contexts can we hope to develop an understanding of more effective and empowering policy strategies. To advance a more just and inclusive form of scholarship, we must continually ask: what is therapeutic, and for whom?

References

Alaazi, D. A., Masuda, J. R., Evans, J. and Distasio, J. (2015). Therapeutic landscapes of home: exploring Indigenous peoples' experiences of a Housing First intervention in Winnipeg. *Social Science & Medicine*, 147, pp. 30–37.

Baer, L. D. and Gesler, W. M. (2004). Reconsidering the concept of therapeutic landscapes in J D Salinger's *The Catcher in the Rye*. *Area*, 36, pp. 404–413.

Bignante, E. (2015). Therapeutic landscapes of traditional healing: building spaces of well-being with the traditional healer in St. Louis, Senegal. *Social & Cultural Geography*, 16(6), pp. 698–713.

Buzinde, C. N. and Yarnal, C. (2012). Therapeutic landscapes and postcolonial theory: a theoretical approach to medical tourism. *Social Science & Medicine*, 74(5), pp. 783–787.

Collins, D. and Kearns, R. (2007). Ambiguous landscapes: sun, risk and recreation on New Zealand beaches. In: A. Williams, ed., *Therapeutic landscapes*. Surrey: Ashgate Publishing, pp. 15–32.

Conradson, D. (2005). Landscape, care and the relational self: therapeutic encounters in rural England. *Health & Place*, 11(4), pp. 337–348.

Curtis, S., Gesler, W., Fabian, K., Francis, S. and Priebe, S. (2007). Therapeutic landscapes in hospital design: a qualitative assessment by staff and service users of the design of a new mental health inpatient unit. *Environment and Planning C*, 25, pp. 591–610.

Doughty, K. (2013). Walking together: the embodied and mobile production of a therapeutic landscape. *Health & Place*, 24, pp. 140–146.

English, J., Wilson, K. and Keller-Olaman, S. (2008). Health, healing and recovery: therapeutic landscapes and the everyday lives of breast cancer survivors. *Social Science & Medicine*, 67(1), pp. 68–78.

Evans, J. (2014). Painting therapeutic landscapes with sound: on land by Brian Eno. In: G. J. Andrews, P. Kingsbury and R. Kearns, eds., *Soundscapes of wellbeing in popular music*. London and New York: Routledge, pp. 173–190.

Finlay, J. and Bowman, J. (2017). Geographies on the move: a practical and theoretical approach to the mobile interview. *The Professional Geographer*, 69(2), pp. 263–274.

Finlay, J., Franke, T., Mckay, H. and Sims-Gould, J. (2015). Therapeutic landscapes and wellbeing in later life: impacts of blue and green spaces for older adults. *Health & Place*, 34, pp. 97–106.

Foley, R. (2011). Performing health in place: the Holy well as a therapeutic assemblage. *Health & Place*, 17(2), pp. 470–479.

Foley, R. (2014). The Roman-Irish bath: medical/health history as therapeutic assemblage. *Social Science & Medicine*, 106, pp. 10–19.

Gastaldo, D., Andrews, G. J. and Khanlou, N. (2004). Therapeutic landscapes of the mind: theorizing some intersections between health geography, health promotion and immigration studies. *Critical Public Health*, 14(2), pp. 157–176.

Gesler, W. (1993). Therapeutic landscapes: theory and a case study of Epidauros, Greece. *Environment and Planning D*, 11(2), pp. 171–189.

Gesler, W. (1996). Lourdes: healing in a place of pilgrimage. *Health & Place*, 2(2), pp. 95–105.

Gesler, W. (1998). Bath's reputation as a healing place. In: R. Kearns and W. Gesler, eds., *Putting health into place: landscape, identity, and well-being*. Syracuse, Pittsburgh: Syracuse University Press, pp. 17–35.

Gesler, W. (2003). *Healing places*. Lanham, MD: Rowman & Littlefield.

Gilmour, J. A. (2006). Hybrid space: constituting the hospital as a home space for patients. *Nursing Inquiry*, 13(1), pp. 16–22.

Gorman, R. (2016). Therapeutic landscapes and non-human animals: the roles and contested positions of animals within care farming assemblages. *Social & Cultural Geography*, 18(3), pp. 1–21.

Hoyez, A. C. (2007). The "world of yoga": the production and reproduction of therapeutic landscapes. *Social Science & Medicine*, 65(1), pp. 112–124.

Jiang, S. (2014). Therapeutic landscapes and healing gardens: a review of Chinese literature in relation to the studies in Western countries. *Frontiers of Architectural Research*, 3(2), pp. 141–153.

Kearns, R. A. and Collins, D. C. A. (2000). New Zealand children's health camps: therapeutic landscapes meet the contract state. *Social Science & Medicine*, 51(7), pp. 1047–1059.

Laws, J. (2009). Reworking therapeutic landscapes: the spatiality of an "alternative" self-help group. *Social Science & Medicine*, 69(12), pp. 1827–1833.

Liamputtong, P. and Suwankhong, D. (2015). Therapeutic landscapes and living with breast cancer: the lived experiences of Thai women. *Social Science & Medicine*, 128, pp. 263–271.

Love, M., Wilton, R. and Deverteuil, G. (2012). "You have to make a new way of life": women's drug treatment programmes as therapeutic landscapes in Canada. *Gender, Place & Culture*, 19(3), pp. 382–396.

Mackian, S. C. (2008). What the papers say: reading therapeutic landscapes of women's health and empowerment in Uganda. *Health & Place*, 14(1), pp. 106–115.

Milligan, C., Gatrell, A. and Bingley, A. (2004). "Cultivating Health": therapeutic landscapes and older people in Northern England. *Social Science & Medicine*, 58(9), pp. 1781–1793.

Mokos, J. T. (2017). Stigmatized places as therapeutic landscapes: the beneficial dimensions of river-bottom homeless encampments. *Medicine Anthropology Theory*, 4, pp. 123–150.

Oster, C., Adelson, P. L., Wilkinson, C. and Turnbull, D. (2011). Inpatient versus outpatient cervical priming for induction of labour: therapeutic landscapes and women's preferences. *Health & Place*, 17(1), pp. 379–385.

Palka, E. (1999). Accessible wilderness as a therapeutic landscape: experiencing the nature of Denali National Park, Alaska. In: A. Williams, ed., *Therapeutic landscapes: the dynamics between wellness and place*. Lanham, MD: University Press of America, pp. 29–51.

Rose, E. (2012). Encountering place: a psychoanalytic approach for understanding how therapeutic landscapes benefit health and wellbeing. *Health & Place*, 18(6), pp. 1381–1387.

Sampson, R. and Gifford, S. M. (2010). Place-making, settlement and well-being: the therapeutic landscapes of recently arrived youth with refugee backgrounds. *Health & Place*, 16(1), pp. 116–131.

Smyth, F. (2005). Medical geography: therapeutic places, spaces and networks. *Progress in Human Geography*, 29(4), pp. 488–495.

Van Ingen, C. (2004). Therapeutic landscapes and the regulated body in the Toronto Front Runners. *Sociology of Sport Journal*, 21(3), pp. 253–269.

Williams, A. (ed.) (1999). *Therapeutic landscapes: the dynamic between wellness and place*. Lanham, MD: University Press of America.

Williams, A. (2002). Changing geographies of care: employing the concept of therapeutic landscapes as a framework in examining home space. *Social Science & Medicine*, 55(1), pp. 141–154.

Williams, A. M. (2017). The therapeutic landscapes concept as a mobilizing tool for liberation. *Medicine Anthropology Theory*, 4(1), pp. 10–19.

Wilson, K. (2003). Therapeutic landscapes and First Nations peoples: an exploration of culture, health and place. *Health & Place*, 9(2), pp. 83–93.

18

WELL-BEING IN HEALTH GEOGRAPHY

Conceptualizations, contributions and questions

Meryn Severson and Damian Collins

The World Health Organization's definition of "health" as "a state of complete, physical, mental and social well-being and not merely the absence of disease or infirmity" (WHO, 1948, p. 100), expanded the concept of health beyond medical matters to encompass a wide range of human experiences. Well-being subsequently became a popular idea with varied applications, coming to be understood as a policy goal, a purchasable commodity, a desirable status and a valuable analytic lens for scholars interested in diverse aspects of health and health care (Fleuret and Atkinson, 2007; MacKian, 2009).

The emergence of the sub-field of health geography in the early 1990s, alongside a broader socioecological turn in conceptualizations of health, led geographers to consider holistic ideas of emotional, social, and physical contentment (Kearns and Andrews, 2010). Concurrently, well-being emerged as a focus in geographical research. However, the term "well-being" itself remained loosely defined and understood, both within and beyond disciplinary boundaries. While we know what well-being is not – it is "not merely the absence of disease or infirmity" (WHO, 1948, p. 100) – what does it consist of? And why is it important?

This chapter seeks to answer both of these questions. In the first section, we outline the varied conceptualizations of well-being in health geography and some of the challenges associated with its application. In the second section, we illustrate the relevance and significance of well-being in health-geography research. We then conclude with questions for future well-being research in health geography.

Conceptualizing well-being

Components of well-being

Fundamentally, the concept of well-being entails a positive focus on various dimensions and experiences that contribute to human potential. The strength of the idea derives from its breadth and holistic nature, encompassing various aspects of quality-of-life, both material and subjective. However, diverse experiences and dimensions of "well-being" make the definition of the term difficult to nail down, as highlighted in Table 18.1:

These definitions focus on the combination of physical health and positive subjective or emotional experience (e.g., happiness, state-of-mind) at the individual level. Several also incorporate material aspects (e.g., prosperity, needs, welfare), which imply a connection to social and economic conditions. These conditions arise partly from the experiences of individuals and households as they engage in economic activity and

Table 18.1 Select definitions of "well-being" in health geography

"a good or satisfactory condition of existence; a state characterized by health, happiness, and prosperity. The state of feeling healthy and happy" (MacKian, 2009, p. 235)
"an experienced state of being, in terms of healthiness and happiness" (Kearns and Andrews, 2010, p. 309)
"a longer-term state-of-mind (and body, if not spirit) . . . often regarded as synonymous with quality of life" (Kearns and Andrews, 2010, p. 311)
"a state of positive mental and physical health and welfare, attained or obtained in some way by fulfilling personal needs" (Andrews, Chen and Myers, 2014, p. 214)

perform work in particular places, and partly from larger socioeconomic systems that shape the distribution of opportunities and affluence. These systems play critical roles in the complex mechanisms that give rise to the socioeconomic determinants of health and the uneven distribution of well-being across the population. To the extent that increasing affluence allows for greater satisfaction of needs, it contributes to individual and collective well-being (Fleuret and Atkinson, 2007). However, in some cases it can have detrimental impacts on well-being, as when economic growth is generated through environmental degradation (MacKian, 2009). Additionally, material needs can vary between places, and, as such, the significance of wealth and income is contextual (Fleuret and Atkinson, 2007).

In important respects, quality-of-life is produced and experienced collectively – in terms of both material and interpersonal needs. Adequate income, meaningful employment, social support, and community integration have strong positive effects on health status, as powerfully illustrated by the Whitehall II study of more than 10,000 civil servants working in London, England (Ferrie, 2004). Thus, the concept of well-being necessarily encompasses both personal and collective components; being well is experienced at multiple scales, from the individual, to institutions (e.g., schools and employers) and neighborhoods, to the national state (Kearns and Andrews, 2010; MacKian, 2009).

Dichotomies in well-being

Fleuret and Atkinson (2007) provide three theories for understanding well-being. The *theory of needs* stems from the work of Maslow and the connection between material affluence and well-being. Here well-being is an outcome of consumption, objectively and universally delineated by needs. However, needs are related to context and circumstances, leading to the second approach: *the relative standards theory*. This approach focuses on the contextual components of well-being, positing that "well-being is therefore linked to individual happiness and conditioned by the individual's perception of the context in which he or she is living" (p. 109). However, this approach's focus on the subjective experience neglects consideration of the physical aspects of well-being. A third understanding of well-being is the *theory of capability*, which seeks to integrate the material aspects of the theory of needs with the subjective aspects of relative standards through the idea of capabilities. Capabilities are "a range of attainable and valuable functionings including sets of skills and power" (p. 109). Here, we see a connection between well-being and understandings of health as a resource for everyday life, and a positive set of capacities that enable the fulfillment of human potential (WHO, 2017). However, approaching well-being in this way can be difficult, due to the problem of operationalizing capabilities (Fleuret and Atkinson, 2007).

These three theories illustrate a set of dichotomies that constitute well-being – health and happiness, material aspects and dimensions of human experience, objective and subjective, universal and contextual, and individual and collective. There is a particularly fundamental divide between *subjective well-being*, associated with dimensions of human experiences and context at the individual level, and *general or objective well-being*, associated with universality and material aspects at the collective level (MacKian, 2009). However, such tensions do not necessarily detract from the utility of well-being as a concept. Rather, they illustrate the breadth of the term and "the various linkages to a range of domains of human experience" (Kearns and Collins, 2009, p. 27). Additionally, while it is often simpler – for example, in research – to focus on a single aspect of well-being, operationalizing the concept in this way loses sight of its breadth and holistic nature.

Health and well-being

Health geography's focus on well-being reflects a sub-disciplinary move away from a biomedical focus on individual diseased bodies and spaces of medical intervention. At a more epistemological level, it also indicates a rejection of the Cartesian dualism, which separates body and mind. These shifts, combined with a concern for socio-spatial context, mean that well-being is by no means the sole (or even primary) domain of biomedicine (Kearns and Andrews, 2010; MacKian, 2009). From one perspective, a strength of the term "well-being" as a focus for inquiry, and a goal of interventions, is that it is difficult to medicalize due to its breadth and diversity. From another perspective, "well-being" can be understood to imply "a progression away from being ill or impaired [and thus] the natural positive outcome that medical intervention seeks to achieve" (Kearns and Collins, 2009, p. 19). Evading medical capture of the term is, thus, far from assured. This is particularly evident with regard to common ways in which well-being is measured, which often focus on life expectancy, mortality rates and incidence of disease, rather than more positive subjective dimensions, such as contentment, life satisfaction or capability (Fleuret and Atkinson, 2007; Kearns and Collins, 2009).

One tool for addressing this disconnect is the WHO-5 index, which was developed in 1998 to enable measurement of subjective well-being. It consists of a questionnaire with five positively worded statements, including "In the last two weeks, my daily life has been filled with things that interest me" and "I have felt active and vigorous," which respondents are asked to rank on a scale of 0–5. This index helps operationalize positive notions of well-being, at the same time as it reduces inherently subjective experiences and states to numbers. Moreover, it does not include direct questions about social well-being (WHO, 2017). As well-being is conceptualized at multiple scales, it should also be measured at multiple scales, from individual markers of health to indicators of national well-being. Measures of the latter now extend beyond primarily economic indicators such as gross domestic product to include more social and quality-of-life aspects, as in the Human Development Index and the Happy Planet Index (Fleuret and Atkinson, 2007; MacKian, 2009). Thus, operationalizing and measuring well-being continues to be an area for growth.

With respect to operationalization of the term "well-being," the example of the New Public Health movement is instructive. It focuses on enhancing population health by empowering and expecting individuals to take responsibility for their own well-being (Kearns and Collins, 2009). Through emotionally charged and persuasive messages about nutrition, exercise, sun exposure and smoking, for example, well-being is presented as "both a status and feeling, [which] can be achieved through individuals being responsible for and assisting themselves mentally and physically" (Andrews, Chen and Myers, 2014, p. 212). It also links to neoliberalism via the responsibilization of the individual and the associated rise of a well-being industry to enable healthy choices – at least for those who can afford them (MacKian, 2009). In New Public Health, well-being is an inherently political project, one in which hopes and fears, and visions of dystopia and utopia, are operationalized to propel certain actions and preempt possible health harms (Evans, 2010).

So, what can be said about well-being? At its core, the term points to a sense of completeness and balance in varied aspects of life, including the emotional, social, material and physical (MacKian, 2009). It is also the positive sense of having potential and ability within each of these dimensions of human experience (Kearns and Collins, 2009). It is possible to have a sense of well-being while living with a chronic illness or disability (Coleman and Kearns, 2015), and, as such, physical health is not the dominant component of well-being (Kearns and Andrews, 2010). Well-being also implies an embodied experience of *being well*, which is anchored in place and context. Thus, it invites geographic research into the places and spaces of well-being

(Kearns and Andrews, 2010). In the next section, we explore the significance and influence of well-being in health-geography research.

The significance and influence of well-being in health-geography research

Within health geography, well-being is a significant and influential concept, centered on the place-based resources and experiences that support health and quality-of-life. In this section, we document some of the major contributions of health geography to understandings of well-being, as developed in three areas of research: spaces of well-being; therapeutic landscapes; and emotion, place and well-being. Each of these themes is explored in turn.

Spaces of well-being

The relationship between place and mental and physical health is central to inquiry in health geography (Kearns and Andrews, 2010). Fleuret and Atkinson (2007) identify three key ways in which this focus enables health geographers to contribute to the study of well-being: "a socio-economic focus on spatial and social injustice"; "an environmental approach . . . and the therapeutic virtues of the landscape"; and "a social welfare approach in which the consequences of vulnerability in terms of health and quality of life are studied" (p. 107). Reflecting a broadening of interest beyond institutional sites of medical intervention, health geographers have studied the rise of landscapes of well-being in the context of community and home-based care (Kearns and Collins, 2009). There is increasing recognition of the informal care work and *caringscapes* that contribute to well-being (Kearns and Andrews, 2010). These "spaces of care are therapeutic environments produced by, for, and through the interest of one person in the well-being of another" (DeVerteuil and Evans, 2009, p. 291). They include sites as varied as drop-in centers and homeless shelters, as well as traditional hospital-based settings, provided they offer "refuge, support, and essential resources" (p. 291).

Fleuret and Atkinson (2007) conceptualize spaces of well-being as informed by four other spaces: spaces of capability, integrative spaces, spaces of security and therapeutic spaces. Spaces of capability focus on experiences of well-being and self-fulfillment, which may be hindered by factors such as (dis)ability, aging and stigmatization. Integrative spaces include those domains and experiences with positive links to well-being and health, such as social contacts, while also identifying the processes that produce spatial and social inequalities. Spaces of security include social, spatial and individual supports that contribute to well-being. Lastly, therapeutic spaces, which have been a major focus of well-being research, are those environments that positively contribute to well-being. This four-part conceptualization reflects the multidimensional nature of well-being itself, as well as the way in which it is positioned outside of biomedical knowledge and medical spaces.

Therapeutic landscapes

Therapeutic landscapes positively impact well-being in a myriad of ways, including via stress relief, social connections, security and belonging – an overall sense of restoration and renewal (Duff, 2011). Here, landscapes are understood as "the complex layerings of history, social structure, symbolism, nature, and built environment that converge at particular sites" (Kearns and Collins, 2009, p. 19). Some landscapes – most often those with considerable natural value – are widely understood as enhancing human well-being. For example, bluescapes – which include river, beach and island environments – "are commonly interpreted as possessing restorative benefits since they offer opportunities for stillness, reflection and respite" (Coleman and Kearns, 2015, p. 207). However, as Duff (2011) warns, this idea that there is something innately health-promoting about these natural landscapes contributes to the exclusion of built and urban environments that can contribute to well-being.

Additionally, the idea that it is the essence of places that is restorative neglects the activity that occurs in those places. Coleman and Kearns (2015) write that "therapeutic experiences cannot be presumed but rather are the product of peoples' diverse and complex relations with place" (p. 206). They support this argument in their study of bluescapes and older adults residing on a New Zealand island, highlighting the ways in which well-being is produced via emotional connections to the water and the island. These emotional connections were also expressed in the complex relations around staying or leaving the island as participants' health declined, with many to choosing to stay because of their sense of connection. Critically, it is "the experience of place rather than the place itself that is generative of well-being" (Coleman and Kearns, 2015, p. 216).

Emotion, affect and well-being

Health geographers' engagements with well-being are increasingly concerned with the emotions and affects generated through human encounters with place. In broad terms, emotions refer to the feelings experienced and expressed by individuals, while affects refer to shared, non-cognitive capacities. Both specific emotions and the more diffuse and collective energies and intensities that constitute affects are routinely mobilized in initiatives to promote forms of well-being (Evans, 2010). However, Andrews, Chen and Myers (2014) contend that well-being should be conceptualized as an affective environment, rather than as the result of a given intervention or the product of experience. From this perspective, well-being is "the environmental action, then feeling of that action, prior to meaning" (p. 219). Under this approach, well-being is temporally short, whereas in previous approaches it was understood as having some endurance or state of being (Kearns and Andrews, 2010).

Affect is dynamic and diverse, changing between people and over time (Duff, 2011). This variability is highlighted in Coleman and Kearns' (2015) analysis of how bluescapes impact aging and well-being, in which participants expressed a deep sense of connection with the water and the island that changed over time and with their experiences. For example, one participant commented how after her husband passed away, the island felt isolating but the water helped with her grief.

Emotions and well-being can also be connected through music and soundscapes. In some instances, this connection is explicit, through lyrical focus on the topic or through support for events like LiveAid. More generally, soundscapes and well-being interact primarily through emotion. For example, the internalization of lyrics and creation of soundscapes "transports [individuals] from their current situation, to help them forget, feel better, and hope" (Kearns and Andrews, 2010, p. 185). Listening to music is often a social event, particularly in the shared experience of attending concerts, which can contribute to feelings of community integration and support and, thus, emotional well-being. The intersection of emotional and health geography offers a particularly rich milieu for investigating the meanings and experiences of well-being in diverse places.

Directions for future research

In this section, we identify three areas for future health-geography research into well-being. First, geographers are already undertaking important work linking the places in which we live and work with the social gradient of health (Kearns and Collins, 2009). Implicit in this research is a recognition that well-being – in terms of overall contentment and quality-of-life beyond satisfaction of basic material needs – is something the relatively privileged can afford to work on. Greater income and wealth often bring with them the time and resources to dedicate to nutrition, exercise, supplements, work-life balance and more (MacKian, 2009). At the same time, neoliberal discourses can place the blame for poor health on individual choices, downplaying or denying the influence of socioeconomic factors (MacKian, 2009). In studying well-being, researchers must continue to highlight the inequalities and inequities that impact well-being at both the individual and population levels (Kearns and Andrews, 2010; Kearns and Collins, 2009). This is important, in part, to counteract the individual as the locus of New Public Health imperatives and to inform policy goals of increasing well-being across all sectors of society.

Conceptually, well-being is linked to both happiness and health (MacKian, 2009). However, there are times when these two ideas conflict, and, as such, it is important that there be a "realization that not all that contributes to personal well-being is necessarily good for one's health in a medical sense ... for example, overtraining [or] taking performance-enhancing drugs" (Kearns and Andrews, 2010, p. 315). Moreover, hedonistic activities, such as drug use, smoking and poor eating, can create at least short-term feelings of contentment and satisfaction, often at the expense of long-term physical and mental health. This creates tension in the idea of well-being. Do these activities produce (genuine) well-being? What separates pleasure from well-being? Moreover, some things that contribute to health, such as physical activity, can in certain instances contribute to negative emotional states. So too can New Public Health campaigns, which are often focused on fear (Evans, 2010) and the private well-being industry that promotes well-being as a consumption-based status (MacKian, 2009). Geographical analyses of well-being need to grapple with these conflicts and tensions and, in so doing, develop further accounts of how health benefits and health risks can be co-present in landscapes and place-based experiences.

Lastly, the term "well-being" is nuanced, complex and difficult to translate and study in different languages (Fleuret and Atkinson, 2007). While the holistic, integrated concept of well-being is relatively new (or re-discovered) in the Western tradition, it has long existed in other traditions, such as among the Maori in New Zealand and Indigenous peoples in Canada (Kearns and Andrews, 2010; Kearns and Collins, 2009). However, under neoliberalism, well-being is increasingly linked to individual responsibility and choices (MacKian, 2009). Given these diverse interpretations and histories, well-being appears as "a relative notion that is socially and culturally constructed and as a result takes on different meanings depending on the context, [which] amounts to stating that the WHO definition of "health" cannot be universal as it is subject to the effects of place" (Fleuret and Atkinson, 2007, p. 111). How can well-being research better account for diverse contexts that influence how the idea itself is understood? How can cross-cultural analyses of well-being be undertaken? Is well-being necessarily a universal goal, as the WHO definition of "health" appears to suggest?

Conclusion

In this chapter, we have reflected on the concept of well-being and its rising prominence in health geography. Well-being is a more complete and broad concept than health, combining varied indicators of quality-of-life and aspects of human experience and connecting multiple disciplines and domains of life. However, issues of definition and measurement remain challenging. Here we see an enduring legacy of the Cartesian dualism: the "divorce in Western rational scientific thinking [between body and mind] that has resulted in subsequent confusion about what well-being is and how to comprehend, measure, or enhance it" (MacKian, 2009, p. 235). Recognition of the impacts of place on physical, emotional and social well-being have given health geographers a prominent position in the study of well-being. Research around therapeutic landscapes and the partial relocation of health care from the hospital to the community and the home invite continued geographic research. While the scholarship around well-being has increased with widespread adoption of the socioecological model of health, there are still areas for additional research, particularly around inequalities, contradictions and context.

References

Andrews, G. J., Chen, S. and Myers, S. (2014). The "taking place" of health and well-being: towards non-representational theory. *Social Science & Medicine*, 108, pp. 210–222.

Coleman, T. and Kearns, R. (2015). The role of bluescapes in experiencing place, aging, and well-being: insights from Waiheke Island, New Zealand. *Health & Place*, 35, pp. 206–217.

Deverteuil, G. and Evans, J. (2009). Landscapes of despair. In: T. Brown, S. McLafferty and G. Moon, eds., *A companion to health and medical geography*. Oxford: Wiley-Blackwell, pp. 278–300.

Duff, C. (2011). Networks, resources, and agencies: on the character and production of enabling place. *Health & Place*, 17, pp. 149–156.

Evans, B. (2010). Anticipating fatness: childhood, affect and the pre-emptive "war on obesity". *Transactions of the Institute of British Geographers*, 35(1), pp. 21–38.

Ferrie, J. E. (ed.) (2004). *Work, stress and health: findings from the Whitehall II study*. London: Public and Commercial Services Union, University College.

Fleuret, S. and Atkinson, S. (2007). Well-being, health and geography: a critical review and research agenda. *New Zealand Geographer*, 63(2), pp. 106–118.

Kearns, R. A. and Andrews, G. J. (2010). Geographies of well-being. In: S. J. Smith, R. Pain, S. A. Marston and Jones, J. P. III, eds., *SAGE handbook of social geographies*. Los Angeles: Sage, pp. 309–328.

Kearns, R. and Collins, D. (2009). Health geography. In: T. Brown, S. McLafferty and G. Moon, eds., *A companion to health and medical geography*. Oxford: Wiley-Blackwell, pp. 13–32.

MacKian, S. C. (2009). Well-being. In: R. Kitchin and N. Thrift, eds., *International encyclopedia of human geography*. Oxford: Elsevier, pp. 235–240.

The WHO-5 Website. (2017). *WHO-Five Well-Being Index (WHO-5)*. [online] Available at: www.psykiatri-regionh.dk/who-5/Pages/default.aspx [Accessed 27 Feb. 2017].

World Health Organization. (1948). *Constitution of the World Health Organization*. [pdf] Geneva: WHO. Available at: www.who.int/governance/eb/who_constitution_en.pdf [Accessed 28 Aug. 2017].

World Health Organization. (2017). *The Ottawa Charter for Health Promotion, 1986*. [online] Geneva: WHO. Available at: www.who.int/healthpromotion/conferences/previous/ottawa/en/ [Accessed 28 Aug. 2017].

19

DECENTERING GEOGRAPHIES OF HEALTH

The challenge of post-structuralism

Joshua Evans

Roughly a quarter-century ago, Robin Kearns' (1993) critique of medical geography established a conceptual space for a reformed geography of health focused on the dynamic relationships between health and place. Kearns' (1993) critique sparked responses from across medical geography, one of them by Dorn and Laws (1994), who welcomed Kearns' reformist argument but argued it overlooked recent advances in social theory regarding the body, its material and representational forms, and the embodied subject positions created and resisted in contemporary society. Dorn and Laws' challenge was important because it initiated an engagement between post-structuralism and health geography. The post-structuralist geographies of health that have emerged since share a concern for the dynamic relationship between people, place and health but reject the idea that the meaning of these experiences is centered in an autonomous, rational subject's experience of a physical location. Instead, poststructural geographies of health pursue *decentered* accounts that relocate the production of meaning from the individual to the way individuals are defined as subjects within discourses, the processes of embodiment by which individuals are placed in the world, and the state strategies by which individual health is politicized.

Post-structuralist geographies of health have come into their own over the past quarter-century but still occupy a marginal position within the sub-field of health geography. However, a powerful challenge has been articulated from this marginal position: the challenge of taking the body seriously as an embodied site of power and politics. In much research, this has necessitated a relational approach to space (Cummins et al., 2007), one that is sensitive to contingency, change and difference.

This chapter begins by introducing post-structuralism and reviewing how geographers have applied its various elements in developing notions of relational space. The chapter then assesses what kinds of relational geographies post-structuralism helped map before reflecting upon the wider influence of post-structuralism within health geography.

Post-structuralism and geography

As the *post-* prefix indicates, post-structuralism developed in conversation with structuralism, a theoretical perspective that originated in France in the mid-20th century and spread across the humanities and social sciences in Europe and North America (Murdoch, 2006). Led by proponents such as Claude Lévi-Strauss, Louis Althusser and Jacques Lacan, structuralists contended that structures, such as language, produce rather than reflect reality. Structuralists were labeled anti-foundational, anti-essentialist and anti-humanist because their ideas challenged the existence of a rational, Cartesian subject that discovered objective truths in the real world.

The *post-* prefix was added after a group of French philosophers writing in the 1960s, individuals such as Jacques Derrida, Michel Foucault and Gilles Deleuze, began to question the universality, stability and simplicity of these structures (Woodward, 2017). Jacques Derrida (1930–2004) founded a kind of textual post-structuralism through literary and cultural analysis. He critiqued the way structuralism manufactured stability and coherence by organizing binary relations (i.e., nature-culture, feminine-masculine, non-human–human) in a hierarchical fashion. This centrism betrayed the fact that meaning was itself a product of a much broader, unpredictable and multifaceted system of differences that stretched into the past and future. Michel Foucault (1926–1984) charted a more materialist approach to post-structuralism by turning to history. He emphasized how structures of knowledge are animated by power relations that manifest in locally specific and historically contingent ways. In Foucault's view, power relations manifest, on one hand, as discourses that allow us to speak certain truths about human life (what he called power/knowledge) and, on the other hand, as non-discursive, material practices that target individual bodies, households and populations over time. Foucault was particularly interested in how these coalesce as institutions (i.e., clinic, asylum, prison) promoting normativity and order. Finally, Gilles Deleuze (1925–1995) added a degree of dynamism and vitality into materialist accounts of post-structuralism. He emphasized that structural relations, constituting language, history, economy and spatial organization, are in a constant state of re-composition and re-configuration; hence, culture and social life are better understood in relation to trajectories of change and becoming.

Today, post-structuralism refers to a loose bundle of philosophies, theories and methods that recognize the structural constitution of social and cultural life – that structures such as discourse actively create and constrain our relationship to the world – but nonetheless maintain skepticism regarding the stability, universality and inertness of these structures. Like structuralism, post-structuralism is concerned with the relational constitution of social and cultural life but is drawn toward the situated practices through which meanings are constructed, identities are formed and knowledge is produced (Murdoch, 2006). From the perspective of post-structuralism, these practices are spatially and historically contingent, highly indeterminate and, thus, open to contest. The influence of post-structuralism has been felt across the humanities and social sciences, particularly in feminist, queer studies, and disability scholarship. Adherents value its critical perspective toward power, its skepticism toward the universality of knowledge claims, and its ability to foster alternative narratives (Popke, 2003).

Human geography's engagement with post-structuralism began in the 1980s during the cultural turn, a period of theoretical diversification and engagement focused on issues of culture. Filtered through the lens of post-structuralism, the absolute spaces typical of positivistic approaches and the experiential places typical of humanistic approaches were reimagined as temporary *permanences* produced through relational processes. As Murdoch (2006, p. 19) describes this relational approach:

> space is made not by (underlying) structures but by diverse (physical, biological, social, cultural) processes; in turn, these processes are made by the relations established between entities of various kinds.

The uniqueness of this relational approach is clearer when some of post-structuralism's key analytical insights are applied to these permanences. For example, drawing on Derrida, they should not be mistaken for some kind of essence; rather, permanence is a product of the temporary stabilization of relations between entities of various kinds at a given moment. These stabilizations are always susceptible to unraveling, slippages, and overflows of meaning, often in ways that disrupt the taken-for-granted understandings of the space. Drawing on Foucault, the configuration of these relations is itself a medium of power; moreover, actually existing configurations manifest in locally specific and historically contingent ways. Finally, drawing on Deleuze, relations are always in the process of being made; in other words, spatialization is a process of becoming other or different. Considered together, a relational approach necessitates sensitivity to uncertainty, instability, contingency, change and difference.

Post-structuralist geographies of health: three relational geographies

As noted in the introduction, early in its history, health geography's nascent inclinations prompted an engagement with post-structural interpretations of discourse, identity, embodiment and knowledge. In hindsight, these engagements have added a valuable perspective to the subfield by drawing attention to the role of power and knowledge in the relational constitution of (un)healthy bodies, populations and places as well as the role of these inscriptions in the everyday experience of well-being.

On the whole, health geographers have been influenced more by Foucault than Derrida or Deleuze. Post-structuralist geographies of health have, to date, relied more heavily on concepts such as discourse, power/knowledge, biopolitics and governmentality than deconstruction, de-territorialization, and multiplicity. Drawing on such concepts, geographers have *decentered* geographies of health by turning to discourse as the source of meanings related to health, the discursively and materially constructed body as the place of health experience, and governmentalities as the means by which health is politicized.

To illustrate the breadth of these applications, examples are provided below, under the categories *representational practices*, *embodied practices* and *governmental practices*. Each category highlights a distinct analytical strategy for understanding the relational constitution of bodies, places and health. While these conceptual categories are helpful in a review such as this, in the analyses of health geographers these relational spaces are more often blurred. Hence, they should be read not as mutually exclusive but rather as helpful conceptual devices that overlap in practice.

Representational practices

Scholarship on the representational practices related to health has largely been concerned with discourses that engender meanings about what constitutes a healthy body or a therapeutic place. Bodies and places are treated as objects and/or sites of representation that are inscribed with different social meanings depending upon how they are positioned (or not positioned) within discourses. Discourse, therefore, is itself a kind of relational formation through which individuals or places are rendered normal as opposed to abnormal, well as opposed to sick. In most cases, discourse is approached using interpretive strategies (Gesler, 1999), focused on texts such as literary works (Tonnellier and Curtis, 2005), television (Lepofsky et al., 2006), self-help materials (Ormond and Sothern, 2012), policy documents (Brown and Bell, 2007; Moon, 2000) and architectural forms (Kearns, Barnett and Newman, 2003; Moon, Kearns and Joseph, 2006) that yield insights into how understandings of people-place-health experiences are formed.

In a smaller number of cases, discourse has been approached as a spatial formation in itself. In this work, historically specific relationships between concepts, objects and speaking positions are understood to engender a kind of *medical gaze*. An important touchstone for this particular approach is Foucault's *Birth of the Clinic* (1973) (see Philo, 2000). Health geographers have drawn upon this approach in various ways. For example, Hall (2003) has examined the intensification of the medical gaze through the discourse of genetic mapping; Curtis et al. (2009) have examined the reconfiguration of the medical gaze through discourses of inpatient care; and Brown and Knopp (2014) have examined the extension of the medical gaze to spaces in the community through discourses of urban public health.

Embodied practices

Scholarship on embodied practices of health has largely been concerned with the complexity of lived experience as it relates to the body, health and place. A principal focus has been on how individuals living with illness or impairment negotiate everyday life through their bodies, which are experienced as both material entities *and* sites of sociocultural inscription (Butler and Parr, 1999). This embodied understanding of the health experience, as something simultaneously materialized in the body and socially and culturally constructed, is particularly prominent within post-structuralist geographies of health. Using primarily qualitative methodologies such as discourse analysis and ethnography, health geographers have produced a variety of accounts of exclusion and marginalization linked to the embodied experience of chronic illnesses (Moss and Dyck, 2002), infectious disease (Craddock, 2000) and impairment/disability (Chouinard, Hall and Wilton, 2010), involving a wide range of sites and settings, including the home (Dyck et al., 2005), spaces of care (Crooks and Chouinard, 2006) and nature (Parr, 2008). Of particular concern are how the power of biomedicine is expressed (and resisted) through processes of embodiment and how these processes redefine everyday environments.

A common refrain echoing through much of this scholarship is that the body, as a culturally encoded and lived entity, is itself a kind of place through which health experiences are formed. This emplacement is a modality of power, a medium through which biomedicine, for example, is woven into the fabric of experience. Health geographers have drawn upon this approach in ways that address Dorn and Laws' (1994) call for a *radical body politics*. For example, Moss and Dyck's (2002) examination of the embodied spacings of chronic illness has shed light on how women actively resist the destabilization of their lives; Parr's (2008) examinations of the embodied spacings of mental illness has shed light on how individuals re-author their lives in relation to and in opposition to biomedical inscriptions; and Wilton, DeVerteuil and Evans' (2014) examination of the embodied practices linked to addiction recovery shed light on the complex re-workings of masculinities involved in the pursuit of sobriety.

Governmental practices

Scholarship on governmental practices has focused primarily on state strategies that target *populations*, transforming them into objects of political intervention predominantly in relation to health risks (Brown and Duncan, 2002). This scholarship has found health policy to be a rich empirical site, especially by attending to the representational and embodied practices outlined above (Evans, 2006; Prince, Kearns and Craig, 2006). Foucauldian concepts such as biopolitics and governmentality have also been especially important for mapping these relational geographies. These concepts have been applied to understanding the governance of health risks across a wide range of contexts, including homelessness (Evans, 2012), (un)healthy behaviors (Thompson, Pearce and Barnett, 2007), infectious disease (Burgess Watson and Stratford, 2008), non-communicable disease (Brown and Bell, 2007; Reubi, Herrick and Brown, 2016) and environmental toxins (Mansfield, 2012).

A common pattern characterizing this scholarship is a shared interest in practices of problematization (Reubi, Herrick and Brown, 2016). Moreover, health geographers have demonstrated a particular concern for how risky behaviors, populations, or places are problematized in relation to neoliberal ideologies and techniques that shift the locus of responsibility from states to communities and individuals (Brown, Craddock and Ingram, 2012). For example, Sothern (2007) has provided an account of the neoliberal politics linked to self-help discourses; Evans and Colls (2009) have drawn attention to the neoliberal politics of anti-obesity policy; and Evans (2012) has provided an account of the neoliberal politics linked to the governance of homelessness.

These post-structuralist approaches have not developed in the absence of criticism. Critics have charged post-structuralism with leaving little space for agency, resistance, and change, and with promoting a form of relativism that renders ethical action problematic (Popke, 2003). Moreover, health geographers in particular have questioned how relevant post-structuralist accounts are to the applied activity of policy-making (Philo and Wolch, 2000). These criticisms are important to consider alongside the critical engagements the approach makes possible. Namely, the recognition that the way in which we understand our bodies, as well or unwell, is supplied to us by discourses that simultaneously enable and constrain how we experience our own bodies as well as how we relate to the variety of everyday contexts within which we are placed. Moreover, these embodied subject positions, the behaviors they initiate, and the experiences they generate, are themselves key nodes in wider power geometries that give form to institutions such as the state. Hence the principal challenge of post-structuralism to health geography: to fully appreciate geographies of health, geographers must remain attuned to power relations operating within and across discourses, bodies and populations.

The post-structuralist challenge reconsidered

Arguably, a fully embodied health geography sensitive to the interplay between power/knowledge within and across discourses, bodies and populations has been slow to develop. Nonetheless, it is hard to dismiss the influence and impact of post-structuralist philosophies on the sub-field. Relational geographies of health stand out as important critical engagements insofar as they enhance understandings of place-health linkages while facilitating different pathways to praxis (Parr, 2004). Three outcomes of these critical engagements are notable.

First, post-structuralism reveals additional layers of power relations and their role in structuring health experiences. In doing so, this scholarship has illuminated the role of these practices in the exclusion of marginalized populations and, relatedly, the maintenance of normativities such as ableism and saneism. This has involved taking the body seriously politically – both the individual body and the social body (population) – and developing closer, rather than more distant, engagements with practices of biomedicine.

Second, post-structuralist geographies of health have informed how health geographers problematize their own situatedness and positionality in relation to their research subjects and sites (Parr, 1998). Openness regarding how power relations connecting researchers, research participants, and research sites shape the production of knowledge has fostered more reflexive and, by extension, authentic accounts within health geography.

Finally, post-structuralist geographies of health have helped create openings for engagements with other theoretical frameworks that offer alternative ways of understanding spatial processes and relations. In a way, post-structuralism can be thought of as the *drink* that relaxed theoretical inhibitions and prepared the way for other theoretical ideas. Post-structuralism has served as an important stepping-stone between the humanist accounts that served as the basis of health geography and more recent engagements with post-humanism, new materialism and post-phenomenology.

References

Brown, M. and Knopp, L. (2014). The birth of the (gay) clinic. *Health & Place*, 28, pp. 99–108.

Brown, T. and Bell, M. (2007). Off the couch and on the move: global public health and the medicalisation of nature. *Social Science & Medicine*, 64, pp. 1343–1354.

Brown, T., Craddock, S. and Ingram, A. (2012). Critical interventions in global health: governmentality, risk and assemblage. *Annals of the Association of American Geographers*, 102(5), pp. 1182–1189.

Brown, T. and Duncan, C. (2002). Placing geographies of public health. *Area*, 33(4), pp. 361–369.

Burges Watson, D. and Stratford, E. (2008). Feminizing risk at a distance: critical observations on the constitution of a preventive technology for HIV/AIDS. *Social & Cultural Geography*, 4, pp. 353–371.

Butler, R. and Parr, H. (1999). *Mind and body spaces: geographies of illness, impairment and disability*. New York: Routledge.

Chouinard, V., Hall, E. and Wilton, R. (2010). *Towards enabling geographies: "Disabled" bodies and minds in society and space*. New York: Routledge.

Craddock, S. (2000). Disease, social identity, and risk: rethinking the geography of AIDS. *Transactions of the Institute of British Geographers*, 25, pp. 153–168.

Crooks, V. and Chouinard, V. (2006). An embodied geography of disablement: chronically ill women's struggles for enabling places in spaces of health care and daily life. *Health & Place*, 12, pp. 345–352.

Cummins, S., Curtis, S., Diez-Roux, A. and Macintyre, S. (2007). Understanding and representing "place" in health research: a relational approach. *Social Science & Medicine*, 65(9), pp. 1825–1838.

Curtis, S., Gesler, W., Priebe, S. and Francis, S. (2009). New spaces of inpatient care for people with mental illness: a complex "rebirth" of the clinic. *Health & Place*, 15, pp. 340–348.

Dorn, M. and Laws, G. (1994). Social theory, body politics, and medical geography: extending Kearns's invitation. *The Professional Geographer*, 46(1), pp. 106–110.

Dyck, I., Kontos, P., Angus, J. and McKeever, P. (2005). The home as a site for long-term care: meanings and management of bodies and spaces. *Health & Place*, 11(2), pp. 175–185.

Evans, B. (2006). "Gluttony or sloth": critical geographies of bodies and morality in (anti)obesity policy. *Area*, 38(3), pp. 259–267.

Evans, J. (2012). Supportive measures, enabling restraint: governing homeless street drinkers in Hamilton, Canada. *Social & Cultural Geography*, 13(2), pp. 175–190.

Evans, B. and Colls, R. (2009). Measuring fatness, governing bodies: the spatialities of the Body Mass Index (BMI) in anti-obesity politics. *Antipode*, 41(5), pp. 1051–1083.

Foucault, M. (1973). *The birth of the clinic*. New York: Vintage.

Gesler, W. (1999). Words in wards: language, health and place. *Health & Place*, 5, pp. 13–25.

Hall, E. (2003). Reading maps of the genes: interpreting the spatiality of genetic knowledge. *Health & Place*, 9(2), pp. 151–161.

Kearns, R. (1993). Place and health: towards a reformed medical geography. *Professional Geographer*, 45(2), pp. 139–147.

Kearns, R. A., Barnett, J. R. and Newman, D. (2003). Reading the landscapes of private medicine: Ascot Hospital's place in contemporary Auckland. *Social Science & Medicine*, 56, pp. 2203–2315.

Lepofsky, J., Nash, S., Kaserman, B. and Gesler, W. (2006). I'm not a doctor but I play one on TV: E.R. and the place of contemporary health care in fixing crisis. *Health & Place*, 12(2), pp. 180–194.

Mansfield, B. (2012). Environmental health as biosecurity: "Seafood choices," risk, and the pregnant woman as threshold. *Annals of the Association of American Geographers*, 102(5), pp. 969–976.

Moon, G. (2000). Risk and protection: the discourse of confinement in contemporary mental health policy. *Health & Place*, 6(3), pp. 239–250.

Moon, G., Kearns, R. and Joseph, A. (2006). Selling the private asylum: therapeutic landscapes and the (re)valorization of confinement in the era of community care. *Transactions of the Institute of British Geographers*, 31(2), pp. 131–149.

Moss, P. and Dyck, I. (2002). *Women, body, illness: space and identity in the everyday lives of women with chronic illness*. Lanham, MD: Rowan and Littlefield.

Murdoch, J. (2006). *Post-structuralist geography*. Thousand Oaks, CA: Sage.

Ormond, M. and Sothern, M. (2012). You, too, can be an international medical traveler: reading medical travel guidebooks. *Health & Place*, 18, pp. 935–941.

Parr, H. (1998). Mental health, ethnography and the body. *Area*, 30(1), pp. 28–37.

Parr, H. (2004). Medical geography: critical medical and health geography? *Progress in Human Geography*, 28(2), pp. 246–257.

Parr, H. (2008). *Mental health and social space: towards inclusionary geographies?* Malden, MA: Blackwell.

Philo, C., (2000). The birth of the clinic: an unknown work of medical geography. *Area*, 32(1), pp. 11–19.

Philo, C. and Wolch, J. (2000). From distributions of deviance to definitions of difference: past and future mental health geographies. *Health & Place*, 6, pp. 137–157.

Popke, J. E. (2003). Poststructuralist ethics: subjectivity, responsibility and the space of community. *Progress in Human Geography*, 27(3), pp. 298–316.

Prince, R., Kearns, R. and Craig, D. (2006). Governmentality, discourse and space in the New Zealand health care system, 1991–2003. *Health & Place*, 12, pp. 253–266.

Reubi, D., Herrick, C. and Brown, T. (2016). The politics of non-communicable diseases in the global South. *Health & Place*, 39, pp. 179–187.

Sothern, M. (2007). You could truly be yourself if you just weren't you: sexuality, disabled body space, and the (neo)liberal politics of self-help. *Environment and Planning D*, 25, pp. 144–159.

Thompson, L., Pearce, J. and Barnett, J. R. (2007). Moralizing geographies: stigma, smoking islands and responsible subjects. *Area*, 39, pp. 508–517.

Tonnellier, F. and Curtis, S. (2005). Medicine, landscapes, symbols: "The country Doctor" by Honoré de Balzac. *Health & Place*, 11(4), pp. 313–321.

Wilton, R., DeVerteuil, G. and Evans, J. (2014). "No more of this macho bullshit": drug treatment, place and the remaking of masculinity. *Transactions: Institute of British Geographers*, 39(2), pp. 291–303.

Woodward, K. (2017). Poststructuralism/poststructural geographies. In: D. Richardson, ed., *The international encyclopedia of geography: people, the earth, environment and technology*. New York: Wiley-Blackwell.

20

AFTER POSTHUMANISM

Health geographies of networks and assemblages

Cameron Duff

This chapter explores some of the ways in which health geographers have debated, contested and adapted ideas and concepts drawn from contemporary discussions of posthumanism. I will start by briefly introducing the key features of posthuman, nonhuman and more-than-human geographies (see Castree et al., 2004), before considering how health geographers have contributed to these discussions. My goals are to synthesize the ideas, concepts and controversies of key relevance for health geographers; to highlight the most engaging and innovative of contemporary researches; and to briefly map some possible future directions for health geography after the posthuman turn (see also Rock, Degeling and Blue, 2014).

Reflecting on the reception of posthumanism among human geographers, Castree and Nash (2006, p. 501) argue that posthumanism has been taken to describe both a "historical condition" and a critical "theoretical perspective" on that condition. With respect to the former, posthumanism calls attention to contemporary developments in philosophy, the life sciences and biotechnology, and their blurring of conventional distinctions between a *natural* human subject and the *artificial* cultural and technological contexts in which that subject is embedded. This concern has helped fashion richer understandings of the "complexity and interconnectivity of life" (Panelli, 2010, p. 79) and the ways in which human and nonhuman forces ramify in the articulation of more-than-human subjectivities. The second current is more concerned with questions of ontology and epistemology and with the social, political and historical organization of the categories *human, nonhuman, nature* and *culture*. Drawing theoretical resources from continental philosophy, science studies and feminism, posthumanist research is concerned with tracing the complex relationalities and generative capacity of these ontological categories and the social and political debates they sustain. With each line of inquiry, posthumanism has provided resources for investigating the "recombinations and disruptions" (Castree and Nash, 2006, p. 502) resulting from innovations in philosophy, biotechnology and the life sciences, while refining understandings of how the categories of human, nature and culture are disrupted and reasserted in this innovation.

The lived experience of health and illness, and the scope and methods of the health sciences, have provided especially rich contexts for these investigations (see Andrews, 2014; Duff, 2014, for reviews). Analysis of the implications of posthumanism for the study of health and illness has largely followed the scission noted by Castree and Nash (2006), with important contributions emerging in health geography (Andrews, Chen and Myers, 2014; Foley, 2011; Greenhough, 2011), social and cultural geography (Dilkes-Frayne and Duff, 2017; Conradson, 2005) and political geography (Atkinson, 2013; Rock, Degeling and Blue, 2014). Within health geography more narrowly, scholars have grappled with the implications of posthumanism for thinking about the meaning and experience of health and illness, undermining established ways of investigating person-place interactions, the social determinants of health, even how the notion of health itself ought to be conceptualized and investigated (Andrews, Chen and Myers, 2014). This chapter will explore how health geographers have tackled these challenges, mapping some of the ways in which health geography has been reoriented after the posthuman turn. I will start, however, with a brief review of how the discipline more broadly has engaged with posthumanism, noting the debates and controversies of key relevance for health geographers.

Posthuman geographies

While one might identify some degree of convergence in geographical discussions of posthumanism, it is important to avoid reifying either the term itself or the ways in which geographers have engaged with it. Fifteen years ago, Castree and colleagues (2004) cautioned against treating posthumanism as an internally coherent theoretical position, referring instead to diverse "theorizations of the posthuman" (p. 1341). In the same vein, there has been little agreement among geographers about either the meaning or the significance of the posthuman turn for human geography. It is for this reason that human geographers have referred to "non-human," "more-than-human" and "post-natural" geographies alongside the language of the posthuman (Lorimer, 2009, pp. 344–346). What is rarely contested, however, is the profound challenge that posthumanism presents to figurations of the *human* in and for human geography.

Among geographers, the key issue has been to overcome the naturalization of the human – what Badmington (see Castree et al., 2004, p. 1348) calls the "obviousness" of humanism's "common sense" approximations of the human condition – in order to open up new insights into the ways in which the human emerges as a distinctive structure of feeling in specific collocations of affect, practice, discourse, space and power. Dismissing the dyadic separation of nature/culture, object/subject, inside/outside and agent/structure, geographers interested in the posthuman turn have worked to revise classically humanist concerns, such as agency, desire, responsibility, choice and intentionality, in other-than-human terms (see Greenhough, 2014, for a review). This has mainly involved rejecting the humanist premise of an autonomous, pre-cultural, self-identical subject and replacing it with an assemblage, network or field of human and nonhuman associations and encounters (Atkinson, 2013; Duff, 2014).

The relational ontologies this orientation prefigures have been a mainstay of science and technology studies, feminism, assemblage theory and new materialisms. Each has been prominent in the articulation, analysis and discussion of posthumanism within health geography (see Apostolidou and Sturm, 2016; Davies, Day and Williamson, 2004; Greenhough, 2011). While there is not the space here to review the varied contributions each area of inquiry has made to discussions of the posthuman, some indicative remarks should be helpful. Science and technology studies (STS) has arguably made the most significant theoretical and conceptual contribution to discussions of posthumanism within health geography (see Greenhough, 2014), particularly by way of the reception of actor–network-theory (ANT) and the writings of Bruno Latour and John Law (Duff, 2012; Davies, Day and Williamson, 2004). Also important have been feminist critiques of science and technology, including seminal works by Donna Haraway (1991), Elizabeth Grosz (1994), Rosi Braidotti (2002), Karen Barad (2007) and Jane Bennett (2010).

STS and feminist critiques have been important for the ways in which they overturn earlier geographical understandings of the relationships between human subjects, sexual difference, knowledge, technology and power (see Whatmore, 2006, for a review). Rather than treat subjects as masters of passive technological artifacts that remain always under their authority, or as passive end-users of technologies that ultimately determine by their very structure the manner in which they are deployed, STS and feminist scholars treat science and technology as co-constituted effects of networks and associations linking heterogeneous human and nonhuman actors, bodies, spaces and forces (see Barad, 2007; Latour, 2005). This means that technologies and subjectivities are co-constituted, made or transformed in specific "relational arrays" (Law, 2002, p. 92) in which objects and subjectivities, and the practices, affects and relationalities that bind them, are constantly produced and reproduced.

The challenges these propositions present to conventional, humanist accounts of human subjects, and their experiences of health and illness, are far-ranging. Arguably of greatest significance for health geographers is the way STS and feminist critiques dislodge the human subject from the center of accounts of social change, technological development, progress and politics (Greenhough, 2014; Lorimer, 2009; Whatmore, 2006). Ushering in a key pillar of posthumanist thinking, this work refuses to treat human subjects as the drivers of technological, social and political development, instead seeing this development as an effect of networks of human and nonhuman forces, in which all such forces must be treated "symmetrically" (Latour, 2005, p. 26). This means there can be no *a priori* distinction of subjects and objects and no elevation of active human subjects above passive technological processes. Instead, all forces must be parsed for the contribution they make to the distribution of agencies throughout a socio-technical network. This gesture effectively replaces the humanist subject with a theory of subjectivation and an empirical account of the ways in which subjectivities are expressed, transformed or disrupted in relational arrays linking human and nonhuman forces (Duff, 2014).

Building on generations of scholarship, feminists have focused on the gendered effects of these processes and the ways in which humanist accounts of subjectivity and technology tend to privilege the white male protagonist of technological innovation, marginalizing female, queer and non-white bodies and voices (Barad, 2007; Whatmore, 2006). These critiques of subjectivation have influenced broad swathes of human geography in the last 15 years (Lorimer, 2009), including groundbreaking work in health geography (Andrews, 2014; Atkinson, 2013; Duff, 2014), leading to detailed studies of the role of affects, spaces, events, objects and relations in the lived experience of health and illness (Apostolidou and Sturm, 2016; Davies, Day and Williamson, 2004; Duff, 2012; Dilkes-Frayne and Duff, 2017).

Within geographical discussions of the significance of the critiques of humanism pursued by feminist and STS scholars, assemblage thinking (DeLanda, 2016) and various new materialisms (Coole and Frost, 2010) have offered important refinements and elaborations (see Anderson et al., 2012; Kirsch, 2013; Whatmore, 2006, for reviews). Both approaches abandon the humanist subject in favor of a flatter ontological reckoning of the assemblages and material infrastructures in which subjects emerge (DeLanda, 2016). New materialism is especially attentive to what Jane Bennett (2010, pp. 1–3) calls the "vibrancy" of matter and the ways in which matter itself exerts an active or vital force in the varied networks and assemblages in which it emerges and circulates. Health geographers interested in assemblage thinking and new materialisms have used this work to account for the ways in which capacities, agencies, affects and norms circulate in networks or assemblages of diverse human and nonhuman forces, attaching to human and nonhuman bodies and transforming their variable scope of activity (see Atkinson, 2013; Andrews et al., 2012; Foley, 2011; Greenhough, 2010, for reviews). This research has opened up novel ways of accounting for the positive (or substantive) features of health and well-being, rather than regarding each as registering the simple absence of disorder (Andrews, 2017). In this vein, health geographers have tended to regard posthumanism as a political and ethical project concerned with development and expression of new human/nonhuman hybrids that exceed the fixed social, ontological and political constraints of the humanist subject.

More-than-human health geographies

Among health geographers, the posthuman turn has served to theoretically, conceptually and ontologically ground long-standing interests in the social, spatial and material contours of health and illness (King, 2010). With its focus on the more-than-human aspects of relational life, posthumanism makes human life, subjectivity and identity a function of varied "molecular" or "pre-personal" forces, rather than their cause or origin

(Lorimer, 2009, pp. 347–349). The implications of this approach for thinking about health and illness are profound. For one, overturning the notion of an ahistorical subject brings into question the very idea of health as the normal condition of that subject (Duff, 2014). In doing away with the notion of a bounded, *natural* biological subject, posthumanism also does away with commonsense understandings of the identity and condition of the entity to which *health* and *healthiness* ostensibly attach (Atkinson, 2013; Greenhough, 2011). This is a profound challenge to the humanist ontologies that undergird the health sciences, whereby the human subject is understood to be naturally healthy barring the intervention of agencies, organisms or events external to that subject (King, 2010).

According to these ontologies, humanism installs a series of dualisms at the heart of the health sciences (subject/object, inside/outside, agent/structure, nature/culture) that necessarily remove the subject from its social, affective and material contexts (Fox and Alldred, 2016, pp. 131–149). This dualism is at work, for example, in all accounts of the social determinants of health (Marmot et al., 2008), in which social and structural factors such as power, class, income inequality and gender are said to have a causal effect on the health of populations and the individuals and groups that they comprise. According to this logic, structure and agency, object and subject, outside and inside are ontologically and empirically severed, such that one may make claims about the ways nonhuman structural factors mediate the health of human agents. Posthumanism overturns this schism in favor of an ontological flattening of inside and outside, structure and agent, and a novel focus on networks or assemblages of health and illness (Andrews, 2017; Greenhough, 2011).

This ontological flattening, with its associated focus on networks and assemblages, entails a profound reorientation of the aims and methods of theoretical and empirical inquiry in health geography. From its inception, health geography has prioritized analysis of the relationship between people and environments, health and place (Kearns and Moon, 2002). Yet this focus has mainly cohered with related work in public health that treats human agents as ontologically distinct entities that may be analytically separated from their environments and contexts in ways that make the very logic of person–place interactions intelligible (Rock, Degeling and Blue, 2014). In other words, health geographers have mainly endorsed the humanist abstraction of subjects from their contexts, even if they have worked ever more assiduously to complicate the conceptual and empirical terms of this abstraction. Cognate work after the posthuman turn has, however, sought new ways of accounting for the distribution of health and illness in place (Greenhough, 2011; King, 2010; Whatmore, 2006). Rather than investigate how certain environments either prevent or enable the transmission of infectious conditions within a population, or how such environments promote or inhibit the adoption of health-promoting habits and practices, health geography after the posthuman turn has invoked networks or assemblages as the primary unit of analysis (Atkinson, 2013; Duff, 2014). This has involved a shift away from attempts to isolate and determine the specific causal contribution of discrete factors in the incidence and prevalence of a given disorder – whether those factors be pathogens, attitudes, norms or practices – in favor of renewed interest in the emergent and dynamic complexity of health in its social, affective and material becomings.

Such interests suggest that it is not a space, a pathogen, a practice or an attitude that makes an individual human subject ill; rather, illness emerges within an assemblage of human and nonhuman forces (Andrews, 2017; Atkinson, 2013; Foley, 2011; Duff, 2014). In other words, it is the assemblage, rather than any one individual body (human or nonhuman) within it, that becomes ill. This approach extends the logic of earlier attempts to formulate a political ecology of health (King, 2010) by making the ecology (or the assemblage) itself the focus of analysis (Atkinson and Scott, 2015). Although it may be important for clinicians to concentrate on the habits of individual human bodies, health geographers after the posthuman turn have sought to understand the ways health emerges as a social, affective and material expression of human and nonhuman encounters within an assemblage of heterogeneous forces. Health is less an attribute of individual bodies and more a function of encounters immanent to a specific assemblage.

My own research into young people's experiences of recovery from mental illness has emphasized the varied assemblages in which these experiences unfold (Duff, 2012; 2016). This focus has required me to shift my gaze from the (human) subject of mental illness to the processes of *becoming well* that recovery partially describes as it advances and retreats within an assemblage of human and nonhuman forces. In this respect, I have refused to treat participants as the locus of their own individual recovery journeys, focusing instead on social, affective and material encounters within an assemblage of forces, and the lines of becoming well that some of these encounters express. This does not mean that I have ignored the role of human agency in recovery, as if individuals who experience mental illness have no impact on their own recovery. Yet it has meant that I have sought to understand human agency as an effect of encounters between forces, not as an innate characteristic of discrete human subjects. I have adopted this perspective mainly because I have wanted to account for the force of the myriad nonhuman bodies that have emerged in the varied recovery processes and contexts I have participated in (see Vannini, 2015).

What I have discovered is that while clinicians and psychiatrists may regard mental illness as an exclusive property of (some) human brains, the process of recovery from mental illness far exceeds this locus, branching out to include a host of human and nonhuman forces. This includes the social, affective and material support of clinicians, carers, friends and family, but also the force of nonhuman objects, factors and settings, such as the guitar that helps to settle feelings of anxiety or boredom, the park bench in the sun that affords views of the city center, or the car that opens up a territory to new encounters and opportunities (Duff, 2016). As each of these human and nonhuman forces is drawn into an assemblage of health, novel lines of becoming emerge, tracing the social, affective and material contours of recovery. Recovery, in this respect, is a function of a broader assemblage of health, rather than a condition of any one body or force within the assemblage.

Emerging lines of inquiry

Health geographers are only beginning to grapple with the challenges presented in and by posthumanism for the study of health and illness and its social, affective, material and discursive contexts. One of the key tasks for pioneers like Atkinson (2013), Andrews (2014) and King (2010) has been to determine what is significant about posthumanism for health geographers and how posthumanism differs from, or coheres with, cognate theoretical and conceptual developments in science studies, continental philosophy, feminisms and social theory. One might reasonably note that health geographers currently enjoy a surfeit of theoretical resources and orientations. Indeed, it is common for scholars to adopt ideas and concepts from diverse theoretical orientations, blurring distinctions between them (see Kearns and Moon, 2002). For this reason, one of the key challenges for health geographers interested in exploring the implications of posthumanism for their discipline may well lie in determining with greater clarity what specific resources may be derived from discussions of posthumanism for the study of health geography and how these resources may be used either in tandem or in creative tension with conceptual and empirical developments in cognate theoretical discussions. Given recent interest among health geographers in assemblage thinking (Anderson et al., 2012; Atkinson, 2013; Dilkes-Frayne and Duff, 2017; Foley, 2011) and affective atmospheres (Bissell, 2010; Boyer, 2012; Conradson, 2005; Duff, 2016), one particularly fruitful line of inquiry may be found in considering how posthumanism might help sharpen and refine the empirical analysis of assemblages and atmospheres by clarifying how human and nonhuman objects, bodies and spaces interact in particular contexts.

One area of research in which this kind of clarification would be especially useful is ongoing analyses of the social and structural determinants of health. The challenge for health geographers interested in the social, economic and political contexts of health and illness is to carefully document the ways distal social and economic forces come to mediate the proximate lived experience of health and illness in specific settings. The challenge is to explain how remote human and nonhuman forces actually work *at a distance* to socially, materially and affectively shape lived experience. As I have argued elsewhere (Duff, 2014), most social-scientific discussions of these matters simply take social determinants like class, income inequality and gender as given – as aspects of a wider social context – and neglect to indicate how these aspects actually come to matter in a given setting (see also Andrews, 2017; Fox and Alldred, 2016). In this way, the social determinants of health become reified as invariant features of social context, as an indefinable background to the major focus of attention, without ever crystallizing into a coherent, empirically verifiable feature of health and illness as each manifests in a specific time and place.

Posthumanism is uniquely placed to help health geographers overcome this reification of the social determinants of health and to begin, once again, the patient empirical work of tracing how diverse bodies, objects and forces actually come to shape, disrupt and transform experiences of health and illness in place. By helping health geographers conceptually and empirically come to grips with how heterogeneous forces combine in the rhythms and pulses of subjectivity, with the ways in which place itself is made and unmade, and with the ways in which health and illness are expressed in an assemblage of forces, recent discussions of posthumanism might finally help health geographers overcome the false antinomies of inside and outside, subject and object, proximate and distal that continue to dog discussions of the social determinants of health (see also Andrews, 2017; Greenhough, 2010; Whatmore, 2006). Following these lines of inquiry ought to open up new ways of conceptualizing and interrogating the link between health and place, as well as new ways of thinking about the lived experience of health and illness. In pursuing these lines, geographers have begun to assemble the rudiments of a novel posthuman, more-than-human, health geography.

References

Anderson, B., Kearnes, M., McFarlane, C. and Swanton, D. (2012). On assemblages and geography. *Dialogues in Human Geography*, 2(2), pp. 171–189.

Andrews, G. J. (2014). Co-creating health's lively, moving frontiers: brief observations on the facets and possibilities of non-representational theory. *Health & Place*, 30, pp. 165–170.

Andrews, G. J. (2017). "Running hot": placing health in the life and course of the vital city. *Social Science & Medicine*, 175, pp. 209–214.

Andrews, G. J., Chen, S. and Myers, S. (2014). The "taking place" of health and well-being: towards non-representational theory. *Social Science & Medicine*, 108, pp. 210–222.

Andrews, G. J., Evans, J., Dunn, J. R. and Masuda, J. (2012). Arguments in health geography: on sub-disciplinary progress, observation, translation. *Geography Compass*, 6, pp. 351–383.

Apostolidou, S. and Sturm, J. (2016). Weighing posthumanism: fatness and contested humanity. *Social Inclusion*, 4(4), pp. 150–159.

Atkinson, S. (2013). Beyond components of well-being: the effects of relational and situated assemblage. *Topoi*, 32(2), pp. 137–144.

Atkinson, S. and Scott, K. (2015). Stable and destabilised states of subjective well-being: dance and movement as catalysts of transition. *Social and Cultural Geography*, 16(1), pp. 75–94.

Barad, K. (2007). *Meeting the universe halfway: quantum physics and the entanglement of matter and meaning*. Durham, NC: Duke University Press.

Bennett, J. (2010). *Vibrant matter: a political ecology of things*. Durham, NC: Duke University Press.

Bissell, D. (2010). Passenger mobilities: affective atmospheres and the sociality of public transport. *Environment and Planning D: Society and Space*, 28(2), pp. 270–289.

Boyer, K. (2012). Affect, corporeality and the limits of belonging: breastfeeding in public in the contemporary UK. *Health & Place*, 18(3), pp. 552–560.

Braidotti, R. (2002). *Metamorphoses: towards a materialist theory of becoming*. Cambridge: Polity Press.

Castree, N. and Nash, C. (2006). Posthuman geographies. *Social & Cultural Geography*, 7(4), pp. 501–504.

Castree, N., Nash, C., Badmington, N., Braun, B., Murdoch, J. and Whatmore, S. (2004). Mapping posthumanism: an exchange. *Environment and Planning A*, 36(8), pp. 1341–1363.

Conradson, D. (2005). Landscape, care and the relational self: therapeutic encounters in rural England. *Health & Place*, 11(4), pp. 337–348.

Coole, D. and Frost, S. (eds.) (2010). *New materialisms: ontology, agency, and politics*. Durham, NC: Duke University Press.

Davies, G., Day, R. and Williamson, S. (2004). The geography of health knowledge/s. *Health & Place*, 10(4), pp. 293–297.

DeLanda, M. (2016). *Assemblage theory*. Edinburgh: Edinburgh University Press.

Dilkes-Frayne, E. and Duff, C. (2017). Tendencies and trajectories: the problem of subjectivity in posthumanist social research. *Environment & Planning D*, 35(5), pp. 951–967.

Duff, C. (2012). Accounting for context: exploring the role of objects and spaces in the consumption of alcohol and other drugs. *Social and Cultural Geography*, 13(3), pp. 13–26.

Duff, C. (2014). *Assemblages of health: Deleuze's empiricism and the ethology of life*. Rotterdam: Springer International.

Duff, C. (2016). Atmospheres of recovery: assemblages of health. *Environment and Planning A*, 48(1), pp. 58–74.

Foley, R. (2011). Performing health in place: the holy well as a therapeutic assemblage. *Health & Place*, 17(2), pp. 470–479.

Fox, N. J. and Alldred, P. (2016). *Sociology and the new materialism: theory, research, action*. London: Sage.

Greenhough, B. (2010). Vitalist geographies: life and the more-than-human. In: B. Anderson, ed., *Taking place: non-representational theories and geography*. Surrey: Ashgate, pp. 37–54.

Greenhough, B. (2011). Citizenship, care and companionship: approaching geographies of health and bioscience. *Progress in Human Geography*, 35(2), pp. 153–171.

Greenhough, B. (2014). More-than-human-geographies. In: A. Paasi, N. Castree, R. Lee, S. Radcliffe, R. Kitchin, V. Lawson and C. Withers, eds., *The Sage handbook of progress in human geography*. London: Sage, pp. 94–119.

Grosz, E. (1994). *Volatile bodies: toward a corporeal feminism*. Indianapolis: Indiana University Press.

Haraway, D. (1991). *Simians, cyborgs, and women: the reinvention of nature*. New York: Routledge.

Kearns, R. and Moon, G. (2002). From medical to health geography: novelty, place and theory after a decade of change. *Progress in Human Geography*, 26(5), pp. 605–625.

King, B. (2010). Political ecologies of health. *Progress in Human Geography*, 34(1), pp. 38–55.

Kirsch, S. (2013). Cultural geography, materialist turns. *Progress in Human Geography*, 37(3), pp. 433–441.

Latour, B. (2005). *Reassembling the social: an introduction to actor-network-theory*. Oxford: Oxford University Press.

Law, J. (2002). Objects and spaces. *Theory, Culture & Society*, 19(5/6), pp. 91–105.

Lorimer, J. (2009). Posthumanism/posthumanistic geographies. In: R. Kitchin and N. Thrift, eds., *International encyclopedia of human geography*. London: Elsevier, pp. 334–354.

Marmot, M., Friel, S., Bell, R., Houweling, T. A., Taylor, S. and Commission on Social Determinants of Health. (2008). Closing the gap in a generation: health equity through action on the social determinants of health. *The Lancet*, 372(9650), pp. 1661–1669.

Panelli, R. (2010). More-than-human social geographies: posthuman and other possibilities. *Progress in Human Geography*, 34(1), pp. 79–87.

Rock, M. J., Degeling, C. and Blue, G. (2014). Toward stronger theory in critical public health: insights from debates surrounding posthumanism. *Critical Public Health*, 24(3), pp. 337–348.

Vannini, P. (ed.) (2015). *Non-representational methodologies: re-envisioning research*. London: Routledge.

Whatmore, S. (2006). Materialist returns: practising cultural geography in and for a more-than-human world. *Cultural Geographies*, 13(4), pp. 600–609.

21

NON-REPRESENTATIONAL GEOGRAPHIES OF HEALTH

Jennifer Lea

Broadly speaking, non-representational theory (NRT) has enabled a shift of geographical focus to the realms of the practical, the bodily and the material. A way of thinking and doing that largely stemmed from Thrift's writings in the 1990s and 2000s (e.g., Thrift, 1996, 2000), NRT has grown and diversified to be an influential reference point in contemporary human geography. Drawing on a range of post-structural and practice theorists (see Thrift, 1996, for a detailed treatment of these points of reference), NRT has been seen to have consequences theoretically and practically. These include a shift in focus toward the "practical rather than cognitive" (Thrift, 1997, p. 126) aspects of life, marking a turn toward the embodied and experiential aspects of life and refiguring what we might consider to be valid academic knowledges.

As geographers have grappled with the consequences of this, they have begun to question how we might *do* geographical research, and this has led to a set of methodological innovations and interventions (see, for example, Vannini, 2015). This chapter sets out to chart the relationship between NRT and health geographies, considering how NRT has enabled geographers interested in health to refigure their conceptions of the body, place and space, and health itself.

Health geographers have been relatively slow to draw on NRT, even though there might seem to be overlaps, as the body is increasingly foregrounded in both areas. Doel and Segrott (2003) staged what was perhaps the first explicit conversation between questions of health and NRT, suggesting that there would be value in problematizing some of the key terms of reference for health geographers (e.g., health, illness, disease, body, medicine, place, space, experience, need, use and provision) because societal shifts had necessitated a new theoretical vocabulary. Revisiting this intervention more than a decade on, it seems that this situation has shifted somewhat. At the very least, geographers have used concepts aligned with NRT to question what health and the body mean, and this chapter will trace some of the ways in which this has happened.

More recently, geographers have begun to build a more explicit agenda for the inclusion of NRT in geographies of health. This has largely been played out through exchanges between Andrews, Chen and Myers (2014), Kearns (2014) and Hanlon (2014). Andrews, Chen and Myers (2014) start the conversation, suggesting that NRT offers the opportunity to *explain* the notion of well-being, because of the close attention to the "intimate processes through which well-being arises" (p. 213) that NRT offers. Working with the concept of affect, they indicate how "well-being arises initially as an energy and intensity through the physical interaction of human bodies and non-human objects, and is experienced as a feeling state" (p. 211). They argue that what is distinctive about their new perspective is its suggestion that well-being is something

144

that arises prior to the formation of the subject, rather than as something that is solely a subjective feeling state. Similarly, rather than being a quality of place per se, they suggest that well-being might "emerge *as* an affective environment" (p. 219).

Commentators on this paper include Kearns (2014), who, agreeing that NRT has potential in the field of health geography, names three areas he sees to be of particular value (firstly, embodiment; secondly, reducing reliance on anthropocentric or environmental deterministic understandings of well-being; and, thirdly, showing how particular human and non-human assemblages generate experiences of health). He also raises critiques, the most substantial being that a focus on well-being skews analyses away from questions of ill health and disability and also obscures the privilege of the forms of well-being that Andrews, Chen and Myers record. Hanlon (2014) adds another intervention, taking issue with Andrews, Chen and Myers' interpretation of affect and well-being, indicating some further useful concepts from NRT (materiality thought in its broadest sense and the more-than-rational realms of feelings, emotion and affect). In a second intervention, Andrews (2014) outlines the characteristics of thought and practice from NRT that he thinks might reinvigorate health geographies (summarized in Table 21.1).

As a series of papers that open debate around the value of NRT for geographies of health, these explore some fertile areas. Rather than reiterate these areas, however, this chapter offers some further ideas about how concepts drawn from NRT might be of value to health geographers. The chapter looks first at the (re)conceptualization of the body that might happen through engagement with ideas of affect and by foregrounding bodily knowledges, before secondly looking at how relationship between place, landscape and the therapeutic is refigured, and thirdly rethinking the notion of health itself.

Rethinking the body

One of the positive directions that Kearns (2014) suggests NRT might offer to health geographers is the focus on embodiment and the body. Although health geographers and disability geographers have, for many years, focused upon the body and embodied experiences, NRT allows a different articulation of the body. This has, perhaps, been most clearly articulated by Dewsbury (2000), who distinguishes between approaches that take the body as the *object* of knowledge and those that take it as the *subject* of knowledge. Briefly, the tendency in approaches that foreground embodiment, such as those that have emerged in health geographies, is to take the body as the *object* of knowledge. These approaches tend to take the body as a passive object, one

Table 21.1 Characteristics of thought and practice from non-representational theory

Characteristic	Description/emphasis
Onflow	Ongoingness, fluidity, potentiality
Relational materialism	Post-humanist, assembly of bodies and objects, associations and relationalities
Transpersonal sensations	Shared, affective environments, feelings are not personal
Practice and performance	Timings and spacings, attention to bodies
Vitality	Life has spirit
Virtuality and multiplicity	Non-fixed, non-linear, ruptured nature of space and time
Everyday	Rhythms of people's lives, everyday places
Wonderment	Positivity, not just looking at power
Witnessing, changing, boosting	Witness, act into change, boost the active world
Diverse and lively theory	What does theory help one get at?
The irrealis mood	Living, lively style of communication

Summarized from Andrews (2014, p. 165)

that can be inscribed by social and cultural forces. The body is taken to be a self-contained, whole organism whose "matter and energy flows" are bound "neatly within unleaking ends" (Dewsbury, 2000, p. 482).

Dewsbury argues that we need to add to these existing accounts, separating the body out from its embodiment to take it as the *subject* of knowledge. This means that the body is approached less as a pre-existing tangible *object* and more as a *connective space* that is constituted through its relations with other bodies, objects and contexts. This has two consequences for health geographers. The first of these is the imagination of the body that emerges; most often, affect has been used as a concept to enable geographers to develop this connective imagination. The chapter will look at how affect has made a difference in approaches to (dis)ability and chronic pain. The second of these is to take the body as knowledgeable itself, rather than the passive object of knowledge, so that the "biological flow[s] of energy, matter, and stimulating chemical fluids" (Dewsbury, 2000, p. 485) are added into our understandings of the body. The chapter will focus on pre-cognitive and alternative health knowledges as areas where NRT and health geography have come into conversation.

What can a body do?

Although there is no shorthand definition for "affect," largely because ideas of affect can be traced through a number of different theoretical lineages, much of the work on affect in health geography draws on Deleuzian definitions, which themselves have a relationship to Spinoza's writing on affect. In this version, affect is a "prepersonal intensity corresponding to the passage from one experiential state of the body to another and implying an augmentation or diminution in that body's capacity to act" (Massumi, 1988, p. xvii). Geographers with an interest in NRT have found affect to be a useful conceptual device for a number of reasons, including that affect (a) acts as a cipher for an atmospheric realm of forces and feelings that undergirds social life and (b) enables a different way of thinking about corporeality and human capacities.

It is this rethinking of the body that is the focus of this section; from a Deleuzian perspective, affect means rethinking the body through the question of what it can do, or how it can respond to the world, rather than looking at its form or functions. Deleuze and Guattari (1988, p. 257) map out a manifesto for this Spinozist ethological view of the body:

> We know nothing about a body until we know what it can do, in other words, what its affects are, how they can or cannot enter into composition with other affects, with affects of another body, either to destroy that body or to be destroyed by it, either to exchange actions and passions with it or to join with it in composing a more powerful body.

The economy of linkages, experiments, lines and relations that affect shifts us into allows us to think of the body as a "mobile machine" that "may connect to the environment in a variety of different ways" (May, 2005, p. 123), and this machinic ontology has directed geographers' interest in questions of health in a number of ways.

One of these is the questioning of the idea of the essentialized body. One of the strongest articulations of this has been the bringing together of NRT and geographies of disability by Hall and Wilton (2016). In an agenda-setting paper, they use work on affect (among other NRT-related concepts) to herald a new phase in the study of disability, one that aims to "unsettle broader assumptions about the nature of the 'able body'" (p. 2) and to destabilize the "authentic" disabled subject in order to think about the making of dis/ability through "relational connections between heterogeneous bodies, objects and environments" (p. 3). Hall and Wilton argue that asking how *all* bodies might be dis/abled through their everyday geographies is a liberatory question because it removes prior assumptions and means that we are not always already approaching bodies through the lens of disability. It also pluralizes bodies so that they are not just conceptualized in opposition to the normative body. Unsettling *all* bodies is a way of "drawing attention to the shifting and prolific diversity of body sites and forms, sensory activities, cognitive abilities, mental states, illnesses, injuries and other differences" that "begins to undo the taken-for-granted dualism between a valued able-body and its disabled 'other'" (p. 10). Focusing on individual instances of connection and disconnection, via affect, can productively disrupt pre-existing and assumed understandings of disability, capacities and relationships.

Another way that affect has disrupted understandings of health is in Bissell's (2009) work on chronic pain. While Bissell does not frame his writing through geographies of health, his writing has the potential to inform research on suffering and pain in health geographies. By looking at his own chronic pain, Bissell uses affect to undo the centering of pain in the human, with a number of consequences. Affect directs attention to the inter-relationships between bodies (where "bodies" have a broader definition than just the corporeal body), and these relationships are important for understanding pain, the unpredictability and unforeseen ways in which pain comes and goes in intensity across time and space, and the "possible field of action" and "realm of possibility for the body in pain" (p. 82). Taking pain out of the individual and distributing it across broader settings also shifts the point of view beyond the immediate personal intensity of pain, thus offering a different perspective on pain. Bissell argues that thinking pain through the concept of affect also has the effect of disrupting the often-assumed "predictable and fixed trajectory" (p. 84) of pain, thus opening up a (hopeful) potential for change. This is contra the effect of chronic pain, which "can close down possibility so much that it becomes increasingly difficult to anticipate or have *any* sort of future" (p. 88). More broadly, Bissell indicates how health geographers looking at pained or suffering bodies might contribute back to debates in NRT, as he argues that looking at the body in pain brings "the ungraspability, indecipherability and incoherence of the body . . . into sharper focus" (p. 85).

Body knowledges

As indicated above, concepts drawn from NRT have stimulated geographers to explore a range of new knowledges associated with the body. Affect has pointed geographers toward the pre-cognitive realm; as Spinks (2001) argues, the affective realm is made up of a "series of inhuman or presubjective forces and intensities" (p. 24). The suggestion that a whole raft of pre-cognitive processes occur before we come to a conscious grasp of a situation troubles any model of a reflective rational human subject and foregrounds the significance of other knowledges (e.g., intuition, habit, reflexes). Health geographers such as Barnfield (2016a) have used the lens of the pre-cognitive to complexify understandings of physical activities such as running and the "tangled processes that shape health practices" (p. 1123). Arguing that the pre-cognitive could usefully be layered into public-health understandings of physical activity and sport, Barnfield (2016b) makes a clear case that "becoming and staying physically active involves more than just purely mindful and rational decision making" (p. 281). Barnfield draws heavily on the work of McCormack (2003), whose work on the therapeutic geographies of dance movement therapy (DMT) offers an example of a body practice that aims to explicitly work in pre- or non-cognitive registers in ways that "can but do not necessarily need to be brought into a relation of verbal or linguistic signification" (p. 492). McCormack suggests that we might usefully supplement our understandings of the therapeutic by attempting to "apprehend those activities and practices whose sense is not always revealed through an interpretative relation with linguistic signification" (p. 491), such as DMT. He also fleshes out a definition of "therapeutic spaces" that does not specify a "direction" for transformation (e.g., feeling better, or giving rise to wellness), describing them as "those spaces emergent through the enactment of practices that explicitly attempt to facilitate a kind of transformation in awareness, thinking, feeling and relating" (pp. 490–491). This anticipates a later section of the paper, which looks at how geographers influenced by NRT are beginning to problematize normative definitions of "health."

Geographers have also looked at bodily knowledges that do not directly map on the non- or pre-cognitive. One of the key examples of this is energy, which is at the basis of the alternative model of health that exists in many complementary and alternative medicines. Energy, a knowledge that might be anomalous or that may push at the fringes of what is believable or acceptable, is nonetheless integral to many health-care practices. Work has already been done to trace how energy is being mapped out at the intersection of NRT and health geographies. Philo, Cadman and Lea (2015), for example, pick out the threads of some "new energy geographies" wherein an "under-current of concern for 'energies'" (p. 41) can be discerned. Examples given include Lea's (2008, 2009) work on bodies in yoga and massage and Foley's work on Irish wells (2011). Philo, Cadman and Lea (2015) go on to outline some potential areas for future focus, including *chakric geographies* (tracing how energetic lines are experienced as providing key alignments for health), *geographies of environmental connectedness* (where health/ill health is shaped and experienced through the energies of places and spaces), *geographies of mental energy* (which might be cultivated by particular bodily practices, such as meditation), and *atmospheric geographies*, wherein bodies in the pursuit of health – through, for example, a shared yoga practice – might generate a kind of healthful energy.

Affect and energy both provide connective imaginations that indicate how different elements, objects, bodies and surfaces might be brought into inter-relation toward the constitution of health or ill health (in the broadest terms). This has resonances with recent work on place, space and therapeutic landscapes influenced by NRT, and it is to this that the chapter now turns.

Rethinking place and space

NRT has prompted a careful rethinking of place and space, which has had a range of impacts (particularly) on studies of therapeutic landscapes. The relational configuration of the subject in NRT has opened up new lines of questioning around the encounter between self and landscape that lies at the heart of therapeutic landscapes. Conradson's (2005) paper was the first explicit attempt to reconceptualize therapeutic landscapes in relation to work being done under the auspices of NRT, and it drew on Thrift's ecologies of place in order to develop an understanding of the agency of the material environment in the constitution of therapeutic landscape experiences. Attending to the relational encounter between people and place, Conradson (2005) develops an understanding of how therapeutic outcomes might emerge through the coming together of "different entities, jostling and engaging with each other, continually evolving in a collective and distributed fashion" (p. 340). The relational approach set out here problematized the notion that places might be intrinsi-cally therapeutic and suggests that we must extend our analyses beyond landscape qualities alone to consider what emerges through the interaction of person and landscape (p. 338). This work has been extended by researchers who hold on to "therapeutic landscapes" as a useful term (including Lea, 2009), who foreground the active of the natural in the creation of specific therapeutic configurations and advocate an attention to the small spaces of connection between body and world or who have proposed alternatives to, or abandoned altogether, the concept of therapeutic landscapes.

Foley (2011), for example, makes a shift to therapeutic *assemblages* to emphasize the coming together of the therapeutic experience. Dawney (2011) offers an NRT-aligned intervention that interrogates the relationship between landscape and the therapeutic, but which does not refer to the therapeutic-landscapes literature. She brings work by Foucault and Spinoza together to frame a discussion of walking in the English countryside as a (therapeutic) practice of the self, arguing that a therapeutic imaginary (a set of "embodied ideas about, and ways of working on the self, involving ideas about health and healing that suggest one can work or be worked upon in a process of making-better" (p. 546)) is layered into experiences of landscape. She traces a reciprocal relationship between the individuals who walk in the landscape and the landscape they walk in; "therapeutic space is performed by embodied subjects who make it therapeutic" (p. 549). Dawney draws wider social imaginaries about health, and more specifically our healthy bodies in (particular) places, into her account of the therapeutic co-configuration of body and landscape.

Whether explicitly named as work on therapeutic landscapes, or not, the questions of the therapeutic and bodily experiences have been the focus of discussion, and often the *landscapes* of therapeutic landscapes have been left somewhat under-theorized. Hanlon (2014) makes this point in his response to Andrews, Chen and

Myers (2014), and he points toward Macpherson's work on the landscape experiences and knowledges of people with visual impairments, which considers how the body and landscape might be rethought through the lens of (dis)ability. Macpherson (2010) notes that NRT has prompted considerable rethinking of the concept of landscape, so that rather than its being seen as an "inert background or setting for human action" it has an "animating" effect, bringing the body into being; "our sense of being an embodied subject arises out of our sensuous engagement with the world" (p. 6). In this work, Macpherson indicates that health geographers can use NRT to advance understandings of the role that landscape can play in the formation of (healthy/ill) selves and how specific landscapes are brought into being by a wide range of (healthy/ill) selves.

NRT-influenced work on therapeutic landscapes has not just sought to extend the theoretical reach of the concept, but has also extended the kinds of landscapes that are thought to be interesting, beyond the green spaces and health-care spaces that have tended to be foregrounded in the literature. In particular, and perhaps inspired by the sensory and bodily focus of NRT, there has been an increasing interest in therapeutic relations to water. Foley has been instrumental in this turn with his explorations of open-water swimming, which interrogate swimming as an immersive therapeutic practice. Foley and Kistemann (2015) set out a new agenda for the exploration of "healthy blue space" (p. 158), touching on themes and concepts drawn from NRT, such as the relational geographies of inter-subjective encounters, embodiment, and bodily movement and activity.

Rethinking health

The chapter started with Doel and Segrott's (2003) provocation of the need to rethink key terms in health geographies, in line with broader societal and cultural changes. "Health" is perhaps the most significant term here, and one key shift in the terrain of health is the reconfiguration of health in terms of pleasure and seduction, rather than illness and disease (although not, by any means, for everyone in society). This need to reconfigure ideas of health is brought about not only by this hedonic turn, but also by conceptual developments in NRT that question the status of the subject, because this has been perhaps the key reference point for definitions of "health." NRT undoes the idea of the rational subject who is coherent and bounded, replacing it with a "radically constructivist understanding of the 'composition of subjectivities'" (Simpson, 2017, p. 10). These composed subjectivities (wherein "all kinds of things are brought into relation with one another by many and various spaces through a continuous and largely involuntary process of encounter" (Thrift, 2007, p. 7) are confronted by Duff (2014) in his book *Assemblages of Health*, which seeks to rethink the "ontological and epistemological status of health at a time when the 'human subject', to which the attribution of health necessarily refers, seems everywhere in retreat" (p. ix). Duff advocates destabilizing the idea of health through the "empirical study of bodies in their posthuman assembling, in their becoming well or ill" (p. 6; see also Andrews, 2017), in order to give rise to contextually located versions of health and wellness.

In addition to Duff's own work on illegal drugs and psychoactive substances, other geographers have begun to destabilize normative ideas of health through contexts as diverse as smoking and childhood (dis) ability. Tan's (2013) work on smoking is particularly powerful here, because it both reconfigures the broader meaning system within which we might insert notions of health and begins to undo the regime of what geographers of health have traditionally taken to be healthy. Tan questions categories of health and illness through the concept of affect; rather than assuming that smoking is intrinsically unhealthy, she instead asks what smoking can do in a particular assemblage. Smoking emerges, through this perspective, as a resource in people's everyday geographies; "smoking in certain places can draw out one's power to act and affect" (p. 174). Tan rejects "inherent/objective distinctions between good/bad" (p. 180) in favor of a fluid understanding of well-being that is distinct from, and even in opposition to, biomedical definitions of "health." Another (perhaps less controversial) attempt to destabilize health using affect is offered by Stephens, Ruddick and McKeever (2015), who look at children's experiences of (dis)ability. They ask what bodies can do in relation to specific assemblages (of bodies, the built environment, social meanings and the adaptations that are made for the children) that might enhance or constrain the children's capacity to act. They take a broad view on what "capacity to act" might be, discussing physical access, senses of inclusion, and also pleasure. They argue that this contextual attention to the everyday lives of these children, listening to their voices, and using affect as a conceptual framework, allows for the "lessening [of] our grip on our idea of what is 'right' for these children, allowing space for multiple identities, multiple preferences and multiple right ways of doing things" (p. 200). This has the potential to reconfigure ideas of health, and also the broader field of power relations within which ideas and practices of health are situated, challenging the received and pervasive understandings of good and bad that emerge in relation to childhood disability in school and home spaces.

Conclusions

The kinds of shifts that the chapter indicates could be seen as useful and productive, but they could also be seen as unacceptable reconfiguration of the very grounds of health geographies, wherein the very coordinates by which health geographers orientate themselves are shifted too far. Unease could well be felt, particularly in response to the undoing of health, not least because health is one of the key arenas in which we make judgments, seek out authority and configure our lives. If "anything goes" in the name of health, it undoes these sources of authority and meaning. The work drawn upon in the chapter does little to assuage the critique that Kearns (2014) outlined; that NRT shifts the focus of health geographers even further away from illness and suffering. This is not to say that NRT cannot extend these lines of inquiry, and in stretching out what we might consider to be the relationship between NRT and health geographies, the chapter has attempted to sketch out some of the future directions by which these difficult and unsettling terrains might be tackled.

References

Andrews, G. (2014). Co-creating health's lively, moving frontiers: brief observations on the facets and possibilities of non-representational theory. *Health & Place*, 30, pp. 165–170.

Andrews, G. (2017). "Running hot": placing health in the life and course of the vital city. *Social Science & Medicine*, 175, pp. 209–214.

Andrews, G., Chen, D. and Myers, S. (2014). The taking place of health and well-being: towards non-representational theory. *Social Science & Medicine*, 108, pp. 210–222.

Barnfield, A. (2016a). Grasping physical exercise through recreational running and non-representational theory: a case study from Sofia, Bulgaria. *Sociology of Health and Illness*, 38(7), pp. 1121–1136.

Barnfield, A. (2016b). Public health, physical exercise and non-representational theory – a mixed method study of recreational running in Sofia, Bulgaria. *Critical Public Health*, 26(3), pp. 281–293.

Bissell, D. (2009). Obdurate pains, transient intensities: affect and the chronically pained body. *Environment and Planning A*, 41, pp. 911–928.

Conradson, D. (2005). Landscape, care and the relational self: therapeutic encounters in rural England. *Health & Place*, 11(4), pp. 337–348.

Dawney, L. (2011). Social imaginaries and therapeutic self-work: the ethics of the embodied imagination. *The Sociological Review*, 59(3), pp. 535–552.

Deleuze, G. and Guattari, F. (1988). *A thousand plateaus: capitalism and schizophrenia*. London: Continuum.

Dewsbury, J. D. (2000). Performativity and the event: enacting a philosophy of difference. *Environment and Planning D*, 18, pp. 473–496.

Doel, M. and Segrott, J. (2003). Beyond belief? Consumer culture, complementary medicine, and the dis-ease of everyday life. *Environment and Planning D*, 21, pp. 739–759.

Duff, C. (2014). *Assemblages of health: Deleuze's empiricism and the ethology of life*. London: Springer.

Foley, R. (2011). Performing health in place: the holy well as a therapeutic assemblage. *Health & Place*, 17, pp. 470–479.

Foley, R. and Kistemann, T. (2015). Blue space geographies: enabling health in place. *Health & Place*, 35, pp. 157–165.

Hall, E. and Wilton, R. (2016). Towards a relational geography of disability. *Progress in Human Geography*, 41(6), pp. 727–744.

Hanlon, N. (2014). Doing health geography with feeling. *Social Science & Medicine*, 115, pp. 144–146.

Kearns, R. (2014). The health in "life's infinite doings": a response to Andrews et al. *Social Science & Medicine*, 115, pp. 147–149.

Lea, J. (2008). Retreating to nature: rethinking therapeutic landscapes. *Area*, 80(1), pp. 90–98.

Lea, J. (2009). Liberation or limitation? Understanding Iyengar Yoga as a practice of the self. *Body and Society*, 3(15), pp. 71–92.

Macpherson, H. (2010). Non-representational approaches to body-landscape relations. *Geography Compass*, 4(1), pp. 1–13.

Massumi, B. (1988). Translators forward: pleasures of philosophy. In: G. Deleuze and F. Guattari, ed., *A thousand plateaus: capitalism and schizophrenia*. London: Continuum, pp. ix–xvi.

May, T. (2005). *Gilles Deleuze: an introduction*. Cambridge: Cambridge University Press.

McCormack, D. (2003). An event of geographical ethics in spaces of affect. *Transactions of the Institute of British Geographers NS*, 28, pp. 488–507.

Philo, C., Cadman, L. and Lea, J. (2015). New energy geographies: a case study of Yoga, meditation and healthfulness. *Journal of Medical Humanities*, 36, pp. 35–46.

Simpson, P. (2017). Spacing the subject: thinking subjectivity after non-representational theory. *Geography Compass*, 11(12), p. e12347.

Spinks, L. (2001). Thinking the post-human: literature, affect and the politics of style. *Textual Practice*, 15(1), pp. 23–46.

Stephens, L., Ruddick, S. and McKeever, P. (2015). Disability and Deleuze: an exploration of becoming and embodiment in children's everyday environments. *Body and Society*, 21(2), pp. 194–220.

Tan, Q. H. (2013). Smoking spaces as enabling spaces of well-being. *Health & Place*, 24, pp. 173–182.

Thrift, N. (1996). *Spatial formations*. London: Sage.

Thrift, N. (1997). The still point: resistance, expressive embodiment and dance. In: M. Keith and S. Pile, eds., *Geographies of resistance*. London: Routledge, pp. 124–151.

Thrift, N. (2000). Afterwords. *Environment and Planning D: Society and Space*, 18, pp. 213–155.

Thrift, N. (2007). *Non-representational theory: space, politics, affect*. London: Routledge.

Vannini, P. (2015). *Non-representational methodologies: re-envisioning research*. Abingdon: Routledge.

SECTION 3

Groups and peoples

22
INTRODUCING SECTION 3
Groups and peoples

Jamie Pearce, Gavin J. Andrews and Valorie A. Crooks

A recurrent theme in contemporary health geography has been to document and explain the stark social and spatial inequities in health outcomes, experiences and behaviors observed across the world. The gradients in health within and between nation states, regions and cities emerged as a key topic in Section 1 of this book. There is also an abundance of work concerned with inequalities in health between populations differentiated by their demographic characteristics, social identity or position in society (Aldridge et al., 2017; Smith, Bambra and Hill, 2016). Unsurprisingly, it is almost always the most disempowered, disenfranchised and disadvantaged groups that are habitually shown to have the poorest health outcomes and experiences across a range of indicators. This concern with social inequities is consistent with many of the fundamental concerns of other human geography sub-disciplines, and indeed across the social sciences. There is a long tradition in human geography of studying how marginalized, disempowered and potentially vulnerable groups are represented and how this depiction shapes their material circumstances and life chances. It is therefore unsurprising that health geographers have often focused their attention on particular groups of people who, by virtue of their position in society, have been materially disadvantaged in multiple ways, including related to their health and well-being. For decades now, researchers have sought to understand how people and groups identifying, or identified, in particular ways have distinct health-related experiences and outcomes (Curtis and Rees Jones, 1998; Smyth, 2008). Health geographers have lent their expertise to these deliberations through scrutinizing how space and place are implicated. The aim of this section is to systematically interrogate these issues by synthesizing the international research on, and with, various people and groups that have received the attention of health geographers in recent years.

As in other parts of the book, a number of cross-cutting themes emerged in this section (on groups and people), and three are highlighted here. First is the concern that recent social, economic and political changes at the global level have disproportionately affected the health of some of the most marginalized populations. Many of the chapters explore the profound health implications of major political and economic shifts observed in many countries for particular people and groups in recent decades. Geographical work has often considered how these health-related experiences are mediated, or even mitigated, by local particularities. Baker, for example, considers how the rollback of the welfare state and accompanying neoliberal agenda that was enthusiastically adopted in many high-income countries in recent decades has been instrumental in understanding attitudes and approaches to homelessness. These structural changes had profound implications for the types and levels of services available to homeless people, which has detrimentally affected this population's health and well-being. In discussing work in health geography on the role of informal caregivers,

Power's chapter also details the pernicious effects of neoliberalism that have diminished the formal support services available in many localities for people with long-term care needs. Often these populations have few alternatives other than to rely on the assistance and goodwill of family and friends, due to decreasing state support. As the author notes, these are concerns that have accelerated since the global financial crisis in 2007 and 2008 and the subsequent implementation of the politics and policies of austerity (Pearce, 2013). These concerns are highly relevant to other groups, and Chouinard's chapter emphasizes them in the context of people living with disability and chronic illness. Richmond and Big-Canoe stress the importance of paying more attention to the historical context of these concerns. In their chapter on health and well-being among Indigenous populations, Richmond and Big-Canoe emphasize how global processes operating over very long periods continue to be highly pertinent to contemporary health status. Using examples from Canada, the authors reveal how the legacy of colonialism – in particular, the reduction or elimination of access to land – continues to undermine the health of populations including First Nations and Inuit communities.

The second theme that is of central importance in many of the chapters in this section is a concern with the health repercussions of the multiple processes that ensure marginalized groups can be explicitly or implicitly excluded from mainstream society. This includes the way in which some groups are disregarded by virtue of their health as well as those who, because of their social position, are prevented from accessing the care and resources that they require. This concern is apparent in Chouinard's chapter, which shows the subtle but important ways in which people living with disabilities and chronic illnesses are routinely excluded (physically and socially) from public and private spaces – for example, changing rooms or the beach. Research in health geography has been particularly instructive in establishing how place and space are enabling or disabling for people living with impairments and chronic illnesses. Power considers the ways in which informal caregivers can be excluded, forgotten or taken for granted in developing health-care plans for those with long-term care needs. The author documents how this recognition and the burden placed on an informal caregiver can negatively affect his or her health and well-being as well as the levels of distress of the person being cared for. In their chapter on aging, Coleman and Wiles consider how the stigma of old age and the aging body reduces the agency of some older people, with implications for their health, well-being and care needs. Similar concerns about the lack of voice and representation is expressed in the context of children's health (see Yantzi's chapter) and homeless individuals (see Baker's chapter), two groups that are regularly excluded from discussions of health, well-being and health-care needs. It is also apparent from the chapter by Newbold that immigrants regularly face various forms of exclusion, not least from accessing health-care provision. As the author demonstrates, new immigrants are potentially confronted with a multitude of social, cultural and financial barriers to accessing medical treatment, which in turn generates additional health needs. Perhaps most vividly, the chapter by Lewis demonstrates how lesbian, gay, bisexual, transgender, and other queer (LGBTQ) individuals are often made invisible by virtue of the paucity of research into the health concerns of this group, and because they often go uncounted in data-collection regimes that adopt crude identity binaries (e.g., male/female). Further, the lives of LGBTQ people continue to be constructed as non-normative, with implications for social, spatial and institutional exclusion.

The final cross-cutting theme, evident in all the chapters in this section, is the recent and important shift in research practice across studies of disempowered groups to ensure the subjects of the studies are active participants in the research. Many of the chapters emphasize the crucial contribution of health geographers in describing and explaining the health disadvantages often experienced among disempowered and minority groups (see, for example, the chapter by Richmond and Big-Canoe on Indigenous populations and the chapter by Newbold on immigrant health). Nonetheless, there is a general consensus across the chapters that enabling these groups to participate in the research process has been a welcome and revealing evolution in health geography. As the chapter by Coleman and Wiles and the chapter by Chouinard astutely note, this sub-disciplinary transition is related to the broad shift in health geography from an almost exclusive biomedical focus toward an agenda that incorporates broader socioecological accounts and a greater use of critical social theory. For example, in their overview of health geography's contributions to mental-health research, Twigg and Duncan draw a distinction between studies that have *examined people* with diagnosed mental-health issues from research actively seeking to involve people, often using participatory methods, to understand the *lived experience* of mental health. Their chapter considers how this latter approach has helped reveal how places are implicated in understanding new forms of mental-health care and the lived experience of mental health. Baker's chapter discusses how work by health geographers is being used to critique attitudes toward, and services for, homeless people. The central tenet of this chapter is that homelessness tends to be treated as a medical concern, with homeless people being objectified as being unhealthy and requiring clinical intervention. Baker calls for more engagement by health geographers with the concerns of homeless populations, including greater involvement of homeless people, and their advocates, in the research process. Similarly, Yantzi notes in the context of work on children's health the welcome shift in academic practice from research that is *about* children, which treats the child as a passive subject, to an agenda that is researching *with* children and including them as meaningful research participants. As these examples demonstrate, the evolution in health geography toward greater participant involvement in the research is more than a methodological drift. Rather, it reflects a recognition that people are active subjects with insights that are vital to understanding the ways in which space and place actively shape the health of people living in a diverse set of circumstances.

In summary, this section of the book identifies some key populations that have been the focus of attention in health-geography research. It is clear that health geographers have been at the vanguard of international research seeking to identify the barriers to, and facilitators of, positive health experiences among some of the most marginalized groups in society. As with the other sections in this book, a different set of editors would almost certainly have identified an alternative configuration of chapters. There are some themes that we deliberately excluded because they cut across so many other chapters in the book. For example, we do not have a chapter that focused exclusively on gender. This was because work on gender, and feminist perspectives on health geography, are relevant to just about all the chapters in the book. Throughout the book, gender is examined by the chapter authors in a variety of social, geographical and health contexts. We decided that gender was better discussed using the multiple theoretical and thematic prisms adopted by our authors. Similarly, this edited collection does not include a chapter on socioeconomic status, social class or other axes of social status. But, of course, these concerns are central to many of the chapters – for instance, they are embedded in the discussion of health inequalities, homelessness and immigrant health. On the other hand, other potential chapters were not included because they have not to date received a significant amount of attention from scholars in health geography. Perhaps most notably, we do not have a chapter on race and ethnicity. Race and ethnicity and their intersections with health and well-being are surprisingly under-researched themes in health geography. This is despite long-standing international recognition of substantial inequalities in health outcomes between groups differentiated by race and ethnicity (Nazroo, 1998) –in particular, how experiences of racism and discrimination (e.g., in housing, education, employment, income, criminal justice and health care) influence the distribution of resources and, in turn, the health of minority groups (Bailey et al., 2017). Further, issues relating to race and ethnicity have received a great deal of attention by geographers in other sub-disciplines, including social and cultural geography. Other marginalized groups that were not the focus of a chapter due to the lack of attention in health geography include prisoners, sex workers, itinerant groups, and people with substance-use disorders. A more general omission from most of this section is attention to the role of place and space in understanding health-related concerns among different populations in the Global South. Health geography has predominantly focused on the populations of high-income countries. These omissions from the international literature represent fertile areas for future scholarship in health geography.

References

Aldridge, R. W., Story, A., Hwang, S. W., Nordentoft, M., Luchenski, S. A., Hartwell, G., Tweed, E. J., Lewer, D., Vittal Katikireddi, S. and Hayward, A. C. (2017). Morbidity and mortality in homeless individuals, prisoners, sex workers, and individuals with substance use disorders in high-income countries: a systematic review and meta-analysis. *The Lancet*, 391(10117), pp. 241–250.

Bailey, Z. D., Krieger, N., Agénor, M., Graves, J., Linos, N. and Bassett, M. T. (2017). Structural racism and health inequities in the USA: evidence and interventions. *The Lancet*, 389(10077), pp. 1453–1463.

Curtis, S. and Rees Jones, I. (1998). Is there a place for geography in the analysis of health inequality? *Sociology of Health & Illness*, 20(5), pp. 645–672.

Nazroo, J.Y. (1998). Genetic, cultural or socio-economic vulnerability? Explaining ethnic inequalities in health. *Sociology of Health & Illness*, 20(5), pp. 710–730.

Pearce, J. (2013). Financial crisis, austerity policies, and geographical inequalities in health. *Environment and Planning A*, 45(9), pp. 2030–2045.

Smith, K. E., Bambra, C. and Hill, S. E. (2016). *Health inequalities: critical perspectives*. Oxford: Oxford University Press.

Smyth, F. (2008). Medical geography: understanding health inequalities. *Progress in Human Geography*, 32(1), pp. 119–127.

23

THE MEDICALIZATION OF HOMELESSNESS

Tom Baker

Health and homelessness are strongly inter-related. Clinical and demographic studies have long shown that poor health can be a contributing factor to homelessness, can be an effect of homelessness and can impede an exit from homelessness (see Hodgetts et al., 2007; Hwang, 2001; Institute of Medicine, 1988). Poor physical and mental health puts people at greater risk of becoming homeless and, once homeless, they are at increased risk of premature death, are more likely to experience a wide range of physical and mental illnesses and are more likely to be assaulted (Hodgetts et al., 2007; Hwang, 2001). However, beyond the *empirical* connection between health and homelessness, health geographers and other critical social scientists have documented the emergence and expansion of *medicalized understandings* of homelessness. As political interest in addressing the structural causes of homelessness has waned, individualized therapeutic interventions have become a central focus of efforts to address homelessness (Gowan, 2010; Lyon-Callo, 2000; Weinberg, 2008). For several decades now, therapeutic interventions have been commonly positioned as a fundamental cure for homelessness, rather than a topical treatment for the symptoms of structural phenomena, such as housing deprivation.

While the medicalization of homelessness is a transdisciplinary research topic that spans academic disciplines including geography, sociology, anthropology and urban studies, this chapter emphasizes the contributions of health geographers. One of the most significant contributions made by health geographers has been to examine the role of *homeless service spaces* in the medicalization of homelessness. The chapter initially outlines the historical development of medicalized homelessness, drawing on sociologist Teresa Gowan's (2010) key text *Hobos, Hustlers and Backsliders*. The chapter then discusses how homeless service spaces have been a key part of the promotion and reinforcement of medicalized understandings and interventions. Synthesizing the work of a range of health geographers, the chapter highlights how homeless service spaces have enabled two distinct styles of medicalization: first, *authoritarian* medicalization typical of mainstream shelters and transitional housing programs; second, *assertive* medicalization typical of supportive housing programs. Following this, the chapter discusses how recent research within health geography suggests that homeless service spaces serve two key functions in the medicalization of homelessness. On one hand, homeless service *systems* allow for the *objectification* of homeless people as members of a (sub-)population with distinctive health characteristics and therapeutic needs. On the other hand, homeless service *sites* allow for members of those populations to be *subjectified* in accordance with social and political norms relating to the healthy individual. The chapter concludes by identifying research opportunities related to the rationales behind, and resistance to, medicalization and the ways in which medicalized understandings and interventions are reconciled with non-medicalized alternatives.

Systems, sin and sickness

Efforts to address homelessness, such as policies and services, reflect particular understandings of homelessness as a problematic social condition. These understandings or problematizations are historically and geographically variable: there is no universally accepted, timeless understanding of homelessness. Within Western nations, Gowan (2010) claims that homelessness has been understood as a problem of sin, systems and sickness.

The first and oldest of these understandings positions homelessness as a problem of individual sin – or moral deficiency – to be addressed with strategies of exclusion and punishment. The origins of this moral problematization stretch back to the Protestant Reformation of the 16th century when, for "the first time, begging and impoverished wandering were primarily understood as indications of moral weakness or criminality and as major social problems" (Gowan, 2010, p. 31). This remained the dominant understanding of homelessness until the early 20th century, when a rival understanding began to emerge, positioning homelessness as a problem of systems that ought to be addressed with systemic regulation or transformation. In the United States, the Roosevelt Administration's package of New Deal reforms in the 1930s – including job creation, enhanced labor rights and cash assistance to vulnerable people – was an early example of such an understanding being put into practice. As was the case in many countries at the time, the combination of Keynesian economic management and (what later evolved into) the postwar welfare state dramatically reduced the incidence of homelessness for several decades. Systemic understandings of homelessness became normalized, and arguments about the moral deficiency of homeless individuals began to subside.

From the 1980s, Keynesian economic management and the postwar welfare state began to be rolled back with market-based, neoliberal reforms. The economic and social upheaval associated with these changes translated into a widely acknowledged homeless crisis (Mitchell, 2011). With systemic change placed beyond the pale of state intervention, the problem of homelessness was increasingly located within the individual. On one hand, historically enduring moral understandings of homelessness reemerged, and homeless people were exposed to a widening array of civil and criminal sanctions. Urban geographers, in particular, have conducted extensive research on the application of such sanctions and their implications for the rights and lives of homeless people (for an overview, see Mitchell, 2011; von Mahs, 2011). On the other hand, a new understanding of homelessness based on individualized pathology began to emerge with the growth and maturation of specialized homeless services. Given its apparently *medicalized* framing (Lyon-Callo, 2000), health geographers have been primary contributors to a growing geographical literature on this approach to homelessness (for early research, see DeVerteuil, 2003, 2006; for more recent research, see Evans, 2012; Evans, Collins and Anderson, 2016). Across the Anglophone world, the 1980s saw activist groups and charitable organizations create stopgap responses to the homeless crisis, in the form of soup kitchens and basic shelters. "While much immediate hardship was mitigated," Gowan (2010, p. 46) notes, "the soup kitchen or emergency shelter tended, as ever, to institutionalize the problem of homelessness." Initially uncoordinated and localized, these efforts became formalized into the homeless service sector through the introduction of specialized public-funding streams. What were conceived as remedial services offered in anticipation of systemic reform, over time, became focused on therapeutic reform of the homeless themselves. "Somewhere along the way," Gowan (2010, p. 51) concludes, "systemic [understandings of homelessness] . . . had become lost in a sea of practices that implied very different ways of understanding the problem." Homelessness was understood less as a disorder of systems and more as a disorder of the self (Lyon-Callo, 2000), with variously moral (a problem of poor behavior) and medical (a problem of poor health) dimensions.

Styles of medicalized homeless governance

Where the justice system focuses on transforming the moral character of the homeless through enforcement of civil and criminal sanctions, the homeless services system has come to focus on transforming their medical character through therapeutic interventions that address individualized mental and physical patholo-gies. By medicalizing a state of housing deprivation, Evans (2012, p. 188) notes, "unhoused individuals are politicized, not simply as 'homeless' citizens but as living beings with biological 'pathologies'." The research of health geographers and other social scientists suggests that there are two distinct but coexistent *styles* of medicalization practiced by homeless services, which stem from two different treatment philosophies: First, the treatment-led philosophy typical of mainstream shelters and transitional housing programs is associated with an *authoritarian* style of medicalization. Second, the housing-led or Housing First philosophy typical of supportive housing programs is associated with an *assertive* style of medicalization. Each of these will be briefly explained.

As localized, volunteer-based, and often radically motivated responses to the homeless crisis became folded into a formalized and professionalized homelessness service sector, "mission creep" set in. The guid-ing philosophy of service provision transitioned from one based around meeting material needs, such as housing, to one based on producing self-sufficient and housing-ready individuals with the aid of therapeutic interventions, such as addiction treatment, counseling and psychiatric treatment. With the prospect of access to permanent housing made contingent on clinical stabilization and treatment adherence, the treatment-led philosophy of homelessness services translated into an authoritarian style of medicalization (Gowan, 2010). Health geographers such as Geoffrey DeVerteuil and Robert Wilton are key contributors in this research area (DeVerteuil, 2003, 2006; DeVerteuil and Wilton, 2009; Wilton and DeVerteuil, 2006), demonstrating how such therapeutic interventions reflect a host of different intentions relating to care and control (for related contributions from urban geographers, see Cloke, May and Johnsen, 2010; Johnsen, Cloke and May, 2005; Johnsen and Fitzpatrick, 2010). Mandated by specialized government funding streams, treatment-led philosophy and practice saw an "army of social work professionals trained in the language of disease and dys-function . . . examining and categorizing their clients' capacities in terms of mental health, substance use, life skills, parenting, budgeting, and overall 'housing-readiness'" (Gowan, 2010, p. 49). In the United States, treat-ment became embedded in federally funded *continuum of care* programs, while in Australia, Canada, France, New Zealand and Sweden the terms "staircase," "linear" and "step-wise" are among those used to describe the underlying treatment-led focus of the homeless services system (Evans, 2015; Houard, 2011; Johnson, 2012; Knutagård and Kristiansen, 2013; Laurenson and Collins, 2007).

While treatment-led approaches remain the orthodoxy for homelessness services in Western countries, since the late 1990s an alternative philosophy of housing-led service provision has gained momentum. Com-monly called Housing First, this philosophy, which health and urban geographers have begun to research in recent years (Baker and Evans, 2016; Evans, 2015; Evans, Collins and Anderson, 2016; Hennigan, 2016; Klodawsky, 2009), is associated with supportive housing programs that target chronically homeless individu-als (for an overview, see Baker and Evans, 2016). The "chronically homeless" – a term with strong medi-cal suggestions – are a statistically small portion of the overall homelessness population who are unable or unwilling to submit to the authoritarian medicalization of treatment-led homelessness services but unable to independently overcome their homelessness. Contrasting with the authoritarian style of treatment-led service programs, these programs engage in assertive medicalization, using persistence and persuasion to *convince* their clients of the need for treatment. Permanent supportive housing programs require that home-less individuals be housed "as quickly as possible with ongoing, flexible and individual support as long as it is needed, but on a voluntary basis" (Busch-Geertsema, 2013, p. 4). Rental lease agreements between clients and landlords allow clients to maintain their accommodation without abstaining from alcohol and drugs and without necessarily engaging in treatment. Rather than mandating treatment uptake as a condition of housing retention, case workers assertively offer treatment services and suggest ways to reduce mental and physical harms associated with problematic alcohol and drug use (see Aubry et al., 2016, regarding the central place of assertive community treatment in Housing First programs). The homelessness services sys-tem has thus become bifurcated: authoritarian, compulsory medicalization is applied to the non-chronic, majority homeless population, and assertive, voluntary medicalization is applied to the chronically homeless.

These ambivalent styles of medicalization align with a broader sensitivity, within geographical scholarship, to diverse practices and motivations implicated in homeless interventions (see May and Cloke, 2014; Murphy, 2009; Sparks, 2012). To understand how these two styles of medicalization have been enabled, the next section draws attention to the role that spaces of homeless service provision play in objectifying and subjectifying homeless populations.

Homelessness services as objectifying and subjectifying spaces

In both authoritarian and assertive modes, the medicalization of homelessness is achieved through spaces of homeless service provision. Indeed, health and urban geographers have advanced the broader field of homelessness research by providing detailed accounts of the ways in which homeless service spaces are implicated in medicalizing homelessness (Evans, Collins and Anderson, 2016). Their work also provides insights into the impacts of homeless service spaces at a number of scales – from the scale of the individual homeless person through to city-scale and national-scale apparatuses of poverty management (May and Cloke, 2014). Homelessness services are impacted by a diverse range of understandings, agendas and motivations, including those of government, service organizations, practitioners and service recipients. As spaces of "regulatory richness," argue DeVerteuil and Wilton (2009), social services may act, variously, as "a platform for operators' therapeutic aims," while at the same time they "may act to contain and control people deemed disruptive and unproductive" by society at large, and, for service recipients, they "may act as a critical node of survival" (p. 463). Even a single homelessness service program "can be read as an entanglement of strategic policy efforts, therapeutic impulses, and personal hopes and desires, all invested in a very unique socio-spatial intervention" (Evans, 2012, p. 186). In order to illustrate how homeless services have been crucial spaces within and through which the medicalization of homeless has been enabled, health and urban geographers tend to distinguish between (1) the roles of homeless services *systems* in governing homeless *(sub-)populations* as political *objects* with distinctive health characteristics and therapeutic needs and (2) the role of homeless service *sites* in governing homeless *individuals* as political *subjects* to be reformed in accordance with social and political norms relating to the healthy individual. In some research accounts, an analogous distinction is made when authors refer to the political and personal dimensions of social service spaces (e.g., Evans, 2012) or to technologies of power and technologies of the self (e.g., Wilton and DeVerteuil, 2006). The chapter will now briefly discuss how spaces of homeless service provision have been instrumental to the medicalization of homeless governance insofar as they have facilitated the objectification and subjectification of the homeless on medicalized terms.

Individual sites of homeless service provision are joined together – formally and informally – as part of a homeless services system, which itself is a crucial component of municipal and national poverty-management apparatuses (DeVerteuil, 2003). Knowledge produced within the homeless services system shapes understandings of the homeless population – its features, its needs, its impacts – which helps turn that population into an object to be governed (Evans, 2012; Hennigan, 2016; Willse, 2010). In the early decades of the formal homelessness services system, it became possible to aggregate administrative data collected from individual homeless clients across the system and, using those data, represent the homeless as a coherent, knowable population that was demographically distinct from the general public, particularly in relation to their health characteristics and needs. In the United States, for example, a prominent report titled *A Nation in Denial: The Truth about Homelessness* (Baum and Burnes, 1993) used administrative data to advance a medicalized understanding of homelessness, arguing that the "primary issue is not the lack of homes for the homeless; the homeless need access to treatment and medical help for the conditions that prevent them from being able to maintain themselves independently in jobs and housing" (p. 3). Data collected by newly formalized homelessness services were crucial in objectifying the homeless as a population in need of therapeutic intervention. Using administrative data from homelessness services, researchers "turned out hundreds of studies of the pathologies of the homeless, establishing their high levels of addiction, depression, and family dysfunction" (Gowan, 2010, p. 50). These findings began to shape and reinforce the views of health professionals, political decision-makers and the public. An authoritarian, treatment-led approach to homelessness was not only feasible and logical but seemingly necessary.

Health geographers and other critical social scientists have shown that, in the late 1990s, analysis of administrative data collected by homelessness service providers suggested a new way of understanding the homeless population as a governable object (Baker and Evans, 2016; Evans, 2012; Hennigan, 2016; Willse, 2010). Applied research began to discern differences in the length of time spent homeless and the extent of poor health within the homeless population (see Culhane and Kuhn, 1998). They found that approximately 10% of the homeless population – termed the chronically homeless – consumed approximately half of the financial resources available to the shelter system, and that same small proportion of the homeless population were disproportionate consumers of public services, such as emergency medical services and psychiatric services (Culhane and Kuhn, 1998; Kuhn and Culhane, 1998).

These findings have had at least two impacts on the way in which the homeless population is understood and governed. First, they suggested that the chronically homeless were fundamentally different from the remaining 90% of the homeless population and, therefore, in need of a different type of homelessness service intervention. Second, the findings provided an economic justification for new interventions that addressed the chronically homeless segment of the homeless population. As the chronically homeless were resistant to authoritarian, treatment-led homelessness services, room was made for assertive, housing-led homelessness services to emerge as a more effective and economical way of addressing the chronically homeless (see Baker and Evans, 2016; Evans, 2015; Hennigan, 2016). In recent years, an expansive literature has developed on the topic of housing-led services for chronically homeless people, including large-scale randomized control trials (e.g., Goering et al., 2011), which has allowed for more fine-grained understandings of the chronically homeless population as a medicalized object to emerge within health and urban geography (see Evans, 2015; Hennigan, 2016; Klodawsky, 2009).

Although homelessness service provision plays an important role in constituting the homeless population as a governable object, service provision activities have individualized "subject-making effects" (Lyon-Callo, 2000, p. 328). Homelessness services, in this sense, are engaged in attempts to produce particular kinds of people – in particular, people who align with social and political norms relating to the healthy individual. "The social relations that exist between staff and clients and among clients are not neutral or devoid of power," argue Wilton and DeVerteuil (2006, p. 660); they are "designed to shape the conduct of individual clients." In the context of the treatment-led shelter system, Lyon-Callo (2000, p. 328) notes how "[t]reatment focuses on reforming and governing the self," such that homelessness service activities "produce homeless subjects who learn to look within their selves for the 'cause' of their homelessness." Housing-led services are also engaged in subject-making. Commenting on a housing-led program for chronically homeless people with alcohol-abuse problems, Evans (2012) observes a "continuous and intense type of medicalized, micro-logistical supervision of drinking and daily living." He describes how "residents are invited to problematize their lives in terms of alcohol dependence. . . [which] provides a new grid of intelligibility for individual desires and actions" (p. 196). Yet, unlike treatment-led services, where clients are required to engage in treatment, housing-led service providers must convince their clients of the need for self-reform. Case managers "must 'sell' participation," Hennigan (2016, pp. 14–15) points out, because the "rehabilitative techniques of [Housing First programs] . . . become considerably (if unevenly) obstructed by the contractual relationship" between the client-cum-tenant and the landlord. The tenant-landlord relationship, while offering protection from compulsory treatment, applies a subtle form of compulsion to participate in treatment: clients must learn to budget effectively on minimal income, lest they fall behind on their rent and bills and find themselves homeless again as a result (see Hennigan, 2016).

Summary and future research

This chapter discussed how homelessness has been medicalized as a social problem, focusing in detail on the insights of health geographers (often in conversation and/or collaboration with urban geographers and critical social scientists from other academic disciplines). It has shown that, in recent years, health geographers have been part of efforts to trace the emergence of an assertive style of medicalized homeless governance, associated with housing-led services for the chronically homeless, alongside the longer-standing authoritarian style, associated with treatment-led services for the general homeless population. Synthesizing the insights of health and other critical social scientists, the chapter suggested that the spaces of homelessness service provision are crucial to the medicalization of homeless governance in general, and to its authoritarian and assertive variants in particular, insofar as they enable the homeless to be objectified as an unhealthy population and subjectified as unhealthy individuals in need of therapeutic reform. Although the chapter has emphasized the contributions of health geographers to a better understanding of the medicalization of homelessness, it is worth noting that research debates have generally occurred outside the formal venues (such as specialized journals) where health geography, as a sub-discipline, is made and remade. This (sub-)discipline-spanning characteristic of homelessness research, combined with the continuing prominence of medicalized understandings and interventions, puts health geographers in a strong position to steer and make key contributions to the transdisciplinary research agenda on homelessness. With the issue of health occupying a central position in contemporary transformations to the global landscape and localized spaces of homeless governance, there are opportunities for health geographers to explore (1) the potentially diverse *rationales* behind medicalized intervention, (2) the ways in which homeless people, advocates, service providers and others *resist* or augment medicalized intervention through a range of practices and (3) the *reconciliation* of medicalized intervention with alternative understandings of homelessness – based, for example, on structural injustices. Studies that address these topics would enable researchers to advance beyond the factual, empirical nexus of homelessness and health – something that health researchers have long grasped – to better appreciate the *political* nexus of homelessness and health that is increasingly implicated in decisions about the availability and type of support provided to those experiencing housing deprivation.

References

Aubry, T., Goering, P., Veldhuizen, S., Adair, C., Bourque, J., Distasio, J., Latimer, E., Stergiopoulos, V., Somers, J., Streiner, D. and Tsemberis, S. (2016). A multiple-city RCT of Housing First with assertive community treatment for homeless Canadians with serious mental illness. *Psychiatric Services*, 67(3), pp. 275–281.

Baker, T. and Evans, J. (2016). "Housing first" and the changing terrains of homeless governance. *Geography Compass*, 10(1), pp. 25–41.

Baum, A. and Burnes, D. (1993). *A nation in denial: the truth about homelessness*. Boulder, CO: Westview Press.

Busch-Geertsema, V. (2013). *Housing First Europe: final report*. Bremen: GISS.

Cloke, P., May, J. and Johnsen, S. (2010). *Swept up lives? Re-visioning the homeless city*. Malden, MA: Wiley-Blackwell.

Culhane, D. and Kuhn, R. (1998). Patterns and determinants of public shelter utilization among homeless adults in Philadelphia and New York City. *Journal of Policy Analysis and Management*, 17, pp. 23–43.

DeVerteuil, G. (2003). Homeless mobility, institutional settings, and the new poverty management. *Environment and Planning A*, 35(2), pp. 361–379.

DeVerteuil, G. (2006). The local state and homeless shelters: beyond revanchism? *Cities*, 23(2), pp. 109–120.

DeVerteuil, G. and Wilton, R. (2009). Spaces of abeyance, care, and survival: the addiction treatment system as a site of "regulatory richness." *Political Geography*, 28(8), 463–472.

Evans, J. (2012). Supportive measures, enabling restraint: governing homeless "street drinkers" in Hamilton, Canada. *Social & Cultural Geography*, 13(2), pp. 185–200.

Evans, J. (2015). A paradox of plenty: ending homelessness in Alberta. In: M. Shrivastava and L. Stefanick, eds., *Alberta oil and the decline of democracy in Canada*. Edmonton, AB: University of Athabasca Press, pp. 313–331.

Evans, J., Collins, D. and Anderson, J. (2016). Homelessness, bedspace and the case for Housing First in Canada. *Social Science & Medicine*, 168, 249–256.

Goering, P. N., Streiner, D. L., Adair, C., Aubry, T., Barker, J., Distasio, J., Hwang, S. W., Komaroff, J., Latimer, E., Somers, J. and Zabkiewicz, D. M. (2011). The At Home/Chez Soi trial protocol: a pragmatic, multi-site, randomised controlled trial of a Housing First intervention for homeless individuals with mental illness in five Canadian cities. *BMJ Open*, 1, 2e000323.

Gowan, T. (2010). *Hobos, hustlers, and backsliders: homeless in San Francisco*. Minneapolis: University of Minnesota Press.

Hennigan, H. (2016). House broken: homelessness, Housing First, and neoliberal poverty governance. *Urban Geography*, 38(9), pp. 1418–1440.

Hodgetts, D., Radley, A., Chamberlain, K. and Hodgetts, A. (2007). Health inequalities and homelessness: considering material, spatial and relational dimensions. *Journal of Health Psychology*, 12, pp. 709–725.

Houard, N. (2011). The French homelessness strategy: reforming temporary accommodation, and access to housing to deliver "housing first": continuum or clean break? *European Journal of Homelessness*, 5(2), pp. 83–98.

Hwang, S. (2001). Homelessness and health. *Canadian Medical Association Journal*, 164(1), pp. 229–233.

Institute of Medicine. (1988). *Homelessness, health, and human needs*. Washington, DC: The National Academies Press.

Johnsen, S., Cloke, P. and May, J. (2005). Day centres for homeless people: spaces of care or fear? *Social & Cultural Geography*, 6(6), pp. 787–811.

Johnsen, S. and Fitzpatrick, S. (2010). Revanchist sanitisation or coercive care? The use of enforcement to combat begging, street drinking and rough sleeping in England. *Urban Studies*, 47(8), pp. 1703–1723.

Johnson, G. (2012). Housing first "down under": revolution, realignment or rhetoric? *European Journal of Homelessness*, 6(2), pp. 183–191.

Klodawsky, F. (2009). Home spaces and rights to the city: thinking social justice for chronically homeless women. *Urban Geography*, 30(6), pp. 591–610.

Knutagård, M. and Kristiansen, A. (2013). Not by the book: the emergence and translation of Housing First in Sweden. *European Journal of Homelessness*, 7, pp. 93–116.

Kuhn, R. and Culhane, D. (1998). Applying cluster analysis to test a typology of homelessness by pattern of shelter utilization: results from the analysis of administrative data. *American Journal of Community Psychology*, 26, pp. 207–232.

Laurenson, P. and Collins, D. (2007). Beyond punitive regulation? New Zealand local governments' responses to homelessness. *Antipode*, 39(4), pp. 649–667.

Lyon-Callo, V. (2000). Medicalizing homelessness: the production of self-blame and self-governing within homeless shelters. *Medical Anthropology Quarterly*, 14(3), pp. 328–345.

May, J. and Cloke, P. (2014). Modes of attentiveness: reading for difference in geographies of homelessness. *Antipode*, 46(4), pp. 894–920.

Mitchell, D. (2011). Homelessness, American style. *Urban Geography*, 32(7), pp. 933–956.

Murphy, S. (2009). "Compassionate" strategies of managing homelessness: post-revanchist geographies in San Francisco. *Antipode*, 41(2), pp. 305–325.

Sparks, T. (2012). Governing the homeless in an age of compassion: homelessness, citizenship, and the 10-year plan to end homelessness in King County Washington. *Antipode*, 44(4), pp. 1510–1531.

von Mahs, J. (2011). Introduction – an Americanization of homelessness in post-industrial countries. *Urban Geography*, 32(7), pp. 923–932.

Weinberg, D. (2008). *Of others inside: insanity, addiction, and belonging in America*. Philadelphia: Temple University Press.

Willse, C. (2010). Neo-liberal biopolitics and the invention of chronic homelessness. *Economy and Society*, 39(2), pp. 155–184.

Wilton, R. and DeVerteuil, G. (2006). Spaces of sobriety/sites of power: examining social model alcohol recovery programs as therapeutic landscapes. *Social Science & Medicine*, 63, pp. 649–661.

24

INFORMAL CAREGIVERS

People, place and identity

Andrew Power

This chapter is about family members and friends who provide care and the spaces and places they occupy as a result. These so-called *caringscapes* (McKie, Bowlby and Gregory, 2001) are produced by the complex interplay of practices, relationships, performances and politics of care that unfold between people who provide or receive care (often interchangeably). Since the early 1990s, Christine Milligan, Allison Williams, Mark Skinner, Valorie Crooks, Andrew Power, Melissa Giesbrecht, Tamara Daley and other geographers have contributed to our understanding of this constituency. Since this time, the question of *who cares?* has been posed many times (Barnett, 2005; Milligan and Power, 2009). Those who provide significant levels of care are known sometimes collectively as informal caregivers/carers, although they are quite often ambiguously defined, relative to other groups. Usually women, caregivers within the family occupy distinct roles, depending on their relationship within the family structure and the caregiving roles that are needed/expected: from dutiful daughter looking after an aging parent, to a self-identified *expert by experience*, supporting an adult with complex and multiple impairments.

Alongside family members, friends, neighbors, and community allies can also exchange small acts of emotional and practical care and support, such as lifts to the doctor, pet care, assistance with small DIY tasks and shopping, as well as sharing convivial encounters with those in need (Bowlby, 2011). Taking this theme further, Lawson (2007) asks that we not forget about these broader acts of caregiving that occur in everyday spaces. Such work has advanced the notion of an *ethics of care* – a guiding principle that all relational practices should be done in a more *care-full* way – with an emphasis on interdependence and the values of caregiving (Lawson, 2007).

For family members, caregiving can involve more intense acts of care work, such as providing physical support to a person who is lacking full independence, including activities of daily living (e.g., eating, getting into bed), as well as cognitive support and help with instrumental tasks of daily living (e.g., paying bills, making decisions, managing bank accounts). They can also be expected to provide moral, psychosocial, emotional and spiritual support. Given the range of these tasks, often caregivers must manage the complex emotional demands of caregiving themselves, which can sometimes result in high levels of personal stress. Finch and Groves (1983) proposed the concepts *caring about* and *caring for* another person to help distinguish these various roles. Caring *about* refers to the corresponding feelings of concern about someone, commonly shared among friends, whereas caring *for* describes the actual labor of performing tasks of care. This distinction has been challenged, however, as the two are argued to coalesce and boundaries are blurred (Ungerson, 2005).

Importantly, caregiving tasks do not occur in a vacuum; they take place in and through space alongside other people on an everyday basis. This chapter examines the people and places of caregiving; those who take up various care roles, and the specific geographies that they occupy and shape. Starting with the international scale, this chapter will trace some of the different sites and practices that encompass the caringscape.

Informal caregivers: people and identity

The origins of caregiving scholarship derived largely from feminist work on women who provide domestic labor (for children) in a marital context (e.g., Finch and Groves, 1983). While this work paved the way for other geographers to explore caregiving, it ignored, for example, the provision of care in a disabilities context (Graham, 1993), as well as an older person's context (Milligan, 2000). It also ignored the context of women from different class and ethnic backgrounds (hooks, 1982). Recognizing that care takes place in varied contexts and for many different constituencies has meant, as indicated in the introduction, that caregivers share a relatively ambiguous identity: many caregivers do not see themselves as such, while others strongly associate with the notion of being a caregiver.

Their ambiguous identity also stems from their close proximity to different care recipients (with chronic illness, frailty and cognitive decline in old age, and disability). This proximity positions them alongside different identity groups, where they must contend with debates over the independence and rights of the care recipient, on one hand, from disability rights or active-aging campaigns (Oliver and Barnes, 1998), and the support of women's rights and the valuing of care, from feminist politics, on the other hand (Finch and Groves, 1983). These complex subject positions are often portrayed as in tension with each other and play out in different ways. The outcomes of these positions readily involve tangible effects on how well caregivers and care recipients are recognized and supported (Power, 2008). Giesbrecht et al. (2016) also helpfully identify the intersectional nature of caregiving that can shape experiences and meanings of place during the caregiving process, where a multitude of personal identities – including gender, age, physical capacity, culture, socioeconomic status, marital status, housing status and social connectedness – intersect.

Despite this diversity, common to all caregivers is the growing profile of and demands on this constituency, due to increasing aging and demographic pressures and cutbacks to social-care services in the Global North (Levine, 2007). There is an increasing reliance on families to care for and support family and community members who are aging, sick or disabled, with less statutory support available (England, 2010). This context particularly impacts those in the *sandwich generation*, responsible for providing care to both their aging parents and their young children. This growing reliance has, in turn, led to a growing awareness of a more politicized informal caregiver identity within this constituency (Levine, 2007); since the mid-1990s, family members have begun demanding to be recognized as co-clients in need of care and rights themselves.

Despite this growth in the profile of caregivers, health-care-agency staff often have divergent views of this group. Twigg's (1989) work identified three accounts of caregivers constructed by professionals: as resources (seen as a constant source of care), as co-workers (positioned alongside professionals as equals), and as clients (presented as a group needing support in their own right). As informal caregiving largely goes unaccounted for by the health-care industry, informal caregivers unsurprisingly have to juggle complex narratives of responsibility, vulnerability and mutuality in their work (Whitmore, Crooks and Snyder, 2015). The narratives also vary in a way that depends on geographical location, the degree of formal care and specific cultural and social factors of the area (as identified by Di Novi, Jacobs and Migheli (2015) at the European transnational level). Elsewhere, Power (2010) examined how caregivers are perceived across the United States, England and Ireland, tracing the way that different cultural expectations have shaped radically different care policies.

One of the dominant recurring themes to emerge from the caregiving literature in health geography has been *caregiver burden*; how care work and often the lack of social and financial support can negatively impact caregivers' health and well-being. This work has identified how models of burden and distress differ between women and men (Stewart et al., 2016) and how burden can evolve over time with a care recipient's illness trajectory (Grunfeld et al., 2004). Caregivers also must manage the obligations and responsibility across a changing landscape of state care provision (Egdell, 2013). In Afram et al. (2014), caregiver burden and the inability of the informal caregiver to care for the patient were stated as key reasons for institutionalization of people with dementia to a long-term care facility. This was more commonly expressed by spouse-caregivers than child-caregivers. McKie, Bowlby and Gregory (2001) have, however, sought to critique the dyadic caregiver/care recipient relationship typically presented in the burden literature. They argue that this assumed (unequal) dyadic relationship can reinforce the binary of *burdened/dependent*. They argue for more recognition of the interdependency between all people and of the multiple care strands that can exist within and between families and communities.

It is also worth highlighting the complex emotionally charged nature of caregiving. This is a point that Casey, Crooks, Snyder and Turner (2013a; 2013b) make in relation to supporters and companions of medical tourists. Medical tourists travel abroad for care, typically with friends and family members who provide support and assistance. The authors argue that caregiver-companions can assist in enhancing best care and offering meaningful support to medical tourists, through three different roles: knowledge broker, companion, and navigator (2013a). However, they also point out that caregiver-companions can also become additional clients themselves if they do not manage the emotional demands of these roles effectively (2013b). In these cases, they can end up requiring more time, attention and resources of professionals, and as a result disrupt the provision of quality care to the medical tourist.

Spaces of family caregiving

While the discussion in the previous section foregrounded the people involved in care work, the following section explicitly examines the complex geographies that family caregivers occupy and shape. There has been considerable interest in how caregiving experiences both shape and are shaped by geographical location, both within rural and urban contexts and within the home itself.

In the rural context, work has examined farmwomen's emotional geographies of care, highlighting the emotional overlay involved in negotiating aging and care in rural areas (Herron and Skinner, 2013). It is evident from this and other geographic work that the absence of formal supports in rural areas creates a higher burden on informal caregivers. McCann, Ryan and McKenna (2005), for instance, pinpoint the difficulties of recruiting care assistants, isolation, travel and distance between clients and their care assistants, and poor housing conditions as relevant factors that place more demand on families. Changing demographic trends and the predicted shortfall in the number of formal care-workers to support families were also considered key future issues for service planners.

This theme is picked up by Joseph and Skinner (2012) and others who identify the importance – and precarity – of the local voluntary sector for family members who care or are cared for in rural and small-town areas. As a result of neoliberal policies, many rural and small-town services have downsized or closed, forcing families to embrace forms of *self-help* (Sherwood and Lewis, 2000) and *stealth voluntarism* by local professionals (Hanlon, Halseth and Ostry, 2011). Indeed, the effectiveness of rural community care was perceived to be reliant upon the goodwill of the community (Ehrlich et al., 2015). How the voluntary sector adapts and reacts ultimately affects how families and communities care and are cared for.

As neoliberalism has evolved and accelerated in the wake of austerity in the Global North, services in urban areas have also not been immune to such precarity (Munro and Morse, 2017). There are parallels to the self-help ethos in Power and Bartlett's (2017) study, who identify how adults with learning disabilities are *self-building* peer-support opportunities in local urban neighborhoods. The participants had negotiated safe havens among care-full and convivial places in the community, such as a local bingo hall, an allotment (community garden), a fish-and-chip shop, and a marina. Behind these safe havens, Power, Bartlett and Hall (2016) found that peer-support groups were key to giving adults with learning difficulties more opportunities for social encounter and helping them acquire skills. In the same vein, Munro and Morse (2017) identify the growing importance of museums as spaces where caregivers and care recipients can visit and receive support through local community-engagement schemes (e.g., storytelling sessions with mental-health-service users). Through such activities, museums are becoming another example of a *space of care* that is unfolding in times of austerity and government retrenchment of services.

Other work that has examined caregiving in urban contexts has revealed the stigma that can emerge when caring for others in public life. Ryan (2005) examined the experiences of having to provide support to young people with learning disabilities in public spaces, such as the local supermarket. When behavioral norms are transgressed (e.g., a person shouts in public), this can often give rise to additional emotional tasks, such as managing expectations with members of the public. Tarrant's (2013) work on grandfathering similarly reveals how certain identities get performed and blurred through public acts of caregiving in urban areas, such as the performance of aging masculinities in care roles that have been historically feminized. As a result, McKie, Bowlby and Gregory's (2001) notion of caringscapes helps us understand the spatial layering of complex practices, emotions and performances associated with care.

Within the home, geographers have also been attuned to how the intimate body work of care (such as bathing, toileting, and catheter management) intersects with the material and discursive site of the home, with its spatial arrangements, location, amenities and furnishings (Dyck and England, 2011). The experience of the home is changed as a result of the practices of caregiving, including daily routines, monitoring of noise, and regulation of visitors (Milligan, 2005; Shannon, 2015). Milligan's (2000) study on the institutionalization of the home is helpful in informing our understanding of how the home space can change in these contexts, with the presence of caregiving routines, practices and formal care visits blurring the distinction between the private and the public domain. This theme is further developed by Yantzi and Rosenberg (2008), who examined the experiences of women caring for children with long-term-care needs in Ontario, Canada. They found that women do not want their homes to be completely defined by long-term-care activities, because many other types of activities are situated in their homes. Second, long-term-care activities and schedules are not segregated but become deeply embedded and enmeshed within the spatial and temporal practices and processes of family life. Third, the meanings, characteristics and ideal of "home" portrayed in popular culture and the academic literature often clashed with what the women experienced on a daily basis. As a result, Wiles (2003) argues the meaning of the home can change from a carefree, relaxing, sociable environment to one where there is careful monitoring of visitors, noise and behavior.

Conclusion

Informal caregiving is a theme that geographers have consistently engaged with since the late 1980s. This scholarship has sought to understand and contribute to the complexity of debates associated with this ambiguous role, tracing the multiple forms of care work that take place and the diverse range of people and identities that become associated with these roles. Geographers have been attuned to the experiences of caregivers and the changing spaces and places that they use and occupy – and ultimately create through their practices and performances. As the global politics of austerity further withers the welfare states of nations in the Global North, the expectations and demands on families and communities will no doubt continue to grow. Simultaneously, a rising awareness of the needs of this group as a client group in and of themselves will likely create greater pressures (e.g., on workforce participation) for the state and private sectors. Geographers are well placed to continue leading and informing debates over how these issues will play out on the ground and to discover how we might best support families in overcoming these challenges.

References

Afram, B., Stephan, A., Verbeek, H., et al. (2014). Reasons for institutionalization of people with dementia: informal caregiver reports from 8 European countries. *Journal of the American Medical Directors Association*, 15(2), pp. 108–116.

Barnett, C. (2005). Who cares? In: P. Cloke, P. Crang and M. Goodwin, eds., *Introducing human geography*, 2nd ed. London: Arnold, pp. 588–601.

Bowlby, S. (2011). Friendship, co-presence and care: neglected spaces. *Social & Cultural Geography*, 12, pp. 605–622.

Casey, V., Crooks, V. A., Snyder, J. and Turner, L. (2013a). "You're dealing with an emotionally charged individual . . .": an industry perspective on the challenges posed by medical tourists' informal caregiver-companions. *Global Health*, 9(31), pp. 1–12.

Casey, V., Crooks, V. A., Snyder, J. and Turner, L. (2013b). Knowledge brokers, companions, and navigators: a qualitative examination of informal caregivers' roles in medical tourism. *International Journal for Equity and Health*, 12(94), pp. 1–10.

Di Novi, C., Jacobs, R. and Migheli, M. (2015). The quality of life of female informal caregivers: from Scandinavia to the Mediterranean Sea. *European Journal of Population*, 31, pp. 309–333.

Dyck, I. and England, K. (2011). Homes for care: reconfiguring care relations and practices. In: C. Ceci, K. Bjornsdottir and M. E. Purkis, eds., *Home, care, practices: critical perspectives on care at home for older people*. New York: Routledge, pp. 81–100.

Egdell, V. (2013). Who cares? Managing obligation and responsibility across the changing landscapes of informal dementia care. *Ageing & Society*, 33, pp. 888–907.

Ehrlich, K., Bostrom, A. M., Mazaheri, M., Heikkilä, K. and Emami, A. (2015). Family caregivers' assessments of caring for a relative with dementia: a comparison of urban and rural areas. *International Journal of Older People Nursing*, 10, pp. 27–37.

England, K. (2010). Home, work and the shifting geographies of care. *Ethics, Place, and Environment*, 13(2), pp. 131–150.

Finch, J. and Groves, D. (1983). *A labour of love: women, work and caring*. London: Routledge.

Giesbrecht, M., Williams, A., Duggleby, W., Ploeg, J. and Markle-Reid, M. (2016). Exploring the daily geographies of diverse men caregiving for family members with multiple chronic conditions. *Gender, Place & Culture*, 23(11), pp. 1586–1598.

Graham, H. (1993). Social divisions in caring. *Women's Studies International Forum*, 16(5), pp. 461–470.

Grunfeld, E., Coyle, D., Whelan, T., Clinch, J., Reyno, L., Earle, C. C., Willan, A., Viola, R., Coristine, M., Janz, T. and Glossop, R. (2004). Family caregiver burden: results of a longitudinal study of breast cancer patients and their principal caregivers. *Canadian Medical Association Journal*, 170(12), pp. 1795–1801.

Hanlon, N., Halseth, G. and Ostry, A. (2011). Stealth voluntarism: an expectation of health professional work in under-serviced areas? *Health & Place*, 17, pp. 42–49.

Herron, R. and Skinner, M. W. (2013). The emotional overlay: older person and carer perspectives on negotiating aging and care in rural Ontario. *Social Science & Medicine*, 91, pp. 186–193.

hooks, b. (1982). *Feminist theory: from margin to center*. Boston: South End Press.

Joseph, A. E. and Skinner, M. W. (2012). Voluntarism as a mediator of the experience of growing old in evolving rural spaces and changing rural places. *Journal of Rural Studies*, 28, pp. 380–388.

Lawson, V. (2007). Geographies of care and responsibility. *Annals of the Association of American Geographers*, 97, pp. 1–11.

Levine, C. (2007). *Always on call: when illness turns families into caregivers*. Nashville: Vanderbilt University Press.

McCann, S., Ryan, A. A. and McKenna, H. (2005). The challenges associated with providing community care for people with complex needs in rural areas: a qualitative investigation. *Health & Social Care in the Community*, 13, pp. 462–469.

McKie, L., Bowlby, S. and Gregory, S. (2001). Gender, caring and employment in Britain. *Journal of Social Policy*, 30(2), pp. 233–258.

Milligan, C. (2000). Bearing the burden: towards a restructured geography of caring. *Area*, 32, pp. 49–58.

Milligan, C. (2005). From home to "home": situating emotions within the caregiving experience. *Environment and Planning A*, 37, pp. 2105–2120.

Milligan, C. and Power, A. (2009). The changing geography of care. In: T. Brown, S. McLafferty and G. Moon, eds., *A companion to health and medical geography*. London: Wiley-Blackwell, pp. 567–586.

Munro, N. and Morse, E. (2018). Museums' community engagement schemes, austerity and practices of care in two local museum services. *Social & Cultural Geography*, 19(3), pp. 357–378.

Oliver, M. and Barnes, C. (1998). *Disabled people and social policy: from exclusion to inclusion*. London: Longman.

Power, A. (2008). Caring for independent lives: geographies of caring for young adults with intellectual disabilities. *Social Science & Medicine*, 67(5), pp. 834–843.

Power, A. (2010). *Landscapes of care: comparative perspectives on family caregiving*. Farnham: Ashgate.

Power, A. and Bartlett, R. (2018). Self-building safe havens in a post-service landscape: how adults with learning disabilities are reclaiming the welcoming communities agenda. *Social & Cultural Geography*, 19(3), pp. 303–313.

Power, A., Bartlett, R. and Hall, E. (2016). Peer-advocacy in a personalised landscape: the role of peer support in a context of individualised support and austerity. *Journal of Intellectual Disabilities*, 20(2), pp. 1–19.

Ryan, S. (2005). "People don't do odd, do they?" Mothers making sense of the reactions of others towards their learning disabled children in public places. *Children's Geographies*, 3, pp. 3291–3305.

Shannon, C. S. (2015). "I was trapped at home": men's experiences with leisure while giving care to partners during a breast cancer experience. *Leisure Sciences*, 37, pp. 125–141.

Sherwood, K. B. and Lewis, G. J. (2000). Accessing health care in a rural area: an evaluation of a voluntary medical transport scheme in the English Midlands. *Health & Place*, 6, pp. 337–350.

Stewart, N. J., Morgan, D. G., Karunanayake, C. P., Wickenhauser, J. P., Cammer, A., Minish, D., O'Connell, M. E. and Hayduk, L. A. (2016). Rural caregivers for a family member with dementia: models of burden and distress differ for women and men. *Journal of Applied Gerontology*, 35, pp. 150–178.

Tarrant, A. (2013). Grandfathering as spatio-temporal practice: conceptualizing performances of ageing masculinities in contemporary familial carescapes. *Social & Cultural Geography*, 14(2), pp. 192–210.

Twigg, J. (1989). Models of carers: how do social care agencies conceptualise their relationship with informal carers? *Journal of Social Policy*, 18, pp. 53–66.

Ungerson, C. (2005). Care, work and feeling. *The Sociological Review*, 53(2), pp. 188–203.

Whitmore, R., Crooks, V. A. and Snyder, J. (2015). Ethics of care in medical tourism: informal caregivers' narratives of responsibility, vulnerability and mutuality. *Health & Place*, 35, pp. 113–118.

Wiles, J. (2003). Daily geographies of caregivers: mobility, routine, scale. *Social Science & Medicine*, 57, pp. 1307–1325.

Yantzi, N. M. and Rosenberg, M. W. (2008). The contested meanings of home for women caring for children with long-term care needs in Ontario, Canada. *Gender, Place and Culture*, 15(3), pp. 301–315.

25

MAPPING LIFE ON THE MARGINS

Disability and chronic illness

Vera Chouinard

Over the past three decades, geographers have made increasingly sophisticated contributions to understanding the lives of disabled people and people with chronic illnesses and the socio-spatial forces shaping those lives. In fact today it may be hard to believe that as late as the mid-1990s it was still possible to criticize geographers for ignoring the lives of these and other marginalized groups (Chouinard and Grant, 1995). And it was not until 1999 that the first edited collection of geographic accounts of disability and chronic illness was published (Parr and Butler, 1999). The second such collection did not appear until 2010 (Chouinard, 2010).

This chapter discusses the emergence and development of research into the socio-spatial lives of disabled people and people managing chronic illness as a sub-disciplinary field – a field informed by both critical social geography and health geography. The chapter begins by outlining key contributions to geographies of disability and chronic illness, defined by the US National Center for Health Statistics as an illness "lasting 3 months or more" (Medicine.net, 2017). Next, work in this sub-disciplinary area that contributes to health geography and other disciplines is discussed. This is followed by a consideration of major criticisms of research on geographies of disability and chronic illness. The chapter concludes with reflections on the future development of this vitally important area of inquiry.

Contributions to geographies of disability and chronic illness

Some of the earliest efforts by geographers to address disability and chronic illness were informed largely by a biomedical conception as given biophysical and/or mental states. In this approach, impairment and illness were seen as necessarily disabling an individual. It was assumed, moreover, that the impaired or ill individual was the problem – a problem to address using biomedical science to cure or at least mitigate symptoms. It was this approach that informed the postwar medical geography literature on the spatial distribution and correlates of disease, the spatial distribution and use of medical facilities and statistical accounts of the characteristics and geographic distribution of different groups of disabled people (see Park, Radford and Vickers, 1998; Wolch and Philo, 2000).

While still largely informed by a biomedical approach, some of these early studies did begin to touch on certain issues relating to access to the environment. Perle (1969) wrote on needs for accessible transit. This topic was not taken up again until the early 1980s (Gant and Smith, 1984; Kirby, Bowlby and Swan, 1983). From the 1970s onward, Golledge examined cognition and way-finding among so-called special-needs groups. His primary interest was in behavioral differences in how persons with and without intellectual and visual impairments navigate their environments (e.g., Golledge et al., 1972; Self et al., 1992; Loomis et al., 1998, 2005). Although critics would later point out that impairment and disability were treated as an individual biomedical condition, Golledge's work was important in providing a basis for the development of way-finding aids (e.g., Loomis, Golledge and Klatzky (1998) on auditory guidance systems for the visually impaired) and in highlighting cognitive and behavioral differences in navigating everyday environments.

The 1980s saw the emergence of what can be described as the beginnings of critical geographic studies of socio-spatial forces shaping the lives of persons with mental illnesses. This included work by geographers such as Dear and Taylor (1982) on community opposition to community care facilities and by Dear and Wolch (1987) on how deinstitutionalization and the political economy of urban development were forcing homeless people, many of whom suffered from mental illness, into service-dependent ghettoes in North America. In the United Kingdom, Philo (1987) was exploring the historical geography of asylums for the mentally ill.

It was not until the mid-1990s, however, that a shift toward critical geographies of impairment, chronic illness and disability, inspired at least in part by the social model of disability then being championed by disability-studies scholars and activists (e.g., Oliver, 1990), took hold. According to the social model, disability was the result of barriers to the inclusion of people with impairments and illnesses in society and space. It was thus no longer seen as an individual biomedical problem but seen as a social problem rooted in ableist attitudes, practices and environments that devalued and excluded disabled people. This shift was evident in critiques by Butler (1994), Gleeson (1996) and Imrie (1996) of the implicitly biomedical perspective that informed Golledge's work. It was also evident in calls for geographers to address disability in more critical ways – including probing how persons with impairments and illnesses were marginalized in academic environments (Chouinard and Grant, 1995).

The 1990s also saw the emergence of a new stream of socioculturally informed medical geography that came to be known as health geography (Kearns, 1993; Kearns and Moon, 2002). This stream differed from traditional medical geography in that it was open to the use of critical social theory (e.g., concepts of structure and agency) to understand people's experiences of health, illness and care and differences in power and oppression that helped shape those experiences. Research moved away from traditional conceptions of place and space as mere containers for phenomena, such as disease patterns, and embraced the idea that place and space mattered in a causal sense in health and well-being. It also tended to be more qualitative than quantitative in its research. Although, as Kearns and Moon (2002) point out, not all geographers concerned with disability necessarily identified as health geographers, there was now a basis for synergy and overlap between the two sub-disciplines. A case in point was Isabel Dyck's (1995) study of the changing lifeworlds of women with multiple sclerosis. She was able to show how the changing bodily abilities of women with this chronic illness combined with increasing difficulty in meeting expectations in places such as the workplace to force their lifeworlds to shrink primarily to the home. Here, as is the case in both geographies of disability and health geography, the emphasis was on how structures, such as workplace organization, as well as the agency of chronically ill women, shape the degree to which ill women are included or excluded in society and space.

Since 2000, there have been many important contributions to geographies of disability and chronic illness. These include studies that broaden our understanding of bodily conditions that can be disabling, such as Longhurst's (2010) study of women who identify as large or fat. She demonstrated that these women experienced physical and social barriers that were disabling; for example, clothing-store changing rooms that were too small and fears of negative attitudes toward their embodied selves if they frequented places such as the beach. For at least some large women, this meant avoiding such places.

Thanks to efforts by geographers such as Philo and Metzel (2005) and Hall and Kearns (2001), more attention is being paid to intellectual impairments and ability/disability – broadening the range of mind differences being considered beyond geography's traditional emphasis on mental illness (e.g., Metzel, 2010). Davidson's work on topics such as the lives of people with autism has also contributed to this project (e.g., Davidson, 2008). This is also true of work on critical geographies of addiction (Wilton and Moreno, 2012).

Other important contributions to geographies of disability and chronic illness have included studies exploring how the meaning of the home reflects links between physical impairment and spatial design (Imrie, 2010) and on how the home looms larger in the lives of women who acquire fibromyalgia syndrome (Crooks, 2010).

Feminist geographers have made important contributions to understanding disability and chronic illness. A case in point is Valentine's (1999) now classic study of how the onset of physical impairment altered the lifeworld and masculine identity of a working-class man. More recently, she has made a case for more inter-sectional approaches to understanding how disability and other embodied differences, such as sexuality, shape people's experiences of different spaces of everyday life (Valentine, 2007). Other examples include Moss and Dyck's (2003) efforts to better understand how chronic illness shapes women's embodied lives and lifespaces.

Contributions to health geography and other disciplines

Geographic research such as that cited above has made important contributions to health geography and other disciplines. For health geography, such research has meant that the socio-spatial implications of an increasingly wide array of mind and body differences are being considered. This research has also helped make processes of disablement, disempowerment and marginalization among those with impairments and/or chronic illnesses more central concerns in health geography. Or, in other words, it has helped, at least to a degree, increase synergy between health and critical social geography. This research has also encouraged consideration of the qualities of place that help make it more therapeutic or more conducive to well-being. Examples include Parr's work on mental health, gardening and the arts (e.g., Parr, 2011).

Geographic research concerned with disability and chronic illness has also made important contributions to other disciplines. Perhaps most importantly, it has helped demonstrate that place and space matter in determining how enabling or disabling the lives of people with impairments and chronic illnesses are. This has enriched the interdisciplinary field of disability studies and disciplines such as the sociology of health and illness and public health.

Geographers, as well as scholars in other disciplines, have helped champion a critical social perspective on disability as a societal rather than an individual medical problem. This is important in focusing attention on the role of oppression in disabled and ill people's lives. At the same time, however, geographers such as Hall (2000) have insisted on taking the body seriously (i.e., as a material entity) in accounts of experiences of impairment and illness. Thus geographers have helped develop what we might term "an embodied critical social understanding" of impairment, illness and disability. This has been a very important contribution to critical disability studies.

Major criticisms of geographies of disability and chronic illness

As noted above, a major criticism of research in geographies of disability and chronic illness in the mid-1990s was that researchers such as Golledge adhered to a biomedical model of disability, impairment and illness. Inspired in part by the social model being advocated by disability scholars such as Oliver (1990), critical geographers called for a shift away from conceiving disability, impairment and illness as an individual bio-medical problem and toward an understanding that it was attitudinal and social barriers that disabled people with impairments and illnesses. Gleeson (1998), for example, proposed a historical materialist geographic approach to understanding how the transition from feudal to capitalist societies worked to marginalize disabled people (e.g., as a result of new regimes of factory production incompatible with non-able working bodies). Efforts to bring the body back in to studies of disability and chronic illness were a response to the criticism, within and outside of geography, that social model–type perspectives neglected bodily realities of impairment and illness such as pain and fatigue (e.g., Hall, 2000; Moss and Dyck, 2003). Examples of such studies are Barns et al. (2015) and Scambler and Scambler (2010). Geographers such as Smith and Davidson

(2006) and Crooks (2010) have also contributed to more embodied conceptions of life with disability and chronic illness.

A more recent major criticism of geographic research on impairment, illness and disability is that it has been conducted almost exclusively in countries of the Global North. This is despite the fact that the vast majority of disabled people (80%) live in the Global South (Meekosha and Soldatic, 2011). One exception to this has been my work on disabled people's lives in Guyana (e.g., Chouinard, 2012, 2014; Chouinard et al., 2016). This work helps demonstrate the importance of understanding experiences of impairment, chronic illness and disability in the context of an uneven global capitalist order that disadvantages countries in the Global South. In Guyana, for example, the outmigration of trained medical specialists (as well as the high costs of importing aids or materials for them) has severely constrained access to aids such as prosthetic arms and legs.

Wilton, DeVerteuil and Evans (2014) criticize health geography for largely failing to address issues of masculinity. They show how successful treatment for drug and alcohol addictions involves a reworking of macho masculinities emphasizing heavy consumption of drugs and alcohol into healthier masculine identities. While scholars do not always agree on how addiction should be defined (e.g., as a mental health condition versus a habit involving compulsive behavior), it is clear that this physical and mental state can have disabling consequences (such as losing one's job or family) and heighten risks of chronic illnesses (e.g., liver diseases).

There are some signs of growing interest in making academic geography more inclusive of persons with mental and physical impairments and chronic illnesses. Feminist geographers such as Linda Peake and Beverly Mullings have, for example, organized conference sessions in Canada and the United States that consider the impacts of neoliberal university life on mental health. This, in turn, has sparked interest in concepts such as slow scholarship. These are welcome developments, but much work remains to be done. I was reminded of this recently when responding to invitations to speak at the 2017 American Association of Geographers (AAG) meeting in Boston. One invitation was from members of the Disability Specialty Group of the AAG. They invited me to give a plenary talk on the geography of disability. As I can no longer travel to conferences by myself due to physical impairment and illness, I responded by offering to give a virtual talk, and, to the organizers' credit, they made this possible. The second invitation was from an organizer of a high-profile conference panel addressing human-rights and diversity issues. Again, I offered to participate virtually, but, in this case, I received absolutely no response. Such experiences are disheartening and help isolate at least some disabled geographers from events such as conferences. I would have liked to have connected, for instance, with the two disabled female geographers who were able to attend in person and were included in the panel. Although I never received an explanation for the lack of response to my offer, one reasonable hypothesis is that virtual presentations are regarded as less valuable than conventional in-person ones. A related hypothesis is that at least some people regard it as too much bother to do things differently than they are used to, even if it would make events such as conferences more inclusionary. Although there has been much progress in geographic scholarship on impairment, disability and chronic illness, there remains a long way to go in order to make the discipline of geography fully inclusive of disabled and ill people. A related challenge, of course, is to ensure that disabled and ill faculty and students are represented and accommodated/supported in our academic units.

Over the years in geography, as well as disability studies, there have been calls for greater attention to ableism in our societies (e.g., Campbell, 2009; Chouinard, 1997). Similar to the concept of whiteness that values and privileges white persons over persons of color, ableism can be thought of as a vantage point and set of attitudes and practices that privileges and values able minds and bodies over their disabled and ill counterparts. Although it is now possible to identify geographic research that helps reveal some of the workings of ableism in relation to specific types of impairment and illness or in specific spaces of everyday life (for instance, Campbell's 2009 work on deafness and cochlear implants and my 2010 account of ongoing struggles for accommodation in academia), there is much still to be learned about why ableism endures and what we can do to help change this. What are the connections, for example, between neoliberal policies and ways of life and ableism? Why are able-bodied people often fearful of and resistant to interaction with disabled and ill people? To what extent is ableism perpetuated by genetic breakthroughs such as identifying genes linked to cognitive disorders (e.g., Northwell Health, 2017) or technological advances (such as bionic prostheses and exoskeletons to allow persons with spinal-cord injury to stand and walk) (*Fixed: The Science/Fiction of Human Enhancement*, 2013)? Ableism thus remains a very fruitful area of inquiry for health geographers and others concerned with how and why persons with impairments and illnesses continue to be disabled in society and space.

Looking ahead: imagining future geographies of disability and chronic illness

From today's vantage point, there are many exciting avenues for geographic studies of disability and chronic illness. Hall and Wilton (2016) have argued that there is considerable potential for more relational understandings of able and disabled bodies and the assemblages of interaction, affect, meaning, objects and environment in which they are caught up. Among other things, such an approach problematizes a binary understanding of able and disabled bodies and minds and instead favors a more fluid conception of people's experiences of ability, disability and chronic illnesses as processes of subjective becoming. These processes unfold through interactions and encounters with different people, places and objects (e.g., mobility aids). This conception also arguably facilitates more critical analyses of the workings of ableist societies and spaces of everyday life.

Thinking back to the challenge of making our academic environments more inclusive of disabled and/or ill people, there is a need to get a better sense of the extent to which they are currently represented in the discipline. We also need to better understand what their experiences of seeking accommodation and inclusion have been. As feminist geographers such as Peake and Mullings have begun to ask, what is it about the neoliberal academic environment that works to undermine mental well-being? Similar questions need to be posed in relation to bodies with physical impairments and illnesses and the extent to which others relate to such bodies in exclusionary ways. Health geographers might want to ask, for example, how neoliberal conditions of life, such as precarious employment and norms and policies favoring individual independence (as opposed to interdependence), are affecting the lives and well-being of persons with physical impairments and illnesses and the ways in which able-bodied persons relate to them.

As noted above, there is a pressing need for geographic research on experiences of disability and chronic illness in the Global South. Although there is a sizeable interdisciplinary literature on these topics, geographers have had little to say about disability and chronic illness in the Global South. In what ways do conditions such as war, famine and extreme poverty foster impairments and illnesses and ensure that those with them are disabled/marginalized in society and space? What needs to change about our current global capitalist order and the place of countries such as Guyana in it to more effectively combat problems such as lack of access to food and income as well as aids such as wheelchairs and prosthetic limbs?

We also desperately need more critical geographies of forces perpetuating ableism in our societies and spaces of everyday life. This could include, for example, analyses of why the in/ability to be physically present at events such as conferences is at least sometimes a basis for exclusionary practices. It also should arguably include analyses of how geographers who embody greater ability are at least sometimes complicit in the oppression of their more disabled counterparts. It also needs to include, as indicated above, further investigation of the links between neoliberalism, global capitalism and the oppression/marginalization of people with impairments and illnesses.

Health and critical social geographers, after almost three decades of research (and, arguably to a lesser extent, activism), are well-positioned to tackle such challenges. What we need is to commit, both in a scholarly and in a political sense, to making the world we share a truly inclusive one.

References

Barns, A., Svanhdm, F., Kellberg, A., Thyberg, I. and Falkmler, T. (2015). Living in the present: women's everyday experience of living with rheumatoid arthritis. *Sage Open*, 5(4), pp. 1–13.

Butler, R. E. (1994). Geography and vision impaired and blind populations. *Transactions of the Institute of British Geographers*, 19(3), pp. 366–368.

Campbell, F. K. (2009). *Contours of ableism: territories, objects, disability and desire*. London: Palgrave Macmillan.

Chouinard, V. (1997). Making space for disabling differences: challenging ableist geographies. *Environment and Planning D*, 15(4), pp. 379–387.

Chouinard, V. (2010). "Like Alice through the Looking Glass" II: the struggle for accommodation continues. *Resources for Feminist Research*, 33(3/4), pp. 161–178.

Chouinard, V. (2012). Pushing the boundaries of our understanding of disability and violence: voices from the global South (Guyana). *Disability & Society*, 27(6), pp. 777–792.

Chouinard, V. (2014). Precarious lives in the global South: on being disabled in Guyana. *Antipode*, 46(2), pp. 340–358.

Chouinard, V., Belle, C., Khan, H. and Adrian, N. (2016). Embodying disability in the Global South: exploring emotional geographies of research and of disabled people's lives in Guyana. In: S. Grech and K. Soldatic, eds., *Disability in the global South: the critical handbook*. Switzerland: Springer International Publishing, pp. 583–598.

Chouinard, V. and Grant, A. (1995). On being not even anywhere near "the project": ways of putting ourselves in the picture. *Antipode*, 27(2), pp. 137–166.

Crooks, V. A. (2010). Women's changing experiences of the home and life inside it after becoming chronically ill. In: V. Chouinard, E. Hall and R. Wilton, eds., *Towards enabling geographies: "disabled" bodies and minds in society and space*. London and New York: Ashgate, pp. 45–61.

Davidson, J. (2008). Autistic futures online: virtual communication and cultural expression on the spectrum. *Social and Cultural Geography*, 9(7), pp. 781–806.

Dear, M. J. and Taylor, S. M. (1982). *"Not on our street": community attitudes towards the mentally ill*. London: Pion Ltd.

Dear, M. and Wolch, J. (1987). *Landscapes of despair: from deinstitutionalization to homelessness*. Princeton, NJ: Princeton University Press.

Dyck, I. (1995). Hidden geographies: the changing lifeworlds of women with multiple sclerosis. *Social Science & Medicine*, 40(3), pp. 307–320.

Fixed: The Science/Fiction of Human Enhancement. (2013). [film] Making Change Media: Regan Brashear.

Gant, R. and Smith, J. (1984). *Spatial mobility problems and the elderly and disabled in the Cotswolds*. Norwich: Geo Books.

Gleeson, B. (1996). A geography for disabled people? *Transactions of the Institute of British Geographers*, 21(2), pp. 387–396.

Gleeson, B. (1998). *Geographies of disability*. London: Routledge.

Golledge, R. D., Brown, L. A. and Williamson, F. (1972). Behavioural approaches in human geography: an overview. *Australian Geographer*, 12 (1), pp. 159–169.

Hall, E. (2000). "Blood, brain and bones": taking the body seriously in geographies of impairment and illness. *Area*, 32(1), pp. 21–29.

Hall, E. and Kearns, R. (2001). Making space for the "intellectual" in geographies of disability. *Health & Place*, 7, pp. 237–246.

Hall, E. and Wilton, R. (2016). Towards a relational geography of disability. *Progress in Human Geography*, pp. 1–18.

Imrie, R. (1996). Ableist geographies, disabling spaces: towards a reconstruction of Golledge's "geography of the disabled." *Transactions of the Institute of British Geographers*, 21(2), pp. 397–403.

Imrie, R. (2010). Disability, embodiment and the meaning of home. In: V. Chouinard, E. Hall, and R. Wilton, eds., *Towards enabling geographies: "disabled" bodies and minds in society and space*. Abingdon: Routledge, pp. 23–44.

Kearns, R. A. (1993). Place and health: towards a reformed medical geography. *The Professional Geographer*, 45(2), pp. 139–147.

Kearns, R. A. and Moon, G. (2002). From medical to health geography: novelty, place and theory after a decade of change. *Progress in Human Geography*, 26(5), pp. 605–625.

Kirby, A., Bowlby S. R. and Swan, N. V. (1983). Mobility problems of the disabled. *Cities*, 3, pp. 117–119.

Longhurst, R. (2010). The disabling affects of fat: the emotional and material geographies of some women who live in Hamilton, New Zealand. In: V. Chouinard, E. Hall and R. Wilton, eds., *Towards enabling geographies: "'disabled" bodies and minds in society and space*. Abingdon: Routledge, pp. 199–217.

Loomis, J. M., Marston, J. R., Golledge, R. D. and Klatzky, R. L. (2005). Personal guidance system for persons with visual impairments: a comparison of spatial displays for route guidance. *Journal of Visual Impairment and Blindness*, 99(4), pp. 219–232.

Loomis, J. M., Golledge, R. D. and Klatzky, R. L. (1998). Navigation system for the blind: auditory display modes and guidance. *Presence*, 7(2), pp. 193–203.

Medicine.net (2017). *Chronic disease.* [online] Available at: www.medicinenet.com/script/main/art.asp?articlekey=33490 [Accessed 1 Nov. 2017].

Meekosha, H. and Soldatic, K. (2011). Human rights and the global South: the case of disability. *Third World Quarterly*, 32(8), pp. 1383–1397.

Metzel, D. (2010). 563 miles: a matter of long-distance caring by siblings of siblings with intellectual and developmental disabilities. In: V. Chouinard, E. Hall and R. Wilton, eds., *Towards enabling geographies: "disabled" bodies and minds in society and space*. London and New York: Ashgate, pp. 123–144.

Moss, P. and Dyck, I. (2003). *Women, body, illness: space and identity in the everyday lives of women with chronic illness.* Lanham, MD: Rowman & Littlefield.

Northwell Health. (2017). *Genetic discovery provides new insight into conditions such as schizophrenia, ADHD.* Science Daily. [online] Available at: www.sciencedaily.com/releases/2017/01/170117084032.htm [Accessed 26 May 2017].

Oliver, M. (1990). *The politics of disability: a sociological approach.* New York: St. Martin's Press.

Park, D. C., Radford, J. P. and Vickers, M. H. (1998). Disability studies in human geography. *Progress in Human Geography*, 22(2), pp. 208–233. Parr, H. (2011). *Mental health and social space: towards inclusionary geographies?* Malden, MA: Wiley-Blackwell.

Parr, H. and Butler, R. (1999). New geographies of illness, impairment and disability. In: *Mind and body spaces: geographies of illness, impairment and disability*. London: Routledge, pp. 1–24.

Perle, E. D. (1969). Urban mobility needs of the handicapped: an exploration. Unpublished PhD dissertation. Department of Geography: University of Pittsburgh.

Philo, C. (1987). "Fit localities for an asylum": the historical geography of the nineteenth-century "mad-business" in England as seen through the pages of the Asylum Journal. *Journal of Historical Geography*, 13(4), pp. 398–415.

Philo, C. and Metzel, D. S. (2005). Introduction to theme section on geographies of intellectual disability: "outside the participatory mainstream"? *Health & Place*, 11(2), pp. 77–85.

Scambler, G. and Scambler, S. (2010). Assaults on the lifeworld: the sociology of chronic and disabling conditions. In: G. Scambler and S. Scambler, eds., *New directions in the sociology of chronic and disabling conditions: assaults on the lifeworld*. London: Palgrave Macmillan, pp. 1–7.

Self, C. M., Gopal, S. Golledge, R. G. and Fenstermaker, S. (1992). Gender-related differences in spatial ability. *Progress in Human Geography*, 16(3), pp. 315–342.

Smith, M. and Davidson, J. (2006). "It makes my skin crawl . . .": the embodiment of disgust in phobias of nature. *Body & Society*, 12(1), pp. 43–67.

Valentine, G. (1999). What it means to be a man: the body, masculinity, disability. In: R. Butler and H. Parr, eds., *Mind and body spaces: geographies of illness, impairment and disability*. London and New York: Routledge, pp. 167–180.

Valentine, G. (2007). Theorizing and researching intersectionality: a challenge for feminist geography. *The Professional Geographer*, 59, pp. 10–21.

Wilton, R., DeVerteuil, G. and Evans, J. (2014). "No more of this macho bullshit": drug treatment, place and the reworking of masculinity. *Transactions of the Institute of British Geographers*, 39(2), pp. 291–303.

Wilton, R. and Moreno, C. M. (2012). Editorial: critical geographies of drugs and alcohol. *Social and Cultural Geography*, 13(2), pp. 99–108.

Wolch, J. R. and Philo, C. (2000). From distributions of deviance to definitions of difference: past and future mental health geographies. *Health & Place*, 6(3), pp. 137–157.

26

THE GEOGRAPHIES OF INDIGENOUS HEALTH

Chantelle A. M. Richmond and Katie Big-Canoe

The geographies of Indigenous health is a relatively new sub-discipline of critical human geography. The fields of health geography and Indigenous geography explicitly influence this emerging research area, as both have identified the need for research to examine the complex and changing relationship between Indigenous peoples' health and the environment. Key questions asked by researchers in this area of health geography include: How is the health and well-being of Indigenous peoples and communities shaped by the environments within which they live? What *processes* are affecting the health and environments of Indigenous peoples, and how so? And can Indigenous engagement in research support positive change for Indigenous well-being and environmental protection? While the field of health geography has provided important theoretical frameworks for exploring the complexity of health and its multiple inter-related determinants, including the ways health may be shaped by current and historical processes (Kearns and Gesler, 1998; Luginaah, 2009), Indigenous geography supports a methodological imperative that places Indigenous communities and their concerns at the forefront of research about Indigenous health determinants, most particularly those related to the environment and processes of environmental change (Coombes et al., 2011; Herman, 2008; Louis, 2007). This relatively new area of study garnering attention among geographers from around the world has its roots in the Canadian experience. This chapter places emphasis on the ways Canadian scholars have informed the development and growth of this field and on the ways it has been taken up globally.

We begin with a short discussion on the importance of Indigenous knowledge frameworks for understanding the connections between Indigenous peoples, their health, and the land, providing a short analysis of Indigenous health disparities as well as the processes that have contributed to these inequities. We then move into a discussion about the origins of the geographies of Indigenous health and its evolution, including a transformation in the conceptualization and measurement of health. This transformation has led to the uptake of more critical approaches that have sought to understand and describe how historical processes, such as colonization and environmental dispossession, have shaped modern Indigenous relationships to land and what this means for health. Indigenous health geographies have also contributed in significant ways to the development of participatory methodologies and approaches that engage Indigenous peoples and communities in research development as a way of allowing them to exert self-determination over their health, their lands and their environmental futures. This chapter concludes with discussion about future directions in the field.

Indigenous knowledge, connection to land and processes of changing health

Across the globe, there are roughly 400 million Indigenous peoples living on all continents and across many diverse ecosystems (Gracey and King, 2009), including some of the harshest and most inhospitable environments (e.g., desert, Arctic). The term *Indigenous* is generally used in a global sense, whereas Indigenous groups and communities typically have geographically contextualized terms that relate to the places they inhabit, as well as the languages they speak. For example, in central Canada, the Ojibway people are often referred to as the *Anishinabe*, a term that reflects the language spoken by this group, *Anishinabemowin*. Around the globe, there are various other terms used to reflect one's Indigeneity, including native, Aboriginal, tribal peoples, original, first, and many others. Perhaps the most important starting place for research at the interface of Indigenous peoples' health and the land is to think about how and why these concepts are related. Among most Indigenous societies, the answer can come from their Indigenous knowledge (IK) systems. In all Indigenous cultures, daily life and continued well-being has traditionally depended on a deep knowledge of local ecosystems. Although there are no commonly accepted definitions of what an IK system is, it typically refers to the culturally and spiritually based ways Indigenous peoples relate both to their local ecosystems and to one another in the maintenance of health and well-being (Battiste and Henderson, 2000; LaDuke, 1994).

Given the special connection between Indigenous peoples and the land, including the social, spiritual, and economic ties it supports, Indigenous peoples are highly vulnerable to processes of environmental change and environmental dispossession (Ford, 2012; Richmond and Ross, 2009), with strikingly similar consequences for health and well-being. Environmental dispossession refers to the processes by which Indigenous peoples' access to their traditional lands and territories is reduced or eliminated (Richmond and Ross, 2009). These processes occur in direct and indirect ways and are generated both historically – mainly through processes of colonialism – and in contemporary times. Direct forms of environmental dispossession (e.g., physical forms) involve processes that physically disable use of the land, such as industrial activities or contamination events that sever ties to traditional foods or resources required for sustaining daily activities. Indirect forms of dispossession (e.g., political) occur because of policies or regulations that lead to the severance of Indigenous peoples' link to the land and the IK it fosters. Indigenous peoples have been subject to massive relocation projects, for example, as well as various assimilationist policies that have included the forced attendance of Indigenous children at church- and state-run residential/boarding schools in Canada, the United States, and Australia.

In the global context, IK systems are based fundamentally on local experiences, connection to and ongoing observation of the local ecosystem. Because of the vast and often rapid changes brought upon by global industrial development, including various types of contamination and climate change, Indigenous peoples are particularly susceptible to these changes. The results of environmental dispossession for Indigenous health, on a global scale, demonstrate remarkable similarities. The reduced access of Indigenous peoples to their traditional lands and territories has had many detrimental impacts for the preservation, practice and intergenerational transmission of IK, leading to significant changes in way of life and a contemporary health and cultural profile characterized by a greater burden of morbidity and early mortality than the non-Indigenous population.

While the last half century has witnessed incredible gains in life expectancy for Indigenous populations, and a considerable reduction in infant mortality, several other patterns of health are quite troubling, including a high prevalence of cardiovascular disease, type 2 diabetes, cancers, mental illness, addictions and violence, and food insecurity, among many others (Adelson, 2005; Gracey and King, 2009). While most Indigenous populations have undergone the epidemiologic transition, wherein the main causes of Indigenous morbidity and mortality have shifted from infectious to long-term chronic disease, infectious disease is common in the health profile of Indigenous populations. The persistence of infectious disease among Indigenous peoples includes many ailments that have been all but eradicated in the general population, such as active tuberculosis and rheumatic fever. International research indicates that these health disparities are linked to alienation from land (Richmond and Ross, 2009), limited opportunities to practice activities that foster Indigenous culture, social connectedness, and traditional health and healing, as well as several social determinants of health, including poverty (Adelson, 2005; Frohlich, Ross and Richmond, 2006).

Despite troubling patterns of health evident in Indigenous populations, there are reasons to be hopeful. As noted above, the wide gap in the health and social profile of Indigenous peoples is slowly closing over time. There are many factors leading to the reduction in health disparities, including increased access to culturally safe and high-quality health care; the development of health and social policy that reflects Indigenous ideals and philosophies; the ongoing resolution of Indigenous land claims; increased funding for Indigenous social development and programming; and steady increases in socioeconomic well-being (e.g., increases in high school completion and post-secondary attendance) (Richmond and Cook, 2016).

Perhaps one of the most important contributions to the closing gap on Indigenous health relates to a growing and hopeful movement of Indigenous community participation in health and social research. Building on Smith's seminal work on decolonizing methodologies (1999), a wide body of Indigenous research has been undertaken, led by a common goal of "centering our own concerns and worldviews, and coming to know and understand theory and research from our own perspectives and for our own purposes" (Smith, 1999, p. 39). Such methods require meaningful research partnerships between Indigenous communities and researchers from universities and various governments and organizations, with the ultimate goal of designing and carrying out research that will lead to improved local conditions. These partnerships are a method of decolonization, as the Indigenous communities involved are placed at the center of the research, thereby empowering local influence on research topics, debate and research design and prioritizing local influence in the development of community programs. It is largely within this context that the geographies of Indigenous health was inspired and has since developed.

The evolution of the discipline

While the field of Indigenous health geography has at its core a focus on the intersection between health and environment, the ways in which these concepts have been both defined and measured in the academic context have evolved significantly in the past 30 years. Some of the earliest research focused on measuring disease prevalence and description of health, social and economic indicators of Indigenous well-being (Newbold, 1998; Thouez, Rannou and Foggin, 1989). These early works focused predominantly on understanding and describing health and social trends, and access to health care, among Indigenous peoples. Mirroring work taking place in the epidemiologic literature, these works focused on large-scale health and social survey data in Canada and elsewhere to describe difference with non-Indigenous peoples and to characterize the spatial distribution of Indigenous peoples' health. Perhaps the earliest application of health-geographic research in the Indigenous context was by Thouez, Rannou and Foggin (1989), who examined the epidemiological and sociocultural determinants of health and indicators of malnutrition among the Cree and Inuit of northern Quebec (in Canada). Drawing from both questionnaires and physical examination, Thouez, Rannou and Foggin (1989) concluded that the shift away from traditional lifestyles, combined with non-native health-service provision, was detrimental to health and well-being in northern Quebec. In 1998, Newbold explored the health status of the Canadian Aboriginal population, along with their perceived community-health problems, and proposed solutions to these issues. Drawing from a geographic analysis of the 1991 Aboriginal People's Survey, Newbold found that geographic location (in particular, on-reserve status) played a significant role in determining self-rated health and perceived community-health problems. Similar to trends occurring within the health-geography literature at the time, Newbold concluded that provision of health services would not be sufficient to improve these troubling health disparities, but that broader social welfare and more comprehensive development programs should be considered.

Beginning in the late 1990s and early 2000s, the research questions and data sources of Indigenous health geographers grew to reflect a wider and more nuanced understanding of Indigenous health. This growth in conceptual definition reflected changes taking place in the wider discipline of health geography – mirroring the recognition of the social determinants of health (i.e., recognizing that health is shaped by many interconnecting determinants (Luginaah, 2009)), and of the necessity of thinking about Indigenous health in a more holistic way (Elliott and Foster, 1995). At the same time, a growing critical emphasis was being placed on the wider social and political forces that frame both Indigenes' health and their environments.

In 2002, Wilson and Rosenberg published a paper that has become one of the seminal pieces in Indigenous health geography. Drawing on data from the 1991 Aboriginal Peoples Survey (APS), they undertook descriptive statistics and logistic regression analyses to explore determinants of health for First Nations people in Canada. While their results indicated that the determinants of health for First Nations are similar to those for the general population, Wilson and Rosenberg's (2002) analysis was among the first to identity the health importance of traditional activities (e.g., harvesting of traditional foods such as berries or animals). While the authors note limitations on the traditional activity measures included in this analysis, this paper called attention to the determinants of Indigenous peoples' health, and especially to other potentially important cultural determinants of health, which they argued were deserving of more thoughtful conceptual development and analysis.

The uptake of these critical perspectives coincided in important ways with a significant methodological reorientation in the ways geographers were undertaking Indigenous health research. Whereas earlier descriptive studies, such as those described above, had relied strongly on quantitative data (e.g., survey, medical and administrative data), research questions about culture and changing relationships with the environment begged for more qualitative and intensive methodologies. It was in 2003, when Wilson published a paper in *Health & Place*, that Indigenous voices themselves began to appear in Indigenous research. Although this important study advanced the theory and uptake of therapeutic landscapes in the Indigenous context, perhaps a more important contribution of this paper was in the way Wilson privileged First Nations voices about their own ideas, their own thoughts on the ways everyday life contributes to health and healing. Wilson (2003, p. 83) concluded that, at the time, "conceptualizations of health and place within the Geography of Health literature were only partial," thereby calling on others in the field to expand their ways of thinking about, and measuring, Indigenous health. Wilson's seminal research set the context for a field that would wholeheartedly embrace this challenge.

In 2005, Richmond et al. investigated First Nations perceptions of the links between environment, economy and health and well-being in 'Namgis First Nation (Alert Bay, British Columbia, Canada), with particular attention on perceived risks and benefits of salmon aquaculture. The results built on Wilson's (2003) holistic conceptualizations of First Nations health (e.g., the mental, social, spiritual and physical dimensions), highlighting the importance of being out on the land and waters for maintaining one's health, as well as the respondents' proclaimed social and moral responsibilities to their communities and as caretakers of the land. These results described the ways that aquaculture development (and other forms of contested development in the traditional territories of First Nations people) had decreased the 'Namgis First Nation's access to local environmental resources and restricted the economic, social, and cultural activities that determine good health and well-being. In their discussion, Richmond et al. (2005) framed these findings around a political-ecological framework (building on Mayer, 1996), that challenged researchers working at the land-health interface, particularly in the Indigenous context, to think more critically about the wider political context within which environmental resource development is occurring and what this means for health. The application of these concepts has since been explored in other Indigenous settings and contexts, including Māori foodscapes in New Zealand (Panelli and Tipa, 2009), the concept of *caring for country* in Australia (Townsend, Phillips and Aldous, 2009), First Nations perceptions of risk in mining expansion (Place and Hanlon, 2011), and contamination of aquatic foods in the American state of Washington (Dona- tuto, Satterfield and Gregory, 2011).

The political turn: environmental dispossession and health

The fields of native studies, anthropology and cultural studies are foundational starting places for many who do research in the geographies of Indigenous health. In these disciplines, a good deal of research has focused on concepts related to identity, culture, and the ways in which processes of colonialism have impacted cul- tural identities and ways of living among Indigenous peoples. In 2000, Peters wrote an article that reviewed the works of Canadian geographers on Indigenous peoples in Canada. Among the many important con- tributions this article highlighted, Peters identified the process of Indigenous peoples' dispossession of their lands and resources as one of the key topics. Within the health context, geographers had been reluctant to use terms like *dispossession* or *colonialism*; however, there had been important discussion around the need for improved conceptualization of place-based notions of well-being that recognize cultural and environmental specificity (Panelli and Tipa, 2009), as well as greater inclusion of Indigenous worldviews in resource devel- opment (Castleden, Garvin and Huuy-ay-aht First Nation, 2009).

In 2009, Richmond and Ross pushed the theoretical boundaries of the land-health relationship when they critically examined the determinants of health through in-depth interviews with First Nations and Inuit community health representatives across Canada. This analysis illustrated holistic connections between health, connection to land and sense of cultural identity. In the wider Canadian literature, up until this point, culture and land had been conceptualized as discrete health determinants. Based on the interviews, Rich- mond and Ross (2009) articulated the concept of environmental dispossession as a historical and enduring process that has significantly altered Indigenous peoples' ties to the land and their abilities to practice their cultures, conceptualizing it as a process that takes both physical and political forms by which Indigenous peoples' access to their traditional lands and territories is reduced or eliminated. These processes affect the health of Indigenous peoples because they have undermined the ability of Indigenous peoples to practice their land-based cultures and IK systems. This framing has provided many scholars in this field, and beyond, a pathway for exampling and further examining changing patterns of Indigenous health and the varied pro- cesses that shape its environmental influences.

The environment as a place of risk

The introduction of the term *environmental dispossession* provided a theoretical framework for understanding, discussing and exploring the historical and contemporary processes that dispossess Indigenous peoples of their lands and livelihoods and shape opportunities for Indigenous peoples' health (de Leeuw, Cameron and Greenwood, 2012; Tobias, Richmond and Luginaah, 2013). On a global level, processes of environmental dispossession – whether operating directly or indirectly – have significant impacts for the health and well- being of Indigenous peoples because they work to destabilize IK systems (Furgal, Martin and Gosselin, 2002). They do so by reducing Indigenous peoples' confidence and ability to know environmental condi- tions (Furgal, 2008; King and Furgal, 2014).

Around the same time that Richmond and Ross (2009) offered the concepts of environmental dispos- session, a base of research was beginning to grow around the possible health risks Indigenous peoples face by living in or near environments undergoing processes of change. That is, while the land had long been seen as a health-enhancing determinant, researchers were beginning to see the health risks associated with environmental change. Perhaps the most important area of research in the Canadian context is climate change and its health impact in affected Indigenous communities (Ford, 2012). While we know that IK systems are dynamic and constantly changing, processes of environmental dispossession can accelerate these changes in ways that makes normal interaction with the land risky for human health – for example, by eating fish or other country food that has been contaminated by heavy metals. The environmental effects of climate change make it more difficult to predict weather patterns and may lead migratory animals to return later in the season than usual (Cunsolo, Shiwak and Wood, 2017; Durkalec et al., 2015; Ford et al., 2009). These changing conditions can lead to people spending less time on the land because the risk of doing so can become too great (Luginaah, 2009). Less time spent out on the land means that opportunities for sharing knowledge and practicing skills are also limited, leading to the weakening of cultural and social ties that are so important for cultural identity, social roles, moral obligations and self-esteem (Durkalec et al., 2015; Tobias and Richmond, 2014).

Methodological advancements: community in research

Indigenous geography supports a methodological imperative that seeks to place Indigenous communities and their concerns at the forefront of research (Louis, 2007; Louis and Grossman, 2009). The geographies of Indigenous health has passionately accepted this important community-centered paradigm in its uptake of research related to Indigenous health and the environment. Perhaps one of the most important contributions of Indigenous health geographers has been in making space for community ideas, concerns and visions throughout the research process (Tobias and Richmond, 2016).

From about 2010 onward, this small field transformed rapidly to embrace a variety of participatory methodologies that included more iterative and culturally relevant data-collection strategies including sharing circles, story, photovoice, digital stories and documentary film. These empirical studies have not only focused on better understanding particular environment-health relationships, but also scrutinized the methodological practice required to carry out this work. This has led to important critiques about the role of friendship in research (de Leeuw, Cameron and Greenwood, 2012), protocol for relationship building (Tobias, Richmond and Luginaah, 2013), and the roles of non-Indigenous scholars (Graeme, 2013; Graeme and Mandawe, 2017).

In 2008, Castleden and Garvin opened the field to the importance of community-based participatory research (CBPR) in their uptake of photovoice methodology with a First Nations community in Western Canada. Interviews with community participants about the photovoice process revealed that this approach to doing environment and health research had balanced power, created a sense of ownership, fostered trust, built capacity in the community and responded to cultural preferences (Castleden, Garvin and Huuy-ay-aht First Nation, 2008). In their conclusion, however, the authors discussed the importance of being flexible in the research methods and analysis and of making appropriate space and time to engage your research partners in various stages throughout the research process, not only in data collection. As environment and health research in the Indigenous community context continues to grow, this study provides a foundational example for researchers to *listen to and discuss community issues*, thereby opening opportunities for scholars to consider ways their research may impart social change at the community level.

In 2012, Castleden was again at the forefront of foundational discussions about the possibilities of CBPR in the environment-health context explored by health geographers. Based on interviews with Canadian university-based geographers and social scientists who engage in CBPR, Castleden, Mulrennan and Godlewska (2012) revealed that its practice often differs quite substantially from its theoretical tenets. Community time and resources, pressures related to academic timelines, restrictions from Research Ethics Boards, and relationship-building are factors that work together in important ways to shape the ways CBPR approaches unfold in practice by health geographers and others. Since 2012, many researchers and institutions have identified the ways in which CBPR can differ across Indigenous and non-Indigenous contexts. Such research has helped identify boundaries around the feasibility of CBPR and the need to respect and balance community timelines and desires with academic timelines and funding structures (Tobias, Richmond, and Luginaah, 2013; de Leeuw, Cameron and Greenwood, 2012).

Future direction: Indigenous research by Indigenous researchers

In the past few years, Indigenous voices have become considerably more prominent in the discipline of Indigenous health geography, as has the commitment to see research used for the benefit of Indigenous communities (Castleden, Shiwak and Wood, 2017; Tobias and Richmond, 2016). This differs from work done in this same discipline a mere 20 years earlier, when individuals and communities were depicted as data points in large surveys. As more Indigenous scholars and communities began to engage in this discipline, however, the motivations for research started to lean considerably toward self-determination (Richmond, 2016), the uptake of Indigenous methodologies and approaches, and the development of real-world strategies for preserving IK and protecting environmental futures. Based on research with Anishinabe youth in Ontario, Big-Canoe and Richmond (2014) proposed the term *environmental repossession*, which refers to the social, cultural and political processes by which Indigenous peoples and communities are reclaiming their traditional lands and ways of life. The concept of environmental repossession is rooted in the idea that Indigenous peoples' health, ways of living, and IK systems are highly dependent on access to their traditional lands and territories. But, as we know, great numbers of Indigenous peoples around the world no longer live in their traditional territories – so how can Indigenous peoples practice their cultures and IK systems in the new spaces and places they occupy? This is an important question for health geographers to address. Existing insights point to the importance of unpacking and understanding residential mobility and rural-to-urban migration by Indigenous people locally and globally (Snyder and Wilson, 2015) and listening to youth voices to understand environmental futures (de Leeuw and Greenwood, 2016; Big-Canoe and Richmond, 2014).

As the field of Indigenous health geographies moves forward, environmental repossession provides both a way of thinking and an applied practice that places Indigenous peoples and their knowledge at the heart of the environment and health research context. Environmental repossession, as a theoretical research area, offers a space wherein Indigenous peoples and researchers can come together to share both their experiences of struggle and their best practices for circumventing these negative experiences. Despite the great geographic, cultural and linguistic diversity of Indigenous peoples across the globe, there are many similarities, including experiences of dispossession, land-based knowledge systems and the contemporary burden of health inequity. Working within a decolonizing methodological framework, environmental repossession provides a promising approach for researchers and communities to collaboratively document the unique and varied processes within which Indigenous communities around the globe are engaging in social, political, economic and many other processes as a means of asserting their Indigenous rights and building healthy futures.

The field of health geography has undergone important transitions in the past three decades. With an expansive and broadening set of theoretical perspectives and methodological tools, the discipline has enabled better descriptions and understandings of the complex relationship between human health and its multiple, inter-related determinants. The discipline's preoccupation with place and space has lent strength to studies focused on Indigenous peoples and the determinants of Indigenous health. Despite the great diversity of the global Indigenous population, there are two important shared characteristics; the first relates to the importance of IK systems for the health of Indigenous communities, and the second relates to their common experiences of colonization and environmental dispossession. Borrowing from both health geography and Indigenous geography, the new field of the geographies of Indigenous health provides a promising discipline from which to explore the ways in which the health and well-being of Indigenous peoples and communities are shaped by the environments within which they live, to critically explore the *processes* that affect the health and environments of Indigenous peoples, and to open more meaningful spaces for Indigenous scholars and communities to engage in research that will support Indigenous well-being and environmental protection.

References

Adelson, N. (2005). The embodiment of inequality: health disparities in Aboriginal Canada. *Canadian Journal of Public Health*, 96, pp. S45–S61.

Battiste, M. and Henderson, J. (2000). *Protecting Indigenous knowledge and heritage: a global challenge*. Saskatoon: Purich Publishing Ltd.

Big-Canoe, K. and Richmond, C. (2014). Anishinabe youth perceptions about community health: toward environmental repossession. *Health & Place*, 26, pp. 127–135.

Castleden, H., Garvin, T. and Huuy-ay-aht First Nation. (2008). Modifying photovoice for community-based participatory Indigenous research. *Social Science & Medicine*, 66(6), pp. 1393–1405.

Castleden, H., Garvin, T. and Huuy-ay-aht First Nation. (2009). "Hishuk Tsawak" (everything is one/connected): a Huu-ay-aht worldview for seeing forestry in British Columbia, Canada. *Society and Natural Resources*, 22(9), pp. 789–804.

Castleden, H., Hart, C., Cunsolo, A., Harper, S. and Martin, D. (2017). Reconciliation and relationality in water research and management in Canada: implementing Indigenous ontologies, epistemologies, and methodologies. In: S. Renzetti and D. Dupont, eds., *Water policy and governance in Canada: Global issues in water policy*. Switzerland: Springer International Publishing, pp. 69–95.

Castleden, H., Mulrennan, M. and Godlewska, A. (2012). Community-based participatory research involving Indigenous peoples in Canadian geography: progress? An editorial introduction. *The Canadian Geographer/Le Géographe canadien*, 56(2), pp. 155–159.

Coombes, B., Gombay, N., Johnson, J. and Shaw, W. (2011). The challenges of and from Indigenous geographies. In: V. J. Del Casino, M. Thomas, P. Cloke and R. Panelli, eds., *A companion to social geography*. Chichester: Wiley-Blackwell, pp. 472–489.

Cunsolo, A., Shiwak, I. and Wood, M. (2017). "You Need to Be a Well-Rounded Cultural Person": youth mentorship programs for cultural preservation, promotion, and sustainability in the Nunatsiavut region of Labrador. In: G. Fondahl and G. Wilson, eds., *Northern sustainabilities: Understanding and addressing change in the circumpolar world*. Switzerland: Springer International Publishing, pp. 285–303.

de Leeuw, S., Cameron, E. and Greenwood, M. (2012). Participatory and community-based research, Indigenous geographies, and the spaces of friendship: a critical engagement. *The Canadian Geographer/Le Géographe canadien*, 56(2), pp. 180–194.

de Leeuw, S. and Greenwood, M. (2016). Geographies of indigenous children and youth: a critical review grounded in spaces of the colonial nation state. *Space, Place, and Environment*, pp. 47–65.

de Leeuw, S., Maurice, S., Holyk, T., Greenwood, M. and Adam, W. (2012). With reserves: colonial geographies and First Nations health. *Annals of the Association of American Geographers*, 102(5), pp. 904–911.

Donatuto, J., Satterfield, T. and Gregory, R. (2011). Poisoning the body to nourish the soul: prioritising health risks and impacts in a Native American community. *Health, Risk & Society*, 13(2), pp. 103–127.

Durkalec, A., Furgal, C., Skinner, M. and Sheldon, T. (2015). Climate change influences on environment as a determinant of Indigenous health: relationships to place, sea ice, and health in an Inuit community. *Social Science & Medicine*, 136, pp. 17–26.

Elliott, S. and Foster, L. (1995). Mind-body-place: a geography of Aboriginal health in British Columbia. In: P. H. Stephenson, S. J. Elliott, L. T. Foster and J. Harris, eds., *A persistent spirit: Towards understanding Aboriginal health in British Columbia*. Victoria, BC: Canadian Western Geographical Series, pp. 94–127.

Ford, J. (2012). Indigenous health and climate change. *American Journal of Public Health*, 102(7), pp. 1260–1266.

Ford, J., Gough, W., Laidler, G., Macdonald, J., Irngaut, C. and Qrunnut, K. (2009). Sea ice, climate change, and community vulnerability in northern Foxe Basin, Canada. *Climate Research*, 38(2), pp. 137–154.

Frohlich, K., Ross, N. and Richmond, C. (2006). Health disparities in Canada today: some evidence and a theoretical framework. *Health Policy*, 79(2), pp. 132–143.

Furgal, C. (2008). Health impacts of climate change in Canada's North. In *Human health in a changing climate, a Canadian assessment of vulnerabilities and adaptive capacity*. Ottawa: Health Canada, pp. 303–366.

Furgal, C., Martin, D. and Gosselin, P. (2002). Climate change and health in Nunavik and Labrador: lessons from Inuit knowledge. In: I. Krupnik and D. Jolly, eds., *The earth is faster now: Indigenous observations of Arctic environmental change.* Fairbanks, Alaska: Arctic Research Consortium of the United States, pp. 266–299.

Gracey, M. and King, M. (2009). Indigenous health part 1: determinants and disease patterns. *The Lancet*, 374(9683), pp. 65–75.

Graeme, C. (2013). Indigenous health research and the non-indigenous researcher: a proposed framework for the autoethnographic methodological approach. *Pimatisiwin: A Journal of Aboriginal and Indigenous Community Health*, 11(3), pp. 513–520.

Graeme, C. and Mandawe, E. (2017). Indigenous geographies: research as reconciliation. *International Indigenous Policy Journal*, 8(2), pp. 1–19.

Herman, R. (2008). Reflections on the importance of indigenous geography. *American Indian Culture and Research Journal*, 32(3), pp. 73–88.

Kearns, R. and Gesler, W. (1998). *Putting health into place: landscape, identity, and well-being.* Syracuse, NY: Syracuse University Press.

King, U. and Furgal, C. (2014). Is hunting still healthy? Understanding the interrelationships between indigenous participation in land-based practices and human-environmental health. *International Journal of Environmental Research and Public Health*, 11(6), pp. 5751–5782.

LaDuke, W. (1994). Traditional ecological knowledge and environmental futures. *Colorado Journal of International Environmental Law and Policy*, 5, pp. 127–148.

Louis, R. (2007). Can you hear us now? Voices from the margin: using Indigenous methodologies in geographic research. *Geographical Research*, 45(2), pp. 130–139.

Louis, R. and Grossman, Z. (2009). *Discussion paper on research and Indigenous peoples.* [pdf] Available at: www.indigenousgeography.net/IPSG/pdf/2009_research_discussion_paper.pdf [Accessed 1 Sept. 2017].

Luginaah, I. (2009). Health geography in Canada: where are we headed? *The Canadian Geographer/Le Géographe canadien*, 53(1), pp. 91–99.

Mayer, J. (1996). The political ecology of disease as one new focus for medical geography. *Progress in Human Geography*, 20(4), 441–456.

Newbold, B. K. (1998). Problems in search of solutions: health and Canadian Aboriginals. *Journal of Community Health*, 23(1), pp. 59–74.

Panelli, R. and Tipa, G. (2009). Beyond foodscapes: considering geographies of Indigenous well-being. *Health & Place*, 15(2), pp. 455–465.

Peters, E. (2000). Aboriginal people and Canadian geography: a review of the recent literature. *The Canadian Geographer/Le Géographe canadien*, 44(1), pp. 44–55.

Place, J. and Hanlon, N. (2011). Kill the lake? Kill the proposal: accommodating First Nations' environmental values as a first step on the road to wellness. *GeoJournal*, 76(2), pp. 163–175.

Richmond, C. (2016). Applied decolonizing methodologies: a community based film project with Anishinabe communities. In: N. Fenton and J. Baxter, eds., *Practicing qualitative methods in health geographies.* London: Routledge, pp. 153–166.

Richmond, C. and Cook, C. (2016). Creating conditions for Canadian aboriginal health equity: the promise of healthy public policy. *Public Health Reviews*, 37(1), p. 2.

Richmond, C., Elliott, S., Matthews, R. and Elliott, B. (2005). The political ecology of health: perceptions of environment, economy, health and well-being among 'Namgis First Nation. *Health & Place*, 11(4), pp. 349–365.

Richmond, C. and Ross, N. (2009). The determinants of First Nation and Inuit health: a critical population health approach. *Health & Place*, 15(2), pp. 403–411.

Smith, L. (1999). *Decolonizing methodologies: research and indigenous peoples.* London: Zed Books Ltd.

Snyder, M. and Wilson, K. (2015). "Too much moving . . . there's always a reason": understanding urban aboriginal peoples' experiences of mobility and its impact on holistic health. *Health & Place*, 34, pp. 181–189.

Thouez, J., Rannou, A. and Foggin, P. (1989). The other face of development: native population, health status and indicators of malnutrition – the case of the Cree and Inuit of northern Quebec. *Social Science & Medicine*, 29(8), pp. 965–974.

Tobias, J. and Richmond, C. (2014). "That land means everything to us as Anishinaabe . . .": environmental dispossession and resilience on the North Shore of Lake Superior. *Health & Place*, 29, pp. 26–33.

Tobias, J. and Richmond, C. (2016). Gimiigiwemin: putting knowledge translation into practice with Anishinaabe communities. *International Journal of Indigenous Health*, 11(1), pp. 228–243.

Tobias, J., Richmond, C. and Luginaah, I. (2013). Community-based participatory research (CBPR) with Indigenous communities: producing respectful and reciprocal research. *Journal of Empirical Research on Human Research Ethics*, 8(2), 129–140.

Townsend, M., Phillips, R. and Aldous, D. (2009). "If the land is healthy . . . it makes the people healthy": the relationship between caring for country and health for the Yorta Yorta Nation, Boonwurrung and Bangerang Tribes. *Health & Place*, 15(1), pp. 291–299.

Wilson, K. (2003). Therapeutic landscapes and First Nations peoples: an exploration of culture, health and place. *Health & Place*, 9(2), pp. 83–93.

Wilson, K. and Rosenberg, M. (2002). Exploring the determinants of health for First Nations peoples in Canada: can existing frameworks accommodate traditional activities? *Social Science & Medicine*, 55, pp. 2017–2031.

27

FROM INEQUITIES TO PLACE ATTACHMENT AND THE PROVISION OF HEALTH CARE

Key concerns in the health geographies of aging

Tara Coleman and Janine Wiles

Early geographies of health and aging emphasized the health trends of older populations, while contemporary geographic research has considered the impact of space and place on older people's health and health care and how older people co-experience their health and place. Health geographies of aging have moved from a biomedical emphasis on health, as the absence of disease, to embrace more holistic socioecological understandings of aging and health in social, physical and symbolic contexts.

In this chapter, we explore spatial, relational and critical approaches to health geographies of aging. We identify three key concerns within these approaches, specifically inequities in health geographies of aging, the health aspects of aging in place and attachment to place, and landscapes of care including the provision of health care and other forms of support and service for aging. We recommend that future research continue to investigate patterns of inequalities and demand for services and supports during aging to anticipate the socioeconomic well-being and health of older people and their future needs. We suggest that this focus be expanded to include consideration of the ways in which developing technologies may be reconfiguring socioeconomic and health care needs, as well as older people's experiences of home and neighborhood. Further, in light of currently increasing numbers of older people in our societies, we recommend that future studies consider the many advantages of an aging population, while challenging the stigmatization and homogenization of aging.

Spatial approaches to health geographies of aging

A *spatial* approach to aging and health geography aligns with the issues and processes used in spatial science, and in human geography more broadly, to focus on the measurement and implications of the changing distribution of aging populations at global, regional, urban and household levels (Davies and James, 2011; Northcott, 2014). Researchers explore how trends in disease, technology and social conditions influence population aging at these different spatial scales (Davies and James, 2011). They investigate spatial variation within and across regions and countries, paying particular attention to inequalities in distributions of older people and in their health outcomes and quality of life (Northcott, 2014). A spatial approach highlights likely demand for a range of services and supports associated with old age (Rosenberg, 2014). Such information is vital to ensure that the future needs of older people are understood and appropriately planned for.

A challenge for spatially focused research projects is the availability of high-quality data; particularly in big national-level datasets. Further, data collection tends to focus on the morbidity and mortality patterns of younger people. Wealthier countries such as New Zealand, Australia, and those in North America and Europe tend to have better data at a much greater level of detail, as well as having proportionally older populations. Researchers in these areas are therefore able to better report on differences and inequalities between older social groups (e.g., by social class, ethnicity and gender) (Isaacson et al., 2017) and in different spaces (such as rural-urban, between cities, or across regions) (Huisman, Kunst and Mackenbach, 2003). Data that are available from less-wealthy nations indicate gross global inequalities and inequities in life expectancy and causes of mortality. People in poorer countries still tend to die younger from injury and infectious diseases (Lloyd-Sherlock, 2000). There are also differences in morbidity, health and quality of life in old age as rates of cardiovascular and similar chronic diseases (Mackay and Mensah, 2004), as well as depressive symptoms and risk-taking behaviors such as alcohol and drug use (Gibson et al., 2016), increase in less-wealthy countries.

More detailed understanding of patterns of inequality highlights the need to improve the socioeconomic well-being of older people as a group in order to improve the health of older populations. For example, comparative cross-national work shows how socioeconomic inequalities shape mortality in old age just as they do in younger age groups (Huisman, Kunst and Mackenbach, 2003) and reveals differences in the rate of development of disability between men and women driving much of the gender difference in health during old age (Deeg, 2015). Cumulative life-course exposure to social and material disadvantage and the current material, social, and health conditions that shape outcomes in later life is also highlighted by cross-national study (e.g., Alvarado et al., 2007). How the accumulated experiences of a lifetime affect health and mortality in old age remains an important question.

Understanding the location and movement of older people, especially as these relate to relevant services and supports for older people, is another key theme in geographical aging and health research (Simpson, Siguaw and Sheng, 2016). Researchers have demonstrated that many young-old move to leisure destinations in early old age, while many older-old return to their original areas, probably to be closer to family and friends as they think about their increasing needs for support (Stoller and Longino, 2001). These patterns are changing in response to a variety of trends, including increasing employment of the elderly, diversifying opportunities for leisure associated with old age, growing constraints on availability of support, and declines in health and health care associated with migration (Hall and Hardill, 2016).

Decisions made by older people and their families about geographic proximity among kin often have implications for the nature and quality of their interactions (Szydlik, 2016). Greater proximity between an older person and his or her family members is typically associated with more provision of care, both in terms of hours and type of support provided, but this is influenced by characteristics such as health status, gender, marital and parental status, and employment of both the older person and potential supporters (Joseph and Hallman, 1996, 1998). Moves motivated by negative life events or circumstances can have depressive effects or be particularly stressful (Bradley and Van Willigen, 2010). Thus, for health geographers, understanding location and movement is central to comprehending the complexities of people's lives, as well as exchanges of care and support and impacts on health.

Increasingly, geographic proximity is mediated by technology (Milligan, Roberts and Mort, 2011). Through remote surveillance and monitoring, developing technologies potentially increase opportunities for people to stay at home as they age, rather than move to residential care (Schillmeier and Domènech, 2010). Such innovations have critical implications for the health and the geographies of older people (Milligan, Roberts and Mort, 2011) and raise many ethical and social issues. Further research is needed to consider how technology might enable older people who are experiencing frailty to retain a greater level of autonomy and independence. Who should have the right to access such private information (e.g., should adult children automatically be able to see their parents' data)? Further, as caring relationships involve complex processes of power and frequently issues of inequality (Bowlby, 2012), how is the cared-for person's ability to make choices and express opinions supported or suppressed in the context of these technologies? Is there potential for greater social and spatial isolation, loneliness, disempowerment or abuse for older people who are monitored remotely rather than in person? Ultimately, might these technologies reconfigure the experience of home and the way care occurs?

Relational approaches to health geographies of aging

Exploring older people's interactions with social and physical environments, and approaching the relationship between people and their environments as reciprocal and indivisible, characterizes a *relational* approach to aging and health geographies (Rowles, 1978; Rubinstein and Parmalee, 1992). This approach often includes the views and voices of older people and emphasizes that every aspect of every older person must always be understood as embedded or situated in socially constructed, dynamic *places* (Cutchin, 2001).

Two important themes in this field of research are sense of place and attachment to place. Geographic researchers exploring these themes have demonstrated that older people with positive connections to place are more likely to feel secure, feel in control, and have a positive sense of self, all of which enhances well-being and the process of aging (Rowles and Ravdal, 2002). Many researchers have illustrated how a sense of attachment to place becomes stronger as people age, with growing investment of meaning in both objects and places over time (Christoforetti, Gennai and Rodeschini, 2011; Wiles et al., 2009). For example, older people are much more likely than younger people to say they like their home or their neighborhood. The reasons this pattern is observed so consistently range from pride and personal investment over time to the immediate environment becoming more important as mobility declines, although recent work suggests the nature and size of older people's social worlds is far more diverse than initially thought (Wiles et al., 2009). Familiarity, emotional connections, networks of friends and family, and a personal sense of contextualized identity or history linked to a place are all likely to contribute to what Rowles refers to as being in place or insideness (Rowles, 1993). There may also be selective recall of positive experiences over a longer period of time (Christoforetti, Gennai and Rodeschini, 2011). Although attachment to place or investment in place is generally understood as having positive effects on health and well-being (Morita et al., 2010), there is evidence that negative associations with place can have a damaging effect on health and well-being (Golant, 2008).

Health geographers have explored positive place effects on health and well-being during aging with reference to the therapeutic landscapes concept – the idea that healing processes can be embedded in places, locales, settings and milieus that develop "an enduring reputation for achieving physical, mental, and spiritual healing" (Gesler, 1993, p. 171). One implication of such research with respect to aging is that particular places (e.g., gardens) may offer comfort and the opportunity for emotional, physical and spiritual renewal among older people (Milligan, Gatrell and Bingley, 2004). Milligan et al. (2004), for example, illustrate how older people may gain a sense of achievement, satisfaction and aesthetic pleasure through gardening activities, thereby experiencing communal gardens as therapeutic and inclusionary spaces that offer social support and access to social networks. This research suggests that a therapeutic landscape may be purposively constructed to positively influence health and well-being among older people and others.

Geographic research in this area has also illustrated the significance of emotions in response to place (Urry, 2007), as well as a tension between emotions and subjective experiences of aging, and the sociocultural expectations of age and emotions (Rose and Lonsdale, 2016). Health geographers investigating the therapeutic qualities of blue spaces, for instance, have demonstrated that many older people engage with blues spaces as part of their daily routines (e.g., interacting with views of the sea on a daily basis) to establish a sense of familiarity and security, as well as an insideness of place (Coleman and Kearns, 2015). In a study of aging-in-place on Waiheke, New Zealand, conducted by Coleman and Kearns (2015), older coastal dwellers reported positive place attachments and restorative benefits, including opportunities for stillness, reflection and respite, in response to the emotional, aesthetic and spiritual qualities elicited by blue spaces. Older people may select particular elements from the landscapes in which they live, such as a view of a calm bay or changing tide, and keep these in the picture in order to cultivate attachment and a restorative experience that helps them cope with ill health or other challenges associated with advanced age (Coleman and Kearns, 2015). These findings show that older people are resourceful, creative, and actively engaged with the settings in which they live. Indeed, researchers and society at large must recognize older people as competent, enterprising, imaginative people who continue to negotiate everyday life and maintain well-being, rather than being merely in decline or increasingly dependent, as negative stereotypes of aging suggest.

Critical approaches to health geographies of aging

Health geographers have argued for much more *critical* understandings of what it means to age in place, both from the perspective of older people, and for urban, regional and global governance (Boyle, Wiles and Kearns, 2015; Cutchin, 2001). A critical approach considers how social forces shape the experience of both place and aging and questions established structures, everyday norms, and taken-for-granted power relationships while employing and expanding social theories to advocate for and work with communities. To date, more critical approaches to aging in place have considered the implications of living in disadvantaged or difficult places (Scharf, Phillipson and Smith, 2005) and current overemphasis on aging in place as a policy mechanism that may prevent people from seeking better alternatives (Golant, 2015).

Research *with* older people suggests that a wide range of factors are important to experiences and perceptions relating to aging in place, such as the quality and adaptability of housing, the availability of social and physical resources in the neighborhoods, and emotional and social and symbolic *perceptions* of these things (Wiles, 2011). Many older people draw on diverse accommodative and adaptive coping mechanisms and repertoires to achieve a sense of residential normalcy in the event that their house or neighborhood becomes uncomfortable (e.g., when stairs become difficult to ascend or as one's ability to engage with the social and physical resources in the neighborhood decreases) (Golant, 2015). Others may try to adapt their housing or move to more accessible or functionally appropriate housing (Peace, Holland and Kellaher, 2011). Studies show that older people's reluctance to move to institutionalized living environments sometimes stems from negative *perceptions* of these environments, though people's actual experiences are frequently more positive and complex (Wiles and Rosenberg, 2003).

A critical geographical perspective on aging and health suggests that we must think more broadly about whether older people have access to appropriate resources for adapting their current living circumstances or relocating. Access to any living environment is influenced by the resources people have available to them. At an individual level, important determining factors include whether older people own or rent their homes and whether they have strong social networks. At the neighborhood scale, important factors are the level of support available (e.g., adequate and good-quality home-based care when needed) and resources such as accessible, adequate and affordable public transport for older people who are no longer driving, or appropriate leisure spaces and opportunities (Pain, Mowl and Talbot, 2000). At even broader scales, the extent to which health services and home-based support are collectively provided (e.g., publicly funded) will shape the degree to which older people must rely on other sources of support, such as family caregiving and the nature of intergenerational exchanges of wealth (Wiles et al., 2012).

More critical geographical approaches to understanding aging and health challenge the stigmatization and homogenization of old age, aging, and aging bodies (Wiles, 2011)in society in general but in particular by research that treats older people as passive and dependent, frail or asexual; that talks *about* older people but not *with* them; or that casts old age as a medical problem to be treated. Critical research recognizes the heterogeneity of older age and considers how aging bodies *relate to* and *shape* a wide variety of spaces and social contexts, from homes and neighborhoods and communities to gardens and sheds, leisure spaces, workspaces, educational spaces, and political spaces. There is growing recognition that older people make significant contributions to their families and the places in which they live (Wiles and Jayasinha, 2013).

Future research

Descriptive tools such as disease atlases, on the rare occasions in which information related to older people is included, show that recognizing the diversity among older people must be a priority (Kerr et al., 2014). This is because there are differences in morbidity and mortality between younger-old and older-old (80 years plus) people or between groups that are diverse on the basis of a range of social differences, including ethnicity, socioeconomic status and gender. Future research that includes multilevel modeling and detailed qualitative research must be prioritized in order to address more complex questions, such as how to understand the relationship between individual and contextual place effects (such as neighborhood socioeconomic status, or rural and urban contexts) on health outcomes. Additionally, continued focus on patterns of inequalities and demand for services and supports during aging is crucial to anticipating the socioeconomic well-being and health of older people and their future needs. Given currently expanding technological developments that mediate everyday environments, this focus could usefully be expanded to include consideration of the ways in which technologies may be reconfiguring socioeconomic and health-care needs, as well as the experience of home and neighborhood. Further, as new opportunities and constraints for old-age employment, support and leisure emerge, and as older people move and adjust to these changes, it is important that research stay abreast of arising impacts on everyday environments, health and health care.

Geographical perspectives on aging and health suggest that we must think more broadly about whether older people have access to appropriate resources for adapting their current living circumstances or relocating. More critical geographical approaches to understanding aging and health are still needed to challenge the stigmatization and homogenization of old age, aging, and aging bodies (Wiles, 2011). This challenge applies both to society in general, and to research that treats older people as passive and dependent, frail or asexual; which talks *about* older people but not *with* them; or, which casts old age as a medical problem to be treated.

Conclusions

Researchers taking a spatial approach to aging and health geography have highlighted demand for a range of services and supports associated with old age, the location and movement of older people and their subsequently anticipated future needs. Geographers taking a relational approach, exploring the relationship between older people and their environments as reciprocal and indivisible, have considered experiences of aging as situated in settings that are socially constructed and dynamic. Critical approaches to aging and health geography have explored the varied meanings of aging from the perspectives of diverse older people, and in the context of urban, regional and global governance, to help us think more broadly about the resources older people may require to maintain daily life. These wide-ranging geographic inquiries have highlighted inequities experienced by older people in their everyday settings, as well as inequities between older people and other groups, and the health aspects of aging in place, including attachment to place, and the provision of health care, caring work and resources.

In all aspects of geographic research on aging and health, there is growing and exciting emphasis on a strengths-based approach to understanding what may help older people achieve and maintain well-being.

Increasing attention is paid to different aspects of aging, including cultural and class differences, and the experiences of those in advanced age as opposed to younger cohorts. A wide range of approaches and methods are deployed, including sophisticated geographic information systems (GIS) and spatial-analytic techniques in a post-positivist theoretical framework and emancipatory methodological frameworks that include participatory and transformative research strategies working *with* older people.

A priority for future research is to achieve an understanding of the multifaceted interactions between individual older people, their everyday settings, and their health and well-being outcomes. In light of rapidly growing proportions and numbers of older people, particularly those in advanced age and from diverse cultural groups, such research must also understand and maximize the many advantages of an aging population while challenging the stigmatization and homogenization of aging.

References

Alvarado, B., Zunzunegui, M., Béland, M., Sicotte, M. and Lourdes, R. (2007). Social and gender inequalities in depressive symptoms among urban older adults of Latin America and the Caribbean. *Journal of Gerontology: Social Sciences*, 62(4), pp. S226–S236.

Bowlby, S. (2012). Recognising the time – space dimensions of care: caringscapes and carescapes. *Environment and Planning A*, 44(9), pp. 2101–2118.

Boyle, A., Wiles, J. L. and Kearns, R. A. (2015). Rethinking aging in place: the "people" and "place" nexus. *Progress in Geography (China)*, 32(12), pp. 1495–1511.

Bradley, D. E. and Van Willigen, M. (2010). Migration and psychological well-being among older adults: a growth curve analysis based on panel data from the health and retirement study, 1996–2006. *Journal of Aging and Health*, 22(7), pp. 882–913.

Christoforetti, A., Gennai, F. and Rodeschini, G. (2011). Home sweet home: the emotional construction of places. *Journal of Aging Studies*, 25(3), pp. 225–232.

Coleman, T. and Kearns, R. (2015). The role of bluespaces in experiencing place, aging and well-being: insights from Waiheke Island, New Zealand. *Health & Place*, 35, pp. 206–2017.

Cutchin, M. (2001). Deweyan integration: moving beyond place attachment in elderly migration theory. *The International Journal of Aging and Human Development*, 52(1), pp. 29–44.

Davies, A. and James, A. (2011). *Geographies of aging: social processes and the spatial unevenness of population aging*. Surrey: Ashgate.

Deeg, D. J. (2015). Gender and physical health in later life. In: *The encyclopedia of adulthood and aging*. Chichester: Wiley-Blackwell, pp. 1–6. http://dx.doi.org/10.1002/9781118521373.wbeaa266

Gesler, W. M. (1993). Therapeutic landscapes: theory and a case study of Epidauros, Greece. *Environment and Planning D: Society and Space*, 11, pp. 171–189.

Gibson, R., Gander, P., Paine, S., Kepa, M., Dyall, L., Moyes, S. and Kerse, N. (2016). Sleep of Maori and non-Maori of advanced age. *The New Zealand Medical Journal*, 129(1436), pp. 52–61.

Golant, S. M. (2008). Commentary: irrational exuberance for the aging in place of vulnerable low-income older homeowners. *Journal of Aging and Social Policy*, 20(4), pp. 379–397.

Golant, S. M. (2015). *Aging in the right place*. Baltimore: Health Professions Press.

Hall, K. and Hardill, I. (2016). Retirement migration, the "other" story: caring for frail elderly British citizens in Spain. *Ageing & Society*, 36(3), pp. 562–585.

Huisman, M., Kunst, A. E. and Mackenbach, J. P. (2003). Socioeconomic inequalities in morbidity among the elderly; a European overview. *Social Science and Medicine*, 57, pp. 861–873.

Isaacson, M., Wahl, H. W., Shoval, N., Oswald, F. and Auslander, G. (2017). The relationship between spatial activity and wellbeing-related data among healthy older adults: an exploratory geographic and psychological analysis. In: T. Samanta, ed., *Cross-cultural and cross-disciplinary perspectives in social gerontology*. Singapore: Springer, pp. 203–219.

Joseph, A. and Hallman, B. C. (1996). Caught in the triangle: the influence of home, work, and elder location on work-family balance. *Canadian Journal on Aging*, 15(3), pp. 392–412.

Joseph, A. and Hallman, B. C. (1998). Over the hill and far away: distance as a barrier to the provision of assistance to elderly relatives. *Social Science & Medicine*, 46(6), pp. 631–639.

Kerr, A., Exeter, D. J., Hanham, G., Grey, C., Zhao, J., Riddell, T., Lee, M., Jackson, R. and Wells, S. (2014). Effect of age, gender, ethnicity, socioeconomic status and region on dispensing of CVD secondary prevention medication in New Zealand: the atlas of health care variation CVD cohort (VIEW-1). *New Zealand Medical Journal*, 127(1400), pp. 39–69.

Lloyd-Sherlock, P. (2000). Population aging in developed and developing regions: implications for health policy. *Social Science & Medicine*, 51(6), pp. 887–895.

Mackay, J. and Mensah, G. (2004). *The atlas of heart disease and stroke*. Geneva: World Health Organization and US Centers for Disease Control and Prevention.

Milligan, C., Gatrell, A. and Bingley, A. (2004). "Cultivating health": therapeutic landscapes and older people in northern England. *Social Science & Medicine*, 58(9), pp. 1781–1793.

Milligan, C., Roberts, C. and Mort, M. (2011). Telecare and older people: who cares where? *Social Science & Medicine*, 72(3), pp. 347–354. http://dx.doi.org/10.1016/j.socscimed.2010.08.014

Morita, A., Takano, T., Nakamura, K., Kizuki, M. and Seino, K. (2010). Contribution of interaction with family, friends and neighbours, and sense of neighbourhood attachment to survival in senior citizens: 5-year follow-up study. *Social Science & Medicine*, 70(4), pp. 543–549.

Northcott, H. (2014). Geographies of ageing: social processes and the spatial unevenness of population aging. *Canadian Studies in Population*, 41(3), pp. 150–151.

Pain, R., Mowl, G. and Talbot, C. (2000). Difference and the negotiation of "old age". *Environment and Planning D: Society and Space*, 18(3), pp. 377–393.

Peace, S. M., Holland, C. and Kellaher, L. (2011). "Option recognition" in later life: variations in aging in place. *Aging & Society*, 31(5), pp. 734–757.

Rose, E. and Lonsdale, S. (2016). Painting place: Re-imagining landscapes for older people's subjective wellbeing. *Health & Place*, 40, pp. 58–65.

Rosenberg, M. W. (2014). Health geography I: social justice, idealist theory, health and health care. *Progress in Human Geography*, 38(3), pp. 466–475.

Rowles, G. D. (1978). *Prisoners of space? Exploring the geographical experiences of older people*. Boulder: Westview Press.

Rowles, G. D. (1993). Evolving images of place in aging and "aging in place". *Generations*, 17(2), pp. 65–70.

Rowles, G. D. and Ravdal, H. (2002). Aging, place, and meaning in the face of changing circumstances. In: R. Weiss and S. Bass, eds., *Challenges of the third age: meaning and purpose in later life*. New York: Oxford University Press, pp. 81–114.

Rubinstein, R. L. and Parmalee, P. A. (1992). Attachment to place and representation of life course by the elderly. In: I. Altman and S. Low, eds., *Human behaviour and environment vol. 12: Place attachment*. New York: Plenum Press, pp. 139–163.

Scharf, T., Phillipson, C. and Smith, A. (2005). Social exclusion of older people in deprived urban communities of England. *European Journal of Aging*, 2(2), pp. 76–87.

Schillmeier, M. and Domènech, M. (2010). *New technologies and emerging spaces of care: cracks in the door?* London: Routledge.

Simpson, P. M., Siguaw, J. A. and Sheng, X. (2016). Tourists' life satisfaction at home and away: a tale of two cities. *Journal of Travel Research*, 55(2), pp. 161–175.

Stoller, E. P. and Longino, C. F. (2001). "Going home" or "leaving home"? The impact of person and place ties on anticipated counterstream migration. *The Gerontologist*, 41(1), pp. 96–102.

Szydlik, M. (2016). *Sharing lives: adult children and parents*. London: Routledge.

Urry, J. (2007). *Mobilities*. Cambridge: Polity Press.

Wiles, J. L. (2011). Reflections on being a recipient of care: vexing the concept of vulnerability. *Social and Cultural Geography*, 12(6), pp. 573–588.

Wiles, J. L., Allen, R. E. S., Palmer, A. J., Hayman, K. J., Keeling, S. and Kerse, N. (2009). Older people and their social spaces: a study of wellbeing and attachment to place in Aotearoa New Zealand. *Social Science & Medicine*, 68(4), pp. 664–671.

Wiles, J. L. and Jayasinha, R. (2013). Care for place: the contributions older people make to their communities. *Journal of Aging Studies*, 27(2), pp. 93–101.

Wiles, J. L., Leibing, A., Guberman, N., Reeve, J. and Allen, R. E. (2012). The meaning of "Ageing in Place" to older people. *Gerontologist*, 52(3), pp. 357–366.

Wiles, J. L. and Rosenberg, M. W. (2003). Paradoxes and contradictions in Canada's home care provision: informal privatisation and private informalisation. *International Journal of Canadian Studies*, 28, pp. 62–89.

28

INCLUDING CHILDREN IN HEALTH GEOGRAPHY

Nicole M. Yantzi

The purpose of this chapter is to show the progress in our understanding of the interrelationships between children's daily spaces and the impacts on their health.[1] This development has moved health-geography researchers from conducting research *about and for children*, in which children are treated as passive research subjects, to research *with children*, in which children are active research participants. Research concerning children's outdoor and neighborhood experiences confirms that a complete understanding of these impacts on children's health must include their perspectives and experiences. The chapter ends by looking at some of the challenging aspects of recognizing children as active research participants and discusses some research areas that would benefit from the inclusion of children's voices and experiences.

The United Nations *Convention on the Rights of the Child* (1989) states that children have the right to express their perspectives concerning their everyday experiences (see Articles 12 and 13). These experiences, whether negative or positive, are contextualized by the policies and organization of three critical spaces that can strengthen or hinder children's health: where they live (homes), where they learn (schools or child-care centers) and where they play (neighborhoods and communities). Although these spaces impact children's health, children can impact and shape their daily spaces; the ability to do so is circumscribed by the rules and policies governing the diversity of spaces in children's lives. This chapter emphasizes the importance of including children and youth as meaningful research participants in order to develop a deep understanding of the interrelationships between their health and their daily environments.

There is a tendency to solely focus on children and youth's chronological age. It is typical when meeting a child to ask his or her age. This assumes that by knowing the child's age it is possible to understand his or her level of maturity, types of experiences and ability to participate. In addition to a child's chronological age, however, it is important to recognize diversity in terms of their development and their social and emotional experiences. *The Convention on the Rights of the Child* applies to any individual under the age of 18 years. This chapter focuses on primary/elementary school-aged children, between the ages of 5 and 12.

The exclusion of children in health geography and health policy

Traditionally children's experiences were not included in geography. It is crucial to realize that children were excluded in both research and public policy. As Davis and Jones (1996, p. 108) state, "[t]he restrictions on children's mobility and access imposed by adult-centred planning, and the assumptions of incompetence and irrationality that justify this approach, become significant health issues, once health is defined in a positive way as a 'resource for everyday life' (WHO, 1984)". In other words, for children to be healthy it is essential that researchers seek children's perspectives and experiences of factors that facilitate or hinder their ability to move around and interact with their daily environments. Having a rich understanding of children's daily environments is essential to advocate for their inclusion in policy.

Traditionally, children were not included as participants in research examining the impacts of spaces and places on children's health. This research, whether qualitative or quantitative, used adult proxies to collect data. In this research, children were not observed or asked – rather, the adults in their lives (e.g., parents/ guardians, teachers, educational assistants) were asked on the children's behalf. A recent systematic review of interventions that have tried to develop healthy urban environments for children shows the diversity of methods used to examine the effectiveness of interventions geared toward road traffic and safety measures, parks and playgrounds, active travel, and multicomponent initiatives (Audrey and Batista-Ferrer, 2015). Many of the studies found in the review feature methods, such as child surveys and participation observation, in which the child is an active research participant. These types of methods are important, as they recognize that children's experiences are often different than those of adults and that they have unique contributions to make in the research process.

The inclusion of children in health geography and health policy

In the 1990s there was a resurgence in geographic research examining children's experiences of the spaces where they live, learn and play (see McKendrick, 2000 for an extended bibliography). The new social studies of children (James and James, 2012) was the catalyst for the paradigm change in how children were included in health-geography research. These changes follow the principles set out by Prout and James (1990), which recognized that children's experiences are diverse, are heterogeneous and intersect with other social positionings such as socioeconomic status, ethnicity, gender and ability. Children are meaningful research participants, who should be asked about their own experiences and observed as they negotiate and interact with their daily environments.

An important distinction in conducting research with children is recognizing children as active agents who can shape and influence their surroundings. According to Freeman (2007): "[A]gents are decision-makers. They are people who can negotiate with others, who are capable of altering relationships or decisions, who can shift social assumptions and constraints. And there is now clear evidence that even the youngest can do this" (p. 8).

In geography, the changing ways of viewing children in research can be seen in the growth of the sub-discipline of the geographies of children and youth. Matthews and Limb (1999) set out seven propositions for their "agenda for the geography of children." The common theme in these propositions is that children's experiences and ways of negotiating their environments are different than those of adults and that "[G]eographers can play a major part in this process of sensitization and [that] there is much work upon which they can reply which highlights that children have different needs, aspirations and behaviour from adults" (Matthews and Limb, 1999, p. 66). The start of the publication of the peer-reviewed journal *Children's Geographies* (in 2003) and Holloway and Valentine's (2000) edited book titled *Children's Geographies* shows this influence. The inaugural editorial for *Children's Geographies* identified health as one of the main areas requiring further exploration (Matthews, 2003).

It is widely accepted that there is a crucial link between children's ability to independently navigate their neighborhoods and their physical, social and psychological health (Carroll et al., 2015). But changes in the built urban environment in many countries, such as Canada (Loebach and Gilliland, 2010) and New Zealand (Carroll et al., 2015); parents' concerns about traffic and strangers (Witten and Carroll, 2016); and children's own fears (Woolley et al., 1999) mean that children's neighborhood activities may be contained and constrained.

Fostering children's health in neighborhoods and communities

Children are spending less time outdoors, and their activities are occurring increasingly in formalized, organized and scheduled settings. In an examination of research published between 1990 and 2011, Gill (2014) found strong scientific evidence of the impact of time spent outdoors and specific health benefits including physical activity, mental and emotional health, healthy eating and motor development. Gill (2014) concluded that the motivation to interact with the environment is for children an intrinsic property of life, but that the quality of the interactions is dependent on richness of the environmental-engagement opportunities.

Hougie's (2010) research reveals the differences in mothers' ideal play spaces and those discussed by their middle-class children living in the United Kingdom. Whereas the mothers emphasized safety, the children felt that "walking in the countryside should be enlivened by going to magical and exciting places, exploring, getting lost and having adventures" (p. 220). For the children, it was the quality of the interaction that they could have with the outdoor environment that was most important. These different spatial perspectives of children and mothers support Rasmussen's (2004) differentiation between places created by adults for children, known as places for children, and places that are engaging, important and meaningful to children. It is only by including children as active research participants that the qualities of healthy children's outdoor environments can be discovered. Children will be much more likely to play and spend time outdoors if they enjoy the space, and they will be more likely to enjoy the space if they are excited about interacting with it.

The use of child-friendly methods such as child-led tours, child travel diaries, children taking pictures and then being asked about the pictures, focus groups and interviews in research have revealed the dynamics and complexities concerning how children and youth feel about and actually experience the places they live in. When asked and observed, children and youth emphasize the importance of spaces that are safe to play in and spend time with friends (Carroll et al., 2015). One of the main challenges in designing a child-friendly city is that planners do not consider the importance of providing children with opportunity structures to encourage independent mobility or include them in the planning process (Björklid and Nordström, 2007). In their work, Carroll et al. (2015) found that children in Auckland were actually playing in streets, driveways, car parks, and common areas in apartment buildings, spaces that are not designed with children's use in mind. This can result in conflict with other users of these spaces and, possibly, safety issues in regard to traffic and accidents (Björklid and Nordström, 2007).

Examining trends, behaviors, experiences and perceptions regarding children's active travel to school and within neighborhoods is a policy priority and active research area. Studies based in Canada, the United States, the United Kingdom and New Zealand all concur that the percentage of children using non-active transportation to school (e.g., cars, buses) is increasing. With this trend, an increasing number of children are losing out on a type of physical activity that they can easily integrate into their daily schedules. Research also shows a difference between how children would like to travel to school and how they actually get to school; generally, children prefer active transportation modes (e.g., walking and biking) (Collins and Kearns, 2001; Osborne, 2005).

Recognizing the challenges of including children in research

While it is important to conduct research in which children are meaningful research participants, there are also some uncomfortable dimensions of children's agency that must be explored (Valentine, 2011). Researchers can tend to neglect the power relations and contexts that influence children's lives, such as parental control in the family context (Holt, 2011). In their homes, schools and neighborhoods, children can experience barriers to their agency including policies and power relationships. Examinations of children's agency must also recognize that demographic and social characteristics, such as gender, ethnicity, ability and family income, can facilitate or hinder children's ability to negotiate their environments (Vanderbeck and Dunkley, 2004). Research on children's experiences and navigations of their city shows the impacts of socioeconomic status, ethnicity and type of neighborhood on children's desires and abilities to move independently around their city (Collins and Kearns, 2001; Witten and Carroll, 2016). Children's experiences of their outdoor environments must be contextualized in terms of geographic location and quality of outdoor environment. For example, while Canadian youth made positive comments concerning the importance of being outside to their health (Woodgate and Skarlato, 2015), interviews with South African youth associated being outside with fear due to crime, violence and pollution (Adams and Savahl, 2015).

At the same time, children may not have equal opportunities to participate in all their daily environments due to context-specific rules and policies (Barker and Weller, 2003; Stephens et al., 2015). There is also a tricky balance between children's lack of physical and social maturity and their ability to exercise agency. As Valentine (2011) states: "[Y]oung children are not recognized as full moral agents because they have an insufficiently developed sense of consequences and the independent lives of others" (p. 350). This statement may lead to the assumption that it is only adults that can truly exercise agency; however, as the author goes on to explain, both adults and children may not make decisions in their best interests.

Including highly vulnerable children in health geography

This section identifies research gaps and discusses two key opportunities for health geographers to examine the impacts of specific contexts and environmental change on children's health. The first gap is that most research focuses on understanding the links between children's health and their negotiations of their urban neighborhoods. In many instances, the rural context may pose even more challenges, due to fewer resources and potentially greater health risks.

The United Children's Emergency Fund's (UNICEF United Nations Children's Fund, 2014) report titled *The Challenges of Climate Change: Children on the Front Line* emphasizes that climate change impacts children not only in terms of mortality and morbidity but also in terms of their physical, emotional and social health. The challenge of climate change is huge; it requires an urgent response from all generations. As the effects of climate change become more visible and extreme, they are likely to affect adversely the lives of children and adolescents all over the world. For example, families that lose their livelihoods to drought will be less able to afford the costs of schooling or health care. Over 99% of deaths already attributable to climate-related changes occur in developing countries (UNICEF United Nations Children's Fund, 2014). Climate change may contribute to the spread of diseases, especially those that threaten children more than adults, such as malaria and diarrhea. There is both a cognitive and an emotional aspect for children in terms of learning about and experiencing climate change. Children can experience a range of emotions, such as worry, despair, anger, guilt and helplessness. In terms of physical health, children undergo rapid developmental and growth changes both before and after they are born. Compared to adults, for every pound of their body weight, children drink more water, eat more food and breathe more air (Etzel and Landrigan, 2014). There are also important differences regarding the ability of children's and adults' bodies to deal with environmental toxins, pollution and temperature change; in addition, children explore their environment in different ways than adults do. Generally, children are more vulnerable to environmental change, but children from low- and middle-income countries are particularly vulnerable, as they "have fewer assets to draw on in every sense of the word, and are more likely to be adversely affected by the various challenges imposed by climate change" (Bartlett, 2008, p. 502).

According to Gibbons (2014), the effects of climate change disproportionately impact children and, yet, children's own concerns are not often heard in the debates or accounted for in policy surrounding the phenomenon. Gibbons demonstrates the effectiveness of using the United Nations Convention on the Rights of the Child (1989) and specifically the principles of "the right to survival and development, the right of participation and to be heard . . . and the right to equality and non-discrimination; and the overarching principle of the best interests of the child" (2014, pp. 20–21) to help ensure climate justice for children.

An important contribution that health geographers can make is examining the health impacts on children due to how climate change disrupts their routines and everyday spaces. Tanner (2010) reveals that children can articulate how environmental hazards impact them, their families and their communities. Children in El Salvador and the Philippines discussed relationships between climate-related risks and other social and economic risks, such as unemployment and violence (Tanner, 2010). Inuit youth from Nunatsiavut, in Labrador, have expressed concern about how the changing sea-ice conditions will impact their abilities to be on the land and connect with their culture and elders (MacDonald et al., 2015). It is also important to realize that children and youth can be assets in terms of the adaptation processes required with climate change (Bartlett, 2008). There is both moral and political justification for collecting the stories and experiences of children and youth who are coping with climate change (Campbell, Skovdal and Campbell, 2013).

Note

1 In this chapter, you will find reference to works both by geographers and non-geographers. It has long been recognized that exploration of children's environments is not the sole domain of geographers (McKendrick, 2000).

References

Adams, S. and Savahl, S. (2015). Children's perceptions of the natural environment: A South African perspective. *Children's Geographies*, 13(2), pp. 196–211.

Audrey, S. and Batista-Ferrer, H. (2015). Healthy urban environments for children and young people: a systematic review of intervention studies. *Health & Place,* 36, pp. 97–117.

Barker, J. and Weller, S. (2003). Never work with children? The geography of methodological issues in research with children. *Qualitative Research*, 3(2), pp. 207–227.

Bartlett, S. (2008). *Climate change and urban children: impacts and implications for adaptation in low- and middle-income countries.* London: International Institute for Environment and Development.

Björklid, P. and Nordström, M. (2007). Environmental child-friendliness: collaboration and future research. *Children, Youth and Environments*, 17(4), pp. 388–401.

Campbell, E., Skovdal, M. and Campbell, C. (2013). Ethiopian students' relationship with their environment: implications for environmental and climate adaptation programmes. *Children's Geographies*, 11(4), pp. 436–460.

Carroll, P., Witten, K., Kearns, R. and Donovan, P. (2015). Kids in city: children's use and experiences of urban neighbourhoods in Auckland, New Zealand. *Journal of Urban Design*, 20(4), pp. 417–436.

Collins, D. and Kearns, R. (2001). The safe journeys of an enterprising school: negotiating landscapes of opportunity and risk. *Health & Place*, 7, pp. 293–306.

Davis, A. and Jones, L. L. (1996). Children in the urban environment: an issue for the new public health agenda. *Health & Place*, 2(2), pp. 107–113.

Etzel, R. A. and Landrigan, P. J. (2014). Children's exquisite vulnerability to environmental exposures. In: P. J. Landrigan and R. A. Etzel, eds., *Textbook of children's environmental health*. New York: Oxford University Press, pp. 18–27.

Freeman, M. (2007). Why it remains important to take children's rights seriously. *International Journal of Children's Rights*, 15(1), pp. 5–24.

Gibbons, E. D. (2014). Climate change, children's rights, and the pursuit of intergenerational climate justice. *Health and Human Rights*, 16(1), pp. 19–31.

Gill, T. (2014). The benefits of children's engagement with nature: a systematic literature review. *Children, Youth and Environments*, 24(2), pp. 10–34.

Holloway, S. L. and Valentine, G. (2000). *Children's geographies: playing, living, learning*. New York: Routledge.

Holt, L. (2011). Introduction: geographies of children, youth and families: Disentangling the socio-spatial contexts of young people across the globalizing world. In: L. Holt, ed., *Geographies of children, youth and families: an international perspective*. New York: Routledge, pp. 1–8.

Hougie, D. P. (2010). Research note: the not-so-great outdoors? The findings of a UK study of children's perspectives on outdoor recreation. *Children, Youth and Environments*, 20(2), pp. 219–222.

James, A. and James, A. (2012). *New concepts in childhood studies*. 2nd ed. Thousand Oaks, CA: Sage.

Loebach, J. and Gilliland, J. (2010). Child-led tours to uncover children's perceptions and use of neighborhood environments. *Children, Youth and Environments*, 20(1), pp. 52–90.

MacDonald, J. P., Willox, A. C., Ford, J. D., Shiwak, I., Wood, M., IMHACC Team, and the Rigolet Inuit Community Government. (2015). Protective factors for mental health and well-being in a changing climate: perspectives from inuit youth in Nunatsiavut, Labrador. *Social Science & Medicine*, 141, pp. 133–141.

Matthews, H. (2003). Coming of age for children's geographies. *Children's Geographies*, 1(1), pp. 3–5.

Matthews, H. and Limb, M. (1999). Defining an agenda for the geography of children: review and prospect. *Progress in Human Geography*, 23(1), pp. 61–90.

McKendrick, J. H. (2000). The geography of children: an annotated bibliography. *Childhood*, 7(3), pp. 359–387.

Osborne, P. (2005). Safe routes for children: what they want and what works. *Children, Youth and Environments*, 15(1), pp. 234–239.

Prout, A. and James, A. (1990). A new paradigm for the sociology of childhood? Provenance, promise and problems. In: A. James and A. Prout, eds., *Constructing and reconstructing childhood: contemporary issues in the sociological study of childhood*. London: Routledge-Falmer, pp. 6–23.

Rasmussen, K. (2004). Places for children – children's places. *Childhood*, 11(2), pp. 155–173.

Stephens, L., Scott, H., Aslam, H., Yantzi, N., Young, N. L., Ruddick, S. and McKeever, P. (2015). The accessibility of elementary schools in Ontario, Canada: not making the grade. *Children, Youth and Environments*, 25(2), pp. 153–175.

Tanner, T. (2010). Shifting the narrative: child led responses to climate change and disasters in El Salvador and the Philippines. *Children & Society*, 24(4), pp. 339–351.

UNICEF United Nations Children's Fund. (2014). *The challenges of climate change: Children on the front line*. Working Paper: eSocialSciences.

Valentine, K. (2011). Accounting for agency. *Children & Society*, 25(5), pp. 347–358.

Vanderbeck, R. M. and Dunkley, C. M. (2004). Introduction: geographies of exclusion, inclusion and belonging in young lives. *Children's Geographies*, 2(2), pp. 177–183.

Witten, K. and Carroll, P. (2016). Children's neighborhoods: places of play or spaces of fear? In: K. Nairn, P. Kraftl and T. Skelton, eds., *Space, place, and environment volume 3 geographies of children and young people*. Singapore: Springer.

Woodgate, R. L. and Skarlato, O. (2015). "It is about being outside": Canadian youth's perspectives of good health and the environment. *Health & Place*, 31, pp. 100–110.

Woolley, H., Dunn, J., Spencer, C., Short, T. and Rowley, G. (1999). Children describe their experiences of the city centre: a qualitative study of the fears and concerns which may limit their full participation. *Landscape Research*, 24(3), pp. 287–301.

World Health Organization. (1984). *Health Promotion: A discussion document on the concept and principles*. Copenhagen: Working Group on Concepts and Principles of Health Promotion. Regional Office for Europe.

29

IMMIGRANT HEALTH

Insights and implications

K. Bruce Newbold

With more than 244 million international migrants in 2015 alone, the number of international migrants continues to grow on a year-by-year basis, reflecting uneven economic opportunities, persecution, and conflict (United Nations, 2016). Fundamentally, international migration includes both voluntary and involuntary (forced) movement, with the reason for the former typically including family reunification as well as economic motivations – including a combination of push factors in the origin country, such as poor employment prospects, large populations and low wages, and pull factors in the receiving country including higher wages and employment prospects. Involuntary or forced migration includes refugees, or individuals who are forced from their home country in order to escape war, persecution or violence. In fact, immigrant and refugee movements have become one of the most pressing social, economic and humanitarian issues faced by countries around the globe. In 2015 alone, more than a million refugees made their way out of Africa and the Middle East into Europe, although this represented just a small part of the estimated 23.3 million refugees globally in the same year (United Nations, 2016).

Globally, the major sending regions include Asia, North Africa and Latin America, while both the developing world and developed world are important destinations. Not surprisingly, international migration (i.e., movement crossing national borders) represents an important research topic that includes disciplines such as demography, economics, sociology and geography, given the large numbers of individuals that cross international borders, with resulting political, economic and demographic implications. Despite the continued interest in the economic implications of immigration, along with the economic integration of new arrivals in their reception country, research has increasingly focused on the health of the immigrant population, particularly in the period immediately after arrival. Known as the *healthy immigrant effect*, the literature suggests that immigrants – at least in the developed world – are typically healthier and are less likely to report chronic conditions or impairments than the native-born of the receiving society (DeMaio and Kemp, 2010). Although most developed countries actively screen immigrants for health issues, such measures do not guarantee that the health of arrivals will be maintained. Indeed, the health advantage demonstrated by immigrants at the time of arrival has been observed to decline within a comparatively short period of time – approximately 5–10 years – after arrival, with health status tending to converge or even fall below the levels of the receiving population. For immigrants in less-developed countries, health screening is typically not used, and access to health-care services may be more precarious or non-existent, particularly for refugees, who are among the most vulnerable immigrants.

Health geographers have engaged in research associated with immigrant health and the healthy immigrant effect through various themes common to health geography, including health-care access and delivery, social models of health and health care, and environment and health. Although often drawing insights from related fields, health geographers have emphasized the role of place and the individual or communal experiences when looking at immigrant and refugee health. Place effects could be reflected in the origin, the journey (particularly for refugees who transit through refugee camps or other countries before their final destination), and the actual destination, as well as everyday places and experiences, such as the local environment, access to employment resources, social and cultural roles, and social networks.

This chapter explores issues associated with the health of immigrants and refugees, including a discussion of the healthy immigrant effect. In addition, the health of the refugee population and other vulnerable immigrant subgroups will be considered, along with consideration of some of the hypothesized reasons why the healthy immigrant effect is observed, including connections to successful acculturation. The chapter concludes by providing insights into the evolving research agenda and key questions that health geographers should consider to advance the research agenda.

The healthy immigrant effect

Despite the abundance of literature that supports the healthy immigrant effect, questions whether changes in health status are real or perceived remain (Barcellos, Goldman and Smith, 2012). In part, these questions arise due to many studies' reliance on self-reported health status. It may be that individuals are re-evaluating their health downward relative to peers within the receiving country as compared to their origin and/or as the reality of immigrant life in the receiving country is revealed. In addition, the healthy immigrant effect may reflect diseases that are not diagnosed in the origin country but discovered after immigration. But such discussions also reflect the emphasis on economic or family class immigrants, as opposed to forced migration, with refugees much more likely to have poorer health because of the nature of their moves and their journey experience of places inclusive of refugee camps and limited access to health care (Newbold and McKeary, 2018), placing them in precarious positions.

Despite these lingering questions over the authenticity of the healthy immigrant effect, it has been observed for a variety of health outcomes beyond self-assessed health (Vang et al., 2015), including disability (Newbold and Simone, 2015), chronic conditions (Newbold, 2006), mental health (Smith et al., 2007; Stafford, Newbold and Ross, 2011), birth outcomes (Farré, 2016), immigrant status (Newbold, 2009) and life expectancy and mortality (Ng, 2011; Trovato, 2003). Additionally, the healthy immigrant effect has been observed in a variety of countries, including Canada, the United States (Jasso et al., 2003), Australia and Europe (Fennelly, 2007; Gotsens et al., 2015; Kennedy et al., 2015; McDonald and Kennedy, 2004).

Although the healthy immigrant effect is widely observed across a number of countries and for different health outcomes, declines in health post-arrival are far from universal and consistent in terms of timing, pace, or degree. For instance, race, origin and/or ethnicity impact health status, with immigrants from non-European origins more likely to experience declining health as compared to those from European origins after controlling for other effects (Veenstra, 2009). Beyond race and ethnicity, Vang et al. (2015) noted significant variations in the healthy immigrant effect across the life course, with the phenomenon strongest during adolescence and adulthood (Kwak, 2016; Vang et al., 2015), while Ng (2011) noted considerable variation (at least for mortality) with respect to geographic scale, gender, duration of residence and specific immigrant origins.

Immigrant status, notably the distinction between economic immigrants and refugees, also impacts changes to health status. Despite evidence that there is also a healthy immigrant effect among refugees (Newbold, 2009), refugees often arrive in the destination country with poorer health than their immigrant counterparts and experience some of the largest declines in health following arrival (relative to immigrants who enter under business or family reunification categories). Such nuances in health status and change among refugees is likely an outcome of their refugee experiences, their greater vulnerability and comparative precarity in society, and the recognized greater physical and mental health needs within this group (McKeary and Newbold, 2010, 2017; Proctor, 2005). Transiting through or residing in refugee camps, individuals may experience violence, lack of access to basic medical care, poor nutrition (Harrison et al., 1999), crowded conditions and food (in)security, resulting in detrimental impacts on health (Koehn, 2005; Newbold and McKeary, 2018; Proctor, 2005). Numerous studies have also noted mental-health issues among refugees as a result of camp conditions, bereavement, separation of family and friends, loneliness, and lack of acceptance in the country where they settle (Whittaker et al., 2005). In particular, post-traumatic stress disorder (PTSD) has been highlighted as a significant concern for refugees, given potential experiences with violence either before being forced to flee or during transit (Beiser, 2009). Post-migration, stressors include the challenge of adapting to a foreign environment and language, economic hardship given limited employment opportunities, discrimination and an altered or absent family and social network (Maximova and Krahn, 2010).

Understanding immigrant health

By definition, immigrants move from one set of health risks, behaviors and constraints to an environment that potentially includes a very different mix, with possible adverse impacts upon health. But why does their health decline? A number of hypotheses have been put forward to explain the healthy immigrant effect. In part, better health at the time of arrival reflects selectivity effects. Immigration is typically self-selective and undertaken by individuals who are young and healthy. Further, countries often impose their own admission requirements that include health screening meant to restrict entry of individuals with health problems that might pose an economic burden to the country.

Some of the earliest work by health geographers interested in immigrant health looked at the health of newcomers as framed by the determinants of health (DoH) framework (Dunn and Dyck, 2000), recognizing that health status reflects social and economic processes as well as access and use of health-care resources. That is, the health of newcomers will be influenced by a range of factors beyond health care per se, with health influenced by lifestyle options, nutrition, housing, work, education and income, as well as social identity, social status, regulatory environments and control over life circumstances. Diminished social networks, poor working conditions, and language barriers may, for example, contribute to declines in health (i.e., Elliott and Gillie, 1998). Consequently, acculturation and integration explanations also figure prominently as explanations for the healthy immigrant effect, with individuals who are socially or economically excluded experiencing poorer health outcomes. Social exclusion has been shown to increase inequalities in health, especially between immigrants and/or ethnic minorities and the general population (Grey, 2003) and has been linked to lower access to appropriate care among ethnic minorities and mistreatment by health professionals (Mclean, Campbell and Cornish, 2003). It is well known that many immigrants experience de-skilling in the workplace and/or are employed in low-skilled jobs. Work by Subedi and Rosenberg (2016), for example, suggests that economic integration is critical for positive health outcomes, with workers engaged in lower-skilled jobs experiencing lower self-esteem, job dissatisfaction and greater work-related stress, resulting in poorer physical and mental health outcomes.

Some health geographers have made significant contributions to the literature and the understanding of changing health status after arrival, while others have contributed to the broader immigrant literature, including economic and social integration that often has implications for the health of new arrivals. Although self-selection and screening ensure good health status among immigrants at the time of arrival, they do not explain the observed rapid declines in health after arrival, with multiple explanations suggested within the literature. Researchers have, for example, hypothesized that changing lifestyles, and more specifically the adoption of riskier health practices among recent immigrants, contribute to declining health. In part, the health advantage experienced by immigrants when they first arrive may be due to cultural buffering effects, whereby immigrants are less likely to participate in activities, such as smoking and drinking, that are associated with poor health outcomes (Hochhausen, Perry and Le, 2010). With duration of residence in the receiving country, alcohol consumption and increased consumption of processed or fast foods have been observed, with both linked to negative health outcomes (Sanou et al., 2014; Subedi and Rosenberg, 2014). Although these studies suggest that there is indeed an uptake of riskier lifestyles resulting in negative health outcomes, Vang et al. (2015) conclude that the overall evidence is limited.

Perhaps the greatest research attention, including in work by health geographers, has focused on access and barriers to health care, which implicitly assumes that faults (barriers) within the health-care system prevent immigrants from accessing care. Work by Wang, for example, has highlighted spatial accessibility to health-care services (i.e., Wang, 2011; Wang and Roisman, 2011). Despite an observed increased need for care (as measured by declining health status), the use of health-care facilities by immigrants does not necessarily increase (Kobayashi and Prus, 2012; Newbold, 2009), potentially reflecting unmet health-care needs and barriers to care (McKeary and Newbold, 2010; Pottie et al., 2008). After arrival, the need to navigate the health-care system highlights the barriers to health, including language, discrimination, cost, insurance, transportation, cultural roles and knowledge. All of these barriers may constrain access and use of health care, resulting in further challenges to health and increased health needs (Edge and Newbold, 2013; Kalich, Heinemann and Ghahari, 2016).

Lack of awareness of health-care opportunities may also play a role in changing health. New arrivals may be unaware of health opportunities. Research has, for example, identified that immigrants are less aware (and consequently less likely to use) preventative health services such as cancer screening (Woltman and Newbold, 2007), which may not be considered essential by the individual. A lack of insurance, distrust or unease of the health system, or a health system that does not provide culturally sensitive and appropriate care may create additional barriers (Woltman and Newbold, 2007), with the literature suggesting that immigrants face difficulties in both finding a physician and maintaining regular contact (Asanin and Wilson, 2008).

Beyond research focused on access to health and health care per se, health geographers have explored a variety of themes related to immigrant health and the broader DoH framework. Williams et al. (2015), for example, considered perceptions of quality of life among immigrant newcomers, finding that immigrants had lower perceptions of quality of life than their Canadian-born counterparts, with potential implications for physical and mental health. Asanin and Wilson's (2008) work on the relationship between neighborhoods and health has resulted in collaborations with local community-health groups and provided a more nuanced picture of what immigrants feel are important health issues, and transnational health-care seeking and behavior are addressed by authors including Wang and Kwak (2015).

Conclusions: where can health geographers add to this research?

Over the past decade, research has increasingly focused on the health of immigrant arrivals and has confirmed the healthy immigrant effect. Although the healthy immigrant effect is not a universal phenomenon, it is commonly observed across a range of health measures (e.g., self-reported health, mental health, and physical health outcomes), immigrant status (immigrant vs. refugee), and other socioeconomic and demographic variables and in various countries within the developed world.

Despite the seeming ubiquity of the healthy immigrant effect, its confirmation comes mainly from developed regions, with the research still needing to broadly consider its role and impact in the developing world. Moreover, concerns with refugee health have become more pressing since the global refugee crisis of 2015. By the end of 2015, the United Nations High Commission on Refugees (UNHCR) estimated that some 59.5 million people had been displaced due to persecution, conflict, violence or human-rights violations. Given their already vulnerable and precarious health, the implications of the healthy immigrant effect among refugees is even greater, as it might be expected that their health will continue to be challenged because of their status and relative vulnerability even after settlement in a new country (Newbold, 2009). For those individuals and families that are internally displaced or are forced to reside in refugee camps, such as the Syrian refugee camps in Jordan and Lebanon, concerns with their long-term physical and mental health are even greater and require attention by researchers. Similarly, questions and attention on receiving countries such as Turkey, Greece and Italy is no less important, with implications for resource availability and delivery, public health and the short- and long-term health of refugees settled in these locations, particularly when their settlement lacks broad public support.

Consequently, there are multiple research avenues for health geographers to pursue, embracing both the more traditional aspects of health geography (i.e., patterns, causes and spread of disease, the planning and provision of health services) and more recent directions, such as the role and meaning of place and the experience(s) associated with health. First, given the ubiquity of the healthy immigrant effect, it is important for research agendas to better answer *why* the healthy immigrant effect occurs. As discussed earlier in this chapter, multiple hypotheses have been forwarded to explain the healthy immigrant effect. Although research has started to explore these more fully, the explanations are undoubtedly complex and overlapping. Importantly, understanding the changing health status and needs of new arrivals, along with how the decline in health can be minimized and the role of place, are important topics for research, particularly when better health status will enable acculturation, and ultimately economic success, within the receiving country. Second, if we are concerned with the impact of barriers to health, research must consider how such barriers can be overcome. That is, what programs or policies can minimize the healthy immigrant effect and ensure that the health advantage of recent arrivals is retained? This agenda would include moving beyond the provision of health and related services and include the role of family, friends and social networks in supporting health. Third, and following Vang et al. (2015), researchers may wish to focus on life-course stages and health outcomes where immigrants are most vulnerable.

If the previous set of research directions captures a more traditional health-geography focus, there is also ample room for health geographers to contribute to the literature with research opportunities that emphasize the role of place and lived experience among immigrants. Research agendas could, for example, pursue questions related to the nuances around the health of refugees, such as the implications associated with the reason for movement (i.e., were they pushed due to war and violence or political reasons and implications with PTSD), and the implications associated with the route, transit time to a host country and settlement in a destination country. Given the role of place, there are further opportunities to understand how the environment influences health and how this differs by context, place, and the meaning of place. Importantly, many of the studies to date have focused on immigrants and refugees in the developed world but have bypassed those in other locations, particularly refugees who either are stuck in camps or have failed to thrive in destination countries. Work in such locations will offer tremendous challenges, but also rewards as new insights are gained. Finally, reflecting a recent call made by Mark Rosenberg (2016), health geographers may (and should) bring different theoretical insights to bear on the research questions associated with international movement and the health of immigrants and refugees. Researchers are able to draw on a range of theories, including (but certainly not limited to) feminist, social and Marxist theories, or new frameworks, such as intersectionality, to frame their work.

Pursuing these research questions could include the use of representative datasets such as the census or other data files that identify immigrants and their health status. Preferably, longitudinal files that track immigrants over a period of time, and hence changes to their health, should be focused on. Although much of the work associated with immigrant health has been based on large, representative data files, qualitative data and research techniques also offer promise, particularly in untangling the explanations for changes in health status by focusing on the lived experiences and outcomes of new arrivals. Finally, there are many opportunities for health geographers to engage with scholars from other disciplines, including economics, sociology and anthropology, and draw upon different frameworks and perspectives to advance the immigrant health research agenda.

References

Asanin, J. and Wilson, K. (2008). I spent nine years looking for a doctor: exploring access to health care among immigrant in Mississauga, Ontario, Canada. *Social Science & Medicine*, 66(6), pp. 1271–1283.

Barcellos, S. H., Goldman, D. P. and Smith, J. P. (2012). Undiagnosed disease, especially diabetes, casts doubt on some of reported health "advantage" of recent Mexican immigrants. *Health Affairs*, 31(12), pp. 2727–2737.

Beiser, M. (2009). Resettling refugees and safeguarding their mental health: lessons learned from the Canadian Refugee Resettlement Project. *Transcultural Psychiatry*, 46(4), pp. 539–583.

De Maio, F. G. and Kemp, E. (2010). The deterioration of health status among immigrants to Canada. *Global Public Health*, 5(5), pp. 462–478.

Dunn, J. and Dyck, I. (2000). Social determinants of health in Canada's immigrant population: results from the National Population Health Survey. *Social Science & Medicine*, 51, pp. 1573–1593.

Edge, S. and Newbold, K. B. (2013). Discrimination and the health of immigrants and refugees: exploring Canada's evidence base and directions for future research in newcomer receiving countries. *Journal of Immigrant and Minority Health*, 15, pp. 141–148.

Elliott, S. J. and Gillie, J. (1998). Moving experiences: a qualitative analysis of health and migration. *Health & Place*, 4(4), pp. 327–339.

Farré, L. (2016). New evidence on the healthy immigrant effect. *Journal of Population Economics*, 29(2), pp. 365–394.

Fennelly, K. (2007). The "healthy migrant" effect. *Minnesota Medical Magazine*, 90(3), pp. 51–53.

Gotsens, M., Malmusi, D., Villarroel, N., Vives-Cases, C., Garcia-Subirats, I., Hernando, C. and Borrell, C. (2015). Health inequality between immigrants and natives in Spain: the loss of the healthy immigrant effect in times of economic crisis. *European Journal of Public Health*, 1–7.

Grey, B. (2003). Social exclusion, poverty, health and social care in Tower Hamlets: the perspectives of families on the impact of the family support service. *British Journal of Social Work*, 33(3), pp. 361–380.

Harrison, A., Calder, L., Karalus, N., Martin, P., Kennedy, M., and Wong, C. (1999). Tuberculosis in immigrants and visitors. *New Zealand Medical Journal*, 112, pp. 363–365.

Hochhausen, L., Perry, D. F. and Le, H. N. (2010). Neighborhood context and acculturation among Central American immigrants. *Journal of Immigrant and Minority Health*, 12(5), pp. 806–809.

Jasso, G., Massey, D., Rosenzweig, M. R. and Smith, J. (2003). *Immigrant health: Selectivity and acculturation.* Santa Monica, CA: RAND.

Kalich, A., Heinemann, L. and Ghahari, S. (2016). A scoping review of immigrant experience of health care access barriers in Canada. *Journal of Immigrant and Minority Health*, 18, pp. 697–709.

Kennedy, S., Kidd, M. P., McDonald, J. T. and Biddle, N. (2015). The healthy immigrant effect: patterns and evidence from four countries. *International Migration and Integration*, 16, pp. 317–332.

Kobayashi, K. M. and Prus, S. G. (2012). Examining the gender, ethnicity, and age dimensions of the healthy immigrant effect: factors in the development of equitable health. *International Journal of Equity Health*, 11(8), pp. 1–6.

Koehn, P. H. (2005). Medical encounters in Finnish reception centres: asylum-seeker and clinician perspectives. *Journal of Refugee Studies*, 18(1), pp. 47–75.

Kwak, K. (2016). An evaluation of the healthy immigrant effect with adolescents in Canada: examinations of gender and length of residence. *Social Science & Medicine*, 157, pp. 87–95.

Maximova, K. and Krahn, H. (2010). Health status of refugees settled in Alberta: changes since arrival. *Canadian Journal of Public Health*, 101(4), pp. 322–326.

McDonald, J. T. and Kennedy, S. (2004). Insights into the "healthy immigrant effect": health status and health service use of immigrants to Canada. *Social Science & Medicine*, 59(8), pp. 1613–1627.

McKeary, M. and Newbold, K. B. (2010). Barriers to care: the challenges for Canadian refugees and their health care providers. *Journal of Refugee Studies*, 23(4), pp. 523–545.

Mclean, C., Campbell, C. and Cornish, F. (2003). African-Caribbean interactions with mental health services in the UK: experiences and expectations of exclusion as (re)productive of health inequalities. *Social Science & Medicine*, 56(3), pp. 657–669.

Newbold, K. B. (2006). Chronic conditions and the healthy immigrant effect: evidence from Canadian immigrants. *Journal of Ethnic and Migration Studies*, 32(5), pp. 765–784.

Newbold, K. B. (2009). The short-term health of Canada's new immigrant arrivals: evidence from LSIC. *Ethnicity and Health*, 14(3), pp. 1–22.

Newbold, K. B. and McKeary, M. (2018). Journey to health: (Re) contextualizing the health of Canada's refugee population. *Journal of Refugee Studies*, fey009, https://doi.org/10.1093/jrs/fey009.

Newbold, K. B. and Simone, D. (2015). Comparing disability amongst immigrants and native-born in Canada. *Social Science & Medicine*, 145, pp. 53–62.

Ng, E. (2011). *Insights into the healthy immigrant effect: Mortality by period of immigration and birthplace.* Canada: Health Research Working Paper Series. Statistics Canada, Catalogue 82–622-X No. 008.

Pottie, K., Ng, E., Spitzer, D., Mohammed, A. and Glazier, R. (2008). Language proficiency, gender and self-reported health: an analysis of the first two waves of the longitudinal survey of immigrants to Canada. *Canadian Journal of Public Health*, 99(6), pp. 505–510.

Proctor, N. G. (2005). "They first killed his heart (then) he took his own life." Part 1: a review of the context and literature on mental health issues for refugees and asylum seekers. *International Journal of Nursing Practice*, 11, pp. 286–291.

Rosenberg, M. (2016). Health geography III: old ideas, new ideas or new determinisms? *Progress in Human Geography*, 41(6), pp. 832–842. DOI: 10.1177/0309132516670054

Sanou, D., O'Reilly, E., Ngnie-Teta, I., Batal, M., Mondain, N., Andrew, C., Newbold, K. B. and Bourgeault, I. L. (2014). Acculturation and nutritional health of immigrants in Canada: a scoping review. *Journal of Immigrant and Minority Health*, 16(1), pp. 24–34.

Smith, K. L. W., Matheson, F. I., Moineddin, R. and Glazier, R. H. (2007). Gender, income and immigration differences in depression in Canadian urban centres. *Canadian Journal of Public Health*, 98(2), pp. 149–153.

Stafford, M., Newbold, K. B. and Ross, N. (2011). Psychological distress among immigrants and visible minorities in Canada. *International Journal of Social Psychiatry*, 57(4), pp. 428–441.

Subedi, R. P. and Rosenberg, M. W. (2014). Determinants of the variations in self-reported health status among recent and more established immigrants in Canada. *Social Science & Medicine*, 115, pp. 103–110.

Subedi, R. P. and Rosenberg, M. W. (2016). High-skilled immigrants – low-skilled jobs: challenging everyday health. *The Canadian Geographer*, 60(1), pp. 56–68.

Trovato, F. (2003). *Migration and survival: the mortality experience of immigrants in Canada*. Edmonton, AB: Prairie Centre for Research on Immigration and Integration (PCRII).

United Nations, Department of Economic and Social Affairs, Population Division. (2016). *International migration report 2015: Highlights* (ST/ESA/SER.A/375). New York: United Nations.

Vang, Z., Sigouin, J., Flenon, A. and Gagnon, A. (2015). The healthy immigrant effect in Canada: a systematic review. *Population Change and Lifecourse Strategic Knowledge Cluster Discussion Paper Series/Un Réseau stratégique de connaissances Changements de population et parcours de vie Document de travail*, 3(1), Article 4.

Veenstra, G. (2009). Racialized identity and health in Canada: results from a nationally representative survey. *Social Science & Medicine*, 69(4), pp. 538–542.

Wang, L. (2011). Analyzing spatial accessibility to health care: a case study of access by different immigrant groups to primary care physicians in Toronto. *Annals of GIS*, 17(4), pp. 237–251.

Wang, L. and Kwak, M. J. (2015). Immigration, barriers to healthcare and transnational ties: a case study of South Korean immigrants in Toronto, Canada. *Social Science & Medicine*, 133, pp. 340–348.

Wang, L. and Roisman, D. (2011). Modeling spatial accessibility of immigrants to culturally diverse family physicians. *The Professional Geographer*, 63(1), pp. 73–91.

Whittaker, S., Hardy, G., Lewis, K. and Buchan, L. (2005). An exploration of psychological well-being with young Somali refugee and asylum-seeker women. *Clinical Child Psychology and Psychiatry*, 10(2), pp. 177–196.

Williams, A., Kitchen, P., Randall, J., Muhajarine, N., Newbold, K. B., Gallina, M. and Wilson, K. (2015). Immigrant's perceptions of quality of life in three second- or third-tier Canadian cities. *The Canadian Geographer*, 59(4), pp. 489–503.

Woltman, K. and Newbold, K. B. (2007). Immigrant women and cervical cancer screening uptake: a multilevel analysis. *Canadian Journal of Public Health*, 98(6), 470–475.

30

ESTABLISHING GEOGRAPHIES OF LGBTQ HEALTH

Nathaniel M. Lewis

Geographies of health have shifted from measuring disease and illness to engaging actively with the social subjectivities that shape health. Sexual and gender identities, including those of lesbian, gay, bisexual, transgender and other queer (LGBTQ) individuals, have gained attention as key axes of social difference contributing to uneven health outcomes and access to care. Yet LGBTQ populations are still studied less frequently in health geography compared to, for example, ethnic minority and immigrant populations.

This chapter first follows the trajectory of geographic research on LGBTQ health from the HIV/AIDS epidemic to broader considerations of health experiences among diverse LGBTQ populations. Next, it considers emerging research on public health as biopolitical regulation, access to care among LGBTQ populations, and sexuality as an intersectional determinant of health. While research in these areas has made a considerable contribution to health geography, there remains ample space in the discipline for quantitative, deductive and ecological research on LGBTQ health outcomes.

Historical and intellectual trajectory

The late 1980s and early 1990s marked the first significant appearances of sexual orientation within geographies of health, then referred to largely as medical geography. Early spatial–epidemiological studies that modeled the HIV/AIDS epidemic pointed to gay men as a key population advancing the spatial diffusion of the disease (Smallman-Raynor and Cliff, 1990; Gould, 1993). This medicalization of gay men, however, was not a new phenomenon. Decades before the advent of HIV/AIDS, urban public health authorities concerned about sexually transmitted infections (STIs) began examining gay men as both an at-risk population and a perceived public-health risk (Brown and Knopp, 2014). Early studies of gay men and the geography of HIV/AIDS were not especially attentive to the ways in which the social construction or stigmatization of non-normative sexualities might affect gay men's risks for HIV infection (Brown, 1995). Rather, they focused largely on the potential for HIV-positive individuals, who also happened to be gay, to infect others or to overwhelm local public-health systems upon returning to family homes (see also Ellis and Muschkin, 1996).

Social, cultural and feminist geographers began studying the lives of LGBTQ individuals more closely during roughly the same period. The cultural turn in human geography, which prioritized the social construction and differential experiences of geographic phenomena, spawned feminist geography in the 1980s and geographies of sexualities (or sexuality and space) in the 1990s. The latter was concerned with how gay men and lesbians made their homes, communities and livelihoods (Bell, 1991; Bell and Valentine, 1995).

While health was not an explicit focus in early geographies of sexualities, many studies dealt with broader issues of well-being, such as fear, harassment and social support (Bell and Valentine, 1995; Valentine, 1998). Much of medical geography, in contrast, continued through the 1980s to treat places as bounded collections of attributes (e.g., deprivation, disease endemicity) that could be modeled to predict health outcomes in a range of populations. As medical geography underwent its own transition to a *reformed* health geography (Kearns, 1993), researchers began employing a wider variety of approaches to understanding the relationship between health and place.

Many geographers sought to examine the role of various therapeutic landscapes (e.g., shrines, spas, camps) in shaping health (e.g., Williams, 1999). While this work typically employed qualitative techniques, it did not always adopt a critical social-theory lens or examine the social relations within places (Kearns and Moon, 2002). Critical and feminist geographers, however, began using detailed interview-based narratives to explore the social relations in places that might perpetuate risk or disadvantage (Dyck, 2003). The first health-geography research to acknowledge the social significance of sexual orientation maintained the focus on HIV/AIDS. Studies of gay men's sexual citizenship during the AIDS crisis (Brown, 1997) and the diminished lifeworlds of those who had contracted the disease (Wilton, 1996) became critical qualitative rejoinders to the disease-diffusion studies from earlier in the decade. Sexuality, however, has remained largely underresearched as an axis of difference in the place-health relationship (Parr, 2004), and sex and gender are still conceived as binaries in most geographic research (Browne, Nash and Hines, 2010). This gap has somewhat narrowed during the past 15 years. Building on research traditions in critical and feminist geography, as well as fruitful collaborations with geographers of sexualities, health geographers have attended to the current and historical regulation of LGBTQ health, access to care for LGBTQ patients and intersectional determinants of health affecting LGBTQ populations.

The current state of the field

Biopolitical regulation has been perhaps the most explored topic within geographies of LGBTQ health. Inspired by Foucault's lectures on sexuality and governmentality (e.g., Foucault, 2008), many health geographers have contributed to a hybrid field of study that examines health as an apparatus through which sexual non-normativity is disciplined. Much of the research in this area outlines the social construction of *good gays* who do their part to uphold the public good and *bad gays* who behave in an irresponsible way that puts society at risk. Brown (2006) counters the notion of sexual responsibility (i.e., safe sex) as a political and moral obligation that must be met to merit full citizenship. He explains that while sex is an act in which individuals naturally lose some of their autonomy, this loss does not stop gay men from taking account of their local epidemiological environments and engaging in strategies (i.e., *safer* sex) to mitigate risk to themselves and others. Public-health agencies often frame gay men's health responsibilities in contradictory ways. Thien and Del Casino (2012), for example, argue that gay men are told to *man up* by getting tested for HIV and disclosing their status to others but also to engage in talk therapy and other traditionally feminized self-care mechanisms to reduce risk.

An additional thread of work on biopolitics explores public-health regimes and the historic regulation of sexual non-normativity by the local state. Using Legg's (2005) framework of biopolitical population control, Brown and Knopp (2014) explain how local health authorities in Seattle and other cities in the United States created gay clinics during the 1970s to gain credibility and control in urban gay communities. While the clinics used surveillance programs and publishable metrics to make male-to-male STI transmission *visible*, they also employed gay staff purposefully to achieve buy-in for testing and safe-sex campaigns. Others have researched the ways in which the governmental regulation of sexuality itself has impacted well-being in LGBTQ populations. In Canada and elsewhere, civil-service expulsions, unwarranted police investigations and closures of social spaces such as bars and bath houses have left many urban LGBTQ communities socially fragmented and lacking resources for the older generation (Andrews and Holmes, 2007; Lewis et al., 2015).

A second body of work focuses on uneven access to health care among LGBTQ populations. The long-term denial of rights and resources for LGBTQ people in most Western countries has institutionalized heterosexuality within health-care provision and limited the availability of LGBTQ-specific services outside the largest cities. Lesbian mothers, for example, are overlooked at most stages of care for both childbirth and parenting. In a study by MacDonnell and Andrews (2006), many of the mothers interviewed observed that birthing-class instructors generally assumed uniform heterosexuality and mother-father relationships. Others noted the there are few supports for children who have lesbian mothers and experience bullying at school. In my own work, I reflect on the implications of institutional invisibility for LGBTQ health across the life course. Gay respondents in my study of HIV risk in Nova Scotia, in Canada, noted that a lack of access to gay-specific sex education during school years, for example, could affect health experiences and outcomes later in life (Lewis, 2015).

Health-care access for LGBTQ people is not only geographically uneven but also geographically imagined. The perception of social disenfranchisement in health care among LGBTQ people may be as damaging as the actual lack of health-care infrastructure for this population. In their study of lesbian, bisexual and trans women in Halifax and Vancouver, in Canada, Baker and Beagan (2016) found that participants alternately imagined LGBTQ people in other cities, towns and countries as better off and worse off in terms of access to care than they were. Building on Knopp and Brown's (2003) observation that queer places are more relational than hierarchical, they find that dense information networks in small, conservative places can enable access while the often diffuse LGBTQ health-care options in large cities can make navigating the health-care system more difficult. Catungal (2013) deftly bridges historical health geographies and access to care issues in his study of ethno-specific AIDS service organizations in Toronto, Canada. He explains that these intersectional safe houses emerged from efforts to replace the whiteness of mainstream gay clinics with HIV prevention messaging that was reparative and culturally affirming.

A third body of work builds upon the established public-health focus in geographies of health by exploring how sexuality coalesces with other place-based social determinants of health. To continue unpacking the black box of place for which medical geography was originally critiqued, many have offered detailed accounts of the social relations within LGBTQ lives. Van Ingen (2004) offers a welcome perspective on LGBTQ sportscapes. She sees running groups and other sports clubs as fluid, socially networked therapeutic landscapes that can be at once health-promoting (i.e., in terms of fitness) and stigmatizing for those who fall outside of white, middle-class notions of gay and lesbian empowerment and self-responsibility. Holmes (2016) considers a broad range of the place-based social relations that threaten the well-being of trans and gender non-conforming individuals in Canada. These relations occur at almost every spatial scale, ranging from intimate partner violence at home to persistent gender segregation in public space and institutional erasure within health-care settings (see also Namaste, 2000).

Other researchers have employed time-space and life-course approaches to understand how health outcomes are shaped for LGBTQ people. Several have used these approaches to problematize the perceived binary of healthy and unhealthy places among gay men (including HIV-positive gay men) and understand the social time-spaces in which health behaviors such as drug use and sex might coalesce (Del Casino, 2007). Nightclubs, for example, can be places that provide social support but are used as contact points for drug use and unprotected sex among men whose lives have been de-routinized by adverse events such as unemployment (Tobin et al., 2013). Conversely, work itself can become a stressful place for HIV-positive gay men who feel the need to conceal both their sexual orientation and HIV status while engaging with customers (Myers, 2010). In some of my previous research, I have attempted to show that movement – like place – is rarely fully emancipatory for LGBTQ people. Health issues are both causes of and consequences of internal and international migration among LGBTQ people navigating uneven landscapes of stigma, stress and social support. While anxiety and depression stemming from stigma, harassment or internalized homophobia might prompt a decision to move, the social upheaval and displacement resulting from migration can result in using drugs or sex as coping mechanisms or ways to fit into new settings (Lewis, 2014, 2016).

Contributions, limitations and future developments

Research on biopolitical regulation, access to care and social time-spaces among LGBTQ people broadens the scope of health geography considerably. The studies outlined in this chapter not only take difference seriously, but also suggest ways in which health geographers can take account of social subjectivities that are not always visible or easily measurable. They also cultivate an understanding of places as sets of social relations shaped by individual subjectivities as well as historical and institutional processes.

The rapidly evolving nature of the social and political landscape affecting LGBTQ people makes work on health and well-being in this population more urgent than ever. Despite advances in legal equalities in some countries, LGBTQ lives are still constructed socially and institutionally as non-normative. They are often punctuated by coming out in multiple settings, uprooting oneself to develop an identity and negotiating social-support systems that extend beyond the heteronormative home, school and community. Although health geographers are ideally positioned to continue examining how the social determinants of health affect LGBTQ lives across space and time, their methods and approaches could perhaps be diversified.

Geographers of LGBTQ health have adopted critical and qualitative methodologies that capture the nuanced understandings of health-place relationships. The lack of quantitative work, in contrast, may owe to the relative lack of statistical data on sexual orientation and non-binary gender identities. This methodological gap may also reflect the perceived shortcomings of quantitative analyses in assessing the flexible, frequently mobile and sometimes traumatic aspects of LGBTQ lives that ultimately lead to adverse health outcomes. Avoidance of the traditional, hypothesis-driven public-health framework may also reflect an understandable aversion to potentially pathologizing identities or spaces associated with LGBTQ communities. The research emphases described in the first section of this chapter consequently tend to deconstruct extant geographies of LGBTQ health rather than measure geographic variations in health risks or highlight preventative interventions.

Other disciplines have begun to produce LGBTQ health research of a geographic *nature*, though not always with the attention to place or social theory for which health geographers have advocated. Research from psychology, epidemiology and even sociology has attempted to quantify the impact of pro- and anti-LGBTQ legislation on mental-health outcomes (e.g., Hatzenbuehler, Keyes and Hasin, 2009); the effect of moving to a gay neighborhood on HIV infection (Egan et al., 2011); and the influence of living in a rural area on life satisfaction among LGBTQ people (Wienke and Hill, 2013). These studies depart from the individual-level psycho-behavioral frameworks more common in these fields to emphasize the place *context* more typical of health geography. At the same time, they may reinforce existing binaries of healthy and unhealthy places for LGBTQ people rather than highlighting the broader health potentials of a range of places. One way that health geographers can contribute to this emerging work is through their expertise in multilevel models that account for the interactions of place context and individual experience (Duncan, Jones and Moon, 1998; see Bauermeister et al., 2015, for an LGBTQ example). Such work could examine, for example, the differential health impacts of gay neighborhood attachment between gay and lesbian individuals of different incomes or with different densities of social-support networks. It might also serve to disentangle place- and individual-level factors, such as living in a place with a high rate of anti-LGBTQ hate crime versus having personally been a victim of such a crime.

A clear area for future development in the geographies of LGBTQ health is the expansion of its socio-demographic scope. Currently, there is lack of account for the full range of LGBTQ identities (as opposed to just gay and lesbian) or for intersections of these identities with social subjectivities, such as race (but see Catungal, 2013; Tobin et al., 2013). Much of the extant work focuses on easier-to-access, typically white and often male populations that might be considered both homonormative (i.e., part of the middle-class main-stream) and metronormative (i.e., located disproportionately in cosmopolitan urban centers). As evidenced in this review, less explicitly geographic work has been conducted on the health of lesbian and bisexual women, and almost none addresses the health of trans individuals. The work of Doan (2007) and Browne, Nash and Hines' (2010) special edited issue of *Gender, Place & Culture* may all provide points of entry to understanding the complexities of trans people's distinct spatial attachments (e.g., to gender clinics and private spaces of social support rather than traditional gay neighborhoods) and the adverse health outcomes they experience.

Finally, there is room to extend the spatial and topical scope of geographic research on LGBTQ health beyond the Anglo-American context. Research on South Africa, China and Eastern Europe has considered how LGBTQ health might be affected by diverse political and epidemiological environments. Tucker's (2009) work on Cape Town, South Africa, underscores the ways in which HIV risk among men who have sex with men is determined by the city's high background prevalence, its racial segregation and individual men's sexual subjectivities. Studies of Chinese male sex workers who have sex with men, or *money boys*, suggests that the growing profession is both rooted in China's capitalist economic transformation and medi-ated geographically by rural-to-urban migration (Kong, 2017). Recent work on the Czech Republic (e.g., Pitoňák and Spilková, 2015) identifies the genesis of homophobia in post-communist Eastern European societies, regional differences in these attitudes and the potential application of educational interventions from abroad.

While this brief review reveals that there is much work to be done on geographies of LGBTQ health, it also catalogs a range of inspiring research on the individual life trajectories that animate health and the institutional structures that have systematically disadvantaged LGBTQ people. The greatest advances can, perhaps, be made by synthesizing key strands of research in LGBTQ health geographies rather than simply accelerating specialized sub-fields such as historical geographies of biopolitics or access to specialized forms of care. Multisite and multifactor ecologies of health that have been popularized in health geography offer models for assessing how both life trajectories and uneven geographies of homophobia, transphobia and health-service infrastructures coalesce in certain places. Similarly, the use of multilevel modeling has the potential to consider the influence of place context on LGBTQ health outcomes, including mental health, in a more nuanced manner than much of the work in the health-science disciplines is currently offering. Such work will not only improve health outcomes in a disadvantaged group, but also help develop the com-plexity and diversity of health geography as a key discipline in health research.

References

Andrews, G. J. and Holmes, D. (2007). Gay bathhouses: transgressions of health in therapeutic places. In: A. Williams, ed., *Therapeutic landscapes: Advances and applications*. Aldershot: Ashgate, pp. 221–232.

Baker, K. and Beagan, B. (2016). Unlike Vancouver . . . here there's nothing: imagined geographies of idealized health care for LGBTQ women. *Gender, Place & Culture*, 23(7), pp. 927–940.

Bell, D. (1991). Insignificant others: lesbian and gay geographies. *Area*, 23, 323–329.

Bell, D. and Valentine, G. (eds.) (1995). *Mapping desire: geographies of sexualities*. London: Routledge.

Brown, M. (1995). Ironies of distance: an ongoing critique of the geographies of AIDS. *Environment and Planning D: Society and Space*, 13, pp. 159–183.

Brown, M. (1997). *Replacing citizenship: AIDS activism and radical democracy*. New York: Routledge.

Brown, M. (2006). Sexual citizenship, political obligation and disease ecology in gay Seattle. *Political Geography*, 25, pp. 874–898.

Brown, M. and Knopp, L. (2014). The birth of the (gay) clinic. *Health & Place*, 28, pp. 99–108.

Browne, K., Nash, C. J. and Hines, S. (2010). Introduction: toward trans geographies. *Gender, Place & Culture*, 17(5), pp. 573–577.

Catungal, J. P. (2013). Ethno-specific safe houses in the liberal contact zone: race politics, place-making and the genealogies of the AIDS sector in global-multicultural Toronto. *ACME: An International Journal for Critical Geographies*, 12(2), pp. 250–278.

Del Casino, V. J. (2007). Flaccid theory and the geographies of sexual health in the age of Viagra ™, *Health & Place*, 13, pp. 904–911.

Doan, P. (2007). Queers in the American city: transgendered perceptions of urban space. *Gender, Place & Culture*, 17(1), pp. 57–74.

Duncan, C., Jones, K. and Moon, G. (1998). Context, composition and heterogeneity: using multilevel models in health research. *Social Science & Medicine*, 46(1), pp. 97–117.

Dyck, I. (2003). Feminism and health geography: twin tracks of divergent agendas? *Gender, Place & Culture*, 10(4), pp. 361–368.

Egan, J., Frye, V., Kurtz, S., Latkin, C., Chen, M., Tobin, K., Yang, C. and Koblin, B. (2011). Migration, neighborhoods, and networks: approaches to understanding how urban environmental conditions affect syndemic adverse health outcomes among gay, bisexual, and other men who have sex with men. *AIDS and Behavior*, 15, pp. S35–S50.

Ellis, M. and Muschkin, C. (1996). Migration of persons with AIDS – a search for support from elderly parents? *Social Science & Medicine*, 43, pp. 1109–1118.

Foucault, M. (2008). *The birth of biopolitics: Lectures at the College de France, 1978–1979*. London: Palgrave Macmillan.

Gould, P. (1993). *The slow plague: a geography of the AIDS pandemic*. New York: Wiley-Blackwell.

Hatzenbuehler, M., Keyes, K. and Hasin, D. (2009). State-level policies and psychiatric morbidity in lesbian, gay, and bisexual populations. *American Journal of Public Health*, 99(12), pp. 2275–2281.

Holmes, C. (2016). Safety, belonging and mental health: exploring the intersections between violence, mental health and place in the lives of trans and gender nonconforming people. In: M. Giesbrecht and V. Crooks, eds., *Place, health and diversity: learning from the Canadian perspective*. New York: Routledge, pp. 53–75.

Kearns, R. A. (1993). Place and health: towards a reformed medical geography. *The Professional Geographer*, 46, pp. 67–72.

Kearns, R. and Moon, G. (2002). From medical to health geography: novelty, place and theory after a decade of change. *Progress in Human Geography*, 26, pp. 605–625.

Knopp, L. and Brown, M. (2003). Queer diffusions. *Environment & Planning D: Society & Space*, 21, pp. 409–424.

Kong, T. S. K. (2017). Sex and work on the move: money boys in post-socialist China. *Urban Studies*, 54(3), pp. 678–694.

Legg, S. (2005). Foucault's population geographies: classifications, biopolitics, and governmental spaces. *Population, Space, and Place*, 11, pp. 137–156.

Lewis, N. M. (2014). Rupture, resilience, and risk: relationships between mental health and migration among gay-identified men in North America. *Health & Place*, 27, pp. 212–219.

Lewis, N. M. (2015). Placing HIV beyond the metropolis: risks, mobilities, and health promotion among gay men in the Halifax, Nova Scotia region. *The Canadian Geographer*, 59(2), pp. 126–135.

Lewis, N. M. (2016). Urban encounters and sexual health among gay and bisexual immigrant men: perspectives from the settlement and AIDS service sectors. *Geographical Review*, 106(2), pp. 235–256.

Lewis, N. M., Bauer, G. R., Coleman, T. A., Blot, S., Pugh, D., Fraser, M. and Powell, L. (2015). Community cleavages: gay and bisexual men's perceptions of gay and mainstream community acceptance in the post-AIDS, post-rights era. *Journal of Homosexuality*, 62(9), pp. 1201–1227.

MacDonnell, J. and Andrews, G. J. (2006). Placing sexuality in health policies: feminist geographies and public health nursing. *GeoJournal*, 65, pp. 349–364.

Myers, J. (2010). Health, sexuality and place: the different geographies of HIV-positive men in Auckland, New Zealand. *New Zealand Geographer*, 66, pp. 218–227.

Namaste, V. (2000). *Invisible lives: the erasure of transsexual and transgendered people*. Chicago: University of Chicago Press.

Parr, H. (2004). Medical geography: critical medical and health geography? *Progress in Human Geography*, 28(2), pp. 246–257.

Pitoňák, M. and Spilková, J. (2015). Homophobic prejudice in Czech youth: a sociodemographic analysis of young people's opinions on homosexuality. *Sexuality Research and Social Policy*, 13(3), pp. 215–229.

Smallman-Raynor, M. and Cliff, A. (1990). Acquired immune deficiency syndrome (AIDS): literature, geographical origins and global patterns. *Progress in Human Geography*, 14, pp. 157–213.

Thien, D. and Del Casino, V. J. (2012). (Un)Healthy men, masculinities, and the geographies of health. *Annals of the Association of American Geographers*, 102(6), pp. 1146–1156.

Tobin, K. E., Cutchin, M., Latkin, C. A. and Takahashi, L. M. (2013). Social geographies of African American men who have sex with men (MSM): a qualitative exploration of the social, spatial, and temporal context of HIV risk in Baltimore, Maryland. *Health & Place*, 22, pp. 1–6.

Tucker, A. (2009). *Queer visibilities: space, identity and interaction in Cape Town*. London: Wiley-Blackwell.

Van Ingen, C. (2004). Therapeutic landscapes and the regulated body in the Toronto Frontrunners. *Sociology of Sport Journal*, 21, pp. 253–269.

Valentine, G. (1998). Sticks and stones may break my bones: a personal geography of harassment. *Antipode*, 30, pp. 303–332.

Wienke, C. and Hill, G. (2013). Does place of residence matter? Rural-urban difference and the well-being of gay men and lesbians. *Journal of Homosexuality*, 60(9), pp. 1256–1279.

Williams, A. (ed.) (1999). *Therapeutic landscapes: the dynamic between place and wellness*. Lanham, MD: University Press of America.

Wilton, R. D. (1996). Diminished worlds: HIV/AIDS and the geography of everyday life. *Health & Place*, 2(2), pp. 68–93.

31

"THIS PLACE IS GETTING TO ME"

Geographical understandings of mental health

Liz Twigg and Craig Duncan

This chapter outlines the ways in which medical and health geographers have contributed to our understanding of mental health. The narrative presented focuses on mental illness and is divided into two main parts. The first part discusses quantitative modeling of mental-health outcomes, detailing how aspects of individual, household and neighborhood socioeconomic context interact to influence risk. Much of this work is interdisciplinary and is produced in collaboration with medical practitioners working in the field of psychiatric morbidity; it addresses both geography and epidemiology audiences. The second part considers the use of qualitative approaches in elucidating the way in which the identities of those experiencing poor mental health are associated with the features and social meaning of places. While these place-based processes may lead to heightened feelings of marginalization, stigmatization, isolation and helplessness, they can also lead to a sense of empowerment, control and more positive identities. Where relevant, we refer to literature highlighting these positive impacts but generally leave *wellness* to take center stage elsewhere in this book.

Defining "mental illness" is a challenging, chaotic and contested exercise and is not attempted here. Sociological work has shown how designations are fluid, often socially and politically constructed, and neither spatially nor temporally universal. We do, however, make a broad distinction between common mental disorders (CMDs) and more serious conditions that can lead to compulsory hospital stays. CMDs include stress, anxiety and low-level depression and, although less disabling than major psychiatric illnesses, can sometimes cause disruption to usual everyday activities, with the potential to lead on to more severe conditions. CMDs' high prevalence is also a cause for concern (in the United Kingdom, around 20% of the adult population may suffer from CMDs at any one time).

Geographers have approached the nexus of place and mental health from several perspectives. They have, for example, attempted to qualify and quantify those aspects of place that may be detrimental to mental well-being, not only focusing on urban penalties but also recognizing the negative impacts of isolation, stigma and poor service provision in remote rural areas. Emphasis has also been placed on determining the geographical scale at which place effects become important, with a focus often centered on socioeconomic inequalities and relative (income) disadvantage. Increasingly, however, there is a realistic recognition that these overarching influences are not necessarily universal. Instead, there is a constant, recursive interplay between people and places, one that is complex and contingent on other geographical and social contexts.

Mental health outcomes: the role and complexity of place

One of the first geographers to capture the influence of place on mental health was John Giggs – a UK-based medical geographer often regarded as one of medical geography's founding fathers. His (now classic) research documented the spatial distribution of people with schizophrenia in Nottingham, in the United Kingdom, during the early 1970s (Giggs, 1973). He built on the idea that inner-city conditions, characterized by poverty, instability and exclusion, act as catalysts to *breed* poor mental health. Although often tabled as a contrasting explanatory framework, the *drift* hypothesis outlined a parallel but more nuanced socio-spatial process, whereby individuals, as a consequence of poor mental health, experienced downward social-class mobility and residential shifts to poorer, marginalized areas of the inner city (see, e.g., Lewis et al. (1992) for a continuation of this debate).

A plethora of research now exists that implicates (to a lesser or greater extent) neighborhood context as part of the explanatory framework for mental-health outcomes. As the drift and breeder hypotheses suggest, the role of place is multifaceted. However, national investments in routine mental-health surveys have facilitated some understanding of this relationship and help us illustrate key themes outlined in the introduction.

Scale

Many studies operationalize socioeconomic context via routine measures of deprivation captured at local scales assumed to be synonymous with neighborhood. In reality, this process is usually data-driven, and *neighborhoods* equate to census areas or administrative/governance-related configurations. In ecological (i.e., aggregate) studies, evidence of deprivation gradients in outcomes is usually present across such areas but reduces significantly, or even disappears, once the compositional mix of people living in these areas is taken into account. This leads us to an obvious question – if geography matters for mental-health outcomes, at what spatial scale is it important (Duncan, Jones and Moon, 1995)?

In essence, contextual factors may operate at much smaller scales than administrative geographies and may interact with individual circumstance and other contexts. Some studies of CMDs in the United Kingdom have recognized the importance of *household* as a meaningful everyday context and show that significant differences remain across households that are not explained by the socioeconomic characteristics of the household or the people within them (Weich et al., 2003a), a finding that suggests that clusters of people within households may confer risk upon one another. Parallels can be drawn with prospective analyses, whereby childhood family psychosocial environment can be associated with mental well-being in early older age, highlighting the influence of family context and the life course (Stafford et al., 2015).

Income inequality and socioeconomic incongruities

As with many other health outcomes, Wilkinson's well-established (but often contested) income-inequality thesis stakes a claim in the explanation of mental-health inequalities (Pickett and Wilkinson, 2010). Within this explanatory paradigm, the high income inequality/poor mental health link is due to the stress response of status anxiety. The causal pathway between poor mental health and income inequality may also be mediated by lower levels of community social capital, social cohesion and trust. The assumption here is that neighborhoods with high levels of inequality are too diverse to sustain strong community bonds, and feelings of isolation, social fragmentation and distrust can lead to more depressive symptoms. Findings regarding Wilkinson's inequality thesis remain inconclusive, and the importance of absolute, rather than relative, disadvantage is often documented. In Wales, Fone et al. (2013) show that absolute income deprivation at the neighborhood level is more important than income inequality as a risk factor for CMD. The authors suggest that these null findings for relative income may be associated with the low values for inequality found across the study area. Likewise, but at the other end of the income-inequality scale, Adjaye-Gbewonyo et al. (2016) find that universally high district-level levels of income inequality in South Africa fail to register an influence on mental-health outcomes, but the authors surmise there may be a time lag before inequality effects are felt.

Flint et al.'s (2013) longitudinal work also delves into ideas about relative disadvantage. Their research across England and Wales confirms the substantial link between individual unemployment and poor mental health but also notes that this is attenuated when the residential context is also characterized by high unemployment. This idea that a socioeconomic alignment between an individual and his or her neighborhood is conducive to good mental health has also been pursued elsewhere. In Auckland, New Zealand, higher treatment rates for anxiety/mood disorder are reported in socioeconomically isolated deprived areas – that is, areas surrounded by more advantaged neighborhoods (Pearson et al., 2013). Albor et al. (2014) show that high-status mothers' risk of depression or anxiety in the United Kingdom is reduced by 41% if they reside in high-status rather than low-status neighborhoods. This was explained by the improved chances of friendship when mothers were living in *socioeconomic congruity* with their neighbors.

Contingencies and the challenge of capturing context

Contextual influences are not universal; they are contingent and may change through time and across the life course. The Whitehall II cohort study of British civil servants has shown that mental-health gradients relating to housing tenure (i.e., owners versus renters) have diminished over time, but significantly higher rates of poor mental health were found across those who continue to experience long-term housing and financial problems (Howden-Chapman et al., 2011). Similarly, Weich et al. (2003b) found that area-level deprivation significantly increases risk of CMD, but only for those who were economically inactive, suggesting that time at home may increase risk exposure. The contingent nature of other environmental contexts has also garnered attention. For example, in a focused study investigating the influence of ethnic density on mental health in the United States across Latino sub-groups, it was shown that the association is dependent on the different migration histories (including reasons for migration and arrival rights) and current contexts across sub-groups (Becares, 2014). It has also been found in the United States that members of ethnic minority groups have reduced risk of poor mental health when they reside in areas with a high density of the same ethnicity (Shaw et al., 2012).

It is not only socioeconomic or socio-demographic context that plays a part in determining mental illness – or wellness. Other chapters in this book discuss how natural surroundings (e.g., blue/greenscapes) or the built environment can play out as *therapeutic*. Sometimes, there may be an interaction between the potentially health-enhancing aspects of neighborhood and socioeconomic context. A Canadian study found that living near a park had a protective effect against depression for people living in crowded conditions, and proximity to health services was also particularly beneficial for those living in deprived neighborhoods. Places classed as socially advantaged and offering cultural services were also protective (Gariepy et al., 2014). The urban penalty associated with mental-health outcomes and the protective effect of rural environments largely disappear when routine measures of population density are used to capture levels of urbanicity. However, when additional, more nuanced aspects of place (e.g., the proportion of residents working in different occupations including agriculture, forestry and fishing) are used to capture rurality, protective effects are detected (Weich, Twigg and Lewis, 2006), thus highlighting the need to capture context in as much detail as possible.

Severe mental-health outcomes

Apart from a brief mention of schizophrenia, our discussion so far has centered on CMDs – a focus largely reflecting data opportunities. More serious conditions usually involve health-care interventions, and many studies investigate service-related administrative data. A recent policy-pertinent example is the English National Study of Compulsory Admissions (ENSCA), looking at geographical inequalities in involuntary hospital admission (Weich et al., 2014). The policy backdrop for this project links to *reinstitutionalization*, a reversal of the trend toward community care (Moon, 2000) and an issue which we return to later in this chapter. ENSCA involved the analysis of approximately 1.2 million individuals who came into contact with secondary mental-health-care services during 2010 and 2011. Cross-classified multilevel models showed that compulsory admission was greater in more-deprived areas and in areas with more non-white residents after adjusting for confounders, a finding replicated in ecological studies of compulsory admission (Keown et al., 2016). ENSCA also provides evidence of the stark inequalities found across different ethnic groups, with black patients almost three times as likely to experience compulsory admission as white people. Although area-level deprivation and individual- and area-level ethnicity are shown to be important, they do little to explain the residual geography and the large variations in risk of compulsory admission remaining across hospitals and small areas (Weich et al., 2017). This possibly reflects differences in community treatments, bed shortages and *revolving door* practices (i.e., continuous ineffective treatment, referral of patients between inpatient and outpatient health-care-delivery sectors). Future work needs to seek out more appropriate longitudinal measures of residential and treatment context to explain these disparities. Quantitative health geographers, using long runs of geocoded patient records, are well-placed to discern nuanced interactions between individuals, residential geographies and treatment contexts in order to address this gap. Further- more, qualitative contributions (such as those outlined below) have the potential to enrich these findings and highlight the detailed, dynamic processes linking places to mental-health outcomes.

From outcomes to care

Geographers have been not only interested in describing and explaining mental-health outcomes, but also concerned with the care and experiences of people with mental-health problems. As we shall see, this second strand of geographical work on mental health is differently orientated in several ways: space and place do not feature as a container for organizing and representing data but feature as an active constituent of daily life; people are active subjects whose voices are heard; and theory and methods become multiple and diverse, focused on understanding and meaning. In short, an epidemiological model of mental health centered on medical definitions gives way to a social model in which historical and cultural interpretations take prec- edence. The idea of mental health, and the care that may be associated with it, is, then, no longer fixed and inevitable.

Confinement

Geographers' attention has often concentrated on the shifts that have occurred in the organization of care for people with mental-health problems. In the United Kingdom, for example, before the 18th century, people with mental-health problems were everywhere, living alongside others. Over the course of the 18th and 19th centuries, however, they came to be placed first in poorhouses, then in private charitable institutions, before finally being confined in large-scale publicly funded hospitals, usually known as asylums and, more often than not, purposefully set apart from urban centers.

Philo (2004) has produced the most detailed history of this intrinsically geographical shift. His work also emphasizes the way in which mental health moved from being a moral concern – disorder, unruliness – to being a medical one – symptoms, conditions. Thus, not only were the *mad* placed in separate physical, mate- rial spaces, but also they were put in new conceptual spaces – the term *mad* being used deliberately by Philo and others to highlight, and make a little less natural and certain, the rise to dominance of medical science's way of knowing irrationality as *illness*.

Release

Asylums have now, largely, disappeared due to a process of deinstitutionalization through the second half of the 20th century in which the *mad* were released back into society – most often, though not exclusively, into particular urban spaces.[1] The work of Dear and Wolch (1987) figures prominently here, though many others (e.g., Knowles, 2000) have added to their framing concept of (inner-city) *landscapes of despair*. Through documenting this shift, people with mental-health problems start to appear as active participants in their lives whose voices are heard. Many studies showed how they had to deal with defensive *not in my backyard* attitudes rather than the warm welcome implied by the standard sobriquet of the time, *care in the community*. Lived experiences proliferate, both within specific communities but also, importantly, within society at large. Drawing on ideas of difference (Who is same? Who is other?), and making connections with studies of disability and chronic physical illness such as the work of Moss and Dyck (2003), workers came to see the treatment of the newly released *mad* as one specific instance of the general way in which different bodies were made invisible in space.

Recapture

While nearly all subsequent health-geography work has concentrated on exploring the implications of the release, it needs to be noted that one further development has been charted in more medically inclined studies of mental-health care (Chow and Priebe, 2013). Although the number of conventional psychiatric beds in hospitals has continued to fall in most Western industrialized countries, there has been a dramatic growth in forensic (criminal) psychiatry places, supported housing places and prison populations. Involuntary hospital admissions, discussed earlier in this chapter, have also risen, and this is considered another part of the process of *reinstitutionalization*. It is, though, perhaps the last setting in the previous list that geographers of mental health need to discover most, at least in some countries. For example, as Scull (2015) has recently pointed out, the largest concentration of seriously mentally ill people in the United States is in the Los Angeles County jail system.

Escape

Work by geographers has, though, remained focused on the escape from confinement, and this is worth a closer look. Parr is the geographer who has done the most work here, and through a series of papers drawn together and expanded upon in a book (Parr, 2008), we have what is, in our view, the fullest, and most subtle and nuanced, analysis of the lived experience of people with mental-health problems by geographers to date. It is life, not as the individual experience of being in the world or the universal human of humanistic geography, but life seen through ideas of socially marked difference and the politics of identity associated with this. Who is included? Who is excluded? Where? Who is valued? Who is devalued? Where? Through close ethnographic, participant-observation-based investigation of a range of mundane, everyday community spaces in both urban and rural locations, Parr remains positive and hopeful that such spaces can provide enough possibility for people with mental-health problems to take some meaningful control of their lives and for others' fear of them to if not subside, then at least be subdued.

Later work by Parr and others has, though, drawn on another way of understanding the lived experience of mental health (Parr and Davidson, 2015).[2] Rather than concentrate on what is going on *out there*, beyond ourselves, attention focuses on what is going on or, perhaps more accurately, has gone on *in here*, inside ourselves. We need, then, to first see interior geographies (how the outside world is organized inside us and how our personality and sense of self is itself organized), especially those that we are not immediately, if at all, consciously aware of, for it is these that affect and shape our external lives and the physical spaces they take place in. In this way, mental life and the problems that can be attached to it are opened up beyond medically defined conditions to include intimate psychic spaces and the full range of human emotions that they encapsulate. Stemming from the work of Freud, and drawing on those who have followed on from him, this psychoanalytical approach has stimulated novel methodological approaches drawing on psychotherapeutic techniques (Bingley, 2003) and provided valuable insights on disruptive and difficult emotions and the debilitating conditions they can give rise to (Davidson, 2003).

Most recently, this desire to investigate unconscious life has led some geographers to take a third, and rather different, approach. Instead of going inward to reveal the long-lasting effects of the unconscious, or outward to expose the enduring differences and identities imposed by social life, Andrews, Chen and Myers (2014) wish to capture all that is going on in the immediacy of the present: lived experience as it takes (and makes) place – before it is understood, before it is represented in understanding. This non-representational approach can be thought of as offering an ecology of mental health in the moment, its metaphors of atmospheres and assemblages seeking to articulate the spontaneous, fluid environments through which, and by which, people with mental-health problems become well (Duff, 2016). In this way, nothing is innate, structured or fixed; instead, recovery is continually produced through the multitude of human, and non-human, resources that come together as the lives of people with mental-health problems quite literally *take place*.

Danger

There is much to celebrate in this second strand of work, and many valuable insights have been gained. The distant, detached calculations of epidemiology have been complemented with knowledge that is *up-close and personal*. Biomedical causes, cures and associations have been augmented with notions of care and recovery. A multitude of often mundane and ordinary places – from Parr's (2008) range of participatory community spaces to Wilton and Evans' (2016) social enterprises – have come into focus. Methodological variety and novelty has burst through, with go-alongs, video-elicitations, autobiographies, and autoethnographies becoming prominent alongside more well-established qualitative research techniques (Chouinard, 2012; Liggins, Kearns and Adams, 2013; Söderström et al., 2016).

Yet, somehow, there is also, at least for us, some danger here. While consideration of the abject poverty of many of those with mental-health problems is acknowledged as missing by Parr, there is no mention of it in the later work by those taking a non-representational approach. Psychoanalytical impulses often seem to lead to playful encounters with urban living more than the harsh, painful reality of city lives disrupted by mental ill health. What is also striking is that so much recent work has talked about *bodies* more than, or even to the exclusion of, *minds*. One correction to this tendency for *mindless bodies* would appear to be recent British work attempting to get to grips with the current vogue for *mindfulness* therapies (Lea, Cadman and Philo, 2015). As the authors make clear, however, and numerous self-help publications now available attest, such techniques aim to move us from mindless to mindful lives by taking us *out* of our heads and *into* our bodies. With their long-standing interest in human-environment interactions, geographers would seem to be well-placed to elicit the ways in which our mental experiences constitute, and are constituted by, the day-to-day material circumstances in which we lead our lives.

Concluding thoughts

This chapter has provided a brief overview of the role of place in understanding mental-health outcomes and care. The part on health outcomes alerts us to be ever mindful of what is relevant as context. Recent political upheavals on global and national stages may have far-reaching impacts on feelings of inclusion, belonging and safety. First World challenges associated with austerity are likely to damage the fabric and levels of community support in our local neighborhoods. Our measures of context must therefore be innovative and sensitive to contemporary potential stressors. The part on health care reflects the move in the broader discipline from social geographies arising from the material structures of society to cultural geographies attached to the workings of power through identity, meaning and representation. In this way, the two strands of work covered in this chapter seem to be moving farther apart, and, for us, it is striking and concerning that currently there is such little *up-close and personal* geographic work examining the mental-health implications for those most affected by the economic and political vicissitudes of recent times.

Notes

1 The trace of asylums can, though, still be found, as work by Moon and colleagues has explored (Moon, Kearns and Joseph, 2006).
2 This approach is referred to in Parr's (2008) earlier work, but only briefly, and is not developed.

References

Adjaye-Gbewonyo, K., Avendano, M., Subramanian, S.V. and Kawachi, I. (2016). Income inequality and depressive symptoms in South Africa: a longitudinal analysis of the National Income Dynamics Study. *Health & Place*, 42, pp. 37–46.

Albor, C., Uphoff, E. P., Stafford, M., Ballas, D., Wilkinson, R. G. and Pickett, K. E. (2014). The effects of socioeconomic incongruity in the neighbourhood on social support, self-esteem and mental health in England. *Social Science & Medicine*, 111, pp. 1–9.

Andrews, G. J., Chen, S. and Myers, S. (2014). The "taking place" of health and wellbeing: towards non-representational theory. *Social Science & Medicine*, 108, pp. 210–222.

Becares, L. (2014). Ethnic density effects on psychological distress among Latino ethnic groups: an examination of hypothesized pathways. *Health & Place*, 30, pp. 177–186.

Bingley, A. (2003). In here and out there: sensations between Self and landscape. *Social & Cultural Geography*, 4(3), pp. 329–345.

Chouinard, V. (2012). Mapping bipolar worlds: lived geographies of "madness" in autobiographical accounts. *Health & Place*, 18(2), pp. 144–151.

Chow, W. and Priebe, S. (2013). Understanding psychiatric institutionalization: a conceptual review. *BMC Psychiatry*, 13(1), pp. 169.

Davidson, J. (2003). *Phobic geographies: the phenomenology and spatiality of identity*. Aldershot, Hants: Ashgate.

Dear, M. and Wolch, J. (1987). *Landscapes of despair: from deinstitutionalization to homelessness*. Princeton: Princeton University Press.

Duff, C. (2016). Atmospheres of recovery: assemblages of health. *Environment and Planning A*, 48(1), pp. 58–74.

Duncan, C., Jones, K. and Moon, G. (1995). Psychiatric morbidity: a multilevel approach to regional variations in the UK. *Journal of Epidemiology and Community Health*, 49(3), pp. 290–295.

Flint, E., Shelton, N., Bartley, M. and Sacker, A. (2013). Do local unemployment rates modify the effect of individual labour market status on psychological distress? *Health & Place*, 23, pp. 1–8.

Fone, D., Greene, G., Farewell, D., White, J., Kelly, M. and Dunstan, F. (2013). Common mental disorders, neighbourhood income inequality and income deprivation: small-area multilevel analysis. *The British Journal of Psychiatry*, 202(4), pp. 286–293.

Gariepy, G., Blair, A., Kestens, Y. and Schmitz, N. (2014). Neighbourhood characteristics and 10-year risk of depression in Canadian adults with and without a chronic illness. *Health & Place*, 30, pp. 279–286.

Giggs, J. A. (1973). The distribution of Schizophrenics in Nottingham. *Transactions of the Institute of British Geographers*, 59, pp. 55–76.

Howden-Chapman, P. L., Chandola, T., Stafford, M. and Marmot, M. (2011). The effect of housing on the mental health of older people: the impact of lifetime housing history in Whitehall II. *BMC Public Health*, 11, p. 682.

Keown, P., McBride, O., Twigg, L., Crepaz-Keay, D., Cyhlarova, E., Parsons, H., Scott, J., Bhui, K. and Weich, S. (2016). Rates of voluntary and compulsory psychiatric in-patient treatment in England: an ecological study investigating associations with deprivation and demographics. *British Journal of Psychiatry*, 209(2), pp. 157–161.

Knowles, C. (2000). Burger King, Dunkin Donuts and community mental health care. *Health & Place*, 6(3), pp. 213–224.

Lea, J., Cadman, L. and Philo, C. (2015). Changing the habits of a lifetime? Mindfulness meditation and habitual geographies. *Cultural Geographies*, 22(1), pp. 49–65.

Lewis, G., David, A., Andréassson, S. and Allebeck, P. (1992). Schizophrenia and city life. *The Lancet*, 340(8812), pp. 137–140.

Liggins, J., Kearns, R. and Adams, P. (2013). Using autoethnography to reclaim the "place of healing" in mental health care, *Social Science & Medicine*, 91, pp. 105–109.

Moon, G. (2000). Risk and protection: the discourse of confinement in contemporary mental health policy, *Health & Place*, 6(3), pp. 239–250.

Moon, G., Kearns, R. and Joseph, A. (2006). Selling the private asylum: therapeutic landscapes and the (re) valorization of confinement in the era of community care. *Transactions of the Institute of British Geographers*, 31(2), pp. 131–149.

Moss, P. and Dyck, I. (2003). *Women, body, illness: space and identity in the everyday lives of women with chronic illness*. Lanham, MD: Rowman & Littlefield Publishers.

Parr, H. (2008). *Mental health and social space: towards inclusionary geographies?* Oxford: Blackwell Publishing.

Parr, H. and Davidson, J. (2015). Geographies of psychic life. In: P. Kingsbury and S. Pile, eds., *Psychoanalytic geographies*. Oxon: Routledge, pp. 119–134.

Pearson, A. L., Griffin, E., Davies, A. and Kingham, S. (2013). An ecological study of the relationship between socioeconomic isolation and mental health in the most deprived areas in Auckland, New Zealand. *Health & Place*, 19, pp. 159–166.

Philo, C. (2004). *A geographical history of institutional provision for the insane from medieval times to the 1860's in England and Wales*. Lewiston, UK: Edwin Mellen Press.

Pickett, K. and Wilkinson, R. (2010). Inequality: an underacknowledged source of mental illness and distress. *British Journal of Psychiatry*, 197(6), pp. 426–428.

Scull, A. (2015). *Madness in civilization: a cultural history of insanity, from the Bible to Freud, from the madhouse to modern medicine*. Princeton, NJ: Princeton University Press.

Shaw, R. J., Atkin, K., Becares, L., Albor, C. B., Stafford, M., Kiernan, K. E., Nazroo, J. Y., Wilkinson, R. G. and Pickett, K. E. (2012). Impact of ethnic density on adult mental disorders: narrative review. *British Journal of Psychiatry*, 201(1), pp. 11–19.

Söderström, O., Empson, L., Codeluppi, Z. and Söderström, D. (2016). Unpacking "the City": an experience-based approach to the role of urban living in psychosis. *Health & Place*, 42, pp. 104–110.

Stafford, M., Gale, C. R., Mishra, G., Richards, M., Black, S. and Kuh, D. L. (2015). Childhood environment and mental wellbeing at age 60–64 years: prospective evidence from the MRC National Survey of Health and Development. *PLoS One*, 10(6), e0126683.

Weich, S., McBride, O., Twigg, L., Duncan, C., Keown, P., Cyhlarova, E., Crepaz-Keay, D., Parsons, H., Scott, J. and Bhui, K. (2017). Variation in compulsory psychiatric in-patient admission in England: a cross-classified, multilevel analysis. *Lancet Psychiatry*, 4(8), pp. 619–626.

Weich, S., Holt, G., Twigg, L., Jones, K. and Lewis, G. (2003a). Geographic variation in the prevalence of common mental disorders in Britain: a multilevel investigation. *American Journal of Epidemiology*. Oxford University Press, 157(8), pp. 730–737.

Weich, S., McBride, O., Twigg, L., Keown, P., Cyhlarova, E., Crepaz-Keay, D., Parsons, H., Scott, J. and Bhui, K. (2014). Variation in compulsory psychiatric inpatient admission in England: a cross-sectional, multilevel analysis. *Health Services and Delivery Research*, 2(49), pp. 1–90.

Weich, S., Twigg, L., Holt, G., Lewis, G. and Jones, K. (2003b). Contextual risk factors for the common mental disorders in Britain: a multilevel investigation of the effects of place, *Journal of Epidemiology and Community Health*, 57(8), pp. 616–621.

Weich, S., Twigg, L. and Lewis, G. (2006). Rural/non-rural differences in rates of common mental disorders in Britain: prospective multilevel cohort study. *The British Journal of Psychiatry*, 188, pp. 51–57.

Wilton, R. and Evans, J. (2016). Social enterprises as spaces of encounter for mental health consumers. *Area*, 48(2), pp. 236–243.

SECTION 4

Places and spaces

32
INTRODUCING SECTION 4
Places and spaces

Valorie A. Crooks, Gavin J. Andrews and Jamie Pearce

The study of place and space is fundamental to all facets of human geography. Space is often thought of as a geographic point or a segment of territory. There is no single way to understand place. In a broad sense, Cresswell (1996) notes that places are both physical and social. In the physical sense, we can think of place as a location, a setting for everyday life, and a material artifact. Place becomes a social construct when we think about ideas such as having a literal and/or metaphorical place in a social hierarchy, undertaking place-making activities such as community-building, and developing an emotional attachment to or relationship with a place (Castleden et al., 2010). Not surprisingly, health geographers are at the forefront of examining and articulating the connections between space, place and health. Health geographers' engagement with these foundational geographic concepts in new ways helps show that place and space continue to be relevant to contemporary health research questions (Crooks and Winters, 2016).

Health geographers' research regarding space-place-health connections has been important for a number of reasons. First, it has raised awareness among clinicians, public-health officials, health-care administrators, social-care providers and others of the profound influence that space and the places in which we live, play and work have on health. Not only can places make us sick – through, for example, exposure to toxins in particular locations (e.g., Kershaw et al., 2013); they can affect our sense of well-being – through, for example, simply shaping how we feel about ourselves in a specific setting (e.g., Mair, Kaplan and Everson-Rose, 2012). Second, health geographers' research has consistently demonstrated that the relationship between health, space and place is complex, context-specific, and messy (e.g., Castleden et al., 2010). This reality has made it difficult to make sweeping generalizations about, for example, how best to design healthy or health-promoting places. If health is a social construct and understandings of place also have a socially constructed nature, then universal truths about bidirectional impacts between the two become extremely rare. Third, health geographers have played an important role in identifying new places/spaces, invisible places/spaces, and ignored places/spaces that warrant our attention because of their implications for health, whether good or bad (e.g., Dyck, 2005). Studies led by health geographers have brought such places and spaces to the fore, making them visible and part of the larger dialogues that surround health and health care in cognate disciplines.

The chapters in this section both implicitly and explicitly examine place and space and their role in health, wellness, well-being, daily life, and health-care delivery and receipt. The diverse places and spaces examined in this section are not intended to reflect the full range of those studied by health geographers. Instead, they are key exemplars of the types of places and broad spaces that geographers engage and interact with that have been particularly revealing and provide a sense of how and why geographers study particular types of places and spaces. There are a number of important themes that run throughout this section. These themes help show the types of places and spaces often studied, and they reveal important aspects of the sub-discipline of health geography. In other words, these themes help identify particular points of note in health geography's evolutionary spectrum. In the paragraphs that follow, we identify three such cross-cutting themes.

Health geographers have long-standing interests in examining the spaces in which health care is delivered and received. Such spaces vary greatly in size, ranging from the intimate setting of a home or apartment in which home-care services are given, to the large multispecialty hospital that treats hundreds of patients every day, and everything in between. A number of chapters in this section take us to particular places of care. Milligan situates her chapter within the changing landscape of health and social care that has shifted the site of care in much of the Global North out of institutional settings, such as hospitals and long-term-care facilities, and into the community. She then critically examines the idea of the home as a private site of life and a public space of care, particularly for persons aging in place. Ormond and Toyota's chapter also considers care for aging persons. They explore a number of movements across international borders (or transnational movements) for health-care delivery or receipt, including the practice of international retirement migration. As they contend, some older people participate in international retirement migration in order to access affordable health care abroad, including in the home and in residential care settings. They question the equity implications of this particular movement, suggesting it may place burdens on the destination country's health system. In his chapter, Kearns takes us to the formal space of the primary care clinic. He contends that these spaces are not only sites of care and components of larger health systems, but also places of social engagement that have significance for community well-being.

While some places are exceptional, in that they are rarely – if ever – visited (e.g., a site of pilgrimage), or they hold a special meaning (e.g., a relative's home), many are taken-for-granted parts of everyday life. Many chapters in this section highlight these daily, sometimes even mundane, places that shape our health and well-being. As noted above, chapters by Milligan as well as Ormond and Toyota engage with the home – a place that is often thought of as a daily site of comfort and refuge. Foley examines green and blue spaces, including parks, gardens, and pools. As he points out, although we may not interact with these spaces on a routine basis, they are often part of our daily visual landscape. Witten and Ivory's chapter also examines parks as spaces of play for children that offer opportunities for building health in city settings. They also touch on other everyday city spaces, ranging from libraries to streets, that include some and marginalize others and thus have complicated roles in promoting well-being in urban settings. Moving from the city to the countryside, Herron and Skinner explore rural spaces and their study by health geographers. They point to the complex nature of health and social care in rural settings and call for a relational approach to future rural-health research in the sub-discipline.

Many social-science disciplines have been influenced by the *mobilities turn* in research (Gatrell, 2011). We are surrounded by movements that hold deep implications for both health and place, many of which are being brought to the fore by health geographers. As mentioned previously, Ormond and Toyota examine transnational health-care mobilities, including medical tourism. Widener provides an overview of transportation research in health geography, including the methodological and theoretical contributions made by the sub-discipline. The chapter explores a number of distinct mobilities and movements and illustrates their connection to individual and population health. Cinnamon and Ronquillo introduce readers to mobile health, or mHealth. mHealth is a domain of research and practice that involves using mobile technologies to gather and transmit health data or health-care information. They explore the uptake of mHealth in the Global South and call for health geographers to play a greater role in examining the intersection of technology and health and its implications for society and social change. Walkability has emerged as a significant area of study in health geography that considers walking as a fundamental form of human mobility. Hirsch and Winters provide a summary of this area of the sub-discipline and also identify some of the challenges about the reliability of the data gathered and used in walkability research.

From local primary care clinics to sizeable and distant transnational movements, this section of the edited collection covers a range of places and spaces. As noted above, however, the chapters offer only a sampling of the places and spaces that occupy health geographers' imaginaries. We sought, unsuccessfully, to include chapters on the body, workplaces and hospitals as key sites of health-geographic inquiry. While these are important places studied by health geographers, they unfortunately are not brought to the fore here. Places and space also exist at multiple scales, and the contributions to this section do not address the full scope of their scalar existence. Micro-spaces as small as cells and body parts, hospital waiting rooms, and toilets have all received attention by health geographers. Meanwhile, macro-spaces are also significant to understanding health-in-place, including global and regional policy environments. It was never our intention to review the full scope of the places and spaces that health geographers examine across all scales. Instead, the chapters in this section serve as a valuable starting point for understanding this area of the sub-discipline, and they offer important insights regarding pressing future research directions that will redefine how we understand the complex and intricate relationships between place, space and health in years to come.

References

Castleden, H., Crooks, V. A., Schuurman, N. and Hanlon, N. (2010). "It's not necessarily the distance on the map . . .": using place as an analytic tool to elucidate geographic issues central to rural palliative care. *Health & Place*, 16(2), pp. 284–290.

Cresswell, T. (1996). *In place/out of place: geography, ideology, and transgression*. Minneapolis: University of Minneapolis Press.

Crooks, V. A. and Winters, M. (2016). 16th International Medical Geography Symposium special collection: a current snapshot of health geography. *Social Science & Medicine*, 168, pp. 198–199.

Dyck, I. (2005). Feminist geography, the "everyday", and local-global relations: hidden spaces of place-making. *Canadian Geographer*, 49(3), pp. 233–243.

Gatrell, A. C. (2011). *Mobilities and health*. London: Routledge.

Kershaw, S., Gower, S., Rinner, C. and Campbell, M. (2013). Identifying inequitable exposure to toxic air pollution in racialized and low-income neighbourhoods to support pollution prevention. *Geospatial Health*, 7(2), pp. 265–278.

Mair, C., Kaplan, G. A. and Everson-Rose, S. A. (2012). Are there hopeless neighborhoods? An exploration of environmental associations between individual-level feelings of hopelessness and neighborhood characteristics. *Health & Place*, 18(2), pp. 434–439.

33

HOME TRUTHS? A CRITICAL REFLECTION ON AGING, CARE AND THE HOME

Christine Milligan

Where and how we spend the most significant moments of our lives is an issue of importance to all. This can be particularly so where poor, or declining, physical and cognitive health and mobility may limit our ability to make choices.

In the United Kingdom, as well as other high-income countries, disclosures about the adverse impact of institutional care on the lives of both residents and staff from the 1960s onward (e.g., Goffman, 1968; Johnson, Rolph and Smith, 2010; Townsend, 1962), combined with a widespread shift in thinking around who and where care and support for those experiencing physical or mental disability should be provided, resulted in the emergence of policies and practices designed around community-based care. This sought to shift the provision of care and support away from institutional settings to community and domestic environments. Within health and social geography, this shift manifested in a refocus of interest on the (domestic) home as a key site of care and support. Here, the home was viewed not just as a physical structure within which life, death and care are played out, but as a space imbued with multiple meanings linked to identity, safety and security, privacy, power and control, emotion, nurture and historical memory (e.g., Blunt and Dowling, 2006; Chapman and Hockey, 1999; Collier, Phillips and Iedema, 2015; Imrie, 2004; Langstrup, 2013; Milligan, 2000, 2009; Twigg, 2000). Work in this field has played an important role in helping us understand the meaning of home – particularly for physically and mentally ill, older and disabled people and family caregivers – through its critical engagement with the home/care nexus. It has also illustrated how an understanding of the multifaceted and relational nature of home is key to the development of successful policies and practices designed around health, care and the home. By and large, the focus has been on aging in place, community care for those with physical, mental and intellectual disabilities and the home death.

Drawing on debates within health geography and environmental gerontology, this chapter aims to critically examine some of the paradoxes that surround notions of home and care. Engaging with ideas around the public and private, belonging and isolation, identity and alienation, the chapter both explores and challenges prevailing orthodoxies around the meaning of home and its conceptualization as a site of safety and security, power and identity, privacy and control.

The meaning of home

Conceptual thinking around community care and aging in place was built on the premise that the home is the optimum space in which to provide care and support for older and/or disabled people in ways that will enable them to remain as independent as possible for as long as possible. The home is viewed as a setting that is both familiar and imbued with particular meaning. Implicit within this notion is an assumption that the ongoing and temporal process of inhabiting a familiar place somehow results in the development of a unique sense of attachment that is both supportive and adaptive. Critical to understanding this sense of attachment is the notion of home as haven. To paraphrase Tuan (2004), this represents home as a place of security, familiarity and nurture, where an individual can retreat to a private world away from public scrutiny and where s/he can control decisions about whom to admit or exclude. This ability to control decisions about whom to include or exclude from the home is also viewed as enhancing a person's sense of self and independence. This can be particularly important for those older and/or physically or mentally disabled people who may feel vulnerable negotiating unfamiliar or challenging environments beyond the bounds of their own home, and where the surroundings of the home provide an important buttress to their sense of self (Milligan, Gatrell and Bingley, 2004; Sacco and Nakhaie, 2001). In this way, the home, and attachment to home, is viewed as key to successful aging in place.

From the late 1970s onward, Rowles's (1978, 1993) work on home and its local environs played an important role in revealing how, as people age, the lifeworlds they inhabit contract to become increasingly focused around the home. In large part, this is linked to issues of social and physical access and mobility. As the work of Wiles (2005), Lager (2016) and others has demonstrated, even relatively simple things, including the presence or absence of key design features and services such as even pavements, dropped curbs, toilets, seats and cafes in local settings, or physical changes to once familiar local environments, can severely curtail the geographical reach and wayfinding of people experiencing physical or cognitive limitations. Rowles' (1993) focus on the home, however, led him to suggest that a person's temporal knowledge of the home, combined with his or her physical attachment to it and the routines performed within it, held the potential to facilitate a person's ability to negotiate that space, despite physical, cognitive or sensory loss, without coming to harm. Framing this knowledge and attachment as a *preconscious sense of setting* (p. 66), Rowles argued that it can be employed positively to support an individual's ability to self-manage even as physical or cognitive abilities begin to decline.

The literature on care, home and older people also highlights its importance as a site of historical memory and identity (e.g., Angus et al., 2005; Brickell, 2012; Milligan, 2009). Private possessions and familiar objects within the home can reinforce this sense of self and identity, bestowing personal meaning to the home (Rubinstein, 1989). This anchors people not only within the home, but within a particular locality, as a site of memory and a daily reminder of continuity with past identity and relationships. More broadly, relatively new initiatives such as the development of dementia-friendly communities and attempts to understand what makes an environment dementia-friendly from the perspective of the person living with cognitive decline, are very much underpinned by notions of historical memory and identity (Davis et al., 2009; Mitchell and Burton, 2006). Also of relevance here is the growing focus on the significance of home as the preferred place of death. Williams (2004), for example, maintained that for many people, the familiarity, physical arrangements and habituated routines within the home imbue it with a sense of comfort, security and ease; this can be important not just for the person requiring care and support, but also for family carers who are faced with the physical and emotional upheaval of care work within the home, potential impending death and the sense of helplessness all this can engender. Collier, Phillips and Iedema (2015) further highlighted the importance of home at end of life as an expression of social and cultural identity, including those symbolic and affective connections manifested through loved ones and meaningful artifacts. Yet as Milligan et al. (2016) point out, facilitating a home death can also create an ambiguity of place, especially for family carers, where the issues they face in caring for a dying older person at home, and the home death itself, can fundamentally reshape the meaning and sense of home in ways that are not necessarily positive. This highlights the need to problematize some of the rather uncritical assumptions about the meaning and sense of home that have underwritten policies and practices designed around notions of aging in place, community care and the home death.

Critically reflecting on the paradox of home

People's experiences of home are both relational – co-produced by the key actors, actions and objects within – and temporally situated, in that the complex socio-spatial relations of home can shift and alter over time. So, while the focus on home as a site of ease, comfort and security may well hold true for some people at some points in their lives, as Blunt and Varley (2004), Brickell (2012) and others have pointed out, such interpretations of home are often eulogized; ignoring or overlooking the ways in which home can, at different times, be a site of stress, loneliness, fear, neglect or confinement. This highlights the paradox of any policy or philosophical approach that places home at the center of care. That is, the idea that home is a place where belonging and isolation, intimacy and violence, desire and fear, power and powerlessness, identity and alienation, and so forth, can all sit on contradistinction to each other.

Health and social geographers have illustrated how the home is a deeply nuanced and complex site of (sometimes shifting) social relations; a site of paradox, ambiguity and contradiction, where attributes are conditional, contextual and not necessarily positive (Milligan et al., 2016). Any understanding of the impact of home on both the person requiring help and support and family carers' experiences of the home needs to be alert to these complexities. In particular, we need to be alert to the ways in which the requirements of care and support can shift these socio-spatial relations in ways that alter the meaning and sense of home for those concerned. Hence, while it is important to identify how a sense of home might enhance a person's ability to retain a level of independence within the home despite physical or cognitive decline, it is also important to recognize that increasing frailty over time can lead to a breakdown in that preconscious sense of setting as the requirements for care and support increase. This includes the increased presence of professional care staff and the portable technologies and paraphernalia of care that, while disrupting the everyday rhythms and order of the home, are necessary to the provision of good care. These disruptions range from the need for space to perform care work, to the equipment designed to facilitate it (hoists, ramps, commodes, hospital beds, wheelchairs, nebulizers) – all of which can infiltrate and alter that preconscious sense of setting. Such transformations can manifest as either permanent or temporary rearrangements of the home (for example, from the temporary presence of a hospital bed or equipment to the more permanent installation of stairlifts, wheelchair ramps, wet rooms, granny annexes, etc.). The once familiar layout thus becomes unfamiliar, the ability to negotiate known spaces becomes challenging and the preconscious sense of setting breaks down. This can be particularly important when a person's cognitive and sensory abilities may be failing. In such circumstances, the person's sense of home can become increasingly institutionalized as the paraphernalia and requirements of care and support take over.

As Twigg (2000) and others have demonstrated, a further consequence of the complex and shifting social relations of care is often manifest as a declining ability to exclude others (e.g., care providers) from even those most personal and private areas of the home. Twigg's work, for example, was seminal in its illustration of the ways in which the home is divided into a range of public and private spaces. The public represented spaces such as hallways and living rooms – open to those visitors invited into the home but who were excluded from those deeply private and personal areas of the home, such as bedrooms, to which only the closest of friends and family might be given permission to enter. Yet as health deteriorates, and a person's ability to negotiate the home declines, the requirement to grant others, such as carers and care workers, access to those private spaces of the home increases, resulting in a declining ability to exclude and a blurring of the boundary between those areas that had formerly been deemed public and private spaces of the home.

Fundamental to many of the assumptions about the importance of the sense and meaning of home in supporting home-based care and support is the notion that home-based care takes place within the homes of the older or disabled persons themselves – but that, of course, is only a partial picture. For many (often lone-dwelling) frail and disabled people, the need for increased care and support can result in a move from the familial home to a new home-based setting, such as the home of a close family member (often an adult child or sibling) or a supported or extra-care housing environment. Here, the multiple meanings of home, and those attributes of home viewed as enabling and supportive, can be disrupted for both the recipient of care and the family caregiver (Milligan et al., 2016). Such disruptions to the meanings, objects and routines that are integral to the concept of home can have profound consequences for the emotional and material landscapes of care, which can impact both care recipients and family caregivers' own well-being. At the farthest end of the spectrum, the provision of a home death for a spouse or close family member can irrevocably change the nature and physical engagement with the home for those remaining (Milligan et al., 2016). For some, this can represent a shift from *home* to *house* – a space now empty of those personal and emotional relations that underpinned the meaning and sense of home. For others, this may result in a desire to remove from that setting and recreate a new home and a new life – one that is not imbued with an historical memory that has become shadowed by loss.

These shifting and temporal processes of home-based care are illustrative of the juxtaposition between private and public space and the shifting relationships of power, independence and autonomy that can accrue for older and disabled people across a continuum of physical and cognitive abilities.

The home/care dichotomy

While the (re)domestication of care serves a range of political and professional agendas, it is based on an idealized version of home and care; as such, it runs the risk of over-romanticizing notions of care that privilege the home (Rowles, 1993; Exley and Allen, 2007). Not only is there is a danger of placing too much emphasis on familiarity and emotional attachment to place as a component of residential preference, but it is important to recognize that not all older people are attached to their home or community in the way that policy might assume. The discourse of aging in place, for example, is underpinned by a number of assumptions: firstly, that older or disabled people necessarily have access to safe and stable housing; secondly, that the home is a place that enables self-expression and identity – a social space where individuals can relax and be at ease; and, thirdly, that the home is a space where privacy can be maintained and where choice and autonomy can be augmented – ultimately, a place where an individual's needs can be best met. It privileges the value of caring relationships without acknowledging the interaction of pre-existing social relationships with the actual work of caring. Hence, it is also based on the assumption that the home is synonymous with loving relationships – in other words, it assumes that caring *about* is the foundation for caring *for*. The home is thus privileged over the institution as the preferred site of care, with the latter often being characterized as a place where the needs both of the institution itself and of group living put a strain on caring ideals.

We should not forget that for some, the home represents a site of fear, physical and mental abuse, neglect and/or violence (Blunt and Varley, 2004; Meth, 2003). Inevitably the private nature of the home – and, by extension, policies designed to facilitate aging in place – is likely to impede the detection of abuse. Indeed, one review of studies looking at elder abuse across a range of (mainly, but not exclusively, Western) countries noted that around one in four vulnerable older people who are dependent on family carers are at risk of psychological abuse, with a further 5–6% subject to physical abuse and neglect (Cooper, Selwood and Livingston, 2008, p. 158). A study by Bonomi et al. (2007) in the United Kingdom also found that a small – but significant – percentage of older women requiring care and support were physically or sexually abused by their own spouses. Abuse is not restricted to family carers, however; as Bonomi et al. further noted, one in six professional carers commit psychological abuse, with one in ten committing physical abuse. For an important minority of older people, then, the promotion of policies designed to support aging in place does not necessarily equate to *good care*.

Idealized assumptions about the home/care relationship can thus mask its inherent tensions. It can bring with it contradictions and strains that are often unacknowledged and that create challenges not only for older people, but also for those providing their care and support. For example, tensions exist between (a) home as habitat and as a site of social and emotional expression and (b) home as a place within which the mechanics of care work, and the regulation attached to its performance (particularly formal care), are enacted. The home/workplace dichotomy was, perhaps, exemplified by the introduction of legislation in the United Kingdom in 2006 that prohibited smoking in the workplace. Care providers are now in the position of having to request that care recipients refrain from smoking in the private space of their own home during periods in which formal care work is taking place.[1] How enforceable this legislation is within the private space of the home is debatable, but clearly any suggestion that noncompliance could result in service withdrawal will be of significant concern to both family carers and care recipients. It would, however, be exceedingly difficult to apply and monitor such legislation in relation to the performance of family care – particularly co-resident care. Similarly, while formal care work is subject to health and safety regulation, no such regulation applies to the work performed by family carers. So, while aging in place highlights the increasing porosity of the boundaries between home and work in the field of care, it also highlights the inequities in the conditions of formal and family care work within the home.

In addition to the tensions between the home/work relationship, it is also important to critically evaluate how older people, themselves, perceive the relationship between home and the place of care. Many older people assess the benefits of aging in place in an entirely pragmatic way (Milligan, 2009). That is, their key concerns are focused around issues of cost, comfort and convenience and, as such, have little to do with any sense of physical, social or emotional attachment to place. Rather, for an older person's ability to remain at home in a community that he or she is knowledgeable about, and integrated into, local networks of care and support are meaningful because they enable individuals to call on practical assistance from friends and neighbors rather than having to rely on formal care options. It is also worth pointing out that while there is substantial evidence to suggest obligatory relocation can have adverse health consequences (e.g., Ferraro, 1983; Keister, 2006; Pruchno and Rose, 2000), the negative consequences of relocation will not be the same for all older people. Where home represents a site of loneliness, fear and abuse, the experience of relocation may well manifest in improved well-being.

One final dichotomy around home and care focuses around the family and how to balance the right of the older person to age in place versus the cost to the family carer. Although legislation in the United Kingdom now acknowledges the rights of carers, this is not just a legal matter. For example, whose rights and desires should prevail when a person wants to remain at home but the family carer is concerned about risk and danger or feels unable to provide an adequate level of care? These are ethical dilemmas that, in the United Kingdom at least, have yet to be fully tackled by policy-makers; though new models of care, often supported by technology, may go some way toward addressing them.

Where now?

Policy drivers focused around community care, aging in place, the home death, dementia-friendly spaces, et cetera, in most high-income countries are based on assumptions that home is the best place in which to locate care. For some, it is undoubtedly the case that home provides an environment that is supportive as health and mobility declines, but as the discussion above illustrates, this is not the case for all. In this chapter, I have sought to take a more critical lens to examine the nature and meaning of home in relation to declining health and care. As the literature illustrates, home can also be a site of loneliness or fear; the requirements of care may fundamentally shift the nature of home in ways that disrupt people's sense of home and those very facets of home that were seen to provide a supportive environment in the first place. Interestingly, to date, most of the literature focuses on the nature of home as it relates to the home of the care recipient – there is little recognition that as care needs increase, the care recipient may well be cared for in the home of the family carer – and what the implications of that are in terms of home for the carer and the care recipient. There is also little consideration of the changing nature of home for the family after a loss – through bereavement or through a family member's move to residential care – or how these experiences of home and care may vary within and across cultural settings. With their critical lens on unpacking the multifaceted and relational nature of the home/care nexus, these are avenues of research that health geographers are well-positioned to pursue.

Note

1 Although clearly it would be far better not to smoke at all, the reality is that (at least among current generations) many older people still do – and see little benefit in attempting to stop, at their age.

References

Angus, J., Kontos, P., Dyck, I., McKeever, P. and Poland, B. (2005). The personal significance of home: habitus and the experience of receiving long-term care. *Sociology of Health and Illness*, 27, pp. 161–187.

Blunt, A. and Dowling, R. (2006). *Home*. Abingdon: Routledge.

Blunt, A. and Varley, A. (2004). Geographies of home. *Progress in Human Geography*, 29(4), pp. 505–515.

Bonomi, A. E., Anderson, M. L., Reid, R. J., Carell, D., Fishman, P. A., Rivara, F. P. and Thompson, R. S. (2007). Intimate partner violence in older women. *The Gerontologist*, 28, pp. 361–368.

Brickell, K. (2012). Mapping' and "doing" critical geographies of home. *Progress in Human Geography*, 36(2), pp. 225–244.

Chapman, T. and Hockey, J. (eds) (1999). *Ideal homes? Social change and domestic life*. London: Routledge.

Collier, A., Phillips, J. L. and Iedema, R. (2015). The meaning of home at the end of life: a video-reflexive ethnography study. *Palliative Medicine*, 29(8), pp. 695–702.

Cooper, C., Selwood, A. and Livingston, G. (2008). The prevalence of elder abuse and neglect: a systematic review. *Age and Aging*, 37, pp. 151–160.

Davis, S., Byers, S., Nay, R. and Koch, S. (2009). Guiding design of dementia friendly environments in residential care settings: considering the living experiences. *Dementia*, 82, pp. 185–203.

Exley, C. and Allen, D. (2007). A critical examination of home care: end of life care as an illustrative case. *Social Science & Medicine*, 65, pp. 2317–2327.

Ferraro, K. F. (1983). The health consequences of relocation among the aged in the community. *Journal of Gerontology*, 38(1), pp. 90–96.

Goffman, E. (1968). *Asylums: essays on the social situation of mental patients and other inmates*. London and New York: Doubleday, Penguin.

Imrie, R. (2004). Disability, embodiment and the meaning of the home. *Housing Studies*, 19(5), pp. 745–763.

Johnson, J., Rolph, S. and Smith, R. (2010). *Residential care transformed: revisiting "The last refuge"*. Basingstoke: Palgrave Macmillan.

Keister, K. J. (2006). Predictors of self-assessed health, anxiety and depressive symptoms in nursing home residents at Week 1 postrelocation. *Journal of Aging and Health*, 18(5), pp. 722–742.

Lager, D. (2016). Rhythms, aging and neighbourhoods. *Environment and Planning A*, 48(8), pp. 1565–1580.

Langstrup, H. (2013). Chronic care infrastructures and the home. *Sociology of Health & Illness*, 35(7), pp. 1008–1022.

Meth, P. (2003). Rethinking the "domus" in domestic violence: homelessness, space and domestic violence in South Africa. *Geoforum*, 34(3), pp. 317–327.

Milligan, C. (2000). Bearing the burden: towards a restructured geography of caring. *Area*, 32, pp. 49–58.

Milligan, C. (2009). *There's no place like home: place and care in an aging society*. Farnham: Ashgate.

Milligan, C., Gatrell, T. and Bingley, A. (2004). Cultivating Health': therapeutic landscapes and older people in Northern England. *Social Science & Medicine*, 58, pp. 1781–1793.

Milligan, C., Turner, M., Seamark, D., Blake, S., Brearley, S., Thomas, C., Wang, X. and Payne, S. (2016). Unpacking the impact of older adults' home death on family carers' perceptions of home. *Health & Place*, 38, pp. 103–111.

Mitchell, L. and Burton, E. (2006). Neighbourhoods for life: designing dementia-friendly outdoor environments. *Quality in Aging and Older Adults*, 7(1), pp. 26–33.

Pruchno, R. A. and Rose, M. S. (2000). The effects of long-term care environments on health outcomes. *The Gerontologist*, 40, pp. 422–428.

Rowles, G. (1978). *Prisoners of space? Exploring the geographical experiences of older people*. Boulder: Westview Press.

Rowles, G. (1993). Evolving images of place in aging and aging in place. *Generations*, 17(2), pp. 51–65.

Rubinstein, R. (1989). The home environments of older people: a description of psychosocial processes in linking person to place. *Journal of Gerontology*, 44, pp. 45–53.

Sacco, V. F. and Nakhaie, R. (2001). Coping with crime: an examination of elderly and non-elderly adaptations. *International Journal of Law and Psychiatry*, 24(2–3), pp. 305–323.

Townsend, P. (1962). *The last refuge: a survey of residential institutions and homes for the aged in England and Wales*. London: Routledge and Kegan Paul.

Tuan, Y. F. (2004). Home. In: S. Harrison, S. Pile and N. Thrift, eds., *Patterned ground: the entanglements of nature and culture*. London: Reaction Books, pp. 164–165.

Twigg, J. (2000). *Bathing – the body and community care*. London: Routledge.

Wiles, J. (2005). Home as a new site of care provision and consumption. In: G. Andrews and D. Phillips, eds., *Aging and place: perspectives, policy and practice*. London: Routledge, pp. 79–97.

Williams, A. M. (2004). Shaping the practice of home care: critical case studies of the significance of the meaning of home. *International Journal of Palliative Nursing*, 10(7), pp. 333–342.

34

RETHINKING CARE THROUGH TRANSNATIONAL HEALTH AND LONG-TERM CARE PRACTICES

Meghann Ormond and Mika Toyota

With more people living longer than ever before, populations' medical and long-term care needs are increasingly placing strain on individual and collective resources and capacities. National governments are gradually withdrawing from responsibility for direct welfare provision for their citizenries and adopting neoliberal policies that facilitate and expand market-based involvement in health and social care. This has profoundly (re-) structured care relations by (re-)domesticating, individualizing, and commoditizing responsibility (Huang, Thang and Toyota, 2012; Raghuram, Madge and Noxolo, 2009). Yet, while much earlier research focused on this (re-)distribution of care responsibility *within* countries, a growing literature traces how formal and informal health- and social-care provision extends well beyond the national. The transnational (i.e., spanning of international borders) dimension of health and long-term care has been significantly heightened not only by neoliberal trade policies facilitating transnational flows of people, goods, and services, but also by advancements in communication technologies and biotechnological innovation (Gatrell, 2011; Sparke, 2009).

In this chapter, we draw health geographers' attention to the increasingly diverse transnational flows of caregivers and care seekers as well as the complex networks that they and their movements constitute and in which they get enfolded. To do this, we highlight contributions by scholars working at the crossroads of migration studies, transnational studies, health and social geography and anthropology, and social gerontology. This set of literatures critically engages with how transnational care practices challenge conventional conceptualizations and territorializations of care responsibility and entitlement between (non)citizens and states; individuals, families and communities; and consumers, workers and markets.

Transnationalizing the provision of care

Transnational health- and long-term care provision involves many people: from skilled health workers (SHWs) temporarily deployed by national governments engaging in health diplomacy and volunteers with not-for-profit organizations in countries in humanitarian emergencies (Snyder, Dharamsi and Crooks, 2011), to longer-term movements of SHWs from lower-income countries in search of better salaries and greater professional opportunities in middle- and higher-income countries; or from de-skilled migrant residential care-home workers caring for their elderly charges while at the same time caring from a distance for their own aging parents in their countries of origin, to the intimate and domestic labors of migrant women in commercially arranged marriages caring for their ailing spouses in their spouses' home countries (Piper and Roces, 2004).

In light of the ever-growing movement of foreign health, residential and domestic care workers to many rapidly ageing upper- and middle-income countries, the concept of global care chains (Hochschild, 2000; Parreñas, 2005) stands out as perhaps the best developed and most widely applied framework in research on transnational care in the last two decades. Based on a political economy of care approach, this concept effectively draws attention to the ways in which the care labor of women in globally more advantageous positions is substituted by that of women from more marginalized positions. For instance, a woman migrates from a city in a poor country to a wealthy country to work as a domestic maid. She, in turn, relies on a woman from the countryside to migrate to the city to care for her children in her absence. Along the chain, care deficits are passed downward in distinctly gendered, racialized and classed manners, with care labor – undervalued and sometimes even stigmatized – hierarchically extracted from the periphery to benefit those in the core (Raghuram, Bornat and Henry, 2011).

Yet, while migrants employed in various types of care work ameliorate care deficits in destination countries at the cost of care provision in their home countries, the phenomenon cannot necessarily be understood as neatly unidirectional care-resource extraction (i.e., brain drain) (Raghuram, Madge and Noxolo, 2009). In addition to remittances, migrant care workers can also usher in processes of brain circulation and other types of tacit knowledge transfer through their physical return or by developing long-distance collaboration (Connell and Walton-Roberts, 2016). Some sending countries are taking notice of their population's global care-labor potential and seeking to harness the benefits. In the Philippines, for instance, where 10% of the country's population works abroad, nurses are trained specifically for export as part of the government's national development strategy (McKay, 2016).

Increased domestic and transnational mobility, the literature points out, may significantly reconfigure the family- and community-centered networks at the core of care arrangements. In the case of elder care, for instance, transnational mobility challenges the meaning and feasibility of popular concepts like aging-in-place and community care, since notions of place and community become decoupled from geographic location (Milligan and Wiles, 2010). In acknowledging the diverse and overlapping forms that care assumes (e.g., practical, financial, emotional, embodied), scholarship on migration and transnational families (e.g., Baldassar, Vellekoop-Baldock and Wilding, 2007; Parreñas, 2005) has been especially useful for highlighting the distinctions delineated by ethicist Joan Tronto (1993) between caring *about*, which involves recognition of care needs; caring *for*, which encompasses assuming responsibility to ensure care needs are met; and care*giving*, which entails physical, hands-on care work. Migrant workers' long-distance family-oriented caring *about* and caring *for* practices, for example, may improve the health status of their children, aging parents and alternative caregivers through frequent phone calls and remittances that enable access to better food, housing and health care in the places in which they live. Likewise, migrants are not only *providers* of care but also themselves *recipients*, entwined in global networks through which multidirectional and multifaceted care flows also support and sustain *them* (McKay, 2016). By challenging widely held notions that physical proximity is requisite for providing *good* care, transnational mobilities not only reconfigure the spatialities of care but also urge us to re-examine what constitutes care in the first place.

Transnationalizing the pursuit of medical care

Spurred by perceived medical-care deficits in their countries of residence, more and more people are engaging in temporary and circular movements abroad for the satisfaction of their medical needs. This phenomenon is popularly, if inadequately and inappropriately, dubbed "medical tourism" (Connell, 2011). Taking advantage of the international patchwork of regulation and currency-exchange differentials (Pennings, 2002), transnational care seekers – most of whom hail from lower- and middle-income countries (Ormond and Sulianti, 2017) – use the global market to circumvent national regulations and costs in order to access care, treatments and procedures that are not available to them in their own health systems. Commonly sought treatments include chemotherapy, organ transplantations and fertility treatments that are inexistent, are in limited supply or face legal restrictions on qualification in care seekers' habitual countries of residence (Inhorn and Patrizio, 2009; Kangas, 2002; Scheper-Hughes, 2001). Other reasons that compel the transnational movements of patients include concerns about accessing good-quality pharmaceuticals and diagnostic testing (Ormond and Sulianti, 2017); the desire to take advantage of experimental treatments like cutting-edge surgical procedures and stem-cell therapies (Petersen, Seear and Munsie, 2014); the wish to avoid moral polemics certain practices, such as commercial surrogacy, abortion and euthanasia, may provoke in care seekers' places of residence (Cohen, 2014; Pande, 2010); or the prohibitive domestic costs of routine and elective procedures such as joint replacements and cosmetic surgery (Bell et al., 2011; Connell, 2011).

Many national governments – having already marketized various components of their health and social systems – have identified transnational medical care to be a lucrative engine of national economic growth. This is facilitated by a range of factors. First, the epidemiological transition in many lower- and middle-income countries, coupled with demand by their rapidly growing middle classes and the proliferation of private hospitals, clinics and imaging centers, have led to a global dissemination of specialized biomedical knowledge and equipment (Bookman and Bookman, 2007). This provides crucial socio-technological infrastructure for the development of transnational medical care seeking. Second, free-trade agreements, particularly the General Agreement on Trade in Services (GATS), have worked to eliminate national barriers to the transnational movement of SHWs, health consumers, telemedicine and foreign direct investment in the care sector (Chanda, 2002), thus providing critical legal infrastructure for transnationalization.

The development of a medical-tourism industry, in turn, significantly impacts the general regulation of health care. State regulatory structures on health and care become subsumed by multifaceted and multiscalar governance structures, with regulation of health and care falling increasingly within the remit of both government bodies concerned with trade and industry and (transnational) private-sector actors. The Malaysian government's desire to attract well-heeled, fee-paying foreign patients, for example, led to the relaxation of national regulations on medical advertising and foreign SHW recruitment as well as the widespread promotion of US-based Joint Commission International accreditation for medical facilities on top of meeting national standards (Ormond, 2013). In the same vein, the Indian, Malaysian, and Thai governments have reduced import tariffs and provided other fiscal incentives to hospitals and clinics acquiring cutting-edge medical equipment and internationalizing their wards, lounges, and menus in order to tout world-class credentials (Crooks et al., 2011; Whittaker and Chee, 2015). Even airlines and airports are learning to adjust their spaces and border procedures to travelers who may be reaching their destinations with complicated, delayed, and poorly managed or poorly diagnosed conditions (Ormond, 2015).

The enthusiastic governmental and private-sector uptake of medical tourism exemplifies how market forces and the spread of biotechnological innovation combine to de-territorialize health and care from within the container of the nation-state. This de-territorialization complicates conventional notions of national citizenship, responsibility and entitlement by advancing the development of what Rose and Novas (2004, p. 132) call *biological citizenship*, in which the recognition of one's corporeality becomes increasingly central to one's individual and collective identities. The subsequent transnational re-territorialization and commoditization of health and care through frames of biological citizenship lead people, first, to see their own bodies as malleable sites of lifestyle and medical intervention (Connell, 2011); second, to see themselves as neoliberal subjects, not as passive patients but rather as proactive consumers of care and intervention (Ormond and Sothern, 2012); and, third, to see a diverse array of places around the globe as sites within which their needs and desires can be met, with scant attention to the health-equity impacts of their individual and collective pursuits on health systems either at home or abroad. Indeed, in the name of both economic and health-sector development, destination countries have come to privileging paying foreigners' medical-care needs and interests through the development of specific innovations and spaces out of reach and inappropriate to the average needs of most local people (Brown, Craddock and Ingram, 2012; Snyder, Dharamsi and Crooks, 2011).

Because of the ways in which transnational health-care seeking upsets conventional notions of citizenship rights and entitlements, it has cast in sharp relief some of the far-reaching yet often invisible effects of advancements in life-science technologies. Scholars using political economy and science, technology and society (STS) studies approaches have called attention to radical inequities in the ways in which modern medicine is accessed and used, particularly in the cases of organ transplantation and commercial surrogacy (Pande, 2010; Scheper-Hughes, 2001). Biotechnical advancements, such as immunosuppressant drugs that keep recipients' bodies from rejecting transplanted organs (Cohen, 2005) and the decoupling of sex and reproduction made possible by *in vitro* fertilization (Greenhough, 2011), for example, have enabled the number of bioavailable people (i.e., those with compatible bodily tissues that can be extracted, as in the case of organ harvesting, or temporarily borrowed in the case of surrogacy) to significantly expand beyond a small circle of close relatives and people with similar genetic backgrounds to much more distant others. As a result, the new supply-side of bioavailability looks very different, largely comprising bodies of the marginalized poor in lower-income countries commodified by those with sufficient purchasing power. Consider, for instance, surrogates repeatedly renting out their wombs in Thailand for foreigners with fertility challenges in order to provide for the needs of their own biological children, or impoverished Filipino men whose chances to get married and start families are reduced by the social stigma of having sold a kidney on the black market to foreigners seeking to circumvent their home-country waiting lists for a much-needed transplant.

Transnationalizing the pursuit of long-term care

A more recent transnational care practice produced by perceived *long-term care* deficits involves temporary, circular and permanent migration for the fulfillment of one's long-term care needs. International retirement migrants (IRMs) have generally moved abroad to maintain and maximize their health and quality of life relative to available resources. In turn, comparatively healthy, affluent, and autonomous seniors (the so-called third age), valued for their purchasing power and their non-competition with locals over jobs and other resources, have been welcomed by destination countries the world over (Toyota and Xiang, 2012). Part of this conditional welcome has hinged on an assumption that, upon becoming physically and mentally frail and economically vulnerable, IRMs return to their countries of origin for support from family and home-country health and welfare institutions instead of depending on the destination country's welfare system.

However, access to health services, resources and caregivers is increasingly recognized as both a motivation for international retirement migration and a growing need as IRMs age. Dissuaded by the prospect of burdening their families, reduced access to social benefits, dwindling savings and prohibitive institutional-care costs in their countries of origin, more IRMs have begun to either remain in their IRM destinations or relocate to such destinations upon reaching a stage of declining health and greater, costlier dependency (the so-called fourth age) (Hall and Hardill, 2016; Toyota, 2006). This permits them to outsource some of their everyday assistance and medical needs abroad and partly relieve their families and governments in their countries of origin of their care roles and costs (Ormond, 2014). Many IRMs may not officially register their presence, preferring instead to strategically circulate between their home and destination countries to circumvent tax regimes and to secure access to social benefits. Because these older migrants fall into varied economic and political-legal categories and are often precariously inserted in diverse sociocultural formations of care and support, they risk marginalization and exclusion from both their home and destination countries' health- and social-care systems (Ackers and Dwyer, 2002).

Just as the movement of care providers unsettles the relations between caring *about*, caring *for* and caregiving, the mobility of long-term care seekers challenges the meaning of care. Physically distant from their previous homes, family members, and national health and social welfare systems, these care seekers are perceived to have special needs, desires, problems and resources, thus constituting a new subject of care. Correspondingly, a variety of transnational care spaces has emerged around long-term care, including luxurious gated communities with advanced medical-care facilities, assisted-living facilities and nursing homes for foreigners, and live-in care arrangements enabling dependent seniors to "age (though abroad) in place" (Toyota, 2006; Ormond, 2014). Transnational long-term care has thus become intertwined with other lucrative business sectors, such as real-estate development and tourism, in ways that merit further study.

Conclusion

The transnationalization of health and long-term care on a global scale is a novel and fast-growing phenomenon. Its study arguably constitutes one of the most dynamic fields in health geography today. Recent scholarship has identified the multiple driving forces behind this development, particularly the demographic shifts, (bio)technological advancements and regulatory changes. The literature has also documented the experiences of a range of actors in the transnationalization of care and examined its multifaceted effects.

A strong common theme that emerges from the most recent literature is the question of inequity. In spite of dramatic economic, social and (bio)technological developments, access to and quality of health and long-term care resources within and between countries remain highly uneven. In fact, deeply entrenched inequities constitute a basis from which transnational care spaces emerge. It is in the market, state, civic, familial and individual responses to the uneven distribution of resources that diverse transnational care practices are produced. In many cases, inequities are intensified, rather than ameliorated, through transnational care practices.

The ways in which inequities are entrenched or transformed are highly complex and dynamic, and many mechanisms are not yet properly understood. The production and reproduction of inequities themselves become highly uneven processes. For instance, transnational care can work out well for those with the socioeconomic and political privilege to more easily deploy their biological citizenship (Rose and Novas, 2004): an underinsured middle-class American man in need of lifesaving heart surgery has dramatically more purchasing power in India than back home in the United States, and a middle-aged woman ineligible for fertility treatment in France because she is considered too old may undergo out-of-pocket treatment in Thailand with no questions asked by authorities. But it can also put those with more limited socioeconomic and political privilege at distinct disadvantage: a nurse born and trained in Nigeria may only be able to find poorly paid work at a residential-care home or in domestic-care services in Germany because her professional credentials go unrecognized in that country, or a Pakistani boy with the same life-threatening heart condition as the American man described above may not be able to enter India with his father for a long-awaited scheduled procedure because recent political tensions have temporarily closed border-crossings.

How can we map the distributions of costs and benefits across places and populations in ways that attend adequately to the nuanced intersections of somatic, socioeconomic and political identities involved in transnational care? We need to better understand how inequities are experienced and reflected on by different actors. Through transnational health and long-term care, many aspects of both care seekers' and care providers' individual attributes, such as gender, ethnicity, class and age, may come to acquire new significance and intersect with each other in new ways. The effects of transnational care, as suggested by such concepts as biological citizenship and bioavailability, is at once intrinsically embodied and socially far-reaching. These effects can be so physical that they cannot be properly captured by health-geography concepts, yet they can span not only space but also time, deeply transforming core understandings of social, economic and biological relationality between countries and across generations. Thus, multidisciplinary inquiries and the development of new theoretical language are urgently needed to make adequate empirical and ethical assessments about the growing range of transnational care practices.

References

Ackers, L. and Dwyer, P. (2002). *Senior citizenship? Retirement, migration and welfare in the European Union.* Bristol: The Policy Press.

Baldassar, L., Vellekoop-Baldock, C. and Wilding, R. (2007). *Families caring across borders: Migration, aging, and transnational caregiving*. New York: Palgrave Macmillan.

Bell, D., Holliday, R., Jones, M., Probyn, E. and Taylor, J. S. (2011). Bikinis and bandages: an itinerary for cosmetic surgery tourism. *Tourist Studies*, 11(2), pp. 139–155.

Bookman, M. Z. and Bookman, K. R. (2007). *Medical tourism in developing countries*. New York: Palgrave Macmillan.

Brown, T., Craddock, S. and Ingram, A. (2012). Critical interventions in global health: governmentality, risk, and assemblage. *Annals of the Association of American Geographers*, 102(5), pp. 1182–1189.

Chanda, R. (2002). Trade in health services. *Bulletin of the World Health Organization*, 80(2), pp. 158–163.

Cohen, I. G. (2014). *Patients with passports: medical tourism, law, and ethics*. Oxford: Oxford University Press.

Cohen, L. (2005). Operability, bioavailability, and exception. In: A. Ong and S. J. Collier, eds., *Global assemblages: technology, politics, and ethics as anthropological problems*. Malden and Oxford: Blackwell, pp. 79–90.

Connell, J. (2011). *Medical tourism*. Wallingford: Cabi.

Connell, J. and Walton-Roberts, M. (2016). What about the workers? The missing geographies of health care. *Progress in Human Geography*, 40(2), pp. 158–176.

Crooks, V. A., Turner, L., Snyder, J., Johnston, R. and Kingsbury, P. (2011). Promoting medical tourism to India: messages, images, and the marketing of international patient travel. *Social Science & Medicine*, 72(5), pp. 726–732.

Gatrell, A. C. (2011). *Mobilities and health*. London: Ashgate.

Greenhough, B. (2011). Citizenship, care and companionship: approaching geographies of health and bioscience. *Progress in Human Geography*, 35(2), pp. 153–171.

Hall, K. and Hardill, I. (2016). Retirement migration, the "other" story: caring for frail elderly British citizens in Spain. *Ageing & Society*, 36(3), pp. 562–585.

Hochschild, A. R. (2000). The nanny chain. *American Prospect*, 11(4), pp. 32–36.

Huang, S., Thang, L. L. and Toyota, M. (2012). Transnational mobilities for care: rethinking the dynamics of care in Asia. *Global Networks*, 12(2), pp. 129–134.

Inhorn, M. C. and Patrizio, P. (2009). Rethinking reproductive "tourism" as reproductive "exile". *Fertility & Sterility*, 92(3), pp. 904–906.

Kangas, B. (2002). Therapeutic itineraries in a global world: Yemenis and their search for biomedical treatment abroad. *Medical Anthropology*, 21(1), pp. 35–78.

McKay, D. (2016). *An archipelago of care: Filipino migrants and global networks*. Bloomington: Indiana University Press.

Milligan, C. and Wiles, J. (2010). Landscapes of care. *Progress in Human Geography*, 34(6), pp. 736–754.

Ormond, M. (2013). *Neoliberal governance and international medical travel in Malaysia*. Abingdon: Routledge.

Ormond, M. (2014). Resorting to Plan J: popular perceptions of Singaporean retirement migration to Johor, Malaysia. *Asian & Pacific Migration Journal*, 23(1), pp. 1–26.

Ormond, M. (2015). En route: transport and embodiment in international medical travel journeys between Indonesia and Malaysia. *Mobilities*, 10(2), pp. 285–303.

Ormond, M. and Sothern, M. (2012). You too, can be an international medical traveler: reading medical travel guidebooks. *Health & Place*, 18(5), pp. 935–941.

Ormond, M. and Sulianti, D. (2017). More than medical tourism: lessons from Indonesia and Malaysia on South-South intra-regional medical travel. *Current Issues in Tourism*, 20(1), pp. 94–110.

Pande, A. (2010). Commercial surrogacy in India: manufacturing a perfect mother-worker. *Signs: Journal of Women in Culture & Society*, 35(4), pp. 969–992.

Parreñas, R. (2005). Long distance intimacy: class, gender and intergenerational relations between mothers and children in Filipino transnational families. *Global Networks*, 5(4), pp. 317–336.

Pennings, G. (2002). Reproductive tourism as moral pluralism in motion. *Journal of Medical Ethics*, 28(6), pp. 337–341.

Petersen, A., Seear, K. and Munsie, M. (2014). Therapeutic journeys: the hopeful travails of stem cell tourists. *Sociology of Health & Illness*, 36(5), pp. 670–685.

Piper, N. and Roces, M. (2004). *Wife or worker? Asian women and migration*. Lanham, MD: Rowman & Littlefield Publishers.

Raghuram, P., Bornat, J. and Henry, L. (2011). The co-marking of aged bodies and migrant bodies: migrant workers' contribution to geriatric medicine in the UK. *Sociology of Health & Illness*, 33(2), pp. 321–335.

Raghuram, P., Madge, C. and Noxolo, P. (2009). Rethinking responsibility and care for a postcolonial world. *Geoforum*, 40(1), pp. 5–13.

Rose, N. and Novas, C. (2004). Biological citizenship. In: A. Ong and S. J. Collier, eds., *Global assemblages: technology, politics and ethics as anthropological problems*. London: Blackwell Publishing, pp. 439–463.

Scheper-Hughes, N. (2001). Commodity fetishism in organs trafficking. *Body & Society*, 7(2–3), pp. 31–62.

Snyder, J., Dharamsi, S. and Crooks, V. A. (2011). Fly-by medical care: conceptualizing the global and local social responsibilities of medical tourists and physician voluntourists. *Globalization & Health*, 7(6), pp. 1–14.

Sparke, M. (2009). Unpacking economism and remapping the terrain of global health. In: A. Kay and O. D. Williams, eds., *Global health governance: crisis, institutions and political economy*. London: Palgrave Macmillan, pp. 131–159.

Toyota, M. (2006). Ageing and transnational householding: Japanese retirees in Southeast Asia. *International Development Planning Review*, 28(4), pp. 515–531.

Toyota, M. and Xiang, B. (2012). The emerging transnational "retirement industry" in Southeast Asia. *International Journal of Sociology & Social Policy*, 32(11/12), pp. 708–719.

Tronto, J. C. (1993). *Moral boundaries: a political argument for an ethic of care*. London: Routledge.

Whittaker, A. and Chee, H. L. (2015). Perceptions of an "international hospital" in Thailand by medical travel patients: cross-cultural tensions in a transnational space. *Social Science & Medicine*, 124, pp. 290–297.

35

THE PLACE OF PRIMARY CARE CLINICS

Robin Kearns

Primary care is generally understood to be the set of health-care activities that take place at the first point of entry to a larger health care system that typically includes secondary (hospitals) and tertiary (specialist) levels. As geographers, we can read the phrase *take place* not only in the conversational sense of occurring, but also more literally as occupying a place – in a system or hierarchy as well as located in specific sites. The general practitioner (GP; also known in some countries as family physician) is the gateway to wider health services for the majority of people in the United Kingdom and those countries coming under its historical influence and is also the cornerstone of primary care practice (Phillips, 1979).

At the outset, we need to distinguish primary care (as the foundational level of a medical system) from primary *health* care (PHC), which grew out of a concern for health for all. This concern was codified in the World Health Organization's Alma Ata Declaration in 1978 that has subsequently been adopted and adapted in various countries (e.g., Ministry of Health, 2001). The declaration that emerged from the Alma Ata conference acknowledged a desire to endorse grassroots approaches to health care rather than universally impose Western models of medicine (Crooks and Andrews, 2009). In its call to address the underlying determinants of ill health, this statement amounted to a radical call to address the injustices embedded in societies and economies as well as the power inherent in medical practice itself. What was sought was *doctors on tap, not on top* (Werner, 1973). However, as is invariably the case with such radical calls, the message of Alma Ata has been diluted over time such that, at least in Western nations, most contemporary expressions of PHC are little more than primary *medical* care, with a few nods to inclusiveness and population diversity. Hence, in this chapter I use the term "primary care clinics" to signal an interest in how these sites and components of health-care systems are organized in contemporary Western societies, influenced by, but not fully expressive of, the ideals of PHC.

Despite the persistence of medical dominance (Willis, 1993), contemporary primary care physicians and teams of allied health-care professionals (e.g., nurses, physiotherapists, occupational therapists) nonetheless hold potential to address issues of health inequity by offering services targeting vulnerable populations. In one sense, all patients potentially experience vulnerability through their expressed state of needing care. Yet at a more searching level, vulnerable populations are those facing discrimination (Kilbourne et al., 2006). This view of vulnerability suggests that its contours are shaped through structural bias: a potent combination of not only ethnic but also socioeconomic or cultural differences.

The remainder of this chapter reviews the significance of the locations where primary care is offered, the meanings attributed to these sites, and methodologies adopted in the course of research. New directions that are being pursued in the quest to better understand aspects of community-based clinics are then identified. The chapter argues that, despite laudable intentions, primary care settings potentially express bias in who and how they serve. This bias, in turn, has implications for members of populations who experience dismissal, discrimination and social exclusion in other aspects of their lives, such as education and employment (Browne et al., 2012). I therefore move in this chapter from surveying the roots of a largely locational concern for primary care to more recent inquiries that arguably have cast a more searching light on marginalizing spaces and practices.

Locating research on access to primary care clinics

"Primary care clinics" is used in this chapter as a stand-in term for both conventional general-practice surgeries and more targeted clinics that are nonetheless a first port of call for patients (e.g., a student health service). Sites such as these have conventionally been considered by medical geographers to be spatial nodes. Reduced to point-form, clinic locations can be mapped and analyzed for their relative size and distance from other relevant sites of care, as well as other variables including the residential coordinates of patients and their socio-demographic characteristics. This spatial-analytic perspective held sway until the 1990s, providing a powerful set of tools for addressing concerns such as medical workforce availability as well as patient accessibility and utilization (Joseph and Phillips, 1984).

Research drawing on this tradition has tended to treat the location of doctors and clinics as fixed in space. This assumption is reasonable in an era in which house calls rarely occur and in which it is incumbent on patients to travel to seek medical care. However, this spatial fixing invariably extends to patients as well, through the tendency for datasets to freeze patients' location in time and space. The collection of census data provides one example of how this analytical immobilizing occurs. Census information is gathered only every few years, yet is used by researchers between cyclical counts as if people still resided where they were last enumerated. Analyses in this tradition have illuminated the role of location in influencing the utilization of clinics by patients, a behavior that has been termed "revealed accessibility"; people reveal how accessible a location is by reaching it, rather than a clinic being merely *potentially* accessible by virtue of, say, proximity (Joseph and Phillips, 1984).

At its core, the idea of access in the primary care context focuses on the degree to which people are able to obtain the health services they need to sustain or improve their well-being. Access most literally involves whether a service is available, and consideration of the concept commonly begins by focusing on distance. However, a straight line between a clinic's location and a patient's residence seldom reflects either the journey traveled or the obstacles encountered. Rather, accessibility is multidimensional. It can include the influences of service quality (*Is a clinic regarded as good enough?*), its temporal availability (*Do opening hours suit?*), affordability (*What is the cost?*), continuity (*Is it reliably available?*), connectivity (*Is the clinic connected to other and often higher-order services?*), and acceptability (*Is the care offered acceptable?*) (Hanlon, 2009; Penchansky and Thomas, 1981). This last notion of acceptability is arguably the most nuanced. Indigenous populations, for instance, may well postpone or completely avoid contact with a clinic, even when it is located within their own community, if traditional beliefs are demeaned or overlooked such that services become culturally unsafe (Wepa, 2005). In Joseph and Phillips' (1984) terms, these foregoing challenges to gaining sought-after primary care services can be regarded as issues of effective accessibility; potential barriers that are less immediately tangible than distance, and the experiential aspects of an environment that ultimately influence people's readiness to utilize a service or amenity.

Understanding the perspectives of primary care patients

Studies of clinic utilization have conventionally employed surveys to establish patterns of attendance. In the tradition of spatial analysis, geographical theory suggests that consumers will travel to the clinic nearest to their residence. However, this tendency has been shown to be a weak indicator of actual attendance patterns.

Hays, Kearns and Moran (1990), for instance, showed that nearby clinics were typically attended, rather than the nearest available location, thus indicating that relative proximity to residence is still generally preferred by patients. The same study showed that greater personal mobility enabled members of higher-income households to travel farther to attend a clinic in an urban setting. Hence, the privilege of personal mobility delivers greater choice. Conversely, those who had less personal mobility were more spatially bounded.

In a rural context characterized by a dispersed population, however, it is not realistic to expect that all necessary services, or even a choice of clinics, be available. Indeed, rural areas have distinctive characteristics in terms of primary care and community clinics. They tend to have smaller, more scattered client bases and are often served by health services that are resource-poor and beholden to centralized policies and bureaucracies (Woods, 2005). Yet connections between patients and providers are invariably characterized by familiarity, and opportunities for collaboration among providers are strong. This situation heightens the imperative for those working in rural clinics to engage in coordinating and establishing effective connectivity between clinical centers at different spatial scales (Hanlon and Kearns, 2016).

The deepening understanding of the dynamics of place that preoccupied geographers in the 1990s, along with attentiveness to the imperatives of the Alma Ata Declaration, brought shifts in thinking that increasingly embraced a more broadly defined *health* rather than medical care per se (Kearns and Moon, 2002). An associated insight was the recognition that community clinics serve more than just medical purposes – these sites are potentially of social significance to their constituent communities of interest. In one of the earliest illustrations of this contention, community clinics in Hokianga, a rural region of northern New Zealand, were shown to act as *de facto* community centers, attracting not only patients (i.e., people with self-assessed reasons to see a medical practitioner) but also community members dropping in for conversations. Further, the character of the conversations noted across all clinics indicated a strong tendency toward topics of community interest (Kearns, 1991). Clearly people need to experience a degree of comfort in a clinic waiting room to socially engage in this manner. Hence, while loyalty to particular physicians has been long-noted (Platonova, Kennedy and Shewchuck, 2016), this observational research helped establish that ideas of place attachment (Manzo and Perkins, 2006) can be extended to the micro-geographies of clinics themselves, especially where there is a high level of at least symbolic community ownership.

Internal geographies of the clinic

The Hokianga study represents an early expression within health geography of a more general qualitative turn within the parent discipline. As this turn has been more fully embraced, there has been a closer consideration of the internal spaces of the clinics themselves. Humanistic thinking, informed by a range of philosophies such as phenomenology, has led geographers to focus on the meanings of places through the application of methods such as observation, in-depth interviews and focus groups (Baxter and Fenton, 2016). In tandem with this embrace of a wider range of methods, geographers influenced by the qualitative turn broadened their attention to consider not only formal sites of primary care, but also other places that held primacy within people's lives and health experience, such as the home and other community settings (e.g., schools, workplaces, churches, community centers). These sites have been increasingly seen as key elements of social infrastructure complicit in the reproduction of community life. As such, not only did some of these sites (e.g., school-based health clinics) have explicitly health-related functions, but also their contribution to fostering healthy communities suggested they are part of the fabric of primary health care, as broadly defined in the Alma Ata Declaration.

Clinics themselves have undergone transformations. First, the mix of practices and practitioners has begun to change. As geographers have noted, there are mounting attempts to integrate complementary and alternative medicine (CAM) into contexts of orthodox medicine (Adams and Andrews, 2012), and primary care clinics are among such shared spaces (Andrews, Wiles and Miller, 2004). Second, we have entered an increasingly digital age of medicine. Developments over recent decades have seen a range of technologies labeled "telehealth" begin to dissolve some of the otherwise tenacious barriers to access (e.g., distance, precipitous terrain) between patients and clinics especially in rural areas (Hanlon and Kearns, 2016). These interventions can facilitate clinical, educational, administrative and research-related services so that populations with otherwise limited access to health-care professionals can benefit from expertise that would otherwise have to be encountered in person (Cutchin, 2002). What is moved in such transactions is information rather than people; hence, in the language of the mobilities paradigm (Hannam, Sheller and Urry, 2006), we can witness a relaxing of the ties of both patients and professionals being moored to a clinic.

Yet, can telehealth take the place of a clinic? Studies that have examined the symbolic significance of a clinic within the fabric of community life would suggest not (Kearns, 1998). Indeed, community clinics can serve as de facto social centers, not only sustaining forms of locally generated care but also indirectly building resilience though generating social capital. Like schools, primary care clinics are sites at which formal delivery of services is invariably complemented by informal relationships, local knowledge and outreach (Hanlon and Kearns, 2016). Hence, both schools and primary care clinics can be regarded as more than simply parts of a community's institutional fabric; they are fundamentally relational places whose importance is most acutely felt when under threat of closure (Witten et al., 2003). Building on this recognition, we can say that clinics are more than the buildings they occupy and the services they provide. They are informally constituted by the social relations they produce and support and potentially give life to the community itself (Kearns and Neuwelt, 2009).

Time and work at the clinic

The qualitative turn in health geography has not only moved a research focus from space to place, but also introduced a greater attention to time. In Western capitalist countries, the dominant discourses around time involve notions of efficiency, with phrases like "time is money" deeply embedded in professional vernacular (Griffiths, 1999). Time within clinics is subjected to close planning with primary care physicians – for instance, generally scheduling appointment times of typically 15 minutes' duration. On the other side of the door to the clinical encounter in the waiting room, time can hang heavy when patients are required to pause on their journey to care.

To make a banal observation, it is patients, not doctors, who spend time in the waiting room. This is not universally the case; Feinsilver (1993) describes Cuban doctors intermittently occupying waiting rooms in order to brief patients on health-promoting practices. In Western capitalist countries, however, the distancing of doctors from this and other aspects of patient experience reflects medical power: the patient awaits the doctor, not the other way around. Drawing on Penchansky and Thomas' (1981) concern for acceptability of care discussed earlier, the waiting room has been acknowledged as an under-examined location in the primary care system. For one set of occupants – receptionists – it is a workplace, where they greet arrivals, respond to phone calls, book appointments and receive payments. For other occupants – patients – it is a space of necessary pause on the journey to a medical consultation, a space in which experiences lead to feeling anxious and self-conscious or feeling comfortable and at home (Neuwelt, Kearns and Browne, 2015). This potential for an experience of comfort in the waiting room can be related to the presence of distractions ranging from magazines, aquariums and artworks to friendly staff (Kearns and Barnett, 1997).

Given that a general practice is the main portal of entry to the health-care system in countries such as Canada, the United Kingdom and New Zealand, it is perhaps understandable that most emphasis has been granted to the spatial behavior and preferences of GPs themselves. Yet the work at the front of a primary care clinic is invariably performed by receptionists, and patients may also commonly see a primary care nurse.

The fact that these roles are mainly occupied by women, whereas in some countries most GPs are men, suggests the need to adopt a gender perspective in understanding the dynamics of primary care clinics and the reproduction of health-care practices more generally (Andrews and Evans, 2008).

More broadly, workforce considerations are critical to the stability of service provision at a community clinic, especially in smaller settlements (Andrews and Evans, 2008). In these settings, stability can be at least partly dependent on the degree to which the aspirations of medical providers themselves and their families are met. In rural communities, for instance, the employment prospects for spouses and educational opportunities for children have both been shown to influence the degree of place attachment experienced by medical professionals working at primary care clinics (Kearns et al., 2006). This complex process of place integration involves establishing a sense of being at home in a place for health-care professionals and reconciling place of residence with personal and professional desires (Cutchin, 1997).

New forms of research engagement

The use of sophisticated statistical analysis and application of geographic information systems (GIS) continues to assist health geographers in examining questions such as the equitable distribution of primary care services with respect to the relative vulnerability of patient populations (Shuurman, 2009). The blossoming of qualitative methodologies has also seen new forms of research engagement with patients themselves. There has been creative engagement in the course of accessing patient perspectives on the significance of clinics in their lives, as well as important reflection on the ethical and practical limitations of doing research in these settings (Baxter and Fenton, 2016). Eggleton, Kearns and Neuwelt (2017), for instance, elicited patient feelings about reception processes through a participatory visual methodology in which participants were invited to craft drawings to evoke their feelings as patients spending time in waiting rooms. This activity provided the framework for conversational interviews in which color, symbols and metaphors were explored through a process of shared analysis of the images that participants created. In the course of this process, participants reflected on ideas that spoke to their experiences of disempowerment, vulnerability and racism.

Conclusion

Examining the locational logic, accessibility and utilization of primary care clinics has long drawn on geographers' skills in the domain of spatial analysis and mapping. More recently, a highly metaphorical form of mapping has offered a potent complement: a tracing of the contours of patient experience through the observation of waiting rooms and close attention to how the physical layout within clinic spaces aligns with patients' aspirations. While medical treatment or advice is invariably the prompt for visits to a primary care clinic, it is not only the quality or affordability of this service that is most likely to help a patient feel at ease. Rather, as some of the foregoing research has signaled, for patients (and especially those with high health and social needs), elements of the journey toward the clinical appointment (pre-clinical care) may be critically important in helping them feel at ease. The friendliness of the receptionist and the cultural appropriateness of the waiting room are aspects that have remained largely invisible in research and service planning. To this extent, Penchansky and Thomas's (1981) attention to acceptability of health care is finally receiving the research attention it warrants. Further research directions will fruitfully see geographers continuing to collaborate across disciplinary boundaries and employ creative forms of engagement in the quest to address equity of access, the cultural acceptability of waiting spaces and the power dynamics inherent in the delivery of primary care.

References

Adams, J. and Andrews, G. J. (2012). *Traditional, complementary and integrative medicine: an international reader.* New York: Palgrave Macmillan.

Andrews, G. and Evans, J. (2008). Understanding the reproduction of health care: towards geographies in health care work. *Progress in Human Geography*, 32, pp. 759–780.

Andrews, G., Wiles, J. and Miller, K. L. (2004). The geography of complementary medicine: perspectives and prospects. *Complementary Therapies in Nursing and Midwifery*, 10(3), pp. 175–185.

Baxter, J. and Fenton, N. (Eds.) (2016). *Practising qualitative methods in health geography*. London: Routledge.

Browne, A., Varcoe, C., Wong, S., Smye, V., Lavoie, J., Littlejohn, D. and Lennox, S. (2012). Closing the health equity gap: evidence-based strategies for primary health care organizations. *International Journal for Equity in Health*, 11(59), pp. 1–15.

Crooks, V. and Andrews, G. (2009). Thinking geographically about primary health care. In: V. Crooks and G. Andrews, eds., *Primary health care: people, practice, place*. Aldershot: Ashgate, pp. 1–20.

Cutchin, M. (1997). Physician retention in rural communities: the perspective of experiential place integration. *Health & Place*, 3, pp. 25–41.

Cutchin, M. (2002). Virtual medical geographies: conceptualizing telemedicine and regionalization. *Progress in Human Geography*, 26, pp. 19–39.

Eggleton, K., Kearns, R. and Neuwelt, P. (2017). Being patient, being vulnerable: exploring experience of general practice waiting rooms through elicited drawings. *Social and Cultural Geography*, 18(7), pp. 971–993.

Feinsilver, J. (1993). *Healing the masses: Cuban health politics at home and abroad*. Oakland: University of California Press.

Griffiths, J. (1999). *Pip Pip: a sideways look at time*. London: Harper Collins.

Hanlon, N. (2009). Access and utilization reconsidered: towards a broader understanding of the spatial ordering of primary health care. In: V. Crooks and G. Andrews, eds., *Primary health care: people, practice, place*. Aldershot: Ashgate, pp. 43–56.

Hanlon, N. and Kearns, R. (2016). Health and rural places. In: M. Shucksmith and D. Brown, eds., *Routledge international handbook of rural studies*. London: Routledge, pp. 62–70.

Hannam, K., Sheller, M. and Urry, J. (2006). Mobilities, immobilities and moorings. *Mobilities*, 1, pp. 1–22.

Hays, S., Kearns, R. A. and Moran, W. (1990). Spatial patterns of attendance at general practitioner services. *Social Science & Medicine*, 31, pp. 773–781.

Joseph, A. E. and Phillips, D. R. (1984). *Accessibility and utilization: geographical perspectives on health care delivery*. London: Harper & Rowe.

Kearns, R. A. (1991). The place of health in the health of place: the case of the Hokianga special medical area. *Social Science & Medicine*, 33, pp. 519–530.

Kearns, R. A. (1998). Going it alone: community resistance to health reforms in Hokianga, New Zealand. In: R. A. Kearns and W. M. Gesler, eds., *Putting health into place: landscape, identity and well-being*. Syracuse, NY: Syracuse University Press, pp. 226–247.

Kearns, R. A. and Barnett, J. R. (1997). Consumerist ideology and the symbolic landscapes of private medicine. *Health & Place*, 3, pp. 171–180.

Kearns, R. A. and Moon, G. (2002). From medical to health geography: theory, novelty and place in a decade of change. *Progress in Human Geography*, 26, pp. 587–607.

Kearns, R. A., Myers, J., Coster, H., Coster, G. and Adair, V. (2006). What makes "place" attractive to overseas-trained doctors in rural New Zealand? *Health & Social Care in the Community*, 14, pp. 532–540.

Kearns, R. A. and Neuwelt, P. (2009). Within and beyond clinics: primary health care and community participation. In: G. Andrews and V. Crooks, eds., *Primary health care: people, practice, place*. Aldershot: Ashgate, pp. 203–220.

Kilbourne, A. M., Switzer, G., Hyman, K., Crowley-Matoka, M. and Fine, M. J. (2006). Advancing health disparities research within the health care system: a conceptual framework. *American Journal of Public Health*, 96, pp. 2113–2121.

Manzo, L. C. and Perkins, D. D. (2006). Finding common ground: the importance of place attachment to community participation and planning. *Journal of Planning Literature*, 20, pp. 335–350.

Ministry of Health. (2001). *The primary health care strategy*. [pdf] Wellington, New Zealand: Ministry of Health. Available at: www.health.govt.nz/publication/primary-health-care-strategy [Accessed 1 Apr. 2017].

Neuwelt, P., Kearns, R. A. and Browne, A. (2015). The place of receptionists in access to primary care: challenges in the space between community and consultation. *Social Science & Medicine*, 133, pp. 287–295.

Penchansky, R. and Thomas, J. (1981). The concept of access: definition and relationship to consumer satisfaction. *Medical Care*, 19, pp. 127–140.

Phillips, D. R. (1979). Spatial variations in attendance at general practitioner services. *Social Science & Medicine*, 13D, pp. 169–181.

Platonova, E., Kennedy, K. N. and Shewchuck, R. M. (2016). Understanding patient satisfaction, trust, and loyalty to primary care physicians. *Medical Care Research and Review*, 65, pp. 696–712.

Shuurman, N. (2009). The effects of population density, physical distance, and socio-demographic vulnerability on access to primary health care in rural and remote British Columbia, Canada. In: V. Crooks and G. Andrews, eds., *Primary health care: people, practice, place.* Aldershot: Ashgate, pp. 57–74.

Wepa, D. (ed.) (2005). *Cultural safety in Aotearoa New Zealand.* 2nd ed. Melbourne: Cambridge University Press.

Werner, D. (1973). *Where there is no doctor: a village health care handbook.* Berkeley: Hesperian Health Guides.

Willis, E. (1993). *Medical dominance.* Melbourne: Allen & Unwin.

Witten, K., Kearns, R., Lewis, N., Coster, H. and McCreanor, T. (2003). Educational restructuring from a community viewpoint: a case study from Invercargill, New Zealand. *Environment and Planning C: Government and Policy*, 21, pp. 203–223.

Woods, M. (2005). *Rural geography: processes, responses and experiences in rural restructuring.* New York: Sage.

36

PALETTES OF PLACE

Green/blue spaces and health

Ronan Foley

The value of the wider environment to human health and flourishing is well established, and human geographers have had a substantive hand in this realization (Gatrell and Elliott, 2015). While environmental health studies have traditionally focused on risk associated with specific damaged or polluted locations (as shown in Section 1 of this volume), the potential of natural and built settings to promote and sustain healthier populations has been studied across a range of subjects, including psychology, anthropology, physiology and sports science (Hartig et al., 2014; Korpela et al., 2008). For qualitative health geographers, much of that research has drawn on historical writing on therapeutic landscapes (Gesler, 1992), and the specific development of that subject is succinctly documented in Williams (2010) and in other sections of this handbook. Other health geographers have preferred to work more directly on person/place interactions within primarily urban settings (Mitchell, 2013; Richardson et al., 2013). In both of these connected strands of health geography, green space – both natural and built – has been a significant focus for that research (Groenewegen et al., 2012; Hartig et al., 2014). In the past decade, the intertwined idea of blue space has been additionally extracted for separate study (Foley and Kistemann, 2015; Völker and Kistemann, 2011). "Green space" is broadly defined as encompassing publicly accessible areas with natural vegetation, such as grass, plants or trees, including spaces like urban parks and wilderness (Lachowycz and Jones, 2013). Blue space broadly refers to a range of natural and artificial spaces associated with water, including lakes, rivers, seas, reservoirs and other pooled waters (Foley and Kistemann, 2015). At the heart of all this research lies an intention to uncover how green/blue spaces act as supportive settings to enable health, recognizing that those settings can also be disruptive (Duff, 2011).

Health geographers interested in green and blue space now routinely discuss an increasing range of spaces and places, natural and built, managed and unmanaged, alongside concerns for how they work (Groenewegen et al., 2012; Lachowycz and Jones, 2013; Foley and Kistemann, 2015). In addition, contemporary empirical work builds on theory to consider the significance of both mobility and stillness and the ways in which places are, simultaneously, active and contemplative settings for the maintenance of physical and mental health (Conradson, 2005; Foley, 2017). While recognizing their, at times, contested natures and uses, Williams (2010) notes the ongoing centrality of therapeutic places and spaces to health promotion and illness prevention as a way of recognizing their importance in public-health policy. There remain nagging issues around evidence, preferred methodologies and the difficulties health geographers sometimes have in identifying *dose* effects associated with nature. In truth, such a narrow focus on health outcomes contradicts much of what contemporary health geography – drawn to mobility, difference, affect and experience – is all about. Nonetheless, this chapter identifies and argues for a number of complementary research strands, driven by a range of approaches and evidence bases that document the value of green/blue spaces to human health. More importantly, there has been, since the turn of the millennium, a visible presence for healthy natures in both cultural and public-health discourse and presents an opportunity for health and medical geographers to show how such spaces matter, and how they, as a sub-group, can contribute meaningfully to that elucidation.

Green space: exploring vitamin G

Green spaces have been studied by health geographers at a range of scales and settings, from national parks and woodlands, down to urban parks, gardens and allotments. While the value and benefits of what has been referred to as *vitamin G* are broadly accepted, they are complex and not always entirely consistent across different subject perspectives (Groenewegen et al., 2012; Lachowycz and Jones, 2013). From other subjects, especially psychology, a range of studies built on psychometric and other instrumental measures have identified improvements in stress reduction, attention restoration and general well-being from being in or viewing natural green spaces (Hartig et al., 2014). Such studies were often experimental and set indoors, especially Ulrich's classic studies in hospital settings that measured better health outcomes for patients facing views of nature than a brick wall (Ulrich, 1984). Arguably, one of the ways in which health geographers have advanced these studies is by bringing in a more *situated* (real-world) spatial component.

One strand of this research has focused on quantitative spatial analysis, wherein the associations between measured *good* access to green space and positive health outcomes have been used as directly quantifiable evidence, often developed via the use of GIS and spatial modeling (Lachowycz and Jones, 2013). In recent years, such work has been especially motivated by a rise in perceived public-health concerns around high levels of societal obesity and overweight. As a result, a focus on obesogenic environments and the ways in which green space might be recruited in urban planning to encourage movement and activity-promoting place/space design are central drivers of health-geography research (Mitchell, 2013; Ward-Thompson, 2011). In Shortt et al. (2014), a focus on environmental justice has taken a socioecological approach, using a national self-reported survey to explore relationships between levels of activity, unhealthy environments and associations with deprivation to show how health inequalities persist at a population level across England. A parallel study using behavioral and psychometric data from the Scottish Health Survey 2008 compared the use of natural environments (woods, parks, paths, beaches) with other environments (home gardens, pavements, gyms, pitches, pools) and found some enhanced mental-health effects for natural spaces, but less clear associations in most other cases (Mitchell, 2013). Importantly, the study identified the need for more work on the specifics of both the activities and the places/spaces to uncover more evident relationships.

While always a suspect distinction, a second strand developed the above scoping work via a more specific place-based qualitative focus. Personal/narrative studies of green space have deepened the focus on experiential and emotional aspects of health and well-being in a range of green spaces (Bell et al., 2014). In addition, those studies have paid attention to specific sub-populations and drawn from approaches and methodologies from other subjects, including psychology, anthropology, health promotion and wider public health. Milligan and Bingley's (2007) work focused on allotments and their value for older people in maintaining both physical and mental health capacities. Both Doughty (2013) and Macpherson (2012) focus on walking as an aspect of active well-being in a range of outdoor spaces. Doughty's study focused on five different walking groups of variable ages, using a *walking-and-talking* methodology to emphasize more active components of therapeutic outdoor spaces to, in turn, uncover a relational well-being in place, framed by in-space encounters. Macpherson (2012) worked outdoors with a group of people with visual impairments to uncover voices of difference and track different experiential outcomes in outdoor green space. Dinnie, Brown and Morris (2013) studied how people used two different types of public parks in Dundee. They identified a range of contested outcomes and uses – inevitable in public spaces – that both promoted and constrained the use of green space. They additionally noted specific local topographies (shape, size, organization of paths) that combined with more topological components (movement through, interactions) to further the findings of both Doughty and Macpherson on an increasingly mobile relational focus on how health and well-being emerge in such spaces to support health education and promotion (Gatrell, 2013). Yet equally, the ongoing power of green spaces to act as sites of rest and restoration, within which stillness, escape and *shift-spaces* from everyday stress and pressure remained important tropes (Plane and Klodawasky, 2013; Finlay et al., 2015). Finlay et al.'s study of a group of formerly homeless women in a park in Ottawa uncovered the significance of a solid place connection for people who had always lived relatively unconnected lives, yet was equally realistic about more unhealthy uses and user populations (drug dealers, thieves) who might occupy the same spaces at different times of the day and especially night.

Lachowycz and Jones (2013) usefully developed a theoretical model that mapped the mechanisms by which exposure/inputs (green-space access) become beneficial health outcomes (physical/psychological), with a particular focus on what they called moderators and mediators. Moderators included factors – such as demography, space characteristics, living contexts and climate – that affected or shaped people's experience and use of green space. Mediators, loosely identifiable as side effects, were associated with presence (knowing the park is there), aesthetic pleasure and individualized use experiences, all of which shaped differential and previously unconsidered benefits.

Such studies, with their focus on how individual and sub-groups use and benefit from parks in quite different ways, document the very relational ways in which green spaces can be approached. While recognizing ongoing difficulties with quantifying dose effects, health geographers are now exploring (using complementary quantitative and qualitative methods) more open and connective sets of possible outcomes, linked to more fully inclusive experiential insights from users and planners and an enhanced focus on external factors (structures, income, opportunity, spatial justice, neighborhood design, exclusion and resilience) that a critical health geography might easily recognize.

Blue space: immersive encounters

Research on green space has always included glimpses of the blue, but spaces associated with water have been extracted recently for separate study. "Blue space" is defined by Foley and Kistemann (2015) as "health-enabling places and spaces, where water is at the centre of a range of environments with identifiable potential for the promotion of human well-being" (p. 157). The idea for a separate study of blue space originated in work on riverine and coastal health but recognized the difficulties in complete separation from green space (Völker and Kistemann, 2011; White et al., 2013). As a result, the term "green/blue space" has gained common usage, with health geographers central to the development of a complementary and parallel set of blue-space studies (Bell et al., 2014, Finlay et al., 2015; Thomas, 2015; Völker and Kistemann, 2011, 2015). That interest in blue space – drawn in part from therapeutic landscape traditions – considers how exactly geographies of water shape and improve health and healing outcomes through developing studies of specific settings and practices (Foley and Kistemann, 2015). An interest in environmental health reflects earlier research and guides a number of city-based studies, in which rivers, lakes and other bodies of water have been productively studied for health gain within the "urban blue" (Völker and Kistemann, 2011). Völker and Kistemann's work picks out both riverine and park locations to explicitly compare public response to and preference for blue over green spaces, though it recognizes that the two are often mingled, as in the case of linear green parks that run alongside the River Rhine in Düsseldorf (Völker and Kistemann, 2011). That blurring across green and blue is also reflected in other urban work. Finlay et al. (2015) carried out an innovative walking-and-talking study on older adults in metropolitan Vancouver that identified improved physical, mental and social well-being associated with access to both types of space as well as directly observed and discussed activities in place. Thomas (2015), in a study of how different women used both green and blue space in Copenhagen, identified similar benefits around exercise and anxiety alleviation, but also more nuanced countereffects for women who were overweight and found such settings anxiety-producing.

As a contrast to this urban work, the coast has also become a site of study. While Collins and Kearns (2007) acknowledged its contested nature, they identify a clear psychological preference value, especially in a country like New Zealand with an extensive and proximal coast. Other studies have used a range of measurable forms, such as censuses, cross-sectional surveys and personal histories. White et al. (2013) noted an enhanced proximity-to-coast effect from self-reported health measures from the 2001 UK census, which they augmented with data from the British Household Panel Survey (BHPS) to identify physical and mental health benefits of living near the coast (though, interestingly, life-satisfaction measures did not show any association). In applying Lachowycz and Jones' (2013) theoretical model, the mediating value of the coast as a site of aesthetic and reflective value emphasizes its ongoing importance for older people, which is especially evident in a study of an older community on Waiheke Island, New Zealand (Coleman and Kearns, 2015), in which the potential of the coast for island-dwellers, both for active visits and for passive views, speaks to differential modes of encounter that produced differential forms of well-being.

In using the sub-heading "immersive encounters," one can uncover subtle ways in which blue space differs from green space. While coastal walking is a good example of a shared person/place interaction, specifically immersive and interactive practices, such as swimming, diving and a range of watersports (kayaking, canoeing, sailing), have become more central to health geographers' work. From earlier research by Collins and Kearns (2007), there was a clear sense of both health-promoting and health-endangering behaviors at the beach. Being active outdoors promoted physical health yet also increased exposure to hot sun and potentially dangerous waves. Foley's (2015, 2017) recent writing on outdoor swimming takes a generally positive slant on its function and value for a range of healthy and unhealthy bodies. It also suggests that, as a relational engagement, swimming is important in terms of shared socialization, mobile outdoor settings and physical engagement across the life course. While some of the work on watersports has been undertaken outside of health geography, Bell et. al's (2014) innovative work on coastal interactions using a range of GPS tracking devices and *in situ* narratives, picks out kayakers as one representative group to see how they negotiate blue space for its affective and well-being value. All of these works present new, "vertically embodied" perspectives – with strong sensory dimensions – from above, on, in and under water that augment the more horizontal visual and active perspectives drawn from green-space research.

Shades, shapes and scales

Although the primary focus of this chapter is on green/blue spaces, therapeutic places come in many shades, shapes and scales. While there are multiple shades in the green and blue themselves, other shadings are also present. For example, browns and grays represent built environment spaces, such as allotments, community gardens and abandoned or vacant plots, that act as valuable interstitial micro-spaces for restoration and well-being (Finlay et al., 2015; Pitt, 2014; Völker and Kistemann, 2015). Duff's (2012) work on enabling spaces for mental health in Melbourne identified other shades – and even shapes – in the shadows and corners, as well as unexpected emergent practices in such settings. One of his respondents identified a wall as an object for practicing handstands, but equally saw that wall as absorptive, capable of "soaking up" anxiety. This role for particular places, as "affective resources for . . . recovery" (p. 1392), tallies well with wider psychological studies of green/blue space, wherein attention restoration, stress reduction and mood enhancement were identified health outcomes (Hartig et al., 2014). Similarly, Houghton and Houghton (2015) call for more study of the blurred boundaries between rural and urban, the *Edgelands* in Richard Mabey's writing, that act to enhance a place-mindfulness within previously unconsidered, even counterintuitive therapeutic settings.

One important and slightly underappreciated aspect of recent health geographers' research has been inno-vative thinking around emplaced methodologies. The shapes of place interactions have been often linear: along streams, rivers and coasts and through forests and gardens; within which different shades of brightness and shadow are identified during those events (Doughty, 2013; Völker and Kistemann, 2011). In using a range of *in situ* approaches, including walk-alongs, visual-elicitation and informant-led wanderings, one can identify relational shifts, emergent from both the time- and space-frames of the interactive conversations and subjects that, in turn, opens up space for voices of difference and more place-responsive narratives of con-nection (Bell et. al., 2015; Coleman and Kearns, 2015; Finlay et al., 2015; Macpherson, 2012). This is evident even in treatment settings, where individual descriptions (written and visual) of imagined therapeutic land-scapes clearly identified green and blue spaces, but with important shadings that saw, for example, lake edges and horizons as both welcoming and disturbing, depending on the dark imaginative shading those individu-als applied to them (Lengen, 2015). Finally, any mention of shape and shade might also include a mention of scale, wherein work by health geographers has operated across scales, from landscape level to very localized affective/active settings of a park bench or a rock by the sea. These more nuanced understandings of multiple forms of green and blue can meaningfully identify shadings of place experience and health outcomes that remain hidden in more aggregated quantitative studies.

Contested palettes

As is to be expected, there has been a strong critical component within much of this work. More clinical perspectives focus on difficulties with measurability in terms of data and causality, a factor fully discussed in Hartig et al. (2014) and Lachowycz and Jones (2013). At the heart of this critique, the identification of spatial associations are not transferable, or, in terms of specific causalities, explainable as solely contextual factors. In addition, finding the right scale of analysis remains an issue for some (Lachowycz and Jones, 2013). Most health geographers would not disagree with this, in the sense that there are multiple and complex factors at play and there will always be problematic scalar and data-matching issues within quantitative and aggregated work. Given the wider difficulties of unpacking composition and context, this clearly connects green/blue health geographies to mainstream medical/health geography work. Nonetheless, the specific focus on tack-ling wider societal "problems" – as classified by public-health initiatives – such as obesity/overweight, and the clear identification in many studies of benefits to physical and mental health, suggest the work has value (Shortt et al., 2014). The focus of much ongoing research is still quantitative, and health geographers can be more creative in using spatial statistics and other forms of modeling that connect health indicators with green/blue spaces to show how geography can augment existing measurable evidence.

More recent work has shifted the focus a little more widely by a fuller validation (even revaluing) of sub-jective experiences, identifying other ways of uncovering scalar effects (Bell et al., 2014). Earlier critiques on therapeutic landscapes research considered the uncritical ways in which place, especially unusual or special places, were incorporated, and this has led to the inclusion of a much broader range of everyday encounters, practices and settings in green/blue space (Williams, 2010). While recognizing the ongoing significance of urban parks, coasts, mountains, wilderness and the generally recuperative potential of nature, a focus on what has been referred to as *nearby nature* has shifted the focus to more mundane and always occupied spaces (Duff, 2012; Foley and Kistemann, 2015). In addition, new methodologies and techniques, such as walk-alongs, GPS, accelerometers and photo/video elicitation interviews, uncover site-specific and *in situ* well-being responses of considerable depth and value from such everyday settings (Bell et al., 2015; Coleman and Kearns, 2015; Finlay et al., 2015; Straughan, 2012; Stewart et al., 2016).

A final critical strand, reflecting the work of Conradson (2005) and Collins and Kearns (2007), identi-fies contested effects and affects; differential outcomes by personality, ability, community, condition and identity affiliation. Here, the ways in which green/blue spaces not only contain light and dark shades – but also, crucially, how those shades of darkness and light are affectively identified and experienced – has shifted thinking in both spatial and, indeed, temporal terms. Studies of urban parks, for example, make very specific distinctions between a positive daytime and negative nighttime affective response in such spaces (Plane and Klodawasky, 2013; Thomas, 2015). Similarly, an emphasis on sub-groups – children, older cohorts, people with disabilities or mental illness – uncovers more differentiated understandings of how and why green spaces become therapeutic (Bell et al., 2014; Duff, 2012). One should note, however, that some of the valuable new technological possibilities for data-gathering, via public participation and active citizen science, can also act as invasive and anxiety-creating tools on bodies and in places. Finally, there is still a tendency for "green space" to have an urban focus, the assumption being that anything rural is by definition all green and/or healthy. Here, too, there are multiple blurred and mobile geographies at play – for example, research on asthma sufferers identified a counterintuitive preference for urban space, given high levels of pollen in the countryside (Edgeley, Pilnick and Clarke, 2011).

New relational geographies of green/blue space

In contemporary studies of green and blue space, one can see an evolution of approaches that provide a more relational understanding of the value of therapeutic places within society at large. Health geographers have been instrumental in making place a visible factor in the measurement of health outcomes. At the same time, they have been central to deeper uncoverings of how experiential, embodied and emotional geographies emerge in green/blue space to maintain and promote health and well-being. While tensions still remain in orientations toward either quantitative or qualitative approaches, there is now a plurality in evidence-based approaches across health geographies – they draw on everything from GIS and correlation coefficients to videos and voices from the water. In combining experimental and experiential approaches, a new relational geographical understanding of green/blue space continues to emerge. The explicitly multiscalar and networked nature of such spaces is another starting point. Ranging in size from tiny (pocket parks) to grand (national parks and reserves), green/blue spaces represent connective patchwork assemblages from lived lives that are also socially produced (Hartig et al., 2014). The design and planning of green and blue spaces remains an important consideration, with much of the work of landscape architecture and urban planning explicitly focused on producing active spaces for active bodies (Stewart et al., 2016).

An exciting direction for this new relational geographical work is to further uncover an enabling practices/spaces approach that blends people and place together; while intriguing new methodologies hint at a combined physiological and narrative shift (Bell et al., 2015; Stewart et al., 2016). We need to know that embodied health improves through exposure to healthy natural spaces, but we also need to know what mechanisms – physical and emotional, direct and managed – affect that process and what part places and spaces play in those mechanisms (Lachowycz and Jones, 2013). Time as well as space is important here, and there is an emergent interest in interactions across the life course, drawing from surveys and spatial mapping as well as new forms of direct monitoring and geo-narrative (Foley, 2017; Pearce et al., 2016). Linked to this, a focus on a *life-course-of-health* perspective suggests a value in new research that considers the relational geographies of preventative, behavioral-management and recovery elements of a range of chronic and acute conditions. Developing the work of health geographers working with specific sub-groups has real potential to uncover how specific conditions and, indeed, disruptive life events (injury, the sudden loss of bodily capacities), in turn, condition place-based health benefits (Doughty, 2013; Macpherson, 2012). What may emerge is not so much individualized medicine as individually and communally emplaced health outcomes across a range of geographies.

Health geographers have always considered the potential critical-policy value of their work. Work on green/blue space clearly overlaps with wider research on the value of nature for human health and flourishing, a concept increasingly framed in public policy under terms such as "ecosystems services" and "natural capital" (O'Brien, Morris and Stewart, 2014). Health geographers are in a strong position to critically engage beyond the subject and draw attention to some of the implicit dangers in framing nature in specifically commodifiable ways, yet they can also carefully co-opt such terms. One way of doing this is to bring a range of spatially attuned methods and approaches into interdisciplinary work that informs public-health policy and that draws attention to wider discussions around inequality, public space and access to healthy spaces. One intriguing area for study is access to private green and blue space. Some of this is picked up in spatial modeling from remotely sensed data, but there have, to date, been almost no small-scale studies that map area-based levels of private green/blue space (gardens or swimming pools) against public space to identify differences in area-based health outcomes. In addition, Duff's (2011) reframing of health-enabling spaces as a set of public resources identifies new ways of measuring their value beyond the economic, highlighting the sense of a sustainable resource for future life courses as well as current and past ones. That concern with value remains crucial – an anecdotal comment by a colleague suggested that *sea* and *view* are the two most valuable conjoined words in the English language. Health geographers are centrally concerned with new valuings of the parks and coasts around us, and framing these as new "resources" or "supports" for health and well-being may be less instrumental than seeing them solely in terms of "service" or "capital." As *third spaces*, the shared and social nature of green/blue spaces are at the heart of their value to health and well-being (Foley, 2017). That multiplicity of uses and users matters. In framing green/blue spaces as relational, the combination of a singular focus and shared intents lends coherence, but much of the joy of green/blue space is in its emergent possibilities – as a contrast from everyday gray spaces and a place for managing both stress and distress – that mark it out as a shift-space, both physical and emotional, that continues to enhance the health of all citizens.

References

Bell, S., Phoenix, C., Lovell, R. and Wheeler, B. (2014). Green space, health and wellbeing: making space for individual agency. *Health & Place*, 30, pp. 287–292.

Bell, S., Phoenix, C., Lovell, R. and Wheeler, B. (2015). Using GPS and geo-narratives: a methodological approach for understanding and situating everyday green space encounters. *Area*, 47(1), pp. 88–96.

Coleman, T. and Kearns, R. (2015). The role of blue spaces in experiencing place, aging and wellbeing: insights from Waiheke Island, New Zealand. *Health & Place*, 35, pp. 206–217.

Collins, D. and Kearns, R. (2007). Ambiguous landscapes: sun, risk and recreation on New Zealand beaches. In: A. Williams, ed., *Therapeutic landscapes*. Farnham: Ashgate, pp. 15–32.

Conradson, D. (2005). Landscape, care and the relational self: therapeutic encounters in rural England. *Health & Place*, 11, pp. 337–348.

Dinnie, E., Brown, K. and Morris, S. (2013). Community, cooperation and conflict: negotiating the social well-being benefits of urban greenspace experiences. *Landscape and Urban Planning*, 112, pp. 1–9.

Doughty, K. (2013). Walking together: the embodied and mobile production of a therapeutic landscape. *Health & Place*, 24, pp. 140–146.

Duff, C. (2011). Networks, resources and agencies: on the character and production of enabling places. *Health & Place*, 17, pp. 149–156.

Duff, C. (2012). Exploring the role of "enabling places" in promoting recovery from mental illness. A qualitative test of a relational model. *Health & Place*, 18, pp. 1388–1395.

Edgeley, A., Pilnick, A. and Clarke, M. (2011). "The air still wasn't good . . . everywhere I went I was surrounded": lay perceptions of air quality and health. *Health Sociology Review*, 20(1), pp. 97–108.

Finlay, J., Franke, T., McKay, H. and Sims-Gould, J. (2015). Therapeutic landscapes and wellbeing in later life: impacts of blue and green spaces for older adults. *Health & Place*, 34, pp. 97–106.

Foley, R. (2017). Swimming as an accretive practice in healthy blue space. *Emotion, Space and Society*, 22, pp. 43–51.

Foley, R. and Kistemann, T. (2015). Blue space geographies: enabling health in place. Introduction to Special Issue on Healthy Blue Space. *Health & Place*, 35, pp. 157–165.

Gatrell, A. (2013). Therapeutic mobilities: walking and "steps" to wellbeing and health. *Health & Place*, 22, pp. 98–106.

Gatrell, A. and Elliott, S. (2015). *Geographies of health: an introduction*. 3rd ed. Chichester: Wiley-Blackwell.

Gesler, W. (1992). Therapeutic landscapes: medical issues in the light of the new cultural geography. *Social Science & Medicine*, 34, pp. 735–746.

Groenewegen, P., van den Berg, A., Maas, J., Verheij, R. and de Vries, S. (2012). Is a green residential environment better for health? If so, why? *Annals of the Association of American Geographers*, 102(5), pp. 996–1003.

Hartig, T., Mitchell, R., de Vries, S. and Frumkin, H. (2014). Nature and health. *Annual Review of Public Health*, 35, pp. 207–228.

Houghton, F. and Houghton, S. (2015). Therapeutic micro-environments in the Edgelands: a thematic analysis of Richard Mabey's *The Unofficial Countryside*. *Social Science & Medicine*, 133, pp. 280–286.

Korpela, K., Ylén, M., Tyrväinen, L. and Silvennoinen, H. (2008). Determinants of restorative experiences in everyday favorite places. *Health & Place*, 14, pp. 636–652.

Lachowycz, K. and Jones, A. P. (2013). Towards a better understanding of the relationship between greenspace and health: development of a theoretical framework. *Landscape and Urban Planning*, 118, pp. 62–69.

Lengen, C. (2015). The effects of colours, shapes and boundaries on perception, emotion and mentalising processes promoting health and wellbeing. *Health & Place*, 35, pp. 166–177.

Macpherson, H. (2012). Guiding visually impaired walking groups: intercorporeal experience and ethical sensibilities. In: M. Paterson and M. Dodge, eds., *Touching space, placing touch*. Aldershot: Ashgate, pp. 131–150.

Milligan, C. and Bingley, A. (2007). Restorative places or scary spaces? The impact of woodland on the mental well-being of young adults. *Health & Place*, 13, pp. 799–811.

Mitchell, R. (2013). Is physical activity in natural environments better for mental health than physical activity in other environments? *Social Science & Medicine*, 91, pp. 130–134.

O'Brien, L., Morris, J. and Stewart, A. (2014). Engaging with peri-urban woodlands in England: the contribution to people's health and well-being and implications for future management. *International Journal of Environmental Research and Public Health*, 11, pp. 6171–6192.

Pearce, J., Shortt, N., Rind, E. and Mitchell, R. (2016). Life course, green space and health: incorporating place into life course epidemiology. *International Journal of Environmental Research and Public Health*, 13(3), pp. 331–342.

Pitt, H. (2014). Therapeutic experiences of community gardens: putting flow in its place. *Health & Place*, 27, pp. 84–91.

Plane, J. and Klodawasky, F. (2013). Neighbourhood amenities and health: examining the significance of a local park. *Social Science & Medicine*, 99, pp. 1–8.

Richardson, E. A., Pearce, J., Mitchell, R. and Kingham, S. (2013). Role of physical activity in the relationship between urban green space and health. *Public Health*, 127(4), pp. 318–324.

Shortt, N., Rind, E., Pearce, J. and Mitchell, R. (2014). Integrating environmental justice and socio-ecological models of health to understand population-level physical activity, *Environment and Planning A*, 46(6), pp. 1479–1495.

Stewart, O. T., Vernez Moudon, A., Fesinmeyer, M., Zhou, C. and Saelens, B. E. (2016). The association between park visitation and physical activity measured with accelerometer, GPS, and travel diary. *Health & Place*, 38, pp. 82–88.

Straughan, E. (2012). Touched by water: the body in scuba diving. *Emotion, Space and Society*, 5, pp. 19–26.

Thomas, F. (2015). The role of natural environments within women's everyday health and wellbeing in Copenhagen, Denmark. *Health & Place*, 35, pp. 187–195.

Ulrich, R. S. (1984). View through a window may influence recovery from surgery. *Science*, 224, pp. 420–421.

Völker, S. and Kistemann, T. (2011). The impact of blue space on human health and well-being – salutogenetic health effects of inland surface waters: a review. *International Journal of Hygiene and Environmental Health*, 214, pp. 449–460.

Völker, S. and Kistemann, T. (2015). Developing the urban blue: comparative health responses to blue and green urban open spaces in Germany. *Health & Place*, 35, pp. 196–205.

Ward-Thompson, C. (2011). Linking landscape and health: the recurring theme. *Landscape and Urban Planning*, 99, pp. 187–195.

White, M., Alcock, I., Wheeler, B. and Depledge, M. (2013). Coastal proximity, health and well-being: results from a longitudinal panel survey. *Health & Place*, 23, pp. 97–103.

Williams, A. (2010). Therapeutic landscapes as health-promoting places. In: T. Brown, S. McLafferty and G. Moon, eds., *A companion to health and medical geography*. Chichester: Wiley-Blackwell, pp. 207–233.

37

URBAN PUBLIC SPACES, SOCIAL INCLUSION AND HEALTH

Karen Witten and Vivienne Ivory

The shared spaces of a city – its parks, squares, libraries and streets – are commonly viewed as "symbols of collective well-being" providing opportunities for people to gather together, to "build sociality and civic engagement out of the encounter between strangers" (Amin, 2008, p. 6). These spaces are recognized as supporting health and well-being through encouraging physical activity (Sallis et al., 2012), and fostering mental health through social engagement, sense of belonging and restorative experiences (Berkman and Kawachi, 2000).

While shared spaces offer much potential for health, the causal pathways to health are complex and contested. This chapter draws on literature from health geography and urban design and planning disciplines to examine the role of inclusion/exclusion in explaining the health-promoting attributes of urban public spaces. Child-friendly public-space design and community resilience through public spaces are presented as illustrative examples. The chapter finishes with research approaches to exploring the pathways between public spaces and health.

Historically, formal public spaces – plazas, squares – have been significant sites of civic life and democratic participation (Low, 2000). Zukin (1995, p. 260) writes:

> Public spaces are important because they are where strangers mingle. . . . As both site and sight, meeting place and social staging ground, public spaces enable us to conceptualise and represent the city – to make an ideology of its receptivity to strangers, tolerance of difference, and opportunities to enter a fully socialized life, both civic and commercial.

While portraying public spaces as the sites where public culture takes shape, through experiences of others and social interaction, Zukin makes clear that not everyone has an equal right to be present in these spaces – meaning there is unequal access to their health-promoting properties – or an equal opportunity to shape "constantly changing public culture" (p. 11). Inequalities of access, rights, recognition and influence reflect the power structures of the wider society (Amin, 2008; Zukin, 1995). Marginalized groups, including children and young people, may struggle to have a legitimate presence in public spaces, deterred by the absence of symbols or affordances that signal they are welcome, or, conversely, the presence of those that indicate they are not. Geographers have a long-standing interest in *rights to the city*, both the benefits of inclusion and participation in community life and the detrimental impacts of exclusionary practices that deny rights to certain groups. People who are homeless and/or experiencing addictions or mental illness have been a particular focus of much of this latter work (DeVerteuil et al., 2007; Mitchell, 2003).

Public spaces come in various forms and have been identified as contributing to mental and physical health, through mechanisms such as sense of belonging and stress reduction (Cattell et al., 2008). Streets are the most ubiquitous form of public space, and their role in the social life of cities and neighborhoods was etched into planning consciousness by Jane Jacobs' (1961) seminal work *The Death and Life of Great American Cities*. Jacobs observed the casual street exchanges of everyday life and articulated the contribution they make, as *eyes on the street*, to interpersonal trust, social cohesion and a sense of belonging to place. Her street advocacy is credited with reinstating streets as valued and vibrant public spaces (Larice and Macdonald, 2007). The importance to well-being of informal, public gathering spaces where people casually mix, converse and develop friendships also underpins the *third place* theorizing of Ray Oldenburg's (1989) *The Great Good Place*. Settings such as libraries, cafes and Laundromats, as well as streets and parks, comprise Oldenburg's third places, so named to distinguish them from first places (home) and second places (work or school) (Larice and Macdonald, 2007). Informal, in-between spaces, such as thresholds of public spaces, commute routes and event sites, have also been identified as potential sites of social interaction between strangers. Simões Aelbrecht (2016) termed these *fourth* places to highlight their increasing importance in providing opportunities for social interaction across diverse groups, but also to explore how they come to be used socially by people alongside their primary purpose (e.g., as transport or retail spaces). Compared with the conversational nature of social activity in third places, Aelbrecht proposed that social activity in fourth places included people watching, waiting, and so forth alongside strangers; interactions that are more casual, shorter and less intense than those in first, second and third places and are suggestive of Granovetter's *weak* ties. The importance to people's health of a range of social interactions and social networks, including those formed through everyday neighborhood exchanges, is well established (Berkman and Kawachi, 2003).

Modest features of the built environment can become gestures of inclusion or exclusion and, in turn, promote or constrain health. Park benches, for example, can be a micro-scale inclusionary feature contributing to the health of older people. As resting places, they help maintain mobility, provide places to meet and interact, and enhance overall use and enjoyment of the public realm (Ottoni et al., 2016). Conversely, seats and ledges designed with spikes or sloping surfaces to discourage skateboarding or street sleeping are micro-scale exclusionary features that undermine the health and well-being of already marginalized groups. Aptly named "hostile" or "defensive" architecture (Andreou, 2015), these measures often target homeless individuals – a vulnerable group with no legitimate right to belong, no *place-in-the-world* (Mitchell, 2003).

The design of public space is the purview of architects, urban designers and planners. Jan Gehl, a Danish architect, identifies three dimensions essential for promoting positive civic relations: a meeting place, a democratic place, and a friendly and safe place (Gehl, 2010). Consistent with Gehl's approach, reclaiming the streets for walking and cycling to enliven cities and reduce car dependency and the "greening" of city centers are gaining traction as urban-renewal strategies. Implications of these strategies for health are considerable in light of climate change, declining levels of physical activity and surging rates of obesity in many countries. Jannette Sadik-Khan (2013), as transport commissioner in New York, was a leader in the use of tactical urbanism to rally public support for reallocating road space for pedestrians and cyclists and changing street culture. Festivals and other events that make a temporary claim on public space – for example, gay-pride parades – can, over time, establish a history of spatial rights to the city.

Design features contribute to how public spaces are perceived and sensed; and whether a space feels safe and welcoming, or foreboding, determines its use. Crime prevention through environmental design (CPTED) is an urban-design approach for reducing the incidence of crime and improving perceptions of safety. CPTED involves the application of design principles such as passive surveillance – eyes on the street – natural access control, and layout, orientation and wayfinding strategies (Ministry of Justice, 2005). The thinking is that a space sensed as unsafe will not be well used, whereas flexibility or looseness about a space can prompt inclusive social interaction (Simões Aelbrecht, 2016) by providing places for people to linger safely among strangers.

Children can display a heightened awareness of the affective atmosphere of spaces in ways that influence whether they advance and explore or instead retreat to safety. By way of illustration, in one of our studies in Auckland, New Zealand, children spoke of discomfort and fear in the presence of intoxicated and homeless people in an inner-city street. They were alert to bodily odors and the potential of encounter, and they mentally rehearsed safe exit routes. They developed tactics of avoidance; they did not mingle in the presence of difference (Witten, Kearns and Carroll, 2015).

For all population groups, it is important that cultural values be recognized and affirmed to help engender a sense of belonging and membership in society. Sense of belonging and sense of place are concepts with currency in health geography (Relph, 1976). Sense of belonging implies feeling accepted; social inclusion is a necessary but not sufficient precursor to belonging. Sense of place entails an affective and meaningful connection to a particular setting. In the New Zealand context, Māori architects and designers have developed Te Aranga Principles to support and advocate for the incorporation of indigenous Māori cultural values and practices through urban design. They offer practical guidance to design professionals and decision-makers for achieving "a widely held desire to enhance mana whenua presence, visibility and participation in the design of the physical realm" (Auckland Council, 2016) (*mana whenua* refers to authority over land). The principles have a strong relational base, and establishing partnering relationships with Māori of the local area is a necessary antecedent to the application of all other principles. The integration of visual and historical cultural markers that affirm Māori cultural values are a probable outcome when Te Aranga principles are applied in the built environment.

Children, young people and public space

Despite children's marginal status in the planning area, public spaces – streets, footpaths and verges – remain important to children's everyday play and mobility, as well as their health and well-being. However, over recent decades, children's rights to the city – to walk, cycle, play and have a public presence – have been undermined by automobility. Sheller and Urry (2003) describe the way in which pedestrians, cyclists and public-transport users become marginalized as a *blurring* of public and private space by a "civil society of automobility" transforms "once public space into road space" (p. 115). Where cars are dominant, parents' fears for children's safety intensify and children become invisible, retreating from the public realm to the domestic spaces of home (Karsten, 2005; Freeman and Tranter, 2011). Neighborhood streets become off-limits, and outdoor play becomes increasingly confined to designated settings such as skate parks and playgrounds. This signals a remarkable shift in children's rights to the city: a century ago, children and adults had a comparable presence in city streets and public spaces (Simpson, 1997; Whitzman, Worthington and Mizrachi, 2010). To draw on Soja's (2010, p. 4) exploration of social justice, increasing urbanization has intensified *struggles over geography* (with implications for spatial justice), and children and young people have been among those to lose out.

Children's enforced retreat from the public realm has had significant implications for their health and development. Considerable attention has been given to declining rates of independent mobility and physical activity in children, now reported in a number of countries (Fyhri et al., 2011). But this retreat from the street also has implications for children's social and cognitive development, as playing and socializing in public settings allow children to learn to negotiate relationships with others (Carroll et al., 2015). They observe and engage, or distance themselves from, the people they encounter; learn to read the street; assess risk; gain confidence and make decisions independently (Cahill, 2000; Witten, Kearns and Carroll, 2015). An apt response to the needs of children and a child-friendly city is Latham and Loch's call for geographers to attend more to the "generative capacities of public spaces and to the material and practical affordances they can offer" (p. 516).

Freyberg square: a case study of children as urban-design consultants

In recent years, Auckland's inner city has re-urbanized. Apartment blocks have been constructed, drawing new residents, including families with children, into the central business district (CBD). As the number of households has risen, pressure on public space has increased, and the absence of suitable play space for children has become apparent. Freyberg Square, in the heart of the CBD, was scheduled for redevelopment. With aspirations to acquire Child-friendly City accreditation, Auckland Council staff saw the redevelopment as an opportunity to involve children in a participatory design process. A child-friendly audit was facilitated by the first author and colleagues in partnership with Council staff. A reference group of children, assembled for the audit/consultation, engaged in on-site and off-site workshops with Council staff and researchers. The children, aged 7–13 years, explored all corners of the square and investigated its play potential. Photographs were taken and annotated with likes and dislikes, proposed plans were critiqued and imaginative ideas for how the square could be developed as an exciting and sociable child-friendly space were generated. Children's feedback and suggestions led to a number of design changes, such as the incorporation of an adventure/scavenger trail, trees for climbing, and a water feature that could be touched by children in wheelchairs.

Not all of the children's recommendations were taken up. For example, a lane closure sought by the children and the designer was overruled based on objections from local retailers. Nevertheless the audit/consultation had successes: the children had fun and felt heard; the urban designer's understanding of what was needed to make a public space child-friendly was transformed from a concern about safety barriers and constraints to creating affordances and play opportunities; and, at an organizational level, there is now an intention to incorporate children's audits into design briefs for future redevelopment projects (Bishop and Corkery, 2017; Carroll and Witten, 2017).

Public space and community resilience

The contribution of public space to the resilience of communities to shocks and stresses is increasingly recognized and has been firmly established in the city of Christchurch following the devastating 2010–2011 Canterbury Earthquake Sequence. Many buildings were demolished across the city, and, as a consequence of the scale of damage, ongoing seismic activity and complex insurance arrangements have seen large numbers of sites left vacant for a number of years while rebuild efforts were being planned (or not, in other cases). Recognizing the opportunity to use transitional spaces to aid recovery, bottom-up initiatives such as Gap Filler (2018), Greening the Rubble (2017), and Life in Vacant Spaces (2017) largely supported by local authorities and official recovery agencies) have developed projects and events to enhance social support through engagement in public spaces (Brand and Nicholson, 2016). Such initiatives have "started to re-engage Cantabrians with their public realm" (p. 173), thus ensuring the city's spaces are "a vehicle for the life and interaction of its citizens" (p. 173), not simply sites of exclusive property ownership and economic productivity awaiting development.

Allen et al. (2016) point out that public open spaces can play a critical role in disaster response and recovery because they are familiar, available and used by the public. Public places that promote everyday social interaction (and, as discussed above, health and well-being) can strengthen the ability of a community to continue functioning post-event. They allow people to come together for informal support rather than waiting in isolation and to respond to the crisis and plan for the future (Allen et al., 2016). Learning lessons on the value of open space for recovery from the 1906 earthquake in San Francisco (1906) and the 2010 one in Conception, Chile, Allen et al. (2016) proposed that resilience through public spaces is facilitated where normal use, before the disaster event, contributed to a community's social capital and spaces and where spaces are familiar and comfortable to people. The importance of pre-event planning is also illustrated by the Blue Line projects (Wellington Region Emergency Management, 2017) in which lines are painted on roads to show safe evacuation zones in the event of a tsunami, with the routes selected based on their common usage along with practicalities such as elevation and topography. To increase the familiarity of routes, additional public spaces such as pocket parks, benches and walkways are planned in strategic locations to foster everyday social interactions and, therefore, social capital and community well-being (Allen et al., 2016).

Research approaches

The disciplines of urban planning and design, health geography and epidemiology all have research histories investigating the relationship between public spaces, social inclusion and health, and each discipline is characterized by methodological diversity. We describe three more common approaches: socioecological frameworks using multilevel modeling (e.g., Francis et al., 2012; Sallis et al., 2012); experiential approaches investigating the meaning of public spaces (Carroll et al., 2015; Ivory et al., 2015); and case studies using direct observation of public spaces *in use* (Koch and Latham, 2012; Low, 2000).

Multilevel modeling combined with uptake of geographical information systems (GIS) technology has become a vital tool for researchers wanting to empirically explore place-health mechanisms suggested by socioecological frameworks. In particular, it allows for more robust statistical examination of the relative contribution to health outcomes of individual and/or household factors and, using GIS, neighborhood or small-area factors. Walkability, and the urban design attributes that sustain physical activity and active travel, has been a topic of intense research attention over recent years. Studies applying socioecological frameworks and statistical modeling have established that street connectivity, among other urban design features, creates more walkable neighborhoods and is associated with increased physical activity (Witten et al., 2012). Perhaps due to greater conceptual and methodological challenges quantifying social aspects of neighborhoods and places, studies linking attributes of public spaces to social, health-related outcomes using similar methods are less common. However, as part of the unpacking of complex relationships between public space and health, an Australian study investigated determinants of sense of community and found a significant relationship between (a) the perceived quality of public open spaces and shops and (b) participants' sense of community (Francis et al., 2012).

Research that focuses on the experience of place is helping unpack the *how* and *why* of the place-health relationship. Ivory et al. (2015) used thematic analysis of focus-group interviews to capture how being active in public places as part of everyday life shaped people's sense of place and well-being. Residents demonstrated agency in seeking out good-quality public places that were safe, seemed pleasant and engendered a sense of belonging and identity (in keeping with Oldenburg's third places). The act of regularly being active in local streets and parks seemed to create a sense of familiarity and Heidegger's [1982] *nearness* "[arising] through experiences of amicability, congeniality, harmony and pleasantness: a warm and welcome affective connection between people and places where everyday domestic lives intersect" (Ivory et al., 2015, p. 320). Such familiarity and socialness not only acted as a driver to be physically active, but also was regarded as a benefit in itself by participants.

Perhaps not surprisingly, what works for many adults also works for children. Carroll et al. (2015) used *go along* interviews and photovoice with children living in both suburban and inner-city settings. Talking while walking allowed children to share with researchers how and why public places became significant to them. Children's personal histories and identities were forged through their repeated presence in locations during their play, commute to school, visits to local shops and so on.

Experiential approaches can also help us understand lack of engagement with public places. A lack of easy access to safe streets and quality public places (because of a less walkable urban form), and/or constraints on time available to spend in them (because of parental restrictions or, for adults, lengthy working hours), can hinder development of a sense of place and, for children in particular, perhaps, engender a sense of distrust in both places and the people (strangers) in them (Carroll et al., 2015; Ivory et al., 2015).

Case studies using direct observation have been a common research approach in urban design. Observational methods have included activity mapping, time-lapse photography, film and pedestrian-path analysis, as well as interviews with users (Koch and Latham, 2012; Larice and Macdonald, 2007; Low, 2000). Intangible attributes of public spaces, such as human scale (e.g., designed for people rather than cars) are seldom objectively measured. However, new methods are evolving. For example, Ewing and Clemente (2013) have drawn on constructs from architecture, urban design and planning to develop operational definitions and tools for measuring imageability, visual enclosure, human scale, transparency and complexity. Emerging technology is also providing new ways for researchers not only to make observations at the micro-scale (e.g., sentiment analysis about a place from social media or information elicited through public participatory GIS methods), but also to see broader patterns – for example, flows of pedestrians in and out of public spaces, along routes, and across settlements (You and Tuncer, 2016). Agendas such as Smart Cities and Sensing Cities are driving innovation among designers and decision-makers to use data generated from infrastructure usage to "listen" more closely to what matters for a city's residents and therefore provide services that meet their diverse needs and wants. In many ways, city operators and researchers are addressing the same issues; technology is providing both with data that can provide broad and rich insights.

Conclusions

Global and local mobility is increasing the diversity of cities and the groups of people inhabiting urban neighborhoods. Planning and designing public spaces that are welcoming, offer safety and conviviality to people of all age and ethnic groups, and encourage positive encounters and interaction between them is enormously challenging. Children's experience and use of neighborhood streets and public spaces has been at the fore in this chapter, but just as relevant would have been a focus on older and/or disabled people. Thoughtful, people-friendly public-space design is needed to foster community participation and the health and well-being of all. The chapter also touched on the importance of formal and informal public spaces for community resiliency for bringing residents together in a post-disaster context. Extreme weather events predicted as a consequence of global climate change may well place high demands on the public spaces in our cities to engage citizens and bolster the resiliency and health of communities in future.

Space and place are core concepts in geography and in health geography. They are also central to urban planning, as reflected in the pithy definition of "planning" used by the Royal Town Planning Institute: *mediation of space; making of place* (RTPI, 2017). Geographers tend to have a healthy interest in the theories and methodologies of their own and other disciplines. In working with architects and urban designers and planners – and embracing the concepts and methods of these disciplines alongside those of geography – to explore and understand how physical, social and affective attributes of cities can be generative of new relational possibilities and better population health, health geographers are likely to develop new and rewarding forms of inquiry.

References

Allen, L., Allan, P., Bryant, M., Becker, J., Johnston, D. and Saunders, W. (2016). *Design for resilience*. Christchurch, New Zealand: New Zealand Society for Earthquake Engineering.

Amin, A. (2008). Collective culture and urban public space. *City*, 12(1), pp. 5–24.

Andreou, A. (2015). Anti-homeless spikes: "Sleeping rough opened my eyes to the city's barbed cruelty." [online] *The Guardian*. Available at: www.theguardian.com/society/2015/feb/18/defensive-architecture-keeps-poverty-undeen-and-makes-us-more-hostile [Accessed 28 Mar. 2017].

Auckland Council. (2016). *Auckland design manual: Te Aranga principles*. [online]. Available at: www.aucklanddesignmanual. co.nz/ [Accessed 28 Mar. 2017].

Berkman, L. and Kawachi, I. (2000). *Social epidemiology*. New York: Oxford University Press.

Berkman, L. and Kawachi, I. (2003). *Neighborhoods and health*. Oxford: Oxford University Press.

Bishop, K. and Corkery, L. (eds.) (2017). *Designing cities with children and young people: beyond playgrounds and skate parks*. New York: Taylor & Francis.

Brand, D. and Nicholson, H. (2016). Public space and recovery: learning from post-earthquake Christchurch. *Journal of Urban Design*, 21(2), pp. 159–176.

Cahill, C. (2000). Street literacy: urban teenagers' strategies for negotiating their neighbourhood. *Journal of Youth Studies*, 3(3), pp. 251–277.

Carroll, P. and Witten, K. (2017). Children as urban design consultants: a children's audit of a central city square in Auckland, Aotearoa/New Zealand. In: K. Bishop and L. Corkery, eds., *Designing cities with children and young people: beyond playgrounds and skate parks*. Abingdon and Oxon: Routledge, pp. 105–119.

Carroll, P., Witten, K., Kearns, R. and Donovan, P. (2015). Kids in the City: children's use and experiences of urban neighbourhoods in Auckland, New Zealand. *Journal of Urban Design*, 20(4), pp. 417–436.

Cattell, V., Dines, N., Gesler, W. and Curtis, S. (2008). Mingling, observing and lingering: everyday public spaces and their implications for well-being and social relations. *Health & Place*, 14, pp. 544–561.

DeVerteuil, G., Hinds, A., Lix, L., Walker, J., Robinson, R. and Roos, L. (2007). Mental health and the city: intra-urban mobility among individuals with schizophrenia. *Health & Place*, 13(2), pp. 310–323.

Ewing, R. and Clemente, O. (2013). *Measuring urban design: metrics for livable places*. Washington, DC: Island Press.

Francis, J., Giles-Corti, B., Wood, L. and Knuiman, M. (2012). Creating sense of community: the role of public space. *Journal of Environmental Psychology*, 32, pp. 401–409.

Freeman, C. and Tranter, P. (2011). *Children and their urban environment: changing worlds*. London: Earthscan.

Fyhri, A., Hjorthol, R., Mackett, R., Fotel, T. and Kytta, M. (2011). Children's active travel and independent mobility in four countries: development, social contributing trends and measures. *Transport Policy*, 18(5), pp. 703–710.

Gap Filler. (2018). About Gap Filler. [online] Available at: http://gapfiller.org.nz/about/ [accessed 9 March 2018].

Gehl, J. (2010). *Cities for people*. Washington, DC: Island Press.

Greening the Rubble. (2017). Current Projects. [online] Available at: http://greeningtherubble.org.nz/ [accessed 9 March 2018].

Ivory, V., Russell, M., Hooper, C., Witten, K., Pearce, J. and Blakely, T. (2015). What shape is your neighbourhood? Investigating where people go for physical activity. *Social Science & Medicine*, 133, pp. 313–321.

Jacobs, J. (1961). *The death and life of great American cities*. London: Jonathan Cape.

Karsten, L. (2005). It all used to be better? Different generations on continuity and change in urban children's daily use of space. *Children's Geographies*, 3(3), pp. 275–290.

Koch, R. and Latham, A. (2012). Rethinking urban public space: accounts from a junction in West London. *Transactions of the Institute of British Geographers*, 37, pp. 515–529.

Larice, M. and Macdonald, E. (eds.) (2007). *The urban design reader*. London: Routledge.

Life in Vacant Spaces. (2017). Site brokers for creative Christchurch. [online] Available at: http://livs.org.nz/home/ [accessed 9 March 2017].

Low, S. (2000). *On the Plaza: the politics of public space and culture*. Austin: University of Texas Press.

Ministry of Justice. (2005). Part 1. Seven qualities of safer places. In: *National guidelines for crime prevention through environmental design in New Zealand*. Wellington: Ministry of Justice.

Mitchell, D. (2003). *The right to the city: social justice and the fight for public space*. New York: Guilford Press.

Oldenburg, R. (1989). *The great good place: cafes, coffee shops, bookstores, bars, hair salons and other hangouts at the heart of a community*. New York: Marlowe & Co.

Ottoni, C., Sims-Gould, J., Winters, M., Heijnen, M. and McKay, H. (2016). "Benches become like porches": built and social environment influences on older adults' experiences of mobility and well-being. *Social Science & Medicine*, 169, pp. 33–41.

Relph, E. (1976). *Place and placelessness*. London: Pion Limited.

RTPI. (2017). *Mediation of space – making of place*. [online] Royal Town Planning Institute. Available at: www.rtpi.org.uk/ [Accessed May 2017].

Sadik-Khan, J. (2013). New York's streets? Not so mean anymore. [online] *TED Talks*. Available at: www.ted.com/talks/janette_sadik_khan_new_york_s_streets_not_so_mean_any_more [Accessed 24 May 2017].

Sallis, J., Floyd, M., Rodriguez, D. and Saelens, B. (2012). Role of built environments in physical activity, obesity, and cardiovascular disease. *Circulation*, 125, pp. 729–737.

Sheller, M. and Urry, J. (2003). Mobile transformations of "public" and "private" life. *Theory, Culture and Society*, 20(3), pp. 115–133.

Simões Aelbrecht, P. (2016). "Fourth places": the contemporary public settings for informal social interaction among strangers. *Journal of Urban Design*, 21, pp. 124–152.

Simpson, B. (1997). Towards the participation of children and young people in urban planning and design. *Urban Studies*, 34(5/6), pp. 907–925.

Soja, E. (2010). *Seeking spatial justice*. Minneapolis: University of Minnesota Press.

Wellington Region Emergency Management. (2017). Blue Lines – Tsunami Safety Zones. [online] Available at: https://wremo.nz/about-us/initiatives/blue-lines/ [accessed 9 March 2018].

Whitzman, C., Worthington, M. and Mizrachi, D. (2010). The journey and the destination matter: child-friendly cities and children's right to the city. *Built Environment*, 6(4), pp. 473–486.

Witten, K., Blakely, T., Bagheri, N., Badland, H., Ivory, V., Pearce, J., Mavoa, S., Hinckson, E. and Schofield, G. (2012). Neighborhood built environment and transport and leisure physical activity: findings using objective exposure and outcome measures in New Zealand. *Environmental Health Perspectives*, 120(7), pp. 971–977.

Witten, K., Kearns, R. and Carroll, P. (2015). Urban inclusion as wellbeing; exploring children's accounts of confronting diversity on inner city streets. *Social Science & Medicine*, 133, pp. 349–357.

You, L. and Tuncer, B. (2016). *Exploring the utilization of places through a scalable "Activities in Places" analysis mechanism*. Big Data, 2016 IEEE International Conference, IEEE, pp. 3563–3572.

Zukin, S. (1995). *The cultures of cities*. Cambridge, MA: Blackwell.

38

RURAL PLACES AND SPACES OF HEALTH AND HEALTH CARE

Rachel Herron and Mark Skinner

The study of rural places in health geography is essential to understanding variation in health services, outcomes and experiences. Rural places are distinct in their own right; they are distinct from one another, and they are distinct when compared to urban and metropolitan settings. In general, rural places share particular challenges related to health, including distance and isolation from larger service centers, small populations with specific health needs, lack of secondary and tertiary services, difficulties recruiting and retaining a health workforce and poorer health outcomes and determinants of health (e.g., lower levels of education and income) when compared to their more urban counterparts at the aggregate level (DesMeules et al., 2012). More qualitatively, the health and well-being of people in rural places are shaped by the intersection of particular economic activities, political decisions and sociocultural norms related to gender, race, age and class (Panelli et al., 2009). Looking at these broader trends and the complex forces propelling them, scholars have long suggested that rural places are in need of better health-care services, as well as research attention, advocacy and targeted policies and programs (Gesler and Ricketts, 1992; Hanlon and Kearns, 2017; Kulig and Williams, 2012).

In this chapter, we build a case for the importance of rural places and spaces in the study of health and health care. We outline changing conceptualizations of rural and discuss the major philosophical and methodological shifts in the study of rural health. In doing so, we recognize the contested nature of rurality, the diversity of rural people and places and their vulnerability and resilience. Looking to the future, we call on geographers to engage with intersectional and relational approaches to unpack the processes influencing distinct patterns of rural health and well-being.

Conceptualizing rural

Although rural places have long been both the subject and the context of health-geography research (Gesler and Ricketts, 1992), there is no universal definition of "rural." Rural is a complex socially and politically constructed category with many associated subcategories, including rural and small town, remote and Indigenous communities. Although rurality is a social construction, this should not be confused with a lack of material basis. Historically, and following a positivist tradition, "rural" has been defined based on various statistical indicators (e.g., population size, density, distance to major urban centers) and economic activities (e.g., agriculture, resource development). Although these measures provide a concrete basis for defining rurality for policy, program and infrastructure purposes, they vary significantly from country to country. For example, the Canadian census defines "rural" as a population of less than 1,000, whereas the census definition for "rural" in the United States is a maximum population of 2,500 people. In a global context, the differences between urban and rural also reflect more than differences in the concentration of people in a given area; they reflect significant differences in access to resources (e.g., electricity, piped water, education, health care). Following transformations in the academy (i.e., the postmodern challenge, the cultural turn, growing acceptance of qualitative and ethnographic methods) and changing conditions in rural places themselves, definitions of rurality have expanded to recognize the diverse lived experiences in rural places (Chalmers and Joseph, 2006). Definitions of rurality have become increasingly multifaceted, including cultural aspects of places, descriptive features and local self-identification (Kulig and Williams, 2012). However, rural places cannot be conceived simply as a single, bordered study site with unchanging characteristics. As such, some scholars have suggested that we must study rural health relationally (Herron and Rosenberg, 2017). Relational approaches to rural health acknowledge that people, ideas and goods flow between urban and rural settings, connecting people and places in diverse ways and across a range of scales (Malatzky & Bourke, 2016). As such, we turn to look at how rural health has been examined across different scales.

International perspectives on rural health

International perspectives on rural health issues have a long-standing role in health geography's two traditional streams: disease ecology and access to health and health care. In the past, positivist approaches dominated both streams; however, there has been a proliferation of diverse epistemological (e.g., humanist, feminist, Marxist, political-economic) and conceptual approaches to the study of rural health at all scales. For example, studies of global health inequalities, such as the incidence of infectious disease, have increasingly drawn on political ecology to understand how political and historical structures across a range of scales promote the spread of disease among particular groups in rural places (Bardosh, 2015). Focusing on rural areas in the Global South, this domain of research has identified complex multiscalar relations that influence health behaviors, interventions and outcomes, including labor practices, migration between rural and urban centers, gender norms and other local social and cultural practices (Mkandawire, Luginaah and Baxter, 2014).

The implications of climate change for the health of rural people and places are also of growing concern at the global scale. Indeed, Indigenous people in northern and remote areas of Canada have been severely impacted by climate change. Many groups are experiencing declining access to country foods, loss of cultural practices and loss of autonomy and freedom associated with travel routes on sea ice (Durkalec et al., 2015). In addition, research has demonstrated that climate change has had particularly negative impacts on farmers' health in Australia; Australian farmers experience lower farm production and lower socioeconomic conditions while confronting cultural norms that praise stoicism in the face of hardship, thereby exacerbating their sense of hopelessness (Alston, 2012). Taken together, these examples of rural health issues recognize the implications of global processes for rural health, rural peoples' socially and culturally rooted responses to global processes and their financial and economic capacity (or limitations) to respond. Such global-local analyses of rural health issues are essential to understanding the complexities of rural change and rural health in the context of these changes.

National and regional perspectives

At the national and subnational (regional) scale, studies of rural health and health services (in health geography and other health and social-science fields) have engaged with questions about how to improve access to health care for rural and remote communities. The bulk of research in this area has taken place at the national and regional scales, because these are typically the scales at which health policies and formal services are organized. Research in this area has moved from modeling optimal access to developing more complex understandings of the relationship between physical, socioeconomic and cultural accessibility (Joseph and Phillips, 1984; Kulig and Williams, 2012). Modeling and geographic information systems remain a crucial component of understanding the relationships among, for example, physician supply and demand (Mazumdar et al., 2016). There has also been increasing attention to the use and limitations of technology to reach rural and remote people and places (Schneider et al., 2015).

Since the late 1980s, political-economy approaches have become a common means of understanding how institutional, political and economic structures shape both provision of and access to care in rural areas. The expansion of this approach parallels the widespread endorsement of health-care restructuring in welfare states of the Global North as a means of reducing the costs associated with health-care systems. For rural places, this has had many implications, including the closure of rural hospitals, centralization of services in urban locales and increasing reliance on the voluntary sector. Such changes influence the extent to which health-care decisions reflect the needs and interests of rural people and places (Skinner and Rosenberg, 2006). Alongside political-economy approaches to health and health-service organization, there has been a growing interest in the role of discursive processes in (re)producing rural health inequalities. For example, Malatzky and Bourke (2016) argue that medical students are exposed to common ideas about rural health practice that position it as less exciting, repetitive, low-level and isolated, and these perceptions play a role in the maldistribution of general practitioners. Indeed, education and media play a role in representing and reproducing health inequalities.

Community-based perspectives on rural health

As health-care systems change at the national and regional scale, there is a need to consider how different communities and community actors respond to such changes. Geographers have sought to understand what access to care means to people in different rural communities, what capacities rural places need in order to support their residents' health and well-being and how rural communities can and should be connected to resources elsewhere (Kearns and Joseph, 1997). Community-based perspectives have identified the needs of community-support-service users and formal service providers, as well as the role of the voluntary sector and volunteers in responding to unmet needs. Studies in this area illustrate the dynamic response of individuals and organizations to health-care restructuring and governance. For example, the voluntary sector may acquiesce or resist broader changes to health-care delivery (Joseph and Skinner, 2012). Ultimately, research continues to show that not all rural communities and community actors are serviced equally. There is often an expectation that social ties are stronger in rural communities; however, this expectation benefits some people and places more than others. For example, assumptions about the caring nature of rural communities place the greatest demands on rural women as mothers, daughters and wives to provide the bulk of unpaid care (Herron and Skinner, 2012). Broad economic and social-welfare restructuring (e.g., the closure of hospitals, schools, businesses) and demographic changes (e.g., youth outmigration, aging in place, retirement to rural areas) have transformed many rural communities and social-support networks.

Influenced by the cultural turn in health geography, there has been significant work on the complex social and cultural dynamics of rural communities as they shape provision and access to services. For example, the proximity of rural people to one another often makes it difficult to keep sensitive health issues (e.g., mental health) private, and this has implications for people's willingness to seek certain types of help. Innovative, culturally sensitive approaches to community care are required to address sensitive health issues in ways that are appropriate for particular rural contexts (Zanjani and Rowles, 2012).

Although rural communities are often characterized as unchanging and inward-looking, there has been growing recognition within community-based studies of the dynamism and activity of rural communities in support of health and well-being. Indeed, some rural areas have more integrated holistic care and innovative programs drawing on the voluntary sector, technology and the smaller scale of organizations to produce local solutions (Skinner and Hanlon, 2016). Research at the community scale points to the great diversity of lived experiences across rural communities and in a single rural place. As such, there is a need to understand interpersonal and individual-level factors influencing health and well-being, as well as the relationships of care.

Interpersonal and embodied perspectives on rural health

Health and health care are interpersonal, embodied and emotional; yet these facets of rural health and health care have not been studied to the same extent as larger processes. Much of the research on embodiment and emotion in health geography has been urban in focus. Research that has examined these micro-scale processes in rural contexts has focused on emotional challenges in care relationships, as well as emotional attachments to rural places. Attention to emotional dynamics in care relationships can reveal when and where care relationships are experienced as oppressive, constraining and frustrating or, alternatively, as rewarding, fulfilling and enjoyable. Some research has indicated that loyalty and stoicism, which reinforce the internalization of negative feelings about care and caregiving, are strong cultural values in rural communities. Moreover, these emotional dynamics are integral to understanding the context, consequences and responses of care in rural places (Herron and Skinner, 2013).

Focusing on the rural environment, research has shown that continuity and consistency in one's social and physical environment can foster feelings of security, trust and comfort that are essential to well-being. Some rural places that have achieved a reputation for their healing properties may be experienced as therapeutic landscapes. Conversely, rural places have been associated with higher rates of negative health behaviors, such as smoking and drinking. To understand such embodied habits in relation to rural places, Waitt and Clement (2016) looked at the assemblages of bodies, objects, feelings and technologies associated with drinking in rural Australia. More work is needed to understand how differently positioned individuals engage with rural places in ways that promote, or present risks to, health and well-being.

New approaches

In this chapter, we have captured some of the ways in which rural spaces and places of health and health care have been examined. We have used a scalar approach as an organizing framework while trying to emphasize that rural health problems and solutions are dynamic and networked. There are still, however, many studies of rural health that treat rural as a uniform and unchanging context rather than an analytic variable with a complex relationship to other variables. There is a long-standing tendency to talk about rural people as if they were all the same or, at the very least, to pass off dimensions of difference in favor of creating a more cohesive narrative about rural places and health. Although there is a long tradition of examining gender and rurality in relation to health and aging, and rurality in relation to health, studies of rural health have not engaged critically with concepts such as intersectionality. In particular, research on race and sexuality as they intersect with rurality and health is relatively scant (Panelli et al., 2009). Rather than looking at rural health as a singular experience, intersectionality provides a lens for understanding the different social and geographic locations of rural people in relation to health and wellness. A more robust appreciation of how different identities intersect in the context of rural health will enhance our understanding of health inequalities, vulnerabilities and resilience in rural settings. Geographers are well-positioned to take on this complex analysis, because of their understanding that places can hold very different meanings and experiences for different individuals and social groups.

Finally, it is important to note that research on rural health has, until recently, ignored the stories and knowledge of Indigenous people. The discourse around rural health in countries such as Australia and Canada points to Indigenous people as a major problem rather than recognizing their stories and knowledge as part of the solution. These discourses reflect a complex and continuing colonial power structure that rural-health researchers need to play a role in overcoming (Greenwood, De Leeuw and Reading, 2015). Indeed, rural-health research and Indigenous health research remain relative silos in relation to each other.

Concluding comments

Rural health is an interdisciplinary academic field that involves a diverse range of epistemological, methodological and conceptual tools. Geographic perspectives add to the rigor of rural-health research by examining rural as more than merely the context of our health outcomes, services and experiences. Rather, through health-geography research, rurality has been shown to be a complex and changing analytic category interacting with other factors to shape spaces and places of health and health care worthy of further investigation.

References

Alston, M. (2012). Rural male suicide in Australia. *Social Science & Medicine*, 74(4), pp. 515–522.

Bardosh, K. (2015). Achieving "Total Sanitation" in rural African geographies: poverty, participation and pit latrines in eastern Zambia. *Geoforum*, 66, pp. 53–63.

Chalmers, A. L. and Joseph, A. E. (2006). Rural change and the production of otherness: the elderly in New Zealand. In: P. Cloke, T. Marsden and P. Mooney, eds., *Handbook of rural studies*. London: Sage, pp. 388–400.

DesMeules, M., Pong, R., Read Guernsey, J., Wang, F., Luo, W. and Dressler, M. P. (2012). Rural health status and determinants in Canada. In: J. C. Kulig and A. M. Williams, eds., *Health in rural Canada*. Vancouver: UBC Press, pp. 23–43.

Durkalec, A., Furgal, C., Skinner, M. W. and Sheldon, T. (2015). Climate change influences on environment as a determinant of Indigenous health: relationships to place, sea ice, and health in an Inuit community. *Social Science & Medicine*, 136, pp. 17–26.

Gesler, W. M. and Ricketts, T. C. (eds.) (1992). *Health in rural North America: the geography of health care services and delivery*. London: Rutgers University Press.

Greenwood, M., De Leeuw, S. and Reading, C. (2015). *Determinants of Indigenous peoples' health*. Toronto: Canadian Scholars' Press.

Hanlon, N. and Kearns, R. (2017). Health and rural places. In: M. Shucksmith and D. L. Brown, eds., *International handbook of rural studies*. New York: Routledge, pp. 62–70.

Herron, R. V. and Rosenberg, M. W. (2017). "Not there yet:" Examining community support from the perspective of people with dementia and their partners in care. *Social Science & Medicine*, 173, pp. 81–87.

Herron, R. V. and Skinner, M. W. (2012). Farmwomen's emotional geographies of care. *Gender, Place and Culture*, 19(2), pp. 232–248.

Herron, R. V. and Skinner, M. W. (2013). The emotional overlay: negotiating aging and care in the countryside. *Social Science & Medicine*, 91, pp. 186–193.

Joseph, A. E. and Phillips, D. R. (1984). *Accessibility and utilization: geographical perspectives on health care delivery*. London: Harper & Row.

Joseph, A. E. and Skinner, M. W. (2012). Voluntarism as a mediator of the experience of growing old in evolving rural spaces and changing rural places. *Journal of Rural Studies*, 28, pp. 380–388.

Kearns, R. A. and Joseph, A. E. (1997). Restructuring health and rural communities in New Zealand. *Progress in Human Geography*, 21, pp. 18–32.

Kulig, J. C. and Williams, A. M. (2012). *Health in rural Canada*. Vancouver: UBC Press.

Malatzky, C. and Bourke, L. (2016). Re-producing rural health: challenging dominant discourses and the manifestation of power. *Journal of Rural Studies*, 45, pp. 157–164.

Mazumdar, S., Butler, D., Bagheri, N., Konings, P., Girosi, F., Feng, X. Q. and Mcrae, I. (2016). How useful are primary care service areas? Evaluating PCSAs as a tool for measuring primary care practitioner access. *Applied Geography*, 72, pp. 47–54.

Mkandawire, P., Luginaah, I. and Baxter, J. (2014). Growing up an orphan: vulnerability of adolescent girls to HIV in Malawi. *Transactions of the Institute of British Geographers*, 39(1), pp. 128–139.

Panelli, R., Hubbard, P., Coombes, B. and Suchet-Pearson, S. (2009). De-centring white ruralities: ethnic diversity, racialisation and indigenous countrysides. *Journal of Rural Studies*, 4(25), pp. 355–364.

Schneider, A. H., Mort, A., Kindness, P., Mellish, C., Reiter, E. and Wilson, P. (2015). Using technology to enhance rural resilience in pre-hospital emergencies. *Scottish Geographical Journal*, 131(3–4), pp. 194–200.

Skinner, M. W. and Hanlon, N. (eds.) (2016). *Ageing resource communities: new frontiers of rural population change, community development and voluntarism*. London: Routledge.

Skinner, M. W. and Rosenberg, M. W. (2006). Managing competition in the countryside: non-profit and for-profit perceptions of long-term care in rural Ontario. *Social Science & Medicine*, 63, pp. 2864–2876.

Waitt, G. and Clement, S. (2016). Women drinking alcohol: assembling a perspective from a Victorian country town, Australia. *Gender, Place & Culture*, 23(8), pp. 1121–1134.

Zanjani, F. and Rowles, G. D. (2012). "We don't want to talk about that": overcoming barriers to rural aging research and interventions on sensitive topics. *Journal of Rural Studies*, 28(4), pp. 398–405.

39

TRANSPORTATION AND HEALTH GEOGRAPHIES

Michael J. Widener

Transportation, broadly conceptualized as the movement of people across space, is connected to personal and population health in complex ways. It directly and indirectly influences health outcomes both positively and negatively, simultaneously enabling pathways to improved well-being (e.g., increasing access to health services or inducing physical activity) and producing exposures that result in higher rates of morbidity and mortality (e.g., the production of pollutants or fatal car crashes) (Widener and Hatzopoulou, 2016). Given the inherently spatial nature of transportation infrastructure, networks and behaviors, geographers have played an important role in establishing and expanding the study of transportation and health. Their contributions range from research on the effects of automobility and active transportation on health, international air travel and disease spread, and inequities in access to health care via public transit networks, in diverse regions across the Global North and Global South.

This chapter documents the role that geographers and geography have played in furthering research on the links between health and transportation over the past half-century, noting important theoretical and methodological contributions. While the emphasis is primarily on the theoretical developments of geographers or research published in geography journals, researchers from other disciplines (e.g., public health, engineering, planning) who have improved understandings of transportation and health geographies will also be included. The next two sections review major theoretical and methodological contributions by identifying significant themes within the literature, and the chapter concludes with a section that notes gaps and limitations where future research can drive transportation and health geographies forward.

Theoretical contributions

Geographers have been active in studying the links between transportation and health for the past half-century, with a theoretical emphasis that has evolved alongside the rest of the discipline. However, despite the diverse scope of this domain, much of the research covered in this chapter uses quantitative methodologies, thanks in part to the dominance of this methodological framework in cognate fields like transportation engineering and public health. As discussed later, this methodological bias has potentially left many questions about the links between transport and health unanswered, with researchers often relying on sometimes incomplete data to help generate and answer questions. The following subsections identify some of the major theoretical contributions, along with a selection of example studies.

Health-care-facility location

Some of the earliest work on transport and health geographies comes out of the quantitative revolution in the 1950s and 1960s, in which geographers began using newly developed computers and location-modeling techniques to solve facility-location and demand–allocation problems (Scott, 1970). Early on, researchers discovered that these models, developed with economic geography in mind, were useful when deciding where to locate hospitals. They explicitly accounted for travel times and costs to a facility or service, the relevant demand (i.e., population distributions) and facility capacities (Morrill and Kelley, 1969). Of course, many assumptions and generalizations (e.g., simplifying the topology of road networks or assigning residential locations to polygon centroids) were required when implementing these models, given the limited computing power and data availability of the time. However, as computational resources increased and higher-resolution datasets have become available, this same methodology has been used in more sophisticated ways and expanded to additional types of transportation and health-related locations. Examples are locating ambulances (Church and Meadows, 1979), helicopter emergency medical services (Schuurman et al., 2009) and a range of other health facilities and services (Wang, 2012).

Spatial access to health services and healthy places

Other early work linking transportation and health geographies was related to the previously described research on location modeling. Rather than taking a proactive approach to determining the best distribution of health services (at least from a positivist approach), studies of spatial accessibility focused on the established health facilities and services to which a population was geographically proximate. Like location models, the role of transportation in understanding access to health services relied on the fact that a person's mode of transportation could either enable or constrain his or her ability to get to some relevant health service or healthy place. It is important to note that *spatial* access is seen as just one of many dimensions of access to health care and services, with affordability, accommodation, availability and acceptability also contributing to a person's ability to acquire the needed health-related opportunities (Penchansky and Thomas, 1981).

The ability to measure the role and effectiveness of transportation systems in facilitating access to health care and services was greatly aided by the proliferation of geographic information systems (GIS), which can readily compute accurate travel times and costs on road networks, in the 1990s and early 2000s (McLafferty, 2003). With this methodology established, geographers have examined how various populations' spatial access (estimated via travel times and costs) differs to a range of health services and healthy places, including primary care physicians (Schuurman, Berube and Crooks, 2010), healthy food (Widener et al., 2013), mammography clinics (Dai, 2010) and trauma centers (Lerner et al., 1999), to name only a few. Finally, it is worth noting that much of the work in this area has focused on urban and suburban regions, using either automobile travel or simplified representations of other modes. A smaller literature has begun to investigate how transportation affects access to health care and services in a rural context (Martin et al., 2002), as well as how more accurate representations of urban-transit travel times (e.g., accounting for transfers and departure times) can be computed for those using public transportation (Farber et al., 2014).

Active transportation

Another relatively recent health-related focus of transportation geographers and other transport researchers is on how active modes of transportation (e.g., walking or bicycling), rather than passive modes (e.g., using an automobile), can improve population health (Dill, 2009). This work builds on a vast literature that documents the positive effects of physical activity and active living on a range of health outcomes and considers transportation as a critical dimension of improving population health. Some of the more popular areas of research on active transport are children's use of active modes to get to school (McDonald, 2008), active transport behaviors in different built environments (Buehler and Pucher, 2012) and how transit use increases overall levels of active transportation (Miller et al., 2015). While most of this work is focused on locations in North America and Europe, there is increasing interest in the topic around the world (Sallis et al., 2016).

Transportation and pollution

Generally, air pollutants are of concern to a range of geographers, given the impact they have on human health, ecological systems and climate. Transportation and health geographers tend to focus on transportation modes that directly produce air pollution, typically through combustion engines, and how the spatial distribution of those emissions affect the health of populations exposed to the related particulate matter ($PM_{2.5}$) and ozone (O_3) that are linked to adverse health outcomes (Jerrett et al., 2005). While much work by geographers has focused on pollution originating from automobiles at the intraurban scale, with a particular emphasis on the how the geography of pollution relates to different types of road infrastructure and traffic volume (Bayer-Oglesby et al., 2006), a more multidisciplinary group has examined how transportation facilities, such as airports (Franssen, Staatsen and Lebret, 2002) and seaports (Moretti and Neidell, 2011), impact health. Additionally, recent work has examined these factors in rapidly developing countries, such as China, where there is a growing number of residents with automobiles and well-established issues with pollution from both transport and non-transport sources (Zhang, Huang and Luo, 2014).

Road safety and crashes

The study of road-traffic crashes by geographers is an example of an examination of a very direct relationship between health and transportation (Whitelegg, 1987). In this arena, geographers are able to use their toolkit of spatial-analysis methods, like cluster detection, to help identify any geographic patterns in collisions and, thus, potential environmental or infrastructure-related causes. In addition to the exploratory spatial data analysis of crashes between cars (Edwards, 1996), authors have also explored the geographies of collisions between cars and pedestrians (Ha and Thill, 2011) and cars and cyclists (Vandenbulcke et al., 2009). More recently, authors have begun to look at the geography of road safety and crashes in rapidly urbanizing areas outside of the Global North, where automobile ownership and road infrastructure construction are on the rise (Loo, Yao and Wu, 2011).

New mobilities and health

One of the few examples of qualitative work being done in transportation and health geographies focuses on mobility and movement and its various relationships to healing, inclusion and exclusion, and power (Sheller and Urry, 2006). Of particular note is work by Anthony Gatrell, who has critiqued the current state of the walkability literature and suggests that a more inclusive therapeutic-mobilities framework is appropriate (Gatrell, 2013). Among other things, Gatrell calls on walkability researchers to engage in more participatory methods that are inclusive of all people (e.g., people considered obese, differently abled people or those with a mental-health condition), as well as account for a broader range of movement activities and well-being outcomes (e.g., positive spiritual and emotional experiences). Another example of work that draws from the new mobilities paradigm includes research that used interviews to explore how immigrants to Ireland sometimes return to their country of origin to receive health care (Migge and Gilmartin, 2011). While the motivations for traveling outside of state boundaries for care vary from person to person, the article illustrates how factors such as cultural norms and affordability can inspire people to travel when seeking health-care services.

Other theoretical contributions

Geographers have contributed in a variety of ways to the many other links between transportation and health that are not covered in this chapter. Two noteworthy research topics are briefly referenced here. The first is transportation-related noise, which has been linked to a number of negative health outcomes and is highly variable across space (Kim et al., 2012). The second focuses on the role that transportation networks play in the diffusion of infectious disease; researchers in this area have examined a number of modes, from transit systems (Cooley et al., 2011) to airline networks (Bowen and Laroe, 2006). For more on the various ways in which transportation and health intersect, the reader is directed to the previously cited work by Widener and Hatzopoulou (2016), which presents a causal map of the many relevant dimensions and feedbacks between health and transportation systems and includes a longer list of example papers.

Methodological contributions

Beyond the theoretical contributions described above, some techniques widely applied in transportation research have been used by geographers to better understand health outcomes. A first method is the collection of travel surveys (or travel diaries) from study participants as a way to understand where and by what mode a person travels, usually over a period of days. Although this tool is commonly used by transportation researchers, recently geographers have also harnessed this data-collection technique, realizing its ability to capture useful information about geographic context. With information from travel surveys and diaries, it is possible to compare health outcomes of people who, for example, walk to work versus those who take an automobile (Lachapelle and Frank, 2009). A second common technique extensively used in the world of transportation research and transportation geography is the use of Global Positioning System (GPS)-enabled devices. Similar to travel surveys, data collected from GPS-enabled devices can be used to explore people's daily transportation patterns and understand how they relate to access to a wide range of health services and healthy places. As one example, Zenk et al. (2011) used GPS trackers to better understand how the individual dietary behaviors of urban residents in Detroit were affected by their various spatial trajectories and exposures to the food environment.

Advancing transportation research in health geography

While the scope of health and transportation research in geography is large and diverse, there are a number of limitations to the current body of literature. For example, as previously mentioned, there is a strong bias toward quantitative methodologies. Given the varied methodologies employed by geographers, health geographers are in a unique position to supplement current work with mixed-methods and qualitative approaches, building on the work already underway by those engaging in health research that uses the new mobilities paradigm. Interviews, focus groups and ethnographies will be particularly valuable in uncovering the processes that result in uneven access and use of transportation networks that potentially exacerbate disparities in health outcomes across space. These are important directions for advancing this area of health geography.

Much of the work outlined in this chapter is conducted in the United States, Canada, and Western Europe. While there are a growing number of studies emerging from China, and some research on transportation and health geography in other parts of the world, there is a need for health geographers to collaborate with academics and institutions across the globe, where transportation behaviors and infrastructure can be substantially different from those in the Global North. With many parts of the world rapidly urbanizing and struggling to build transportation infrastructure to accommodate growing populations, there is an acute need for geographers to study how health outcomes and transportation behaviors affect each other. This includes developing attentiveness to differing social environments, such as contexts where populations are rapidly aging, and changing physical environments, such as remote settings where transportation networks are being altered as a result of climate change.

Health geographers continue to be well-positioned to study how a wide range of health outcomes are affected by, and affect, transportation. Geographers' training in diverse methods and content areas, from social theories to atmospheric sciences, as well as their interest in spatial effects, provide them with the tools and perspective to tackle this fundamentally multidisciplinary topic.

References

Bayer-Oglesby, L., Schindler, C., Hazenkamp-Von Arx, M. E., Braun-Fahrländer, C., Keidel, D., Rapp, R., Künzli, N., Braendli, O., Burdet, L. and Liu, L. S. (2006). Living near main streets and respiratory symptoms in adults the swiss cohort study on air pollution and lung diseases in adults. *American Journal of Epidemiology*, 164(12), pp. 1190–1198.

Bowen, J. T. and Laroe, C. (2006). Airline networks and the international diffusion of severe acute respiratory syndrome (SARS). *The Geographical Journal*, 172(2), pp. 130–144.

Buehler, R. and Pucher, J. (2012). Cycling to work in 90 large American cities: new evidence on the role of bike paths and lanes. *Transportation*, 39(2), pp. 409–432.

Church, R. L. and Meadows, M. E. (1979). Location modeling utilizing maximum service distance criteria. *Geographical Analysis*, 11(4), pp. 358–373.

Cooley, P., Brown, S., Cajka, J., Chasteen, B., Ganapathi, L., Grefenstette, J., Hollingsworth, C. R., Lee, B. Y., Levine, B. and Wheaton, W. D. (2011). The role of subway travel in an influenza epidemic: a New York City simulation. *Journal of Urban Health*, 88(5), pp. 982.

Dai, D. (2010). Black residential segregation, disparities in spatial access to health care facilities, and late-stage breast cancer diagnosis in metropolitan Detroit. *Health & Place*, 16(5), pp. 1038–1052.

Dill, J. (2009). Bicycling for transportation and health: the role of infrastructure. *Journal of Public Health Policy*, 30(1), pp. S95–S110.

Edwards, J. B. (1996). Weather-related road accidents in England and Wales: a spatial analysis. *Journal of Transport Geography*, 4(3), pp. 201–212.

Farber, S., Morang, M. Z. and Widener, M. J. (2014). Temporal variability in transit-based accessibility to supermarkets. *Applied Geography*, 53, pp. 149–159.

Franssen, E. A., Staatsen, B. A. and Lebret, E. (2002). Assessing health consequences in an environmental impact assessment: The case of Amsterdam Airport Schiphol. *Environmental Impact Assessment Review*, 22(6), pp. 633–653.

Gatrell, A. C. (2013). Therapeutic mobilities: walking and "steps" to wellbeing and health. *Health & Place*, 22, pp. 98–106.

Ha, H. H. and Thill, J. C. (2011). Analysis of traffic hazard intensity: a spatial epidemiology case study of urban pedestrians. *Computers, Environment and Urban Systems*, 35(3), pp. 230–240.

Jerrett, M., Arain, A., Kanaroglou, P., Beckerman, B., Potoglou, D., Sahsuvaroglu, T., Morrison, J. and Giovis, C. (2005). A review and evaluation of intraurban air pollution exposure models. *Journal of Exposure Science and Environmental Epidemiology*, 15(2), pp. 185–204.

Kim, M., Chang, S. I., Seong, J. C., Holt, J. B., Park, T. H., Ko, J. H. and Croft, J. B. (2012). Road traffic noise: annoyance, sleep disturbance, and public health implications. *American Journal of Preventive Medicine*, 43(4), pp. 353–360.

Lachapelle, U. and Frank, L. D. (2009). Transit and health: mode of transport, employer-sponsored public transit pass programs, and physical activity. *Journal of Public Health Policy*, 30(1), pp. S73–S94.

Lerner, E. B., Billittier, A. J., Sikora, J. and Moscati, R. M. (1999). Use of a geographic information system to determine appropriate means of trauma patient transport. *Academic Emergency Medicine*, 6(11), pp. 1127–1133.

Loo, B. P., Yao, S. and Wu, J. (2011). Spatial point analysis of road crashes in Shanghai: a GIS-based network kernel density method. *Geoinformatics, 2011*, 19th International Conference, IEEE, pp. 1–6.

Martin, D., Wrigley, H., Barnett, S. and Roderick, P. (2002). Increasing the sophistication of access measurement in a rural healthcare study. *Health & Place*, 8(1), pp. 3–13.

McDonald, N. C. (2008). Household interactions and children's school travel: the effect of parental work patterns on walking and biking to school. *Journal of Transport Geography*, 16(5), pp. 324–331.

McLafferty, S. L. (2003). GIS and health care. *Annual Review of Public Health*, 24(1), pp. 25–42.

Migge, B. and Gilmartin, M. (2011). Migrants and healthcare: investigating patient mobility among migrants in Ireland. *Health & Place*, 17(5), pp. 1144–1149.

Miller, H. J., Tribby, C. P., Brown, B. B., Smith, K. R., Werner, C. M., Wolf, J., Wilson, L. and Oliveira, M. G. S. (2015). Public transit generates new physical activity: evidence from individual GPS and accelerometer data before and after light rail construction in a neighborhood of Salt Lake City, Utah, USA. *Health & Place*, 36, pp. 8–17.

Moretti, E. and Neidell, M. (2011). Pollution, health, and avoidance behavior evidence from the ports of Los Angeles. *Journal of Human Resources*, 46(1), pp. 154–175.

Morrill, R. L. and Kelley, P. (1969). Optimum allocation of services: an hospital example. *The Annals of Regional Science*, 3(1), pp. 55–66.

Penchansky, R. and Thomas, J. W. (1981). The concept of access: definition and relationship to consumer satisfaction. *Medical Care*, 19(2), pp. 127–140.

Sallis, J. F., Cerin, E., Conway, T. L., Adams, M. A., Frank, L. D., Pratt, M., Salvo, D., Schipperijn, J., Smith, G. and Cain, K. L. (2016). Physical activity in relation to urban environments in 14 cities worldwide: a cross-sectional study. *The Lancet*, 387(10034), pp. 2207–2217.

Schuurman, N., Bell, N. J., L'Heureux, R. and Hameed, S. M. (2009). Modelling optimal location for pre-hospital helicopter emergency medical services. *BMC Emergency Medicine*, 9(1), p. 6.

Schuurman, N., Berube, M. and Crooks, V. A. (2010). Measuring potential spatial access to primary health care physicians using a modified gravity model. *The Canadian Geographer/Le Geographe Canadien*, 54(1), pp. 29–45.

Scott, A. J. (1970). Location–allocation systems: a review. *Geographical Analysis*, 2(2), pp. 95–119.

Sheller, M. and Urry, J. (2006). The new mobilities paradigm. *Environment and Planning A*, 38(2), pp. 207–226.

Vandenbulcke, G., Thomas, I., De Geus, B., Degraeuwe, B., Torfs, R., Meeusen, R. and Panis, L. I. (2009). Mapping bicycle use and the risk of accidents for commuters who cycle to work in Belgium. *Transport Policy*, 16(2), pp. 77–87.

Wang, F. (2012). Measurement, optimization, and impact of health care accessibility: a methodological review. *Annals of the Association of American Geographers*, 102(5), pp. 1104–1112.

Whitelegg, J. (1987). A geography of road traffic accidents. *Transactions of the Institute of British Geographers*, 12(2), pp. 161–176.

Widener, M. J., Farber, S., Neutens, T. and Horner, M. W. (2013). Using urban commuting data to calculate a spatiotemporal accessibility measure for food environment studies. *Health & Place*, 21, pp. 1–9.

Widener, M. J. and Hatzopoulou, M. (2016). Contextualizing research on transportation and health: a systems perspective. *Journal of Transport & Health*, 3(3), pp. 232–239.

Zenk, S. N., Schulz, A. J., Matthews, S. A., Odoms-Young, A., Wilbur, J., Wegrzyn, L., Gibbs, K., Braunschweig, C. and Stokes, C. (2011). Activity space environment and dietary and physical activity behaviors: a pilot study. *Health & Place*, 17(5), pp. 1150–1161.

Zhang, W., Huang, B. and Luo, D. (2014). Effects of land use and transportation on carbon sources and carbon sinks: a case study in Shenzhen, China. *Landscape and Urban Planning*, 122, pp. 175–185.

40

mHEALTH GEOGRAPHIES

Mobile technologies and health in the Global South

Jonathan Cinnamon and Charlene Ronquillo

The rapid expansion of mobile networks has revolutionized communications worldwide, yet mobile telephony is also having significant effects across a range of issues including education, politics, entertainment, and finance (Steinhubl, Muse and Topol, 2015). This chapter explores one of these effects: the use of mobile technologies in health. Mobility is a fundamentally geographic concept, and thus mobile health (mHealth) is an area that health geographers have and continue to examine.

mHealth refers to the use of wireless and mobile technologies in health and health care and includes a diverse range of activities such as health-information hotlines, telemediated patient care, personal mobile health apps and mobile-health surveillance. mHealth is undertaken actively, by people using all types of mobile phones (basic, feature and smartphones), personal digital assistants (PDAs) and tablet computers, as well as passively, using automated mobile sensing technologies. Whether produced actively by people or passively by sensors, at the core of mHealth is the collection and transfer of digital data; this could be a patient's vital sign information, survey data pertaining to a disease outbreak or simply digital voice or text message data transferred from a rural patient to a medical professional in an urban hospital. Modern digital mobile communication networks are thus central to mHealth, enabling the transfer of data in near real-time over vast distances; however, technologies enabling more local data transfer, such as Wi-Fi, Bluetooth and Near Field Communications (NFC), are also increasingly central to mHealth architectures.

Global interest in mHealth stems from the rapidly increasing availability of mobile technologies and networks, with estimates suggesting that 75% of the world's population will have a mobile phone by 2020 (5.7 billion unique subscribers, and 10 billion total mobile connections) (GSMA, 2017). The decreasing costs and increasing functionality of mobile technologies have captured the attention of the scientific community, governments and health systems, resulting in significant investment of resources into mHealth solutions. Owing to its core characteristics of mobility and portability, mHealth can expand the reach of health-care services to previously underserved populations, including those in rural and remote regions, representing an important step toward universal health coverage, a key goal of the global health agenda (Mehl and Labrique, 2014). Although the evidence is still thin, mHealth is also pursued to meet the objectives of reducing health-care costs (Betjeman, Soghoian and Foran, 2013), influencing positive health behaviors (Gurman, Rubin and Roess, 2012) and improving patient outcomes through more rapid diagnosis of illness and better adherence to treatment regimens (Hamine et al., 2015). Additionally, emerging reasons for attention to mHealth include its potential empowerment of health-care providers, communities and individuals, and the introduction of innovative health-service delivery models that leverage underused resources – for instance, by engaging lay and non-professional health workers and patients (Thondoo et al., 2015).

Across all world regions, people have more access to mobile phones than they do to basic necessities such as toilets and safe drinking water (World Bank, 2013). Growth in mobile-phone ownership presents an important opportunity for the Global South, given their ubiquity in many less-developed settings. Situated within the broader context of limited existing health-care services, mHealth presents the potential to *leapfrog* entrenched health practices and opportunities to develop new ways of delivering health care, with less-developed markets recognized as the key drivers of mHealth (PwC, 2014). Although recent developments in mobile internet and sensor technologies significantly increase the possibilities for mHealth, smartphones are still rare in many less-developed settings, a reality that is tempering some of the enthusiasm regarding mHealth in the Global South. In less-developed settings, many of the more advanced forms of mHealth are currently out of reach, and so, as always, the development of mHealth as a global health agenda must be sensitive to local contexts.

This chapter aims at advancing a contextually sensitive perspective on mHealth in the Global South. Following a review of mHealth in health geography and a call for more empirical and conceptual attention to the field by geographers, the main section of the chapter explores the recent developments around using two-way SMS (short message service, text messaging) platforms in health research and practice in the Global South, as an increasingly ubiquitous and low-cost data-transfer technology available on all mobile phones. The chapter then concludes with a brief discussion of the challenges and future prospects of mHealth, focusing specifically on the need for health geographers to contribute to this global discussion.

mHealth and health geography

mHealth consists of a broad and expanding range of activities and practices deployed for a variety of purposes. As part of the World Health Organization (WHO) Global Observatory for eHealth's (2011) second global survey on electronic health technologies, mHealth was explored for the first time as a sub-focus of eHealth. The resulting report identified 14 types of mHealth *services* within six broader categories of *activities* being undertaken: communication between individuals and health services, communication between health services and individuals, consultation between health care professionals, intersectoral communication in emergencies, health monitoring and surveillance and access to information for health-care professionals at point of care. Based on a meta-analysis of mHealth literature, Olla and Shimskey (2015) developed a taxonomy to specify the overarching *purpose* of various mHealth applications, based on eight categories of *end-uses*: point of care diagnostics, patient monitoring, wellness, compliance, education and reference, behavior modification, efficiency and productivity and environmental monitoring. Given the disparate collection of mHealth practices to date, classification activities represent an attempt to bring clarity, recognizing that further conceptual attention is needed to understand the complexities of mHealth technologies and applications. However, such attempts at discrete classifications may be of limited utility; the continual advancement of mobile technology and the diverse ways it is intervening in a wide range of health and well-being practices make it clear that mHealth represents a moving target.

Taxonomic impulses of medical science notwithstanding, mHealth as a field remains under-developed theoretically and conceptually, and it can benefit considerably by a more thoroughgoing engagement with the social sciences. Geographic concepts are often invoked in mHealth, often with little explanation, and perhaps as an assumed normative aim. For instance, *overcoming geographical distance and barriers* is a central trope (e.g., Steinhubl, Muse and Topol, 2015); indeed, it was deployed as an organizing concept in the WHO's (2011) path-breaking report on mHealth, subtitled *New Horizons for Health through Mobile Technologies*. Health geographers are well-placed to conceptualize the dimensions and concerns of mHealth, given the decades of attention in geography to interrogating concepts such as distance and its social and ethical implications. For instance, human geographers have fruitfully developed a spatial dimension to understandings of *care ethics* (e.g., Lawson, 2007) – a specifically relational approach to moral actions based on mutuality and reciprocity – and have applied it to consider the ethical responsibilities of local and global actions and caring-at-a-distance.

Achieving health equity is ostensibly a core aim of mHealth programs, which aim to overcome geographic barriers to care access, and it is conceivable how this spatially explicit conceptualization of care ethics can be useful here. Indeed, concepts central to contemporary human geography that have been applied to understand developments in information and communication technologies (ICT) – for example, space-time, scale, networks and mobilities – can also be productively deployed to advance mHealth.

It is perhaps surprising that mHealth appears to be an underexplored research topic in health geography. A search of the leading health-geography journals is potentially revealing: the term "mHealth" has never been printed in the pages of *Health & Place* (a top health-geography journal) and has been only briefly referred to in a single article in the *International Journal of Health Geographics* (a leading technology-focused health-geography journal). The interdisciplinary journal and frequent venue for health-geographic research *Social Science & Medicine* does show some interest in mHealth (~10 articles) in the broader social sciences. However, overall it appears that the term is primarily deployed in the health and medical sciences. The lack of attention to the term does belie a substantial amount of research at the intersection of health geography and mobile technologies, which broadly fits the mHealth purview, if not specifically identified as such.

A key area of mHealth research geographers are engaging with is the distributed collection of disease surveillance data using mobile phone–based data-collection platforms (e.g., Robertson et al., 2010; Cinnamon, Jones and Adger, 2016). Another important research theme is the use of passive mobile sensor technologies to collect environmental-exposure data. Geographers have long been interested in the concept of *activity spaces* (Golledge and Stimson, 1997), the areas of influence on our daily lives often bounded by the places in which we live, work, shop and take leisure. Mobile technologies are enabling significant advances in measuring activity spaces and therefore more geographically accurate collection of data on, for example, exposures to toxins and pollutants (Steinle, Reis and Sabel, 2013), unhealthy food (Sadler and Gilliland, 2015), substances such as tobacco and drugs (Lipperman-Kreda et al., 2015; Mason et al., 2015) and urban social and environmental exposures in children (Loebach and Gilliland, 2016). This area of research is advancing our understanding of *salutogenic* exposures: contact with spaces that are therapeutic or health-promoting (Bell et al., 2015). Geographers have also contributed to a growing body of research on the use of location and movement sensors (e.g., GPS, accelerometer, gyroscope) embedded in wearable technologies or smartphones to record and analyze participants' physical activity and mobility (Barratt, 2017; Jestico, Nelson and Winters, 2016). Similarly, geographers are researching how mobile apps are used for personal monitoring, fitness and self-care activities as part of the *quantified self* movement (Boulos and Yang, 2013). There has been some research by geographers on mobile technologies in health-care settings, such as mobile point-of-care platforms for managing patient data in hospitals (Zargaran et al., 2014). Overall, however, geographers have focused more on the health rather than the health-care side of mHealth.

mHealth in the Global South using SMS technologies

mHealth in the Global South is frequently considered within the broader field of information and communication technologies for development (ICT4D), a field with considerable geographical influence (e.g., Kleine and Unwin, 2009). Research and practice in this domain focuses on the use of ICTs to improve opportunities and living conditions in less-developed settings, with particular focus on business, governance, health and education (Hilbert, 2012). Positioning mHealth as part of a more established ICT4D project enables mHealth initiatives to learn from its well-developed knowledge base. This is not wholly unproblematic, however. Despite the clear linkages between economic development and health, the overall objectives of *growth* and *progress* that underpin development, and therefore ICT4D (Unwin, 2009), are not always the best route to health improvement; thus, care should be taken in drawing on ICT4D experiences and frameworks to plan and evaluate mHealth projects.

SMS is a simple and highly limited format for data exchange, restricted to 160 characters. Yet in its simplicity lies its significant potential for improving public-health research and practice in the Global South.

SMS messages can be sent and received between any mobile phones, including basic and feature phones most common in the Global South, and between computers and phones. Messages can be either personally tailored for a single recipient or sent out in bulk to all or a targeted subset of mobile users. It is a largely stable and reliable means of data transfer, enabling instant communications between distributed parties anywhere that mobile networks reach, often working even in emergency situations when networks can be too overloaded to connect mobile-phone calls (Revere, Schwartz and Baseman, 2014). It is also a very-low-cost means of communication; mobile customers typically receive messages at no cost and can send them for very little. For the mobile network operators (MNOs), routing SMS messages over the network is very low-cost – estimated at US$0.03 per message in 2009 (Keshav, 2009). As such, there is considerable potential to partner with MNOs to enable free messaging for senders and receivers in mHealth initiatives, especially given the desire of many mobile operators to engage in corporate social responsibility activities – as demonstrated, for example, by the Safaricom Foundation from the leading mobile operator in Kenya (http://safaricomfoundation.org/).

The use of SMS in mHealth initiatives is relatively well documented, especially in the Global North, where text messaging has been used since the 1990s to communicate directly with health-care patients or the general public. A common goal of these programs has been to modify individual health behaviors through regular interaction with health information and advice. For instance, Franklin et al. (2003) describe an SMS initiative that sent targeted messages and general information to young people with type 1 diabetes, with the aim of increasing adherence to intensive insulin regimens. A similar study focusing on follow-up care for bulimia nervosa patients found that the system was effective in providing support to patients who had completed inpatient treatment but still required ongoing monitoring (Bauer et al., 2003). Although these early uses of SMS have been promising, many examples are small-scale pilot initiatives requiring significant human resources to undertake – especially if targeted or responsive messaging is used – which has limited the potential for scaling up.

There has been some progress toward larger-scale mHealth SMS projects that engage a wider population. They are, however, often quite rudimentary in scope, often used only as a platform for distributing generic health information in settings or with groups that lack widespread access to traditional media. Bangladesh has exploited its high mobile-penetration rate (even in rural areas) to undertake SMS health campaigns targeting the entire (mobile) population. The country's government has used SMS to send alerts and information about national immunization day and breastfeeding week and reported wide acceptance, although their ability to evaluate the program for effectiveness (e.g., in influencing individual health-seeking behaviors) is limited (Khatun et al., 2015). The MAMA (Mobile Alliance for Maternal Action) and MomConnect initiatives in South Africa have used SMS as part of a multichannel strategy to communicate directly with expectant and new mothers. SMS has been an especially important channel in South Africa, a country that has high rates of maternal death in younger women and higher mobile-phone penetration than that of radio and television (Bateman, 2014). The South African government received a 50% discount on SMS messaging costs from the country's MNOs (Bateman, 2014); cost will hopefully be an uncommon deterrent in the future, as more substantial discounts or free messaging should be possible, given the negligible cost to operators.

The potential for scaled-up SMS mHealth initiatives has significantly advanced in recent years, due to the development of computer-based software platforms for creating and operating custom automated SMS messaging services. Platforms such as Frontline SMS (www.frontlinesms.com) and the UNICEF-developed RapidPro SMS (https://community.rapidpro.io) are being used across the Global South by organizations engaging in ICT4D activities. A key advantage here is the ability of these systems to enable automated *two-way conversations*, compared to earlier uses of SMS in these contexts largely based on *one-way information sharing*. In these two-way SMS systems, not only information but also questions can be sent out to mobile-phone users. If receivers respond to the prompt, the answer is recorded in a database on the host computer/website. Depending on the response, the SMS software can trigger further, customized questions (see Figure 40.1). The simple ability to answer an automated message means that anyone with a mobile phone – any health

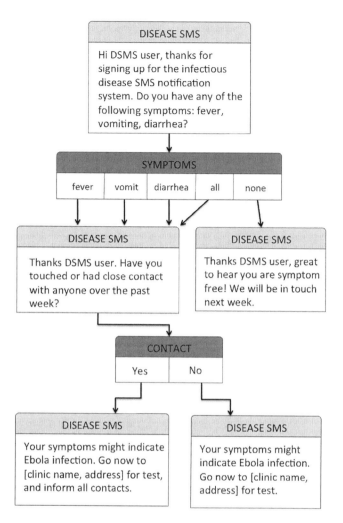

Figure 40.1   An example of SMS two-way conversation architecture

professional, civic official or member of the public – in any part of the country can contribute real-time information on any health issue. The potential is almost limitless. In particular, these developments can enable much more rapid and effective surveillance of health conditions, information on the social and environmental determinants of health and monitoring of health resources and infrastructure.

A particularly promising application of two-way SMS is its use for health surveillance during emergencies. Cinnamon, Jones and Adger (2016) conducted a study to ascertain how two-way SMS systems can be used to enhance informational awareness in the management of communicable-disease outbreaks. Results highlighted the significant potential to enable near-real-time surveillance of disease cases via automated SMS-based surveys directed to health professionals or emergency responders in remote outbreak zones underserved by other communications or transportation links, as was the case in parts of West Africa during the 2013–2015 Ebola outbreak.

Connecting directly with the public via SMS may stop outbreaks before they happen. *Syndromic surveillance* efforts aim to enable early prediction of outbreaks by accessing various data sources that pertain to disease symptoms – everything from Web searches to school/work absentee records, medication sales and weather data (Mandl et al., 2004). Drawing on citizen science and crowdsourcing approaches, two-way SMS systems can also be a source of actively produced syndromic data for identifying new locations and populations to target with prevention resources. The EbolaTracks SMS system (Tracey et al., 2015), while only a proof of concept and undertaken in a developed-country setting (Australia), illustrates the potential use of a basic two-way SMS system for direct symptom monitoring and triage for individuals potentially infected with a communicable disease. Developed during the recent Ebola outbreak, the system was pilot-tested with participants returning to Australia from Ebola-affected countries in West Africa. Participants were provided with a mobile phone and digital thermometer. They were told to take their temperature twice daily for 21 days (maximum Ebola incubation period) and respond via SMS with the reading and any noted symptoms. High readings triggered SMS questions to participants, and then, if there was no response, an SMS and email alert were sent to an on-call medical officer, who would contact and follow up with participants to assess their condition. This type of mHealth application is potentially invasive; however, restrictive control measures are often necessary under conditions of highly contagious disease outbreaks (see also Koch, 2016) and are perhaps more appealing than the alternative of quarantine.

In addition to the use of two-way SMS to collect public health surveillance data, it can also be used to share up-to-date information on health protocols, resources and infrastructure to enhance the functioning of the health-care system itself. A project undertaken in two remote regions of Malawi enrolled 77% of its community health workers (CHWs) in a two-way SMS information sharing and reporting system, which replaced less-effective methods that included sending messages via ambulance or using the radio for one-way communications (Lemay et al., 2012). District health managers used the system to send out messages advising CHWs of available resources, training opportunities and changes in protocol, while the CHWs used the system to ask specific medical questions, to report resource requirements and to organize patient referrals and transfers. Through such a system, there is the potential to better leverage non- and para-professional health-care workers, a vital but often underused health-system resource in the Global South. The key premise of equipping these workers with mHealth lies in its potential to function as additional health-system infrastructure and open the possibility for *task shifting* of health-care tasks and responsibilities traditionally held by professional or specialized health-care providers, which can enable health workers with limited training to receive support and expand their scope of work.

Conclusions: optimism for the future of mHealth

mHealth is still a relatively new field; its relevance and scope are still evolving, paralleling the dynamism of the mobile-technology sector itself. That said, mHealth is a mature enough field to be subject to significant critique. Figure 40.2 illustrates the phases that technologies often progress through over time, from initial introduction to eventual widespread productivity. mHealth consists of a dizzying array of technologies and application areas, and so its current position of technological progression is variable. In fact, few mHealth technologies are likely to have made it even to the *slope of enlightenment* stage of progression, despite some having a history reaching back to the 1990s.

Pilotism is a widespread problem, whereby mHealth technologies never emerge out of testing or are never scaled up. Indeed, as Andreassen, Kjekshus and Tjora (2015, p. 62) note after 20 years of work on ICT projects in health, "most projects remain projects." The ability to confirm health benefits of mHealth initiatives is limited, and so it is unclear whether initiatives improve clinical outcomes (Free et al., 2013), and there is little evidence confirming some of the grander ambitions of cost savings, progress toward universal health coverage and patient and care-worker empowerment. Indeed, some studies have documented negative impacts of mHealth introduction, such as increased costs (Ryan et al., 2012), which suggests that proponents must be wary of technological determinism. In fact, in some cases, initiating these so-called solutions sometimes just strips time and resources away from established health and medical practices (see Higgins, 2014).

Figure 40.2   The technology cycle. This *cycle* describes a generalized pattern of five key stages that technologies often follow over time – from their introduction, to reaching peak hype, followed by a sharp decline in expectations and a slower return to productivity.

Creative commons licensed image from Wikimedia Commons, based on Gartner Hype Cycle (www.gartner.com).

Yet, there is space for optimism about the future of mHealth, especially when considering it as part of the broader social phenomenon of digital citizenship and research on digital lives. The intersection of health, society and technology is a key space in which for health geographers and other social scientists to make an important contribution. For instance, work by Hampshire et al. (2015, p. 97) documents how younger generations in sub-Saharan Africa are engaging in a "digitally-mediated form of therapeutic citizenship" in which they use their mobile phones in informal, but creative and strategic, ways to access health information and health care. As we progress toward an even greater recognition of the social and spatial conditions that shape health and health care, health geographers can and should be at the forefront of research on both the formal and the informal roles that mobile technologies play in health advancement around the world.

References

Andreassen, H. K., Kjekshus, L. E. and Tjora, A. (2015). Survival of the project: a case study of ICT innovation in health care. *Social Science & Medicine*, 132, pp. 62–69.

Barratt, P. (2017). Healthy competition: a qualitative study investigating persuasive technologies and the gamification of cycling. *Health & Place*, 46, pp. 328–336.

Bateman, C. (2014). Using basic technology – and corporate social responsibility – to save lives. *South African Medical Journal*, 104(12), pp. 839–840.

Bauer, S., Percevic, R., Okon, E., Meermann, R. and Kordy, H. (2003). Use of text messaging in the aftercare of patients with bulimia nervosa. *European Eating Disorders Review*, 11(3), pp. 279–290.

Bell, S. L., Phoenix, C., Lovell, R. and Wheeler, B. W. (2015). Using GPS and geo-narratives: a methodological approach for understanding and situating everyday green space encounters. *Area*, 47(1), pp. 88–96.

Betjeman, T. J., Soghoian, S. E. and Foran, M. P. (2013). mHealth in sub-Saharan Africa. *International Journal of Telemedicine and Applications*, [online], pp. 1–7. http://doi.org/10.1155/2013/482324

Boulos, M. N. K. and Yang, S. P. (2013). Exergames for health and fitness: the roles of GPS and geosocial apps. *International Journal of Health Geographics*, 12, pp. 1–7.

Cinnamon, J., Jones, S. K. and Adger, W. N. (2016). Evidence and future potential of mobile phone data for disease disaster management. *Geoforum*, 75, pp. 253–264.

Franklin, V., Waller, A., Pagliari, C. and Greene, S. (2003). "Sweet Talk": text messaging support for intensive insulin therapy for young people with diabetes. *Diabetes Technology & Therapeutics*, 5(6), pp. 991–996.

Free, C., Phillips, G., Watson, L., Galli, L., Felix, L., Edwards, P., Patel, V. and Haines, A. (2013). The effectiveness of mobile-health technologies to improve health care service delivery processes: a systematic review and meta-analysis. *PLoS Medicine*, 10(1), pp. 1–26.

Golledge, R. G. and Stimson, R. J. (1997). *Spatial behavior: a geographic perspective*. New York: Guilford Press.

GSMA. (2017). *The mobile economy 2017*. London: Groupe Speciale Mobile Association.

Gurman, T. A., Rubin, S. E. and Roess, A. A. (2012). Effectiveness of mHealth behavior change communication interventions in developing countries: a systematic review of the literature. *Journal of Health Communication*, 17(Supp 1), pp. 82–104.

Hamine, S., Gerth-Guyette, E., Faulx, D., Green, B. B. and Ginsburg, S. A. (2015). Impact of mHealth chronic disease management on treatment adherence and patient outcomes: a systematic review. *Journal of Medical Internet Research*, 17(2), pp. 1–15.

Hampshire, K., Porter, G., Owusu, S. A., Mariwah, S., Abane, A., Robson, E., Munthali, A., DeLannoy, A., Bango, A., Gunguluza, N. and Milner, J. (2015). Informal m-health: how are young people using mobile phones to bridge healthcare gaps in Sub-Saharan Africa? *Social Science & Medicine*, 142, pp. 90–99.

Higgins, A. (2014). *Fighting Ebola in Liberia with technology*. [online] Al Jazeera. Available at: www.aljazeera.com/indepth/features/2014/11/fighting-ebola-liberia-with-technology-2014111584621373278.html [Accessed 1 Apr. 2017].

Hilbert, M. (2012). Toward a conceptual framework for ICT for development: lessons learned from the cube framework used in Latin America. *Information Technologies & International Development*, 8(4), pp. 243–259.

Jestico, B., Nelson, T. and Winters, M. (2016). Mapping ridership using crowdsourced cycling data. *Journal of Transport Geography*, 52, pp. 90–97.

Keshav, S. (2009). *The cost of text messaging. Testimony at the hearing on: cell phone text messaging rate increases and the state of competition in the wireless market*. Senate Subcommittee on Antitrust, Competition Policy and Consumer Rights.

Khatun, F., Heywood, A. E., Bhuiya, A., Liaw, S. T. and Ray, P. K. (2015). Prospects of mHealth to improve the health of the disadvantaged population in Bangladesh. In: S. Adibi, ed., *mHealth: multidisciplinary verticals*. Boca Raton, FL: CRC Press, pp. 465–484.

Kleine, D. and Unwin, T. (2009). Technological revolution, evolution and new dependencies: what's new about ICT4D? *Third World Quarterly*, 30(5), pp. 1045–1067.

Koch, T. (2016). Fighting disease, like fighting fires: the lessons Ebola teaches. *The Canadian Geographer*, 60(3), pp. 288–299.

Lawson, V. (2007). Geographies of care and responsibility. *Annals of the Association of American Geographers*, 97(1), pp. 1–11.

Lemay, N. V., Sullivan, T., Jumbe, B. and Perry, C. P. (2012). Reaching remote health workers in Malawi: baseline assessment of a pilot mHealth intervention. *Journal of Health Communication*, 17(Supp 1), pp. 105–117.

Lipperman-Kreda, S., Morrison, C., Grube, J. W. and Gaidus, A. (2015). Youth activity spaces and daily exposure to tobacco outlets. *Health & Place*, 34, pp. 30–33.

Loebach, J. and Gilliland, J. (2016). Neighbourhood play on the endangered list: examining patterns in children's local activity and mobility using GPS monitoring and qualitative GIS. *Children's Geographies*, 14(5), pp. 573–589.

Mandl, K. D., Overhage, J. M., Wagner, M. M., Lober, W. B., Sebastiani, P., Mostashari, F., Pavlin, J. A., Gesteland, P. H., Treadwell, T., Koski, E., Hutwagner, L., Buckeridge, D. L., Aller, R. D. and Grannis, S. (2004). Implementing syndromic surveillance: a practical guide informed by the early experience. *Journal of the American Medical Informatics Association*, 11(2), pp. 141–150.

Mason, M., Mennis, J., Way, T., Light, J., Rusby, J., Westling, E., Crewe, S., Flay, B., Campbell, L., Zaharakis, N. and McHenry, C. (2015). Young adolescents' perceived activity space risk, peer networks, and substance use. *Health & Place*, 34, pp. 143–149.

Mehl, G. and Labrique, A. (2014). Prioritizing integrated mHealth strategies for universal health coverage. *Science*, 345(6202), pp. 1284–1287.

Olla, P. and Shimskey, C. (2015). mHealth taxonomy: a literature survey of mobile health applications. *Health and Technology*, 4(4), pp. 299–308.

PwC (2014). *Emerging mHealth: paths for growth*. New York: Price Waterhouse Cooper.

Revere, D., Schwartz, M. R. and Baseman, J. (2014). How 2 txt: an exploration of crafting public health messages in SMS. *BMC Research Notes*, 7, pp. 1–8.

Robertson, C., Sawford, K., Daniel, S. L. A., Nelson, T., A. and Stephen, C. (2010). Mobile phone–based infectious disease surveillance system, Sri Lanka. *Emerging Infectious Diseases*, 16(10), pp. 1524–1531.

Ryan, D., Price, D., Musgrave, S. D., Malhotra, S., Lee, A. J., Ayansina, D., Sheikh, A., Tarassenko, L., Pagliari, C. and Pinnock, H. (2012). Clinical and cost effectiveness of mobile phone supported self monitoring of asthma: multicentre randomised controlled trial. *British Medical Journal*, 344, pp. 1–15.

Sadler, R. C. and Gilliland, J. A. (2015). Comparing children's GPS tracks with geospatial proxies for exposure to junk food. *Spatial and Spatio-temporal Epidemiology*, 14–15, pp. 55–61.

Steinhubl, S. R., Muse, E. D. and Topol, E. J. (2015). The emerging field of mobile health. *Science Translational Medicine*, 6, pp. 79–81.

Steinle, S., Reis, S. and Sabel, C. E. (2013). Quantifying human exposure to air pollution – moving from static monitoring to spatio-temporally resolved personal exposure assessment. *Science of the Total Environment*, 443, pp. 184–193.

Thondoo, M., Strachan, D. L., Nakirunda, M., Ndima, S., Muiambo, A., Källander, K., Hill, Z. and Group, I. S. (2015). Potential roles of Mhealth for community health workers: formative research with end users in Uganda and Mozambique. *JMIR mHealth and uHealth*, 3(3), e76.

Tracey, L. E., Regan, A. K., Armstrong, P. K., Dowse, G. K. and Effler, P. V. (2015). Ebola Tracks: an automated SMS system for monitoring persons potentially exposed to Ebola virus disease. *Euro Surveillance*, 20(1), pp. 1–4.

Unwin, T. (ed.) (2009). *ICT4D: information and communication technology for development*. Cambridge: Cambridge University Press.

World Bank. (2013). *World development report 2014: risk and opportunity – managing risk for development*. Washington, DC: World Bank.

World Health Organization. (2011). *mHealth: new horizons for health through mobile technologies*. Geneva: World Health Organization.

Zargaran, E., Schuurman, N., Nicol, A. J., Matzopoulos, R., Cinnamon, J., Taulu, T., Ricker, B., Garbutt Brown, D. R., Navsaria, P. and Hameed, S. M. (2014). The electronic trauma health record: design and usability of a novel tablet-based tool for trauma care and injury surveillance in low resource settings. *Journal of the American College of Surgeons*, 218(1), pp. 41–50.

41

WALKABILITY AND PHYSICAL ACTIVITY

Jana A. Hirsch and Meghan Winters

Walkability has become an important research area across the fields of geography, public health, urban planning and transportation. Walkable neighborhoods, characterized by density, land-use diversity and well-connected transportation networks, may encourage walking for daily activities among residents. Currently, physical inactivity is the fourth leading underlying cause of mortality worldwide, with an estimated 3.3 million deaths annually due to inactivity (World Health Organization, 2009). Individual characteristics such as age, gender and attitudes are important determinants of physical activity. More recently, neighborhood contexts have also been recognized to promote or discourage physical activity. Since walking is linked with significant health benefits at the population level, such as reduced rates of chronic disease, walkability has emerged as an important policy intervention (Giles-Corti et al., 2016; Sallis et al., 2016).

This area of research has been and remains highly interdisciplinary, though there are important contributions by health geographers. In the late 19th century, efforts to reduce harmful effects of rapid industrialization and urbanization resulted in planning approaches that separated industry and residential areas using zoning (Corburn, 2007). In the 20th century, fields diverged as public health shifted to a biomedical model of disease focusing on individuals, and planning turned toward large infrastructure and transportation projects. These disparate foci had the detrimental consequence of promoting urban sprawl and reliance on the automobile, ultimately impacting health through increased pedestrian injuries, decreased physical activity and environmental costs. By the early 21st century, the fields converged again, in part motivated by rising attention to chronic disease and obesity. This era was associated with increased awareness of the role of context in shaping health behaviors (Corburn, 2007; Sallis et al., 2004) – specifically, how neighborhood walkability, access to healthy foods or other amenities might encourage positive changes in health (Sallis et al., 2004; Schulz and Northridge, 2004). In the most recent decade, walkability has blossomed as a rich and well-investigated field, with hundreds of papers crossing a number of disciplines, such as public health, population health, social epidemiology and urban planning (Andrews et al., 2012).

Numerous reviews summarize the walkability, physical activity and health literature (see, e.g., Lovasi, Grady and Rundle, 2011; Van Cauwenberg et al., 2011). In general, these reviews find that walkable neighborhoods have been linked with increased walking, lower obesity and lower cardiovascular disease. Despite findings that support these links, real challenges remain in terms of the strength of evidence regarding walkability and its links to health. Most of the evidence comes from observational cross-sectional studies that describe the health behaviors of different people living in different settings at a single point in time. Some of these studies are large and capture populations across multiple countries, with diverse built-environment conditions (Sallis et al., 2016). However, cross-sectional studies do not contribute toward building a causal evidence base.

With the majority of people worldwide living in urban areas, cities remain a crucial component of population health. Designing vibrant, walkable urban spaces is an opportunity to engage residents and to influence physical, mental and social health. Walkability is a key pillar of urban futures, and it may be a strategy to help combat air pollution, noise, physical inactivity and isolation. Thus, more research is needed. For the field of walkability in health geography and beyond to advance, there is a need for longitudinal studies and quasi-experimental studies, examining the impacts of changes in walkability on changes in health over time. Notably, there has been an emergence of studies that follow the health behaviors of people who move to areas with different walkability (Giles-Corti et al., 2013; Hirsch et al., 2014a), as well as natural experiment studies that examine how health behaviors change as modifications are made to the built environment (Benton et al., 2016). This work reveals that effect sizes from longitudinal studies may be smaller than observed in cross-sectional studies, reopening the question as to the potential of walkability to promote health after decades of work.

Roadmap for research methods on walkability and physical activity

Research on walkability requires specific attention to how spatial concepts are measured, or operationalized, and health geography has much to contribute. We present a roadmap for geographically relevant considerations in research on walkability and physical activity that may be of value to those new to the area (Table 41.1). While the considerations are presented sequentially, in practice they are always iterative. Further, depending on the research approach, considerations may be concurrent or in a different order. To date, health geographers have been most involved in considerations A and B (around *where* and *what*), and somewhat less so in consideration C (*how much*), relating walkability with health outcomes.

Consideration A: defining the spatial dimension (where)

Researchers have used a variety of methods to operationalize neighborhoods, including administrative units, buffers, activity spaces and self-defined neighborhoods (Thornton, Pearce and Kavanagh, 2011) (Figure 41.1). In this section, we present these broad types, along with their respective advantages and disadvantages.

Table 41.1 Roadmap for research on walkability and health

Methodological considerations	Examples
A. defining the spatial dimension (*where*)	administrative boundaries buffers GPS/activity spaces self-defined boundaries
B. selecting and measuring relevant built environment features (*what*)	walkability indices individual components of the built environment (e.g., residential density, land-use mix, density of parks)
C. obtaining physical-activity data (*how much*)	aggregated data self-report accelerometry GPS

Much early walkability research used administrative boundaries, such as counties, census tracts or zip codes, to define spatial dimensions (Leal and Chaix, 2011). Administrative boundaries are useful if health data (walking, physical activity, outcomes) needs to be aggregated to protect the identities of individuals. Additionally, some administrative boundaries have utility as political, social or jurisdictional areas. For example, if examining how walkability relates to active travel to school, researchers might choose to use school district as a meaningful spatial dimension. Similarly, researchers may use city boundaries to make policy recommendations for city planners.

To generate more individual-specific spatial dimensions for walkability research, many studies geocode individuals' residential addresses in geographic information systems (GIS) and generate distance-based buffers. There are two dominant approaches: Euclidean buffers and network buffers (Figure 41.1). Euclidean (crow-fly) buffers are created by drawing a straight line to a given distance from a home address (Thornton, Pearce and Kavanagh, 2011), whereas network (line-based) buffers are created by drawing a line a given distance along a network, usually a street, and then buffering that line a short distance (e.g., 50 meters) (Oliver, Schuurman and Hall, 2007; Thornton, Pearce and Kavanagh, 2011). In a review of literature using buffers, 65% of studies used Euclidean buffers (Leal and Chaix, 2011). However, as network buffers are widely considered to provide more accurate representations of the spatial context that might influence walking, this trend may change.

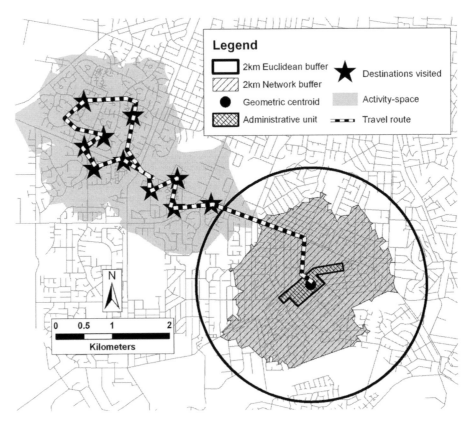

Figure 41.1   Methods of operationalizing neighborhoods

There has been a rise in the use of Global Positioning System (GPS) devices within the walkability field. GPS provides point data and reveals the true spatial context that individuals experience. When linked with accelerometry data, GPS data can provide insights as to where and when individuals are physically active (Jankowska, Schipperijn and Kerr, 2015). GPS has been especially critical in defining spatial extents in a more fluid, individualized way known as *activity spaces*, which reflect a subset of locations that individuals experience during their day-to-day activities. Three major types of activity spaces in walkability research are (1) the standard deviation ellipse, (2) the minimum convex polygon and (3) the daily path area (Hirsch et al., 2014b) (Figure 41.2).

Some research has also relied on spatial dimensions as defined by participants. This work is reminiscent of behavioral-geography work from the 1960s and 1970s, drawing mental maps by defining neighborhoods through the lived experience of residents and humanist-geography work from the 1970s and 1980s, interested in the notion of place as the subjective experiences of a physical location, endowed with meaning and associated with certain activities and social relations. In the handful of published studies on walkability that have asked participants to draw their neighborhood boundaries, it is clear that self-defined boundaries vary between people, as well as differ from other approaches to determining boundaries. For example, one study found substantial differences between participants in the shape and area of neighborhoods, as well as large inconsistencies between the size and amenities (including restaurants, recreation centers, food stores, parks) of self-drawn boundaries as compared to buffer and administrative boundaries (Colabianchi et al., 2014). Nonetheless, self-defined boundaries have the advantage of capturing participants' perceptions of their boundaries and have been very useful as an engagement or qualitative research tool for sparking discussion around neighborhood amenities.

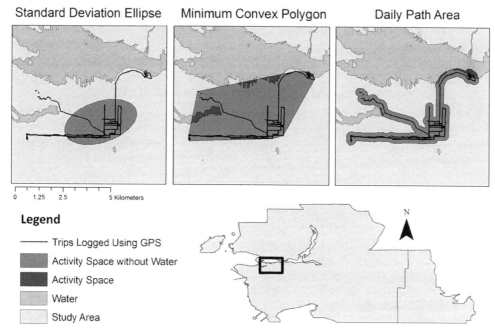

Figure 41.2 Major types of activity spaces in walkability research

Three common geospatial issues occur during the operationalization of spatial extent in walkability research: the Uncertain Geographic Context problem; the Modifiable Areal Unit Problem (MAUP); and the focus on residential environments. First, the Uncertain Geographic Context problem, first put forward by Kwan (2012), relates to the operationalization of neighborhood, specifically uncertainty whether the selected spatial dimension is the relevant space to influence behavior. The diverse and sometimes arbitrary nature of spatial dimensions employed in walkability research highlights this problem. For example, in research on buffer-based neighborhoods, Euclidean buffers range from 100 to 4,800 meters and network buffers range from 640 to 2,000 meters (Leal and Chaix, 2011). This lack of uniformity in size, combined with the lack of uniformity in shape (Euclidean versus network) raises questions around relevance and comparability across studies. Further, the appropriate distance may differ by individual sociodemographic or health characteristics – for example, older adults with mobility issues travel shorter distances. To discern the relevant spatial dimension, researchers require a clear understanding of the underlying causal processes (Diez Roux and Mair, 2010), an area needing much attention.

Second, the MAUP is a widespread issue in walkability research, as with other fields of geography. MAUP is the phenomenon in which changes in either scale or shape of spatial units result in differences in measures (e.g., environment features, walking rates) and/or the observed relationship between them (Openshaw, 1984). MAUP occurs in two different dimensions: scale effect and zoning effect. The scale effect relates to different levels or sizes of spatial aggregation (e.g., census-block groups compared to counties). The zoning effect relates to the patterns of configuration, even with a fixed level of spatial aggregation (e.g., if census-block groups were rotated 180 degrees). The administrative boundaries used in walkability research may be particularly at risk for MAUP, since daily travel is not spatially constrained to census boundaries or zip codes. Investigations within this area have found substantial scale and zoning effects (Clark and Scott, 2014; Haynes et al., 2007; Houston, 2014; Mitra and Buliung, 2012). Clark and Scott (2014) provide guidance, including the use of policy-relevant zones, the use of study-relevant distances to define buffers and the use of disaggregate data, to minimize MAUP in walkability research.

Third, another concern in operationalizing neighborhoods for walkability research is the heavy reliance on residential neighborhoods. Many have argued that relevant spatial dimension should include other common settings, including work, school or leisure-time environments (Chaix, 2009). The bias toward residential environments is ubiquitous: in a review of 131 studies on built environment and chronic disease, 90% of studies investigated exclusively residential exposures (Leal and Chaix, 2011). This typically arises as home addresses are the only available geographic identifier; future studies may want to query work locations or third places. Studies using a self-defined or GPS-based operationalization of neighborhood may alleviate this issue but bring their own issues. For one, despite a long history in space-time geography (Hägerstrand, 1970), the application of activity spaces in walkability research is relatively new. Time-geography frameworks explicitly consider individual constraints (e.g., capability, authority, coupling constraints) that may limit movement and define a time pattern of mobility. These concepts have received little attention in the walkability and health work to date.

Consideration B: selecting and measuring relevant built environment features (what)

Health geography contributes to decisions about relevant environmental features and the translation of these into metrics for walkability. Metrics for walkability may fall into two broad categories: those created using GIS, often called objective; and those generated through surveys of residents or audit tools, sometimes referred to as perceived or subjective. Several recent articles provide a thorough review of this area (Brownson et al., 2009; Eyler et al., 2015; Thornton, Pearce and Kavanagh, 2011).

Walkability is operationalized in GIS using a plethora of metrics. Table 41.2 lists common concepts and metrics.

Table 41.2 Common metrics used in research on walkability and health

Built-environment feature	Description	Examples of operationalization	Conceptual relationship with physical-activity behaviors	Typical data source
Land use	Composition of land use (residential, office, commercial, industrial, recreational, etc.)	Density: Area of (specific land use)/total land area	Lower residential land use, more walking; Higher commercial/ recreational/ industrial land use, more walking	Zoning data, land-use parcel data
Land-use mix	Extent to which there is evenness in the land-use categories across an area	Entropy measures, low values for more homogenous land use	Higher land-use mix, more walking	Zoning data, land-use parcel data
Destinations	Accessibility of various destinations (banks, grocery stores, parks, recreation centers)	Availability: presence of (specific destination) in area Distance: meters to the nearest (specific destination) Density: number of (specific destination)/area	Shorter distance, more walkable; higher density, more walking	Point data from commercial databases
Street connectivity	Degree of route connectivity from a location to a destination – for example, the diversity of route choices	Intersection density Average block length Link-to-node ratio	Higher street connectivity, more walking	Road-network data
Urban design features	Traffic calming, benches, greenness, etc.	Availability: presence of [specific feature] in area Distance: meters to the nearest [specific feature] Density: number of [specific feature]/area	Higher values for urban design, more walking	Municipal files
Walkability	Composite measures that combine built-environment components into a single index	Z-score based walkability index including residential density, street connectivity, land-use mix Walk Score™: availability of common destinations in nearby areas	Higher walkability, more walking	Researcher-derived or commercially available

(Brownson et al., 2009; Thornton, Pearce and Kavanagh, 2011)

Survey tools exist to capture analogous or complementary data to objective measures. These data are subjective or perceived in nature. One of the most popular, the Neighborhood Environment Walkability Scale (NEWS) (Saelens et al., 2003), includes measures related to land-use mix, street connectivity, walking and cycling infrastructure, aesthetics, traffic safety and more. There are also dozens of audit tools for walkability, which can be used by research teams or members of the public (Moudon and Lee, 2003), with a majority examining sidewalks, roads, intersections, vehicles, pleasantness and safety (Maghelal and Capp, 2011). In general, more attention must be placed on perceived measures *in addition to* more objective measures as valid, credible metrics (Andrews et al., 2012).

Consideration C: obtaining physical-activity data (how much)

A third consideration in walkability research is characterizing physical activity – sometimes more specifically by intensity (e.g., moderate to vigorous physical activity, sedentary time), by activity (walking, cycling) or by purpose (commuting, leisure). Unfortunately, at times there seems to be a singular focus on walking, perhaps leading to a lack of attention to other forms of mobility (Andrews et al., 2012).

The physical-activity data may be linked to place in different ways. Some studies use data aggregated to administrative boundaries – for example, commute-to-work data for census tracts, or self-reported physical activity data for health regions from the US Behavioral Risk Factor Surveillance System (BRFSS) or the Canadian Community Health Survey (CCHS). Other studies collect individual-level physical activity, including self-reported or accelerometry data. This disaggregated data may be related to place through residential address and the use of buffers. Finally, some studies use mobile sensing data, linking GPS, accelerometry and built-environment data in GIS to analyze location-specific physical activity. A useful framework for this element of the roadmap is provided elsewhere (Jankowska, Schipperijn and Kerr, 2015).

Future directions

Moving forward, health geographers are poised to inform a number of outstanding questions for walkability research (Andrews et al., 2012; Rosenberg, 2016). For example, do changes in walkability reduce or exacerbate health inequities? Which population groups benefit most from changes in walkability? What is the potential for urban form change across different settings (urban, suburban, rural), and what is the potential for behavior change in each? In studies that focus on behavioral impacts, have other co-benefits and risks been considered (e.g., air pollution, noise, traffic safety, personal safety)? What can the research say about the different motivations for walking and the experiences of different demographic and social groups with different capacities and desires? Are different types of walking (strolling, short trips) sufficient to result in health benefits? Is the innate focus on walking limiting research on other forms of mobility (wheeling, skateboarding, cycling)? Critical health geographers can help shape future research directions to address these questions while engaging with and across all considerations that make up the roadmap we have presented in this chapter.

Methodologically, research on walkability can further benefit from analysis methods emerging from health geography. For example, despite the fact that neighborhood exposures or health behaviors are not randomly distributed in space, but rather naturally clustered, most studies in this area do not recognize spatial autocorrelation (Spielman and Yoo, 2009). Conventional regression methods used in walkability research may introduce bias by not accounting for place (Cerin, 2010). Some studies do use multilevel modeling approaches that account for clustering by administrative units (e.g., individuals in the same city or neighborhood). More nuanced spatial-regression methods, such as spatial auto-regressive models, are beginning to be used, and these may be promising avenues to advance rigor in this field. As well, the patterns of spatial autocorrelation can enrich our understanding of the impact of the built environment, as it is precisely the non-random patterns that help contextualize the role of the built environment relative to other socioeconomic factors, such as education.

An ongoing hurdle in research on walkability is the heterogeneity observed in the size and direction of associations across settings and studies. This heterogeneity may be due to historical contexts, city development patterns, sociodemographic patterning or cultural norms. In recent years, geographically weighted regression (GWR) has gained popularity for exploring these types of spatial heterogeneity. In GWR, the parameter estimates can vary locally, emphasizing the spatial patterning of relationships. GWR is very new to this field, but it could become a cornerstone in walkability research.

Emerging technologies and big data bring new questions. For example, GPS has enabled researchers to generate activity spaces, which provide enhanced specificity of spatial dimension and built-environment measurement (*where* and *what*, on the roadmap) and location-specific activities (*how much*). However, Chaix et al. have warned that these methods may introduce new biases, via selective daily mobility (Chaix et al.,

2013). For example, a GPS/accelerometer study might find that more physical activity happens in parks. Yet if participants who like (and do) more activity are going to the park specifically to be active, then this will result in selective daily mobility bias. Additionally, methodologies that capture a more holistic view of mobility are often more burdensome to participants and researchers (Jankowska, Schipperijn and Kerr, 2015) and invite potential privacy issues, especially when GPS is used to identify frequently visited destinations or with vulnerable populations. Health geographers should be contributing research design and study interpretations to ensure that nuanced issues associated with high-resolution geospatial data are considered.

Health geographers can help lead exploration of the role of place in health inequities by being mindful of avoiding environmental determinism (Andrews et al., 2012). There is an increasing focus on equity across diverse disciplines working on walkability, and findings are conflicting. One of the largest studies, spanning almost 65,000 census tracts in 48 US states, found disadvantaged neighborhoods and those with more educated residents to be more walkable (coined the disadvantaged advantage) (King and Clarke, 2014). Other studies cite opposite trends, suggesting that local context plays a large role (Bereitschaft, 2017). Geographers may have contributions to make in terms of differences by scale and measurement (of both social factors and walkability), as well as including urbanism, history and development patterns in characterizing context.

References

Andrews, G. J., Hall, E., Evans, B. and Colls, R. (2012). Moving beyond walkability: on the potential of health geography. *Social Science & Medicine*, 75(11), pp. 1925–1932.

Benton, J. S., Anderson, J., Hunter, R. F. and French, D. P. (2016). The effect of changing the built environment on physical activity: a quantitative review of the risk of bias in natural experiments. *International Journal of Behavioral Nutrition and Physical Activity*, 13(1), p. 107.

Bereitschaft, B. (2017). Equity in neighbourhood walkability? A comparative analysis of three large US cities. *Local Environment*, 22(7), pp. 859–879.

Brownson, R. C., Hoehner, C. M., Day, K., Forsyth, A. and Sallis, J. F. (2009). Measuring the built environment for physical activity: state of the science. *American Journal of Preventive Medicine*, 36(4), pp. S99–S123, e12.

Cerin, E. (2010). Statistical approaches to testing the relationships of the built environment with resident-level physical activity behavior and health outcomes in cross-sectional studies with cluster sampling. *Journal of Planning Literature*, 26(2), pp. 151–167.

Chaix, B. (2009). Geographic life environments and coronary heart disease: a literature review, theoretical contributions, methodological updates, and a research agenda. *Annual Review of Public Health*, 30, pp. 81–105.

Chaix, B., Méline, J., Duncan, S., Merrien, C., Karusisi, N., Perchoux, C., Lewin, A., Labadi, K. and Kestens, Y. (2013). GPS tracking in neighborhood and health studies: a step forward for environmental exposure assessment, a step backward for causal inference? *Health & Place*, 21, pp. 46–51.

Clark, A. and Scott, D. (2014). Understanding the impact of the modifiable areal unit problem on the relationship between active travel and the built environment. *Urban Studies*, 51(2), pp. 284–299.

Colabianchi, N., Coulton, C. J., Hibbert, J. D., McClure, S. M., Ievers-Landis, C. E. and Davis, E. M. (2014). Adolescent self-defined neighborhoods and activity spaces: spatial overlap and relations to physical activity and obesity. *Health & Place*, 27, pp. 22–29.

Corburn, J. (2007). Reconnecting with our roots: American urban planning and public health in the twenty-first century. *Urban Affairs Review*, 42(5), pp. 688–713.

Diez Roux, A. V. and Mair, C. (2010). Neighborhoods and health. *Annals of the New York Academy of Sciences*, 1186(1), pp. 125–145.

Eyler, A. A., Blanck, H. M., Gittelsohn, J., Karpyn, A., McKenzie, T. L., Partington, S., Slater, S. J. and Winters, M. (2015). Physical activity and food environment assessments. *American Journal of Preventive Medicine*, 48(5), pp. 639–645.

Giles-Corti, B., Vernez-Moudon, A., Reis, R., Turrell, G., Dannenberg, A. L., Badland, H., Foster, S., Lowe, M., Sallis, J. F., Stevenson, M. and Owen, N. (2016). City planning and population health: a global challenge. *The Lancet*, 388(10062), pp. 2912–2924.

Giles-Corti, B., Bull, F., Knuiman, M., McCormack, G., Van Niel, K., Timperio, A., Christian, H., Foster, S., Divitini, M. and Middleton, N. (2013). The influence of urban design on neighbourhood walking following residential relocation: longitudinal results from the RESIDE study. *Social Science & Medicine*, 77, pp. 20–30.

Hägerstrand, T. (1970). What about people in regional science? *Papers in Regional Science*, 24(1), pp. 7–24.

Haynes, R., Daras, K., Reading, R. and Jones, A. (2007). Modifiable neighbourhood units, zone design and residents' perceptions. *Health & Place*, 13(4), pp. 812–825.

Hirsch, J. A., Diez Roux, A. V., Moore, K. A., Evenson, K. R. and Rodriguez, D. A. (2014a). Change in walking and body mass index following residential relocation: the multi-ethnic study of atherosclerosis. *American Journal of Public Health*, 104(3), pp. e49–e56.

Hirsch, J. A., Winters, M., Clarke, P. and McKay, H. (2014b). Generating GPS activity spaces that shed light upon the mobility habits of older adults: a descriptive analysis. *International Journal of Health Geographics*, 13(1), p. 51.

Houston, D. (2014). Implications of the modifiable areal unit problem for assessing built environment correlates of moderate and vigorous physical activity. *Applied Geography*, 50, pp. 40–47.

Jankowska, M. M., Schipperijn, J. and Kerr, J. (2015). A framework for using GPS data in physical activity and sedentary behavior studies. *Exercise and Sport Sciences Reviews*, 43(1), p. 48.

King, K. E. and Clarke, P. J. (2014). A disadvantaged advantage in walkability: findings from socioeconomic and geographical analysis of national built environment data in the United States. *American Journal of Epidemiology*, 181(1), pp. 17–25.

Kwan, M. P. (2012). The uncertain geographic context problem. *Annals of the Association of American Geographers*, 102(5), pp. 958–968.

Leal, C. and Chaix, B. (2011). The influence of geographic life environments on cardiometabolic risk factors: a systematic review, a methodological assessment and a research agenda. *Obesity Reviews*, 12(3), pp. 217–230.

Lovasi, G. S., Grady, S. and Rundle, A. (2011). Steps forward: review and recommendations for research on walkability, physical activity and cardiovascular health. *Public Health Reviews*, 33(2), p. 484.

Maghelal, P. K. and Capp, C. J. (2011). Walkability: a review of existing pedestrian indices. *URISA Journal*, 23(2), pp. 5–20.

Mitra, R. and Buliung, R. N. (2012). Built environment correlates of active school transportation: neighborhood and the modifiable areal unit problem. *Journal of Transport Geography*, 20(1), pp. 51–61.

Moudon, A. V. and Lee, C. (2003). Walking and bicycling: an evaluation of environmental audit instruments. *American Journal of Health Promotion*, 18(1), pp. 21–37.

Oliver, L. N., Schuurman, N. and Hall, A. W. (2007). Comparing circular and network buffers to examine the influence of land use on walking for leisure and errands. *International Journal of Health Geographics*, 6(1), p. 41.

Openshaw, S. (1984). *The modifiable areal unit problem.* (Concepts and Techniques in Modern Geography series). University of East Anglia: GeoBooks.

Rosenberg, M. (2016). Health geography III: old ideas, new ideas or new determinisms? *Progress in Human Geography*, 41(6), pp. 832–842.

Saelens, B. E., Sallis, J. F., Black, J. B. and Chen, D. (2003). Neighborhood-based differences in physical activity: an environment scale evaluation. *American Journal of Public Health*, 93(9), pp. 1552–1558.

Sallis, J. F., Bull, F., Burdett, R., Frank, L. D., Griffiths, P., Giles-Corti, B. and Stevenson, M. (2016). Use of science to guide city planning policy and practice: how to achieve healthy and sustainable future cities. *The Lancet*, 388(10062), pp. 2936–2947.

Sallis, J. F., Frank, L. D., Saelens, B. E. and Kraft, M. K. (2004). Active transportation and physical activity: opportunities for collaboration on transportation and public health research. *Transportation Research Part A: Policy and Practice*, 38(4), pp. 249–268.

Schulz, A. and Northridge, M. E. (2004). Social determinants of health: implications for environmental health promotion. *Health Education & Behavior*, 31(4), pp. 455–471.

Spielman, S. E. and Yoo, E. H. (2009). The spatial dimensions of neighborhood effects. *Social Science & Medicine*, 68(6), pp. 1098–1105.

Thornton, L. E., Pearce, J. R. and Kavanagh, A. M. (2011). Using Geographic Information Systems (GIS) to assess the role of the built environment in influencing obesity: a glossary. *International Journal of Behavioral Nutrition and Physical Activity*, 8(1), p. 71.

Van Cauwenberg, J., De Bourdeaudhuij, I., De Meester, F., Van Dyck, D., Salmon, J., Clarys, P. and Deforche, B. (2011). Relationship between the physical environment and physical activity in older adults: a systematic review. *Health & Place*, 17(2), pp. 458–469.

World Health Organization. (2009). *Global health risks: mortality and burden of disease attributable to selected major risks.* Geneva: World Health Organization.

SECTION 5

Practicing health geographies

42

INTRODUCING SECTION 5

Practicing health geographies

Gavin J. Andrews, Valorie A. Crooks and Jamie Pearce

An important part of what health geographers do, indeed of what any researchers do, is the particular ways in which they practice. In the context of academic disciplines, practice involves the research methods employed, the analytical approaches and techniques adopted, and the ways in which the knowledge produced through studies is disseminated and translated to audiences of various types. Practice, then, is often quite practical and applied, arising as sets of skills to be learned and deployed. Having said this, practice has consequences that go way beyond the processual aspects of research related to its very form and reputation. Indeed, on one level, being highly connected to theoretical orientations and lenses such as those discussed in Section 2 – making it possible to see through these lenses – practice informs and refines research questions. On another level, in terms of standards, *good* practice is absolutely critical in the answering of research questions; it enhances the quality (integrity and rigor), insightfulness, relevance and impact of the findings produced. Ultimately, then, practice not only can help determine the focus, strengths and successes of research, but also has implications for academic identity and the particular type of scholar one becomes and is known to be.

Practice has been a particular strength of health geography. Over the years, scholars have produced numerous journal articles – and a number of journal special editions and books – focused on the *how to* of the sub-discipline (Cromley and McLafferty, 2011; Elliott, 1999; Fenton and Baxter, 2016). Why practice is a particular disciplinary strength is difficult to ascertain, though it might be because health geography, rather than concerning itself with leading theoretical developments in human geography, has sought to make a tangible difference in the worlds of health and health care; realms where strong institutional structures (ranging from funding bodies with focused and applied agendas to broader sectorial movements such as public health and evidence-based practice) exist to improve research practice, with a view to making this difference. At the end of the day, health geographers have paid attention to bettering the ways in which they work because they realize they are dealing with matters of life, death and at the very least human comfort, well-being and happiness; things at the very core of human existence.

Practice is something that needs to be learned and improved at all levels; senior undergraduate, graduate and professional, as one can always know more and do better. Almost all the chapters in this handbook consider or showcase practice to some extent. Indeed, one cannot do health geography without practicing, so this is inevitable. However, the chapters in this section are focused explicitly on particular types of practice in health geography, providing an entry point into, and overview of, each, together providing some breadth and depth on the topic. Because practice is so varied in health geography, there are not really any themes that cross-cut all or most of the chapters in this section of the book. However, three groups of chapters are distinguishable for their particular interests and focuses.

The first two chapters in this section are focused on practices in qualitative health geography. First, Jamie Baxter considers the ascendance of qualitative methods since the 1990s, which are now a mainstream approach in the sub-discipline. He describes how they have been based on, and have articulated, far more expansive and socially aware versions of both health and place; how they have uncovered people's experiences, attachments and identities related to health and place; and how they have provided different kinds of evidence that speaks to the personal, everyday impacts of disease and the impacts or neglects of health services (Kearns, 1993). Indeed, in his chapter, Baxter defines some key terminologies, and some key approaches and debates, in qualitative health geography. Second, Candice Boyd and Michelle Duffy focus on arts-based approaches. These are increasingly popular across the social sciences (in human geography, often under the label *geohumanities*), and this is also the case in both health geography and health research more broadly, where art can be either an intervention or a form of knowledge translation. Specifically, Boyd and Duffy consider the influences of arts-related developments in social and cultural geography on health geography, the specific modalities of therapeutic art practices and music, and how these relate to the health of individuals and communities. Overall, although including only two chapters on qualitative approaches in this section might seem a little light, qualitative methods are in fact explored and conveyed throughout many other chapters in the book.

The next four chapters in this section are focused on practices in quantitative health geography. First, Daniel Lewis takes a general look at quantitative research and the qualitative revolution. Addressing fundamentals, he considers what data is and how scholars understand and use it. He focuses critically, not only on forms of data, but also on the institutions and infrastructures that provide data or give access to it. Second, Alec Davies and Mark Green introduce and describe big data, an overarching concept in social and health research that describes the compilation and use of large datasets. In particular, they review the application of big data in health geography for understanding health patterns. They argue, however, that despite the considerable promotional rhetoric on big data from political and service spheres, much needs to be done, and so they also focus on the challenges and opportunities for future scholarship. Third, continuing the focus on big data, Nina Morris considers ethical issues in the use of large datasets in health geography discussing issues related to consent and re-identification. Fourth and finally, Sara McLafferty and Sandy Wong consider trends, critiques and directions in spatial health modeling, a long-standing approach that is central to health geography's understandings of population health and health-services use. McLafferty and Wong take a historical approach to the subject, looking at innovations using geographic information systems (GIS), but they also look to the future – particularly to how spatial health modeling might be used in conjunction with, or flow to or from, qualitative methods in health geography. This cross-cutting debate nicely previews the final chapters in this section.

The final four chapters in this section are focused on practice concerns that cross-cut both qualitative and quantitative approaches in health geography (i.e., some important considerations regardless of researchers' particular methodological orientations). First, Melissa Giesbrecht discusses the need to acknowledge diversity in the practice of health geography. She explores difference, sameness and forms of solidarity related to various social intersectionalities (such as gender), the ways in which they both are based in and make places, and the ways in which they affect health states and experiences. Second, Mylene Riva and Sarah Mah focus on population-health interventions for public health, describing population health as the science underpinning the practice of public health (which attempts to understand how health is generated and distributed in populations). Using examples, they explore health geography's contribution to the approach in terms of defining the full contexts of interventions. Third, Allison Williams reviews place-based intervention research, affirming that health geographers have long been involved in health interventions at all three key stages (needs assessment, implementation and evaluation), that they have provided place-based knowledge and expertise, and that, in doing so, they have spoken particularly to the impact of deprivation, poverty and caregiving circumstances and relationships. Fourth and finally, Elizabeth Peter and Joan Liaschenko explore practitioner perspectives through the case of nursing. They describe how nurse researchers have undertaken health-geography research to better understand nursing practice – particularly, how the ethics and power of place play a role in this particular health profession. Thus, Peter and Liaschenko describe a form of health geography, practiced by practitioners (non-geographers) to improve their practice. Although this might seem an unusual situation, consider the sheer volume of nurse researchers involved in this endeavor (Andrews, 2016).

In terms of the future, certainly there is a need for health geographers to develop practice that recognizes and helps tackle pressing health issues head-on; the health consequences of aging populations, widening health inequalities, the health impacts of climate change and other global health phenomena all being very important considerations. Indeed, health geographers might consider the development of understandable, accessible, publicly involved – even activist – approaches that work directly to produce change, perhaps borrowing from human geography's idea and agenda of *public geographies* (Fuller, 2008). In addition, there are other challenges at a disciplinary level; a key question being whether health geography is adequately equipped to capture a *new world* of health. In health care, this new world involves services provided almost everywhere people are found – where they live, work, are educated, shop and spend leisure time (McKeever and Coyte, 2002). Moreover, it involves rapid technological developments and innovations in the areas of monitoring, assistance, communication, record-keeping, consultation, triage and even treatment, as well as changing relationships between human bodies and care (McKeever and Coyte, 2002). More generally, beyond health care, as Thrift (2008) argues, this new world involves life moving at an ever-faster pace and with increasing intensity, where public and private interests engineer the core textures and feelings of people's lives – adding multiple aesthetics, attractions and distractions that perform to multiple senses – and where people's technological devices/obsessions give them new forms of awareness and knowledge, adding multiple coexisting timescales and levels of consciousness to their existence. For Thrift (2011), this is an exaggerated humanity and, we argue, one with huge implications for the physical and mental health and well-being of people. Research practices in health geography need to in some way account for and show all this. Indeed, scholars need technologies and techniques to engage, capture and animate health in the fractured, moving, virtual multiplicity of modern life.

References

Andrews, G. J. (2016). Geographical thinking in nursing inquiry, part one: locations, contents, meanings. *Nursing Philosophy*, 17(4), pp. 262–281.

Cromley, E. K. and McLafferty, S. L. (2011). *GIS and public health*. 2nd ed. New York: Guilford Press.

Elliott, S. J. (1999). And the question shall determine the method. *The Professional Geographer*, 51(2), pp. 240–243.

Fenton, N. E. and Baxter, J. (eds.) (2016). *Practicing qualitative methods in health geographies*. Abingdon: Routledge.

Fuller, D. (2008). Public geographies: taking stock. *Progress in Human Geography*, 32(6), pp. 834–844.

Kearns, R. A. (1993). Place and health: towards a reformed medical geography. *The Professional Geographer*, 45(2), pp. 139–147.

McKeever, P. and Coyte, P. C. (2002). Here, there and everywhere. *University of Toronto Bulletin A*, 16–25.

Thrift, N. (2008). *Non-representational theory: space, politics, affect*. Abingdon: Routledge.

Thrift, N. (2011). Lifeworld Inc – and what to do about it. *Environment and Planning D: Society and Space*, 29(1), pp. 5–26.

43

QUALITATIVE HEALTH GEOGRAPHY REACHES THE MAINSTREAM

Jamie Baxter

For decades, qualitative research remained on the fringes of social science, including health geography, but most students today will recognize this broad suite of approaches, methodologies and methods as mainstream. Yet, some tensions within qualitative inquiry that emerged decades ago remain today.

This chapter has two main goals. The first is to define "qualitative research" in a way that not only acknowledges that qualitative approaches are indeed different from quantitative ones, but also recognizes that such differences can be blurry in practice. The second is to trace some key points of tension concerning qualitative approaches, including such issues as counting or quantification, assessing the quality of empirical work (rigor), and determining how to mix qualitative and quantitative methods in the same study (Crang, 2002). Given the vastness of qualitative inquiry, this chapter does not cover the full range of issues relevant to health geographers, which is something that is better covered in a book-length manuscript (e.g., DeLyser et al., 2010; Fenton and Baxter, 2016; Hay, 2016).

Explosive growth in qualitative studies

Though much has changed since Dyck (1999, p. 250) wrote about the value of qualitative research for moving us away from a "medicocentric" approach to health geography, her characterization of the roles for qualitative research remain as relevant as ever:

> Qualitative research approaches allow the examination of variations in the construction of ideas about health, illness, and health care over time and place which unsettle taken-for-granted social categories and expose relations and distributions of power that are involved in the constitution of subjectivities and experiences of being healthy, sick, or disabled.

Analysis of journal-database searches brings to light the exponential growth of qualitative research in health geography in the last two decades. Ironic for the use of numbers, perhaps, this section includes ratios of qualitative to quantitative publications from 1900 to 2000 compared to 2001–2017. A raw Web of Science topic (keyword) search[1] for "health geography qualitative" finds that published studies grew from 12 to 250 across the two periods, while "health geography interview" grew from 18 to 369 journal articles. Yet quantitative research also grew; for example, the comparators "health geography statistic★"[2] and "health geography survey" grew from 31 to 409 and from 56 to 476 articles, respectively (see Table 43.1, column K).

Breaking this down by period, calculating ratios from this same journal-database search gives us a sense of the intensification of qualitative research relative to quantitative research. For example, the ratio of qualitative to statistical health-geographic work has increased from 0.39 to 0.61 (ratio A:B) across these two periods – an increase or intensification of 0.22 (column L); and the ratio of interviews to surveys grew even more dramatically, from 0.32 to 0.78 (ratio C:D) – an intensification of 0.45 (See Table 43.1). Before we health geographers get too proud of the rise of qualitative research in our sub-discipline, though, the bottom half of Table 43.1 shows that we seem to lag behind geography more generally according to these ratio measures.

While qualitative approaches in health geography may not be increasing at the same pace as in the wider discipline, health geographers nevertheless continue to innovate with qualitative approaches. Innovations stem partially from the fact that the traditional use of interview methods has proven unsatisfying for many practitioners, which has prompted two main directions: (1) mixing qualitative and quantitative approaches (e.g., Cutchin, 2007) and (2) pushing qualitative ways of knowing even further (e.g., participant-generated drawing – Coen, 2016; non-representational theory – see Andrews, Chen and Myers, 2014). Thus, many health geographers are constantly critical of the status quo in qualitative research (Fenton and Baxter, 2016).

Table 43.1 Ratio of qualitative to quantitative published in Web of Science journal articles by era

Search keywords	# WOS Journal Articles			
	I 1900–2000	J 2001[#]–2017	K POOLED 1900–2017	L Qualitative Intensification J minus I
HEALTH GEOGRAPHY				
A: (medical OR health) & geography & qualitative	12	250	262	
B: (medical OR health) & geography & statistic★[1]	31	409	440	
Ratio A:B	**0.39**	**0.61**	**0.60**	**0.22**
C: (medical OR health) & geography & interview★	18	369	387	
D: (medical OR health) & geography & survey★	56	476	532	
Ratio C:D	**0.32**	**0.78**	**0.73**	**0.45**
GEOGRAPHY				
E: geography & qualitative	91	1159	1250	
F: geography & statistic★[1]	277	1808	2085	
Ratio E:F	**0.33**	**0.64**	**0.60**	**0.31**
G: geography & interview★	128	1843	1971	
H: geography & survey★	307	1934	2241	
Ratio G:H	**0.42**	**0.95**	**0.88**	**0.54**

★ search-term wildcard – e.g., statistic, statistical, statistically are all found with this wildcard suffix
Initially 1995 was the breakpoint – the year of the inaugural issue of *Health & Place*, a key journal for qualitative health research – but only 1 article appears in the 1900–1994 period for "(medical OR health) & geography & qualitative"

It is essential, then, to engage with qualitative researchers in allied social sciences (e.g., medical sociology, feminist approaches) and health sciences (e.g., nursing) to bring issues facing qualitative researchers in health geography to full light.

What is *qualitative*? Beyond dualistic thinking

Health geographers have a long tradition of anti-dualistic thinking in that we strive to avoid, for example, separating the mind and the body or, similarly, separating social processes from health (Hall, 2000). Yet, what distinguishes qualitative research is not always apparent as we step away from the idea of qualitative and quantitative research as entirely separated polar opposites (Lawson, 1995). Many of us have a pragmatic sense of qualitative research, in particular with reference to method. For example, Asanin and Wilson's (2008) study of immigrant experiences of access to health care in Mississauga, in Canada, would likely be recognized immediately as a qualitative study. That the authors use focus groups is an initial clue, but, even more so, linking to Dyck's quote above, their study concerns deepening our understanding of and redressing the marginalization of immigrants accessing health care. Other methods that might tip us off to the qualitative nature of a study include interviews, discourse analysis, textual analysis, content analysis, photovoice and participant observation. However, what distinguishes qualitative research most from other forms of inquiry is not method but a focus on depth, texture, and nuance, as well as context and social contingencies that accompany phenomena – including health outcomes. Method alone is not sufficient for identifying qualitative work; doing so is a potential source of confusion, as methods like textual analysis, content analysis and observation all have variants that are decidedly quantitative in that they are more focused on counting instances of a phenomenon than on interpretive depth.

Most textbooks describing qualitative methods produce lists something like Table 43.2, which shows the many ways in which qualitative research is presumably the opposite of quantitative research – the potentially misleading dualisms that Lawson (1995) and others criticize (e.g., Longhurst, 1995; Plumwood, 1991). While we may reject dualistic thinking in general, in some cases there may remain stark enough differences (e.g., epistemology, ontology). In other cases, there may be more of a continuum (e.g., analysis-presentation). Such tables still serve as a useful heuristic device for establishing the ways in which qualitative approaches can be both interrelated and complementary with quantitative approaches. The wavy line, then, is meant to represent permeability, and it reminds us that the differences are fuzzy in many cases.

Further, there is a range of approaches on both sides of the wavy line – particularly on the qualitative side. For example, while we typically refer to positivism (or post-positivism – see Sheppard, 2001) as the philosophical basis of quantitative research, the philosophical underpinnings of qualitative research are not that simple. There are several very different philosophies supporting the variety of approaches to qualitative research; for example, post-structuralism, feminism, humanism, symbolic interactionism, realism, critical theory and phenomenology (Kearns and Moon, 2002). Qualitative researchers may disagree somewhat on the philosophical underpinnings of qualitative research, including *epistemology* – what counts as legitimate knowledge – and *ontology* – how the world exists so that we may know it. For example, in terms of epistemology, whereas humanists and feminists with an eye to symbolic interactionism would tend to engage *directly* with the populations they study (for example, through interviews and participant observation), some critical theorists and other sorts of feminists and humanists may prefer to engage *indirectly* (through cultural artifacts such as literature, the media or policy documents) to uncover hidden, implicit, yet insidious social processes we may unconsciously ignore yet reproduce in everyday life (e.g., racism).

The same can be said of ontology: there are varieties within qualitative research. Perhaps the most notable and familiar distinction is the different way critical theorists and social constructionists approach ontology. For example, as there is likely no single *reality* of immigrant access to health care, a qualitative

Table 43.2 Qualitative and quantitative research – the dualistic approach

	Quantitative research	*Qualitative research*
Purpose	Theory testing	Theory development
Epistemology	Measurable and objective is the foundation of knowledge	Interpreted and subjective is the foundation of knowledge
Ontology	Single unchanging reality	Multiple shifting realities
Sampling – type	Random	Purposive
Sampling – size	Large (100+)	Small (10–30)
Methods	e.g., surveys, secondary data	e.g., interviews, discourse analysis
Data type	Variables	Concepts/theory
Analysis – mode	Deduction	Induction
Analysis – presentation	Statistics, numbers	Quotations, text
Analysis – goal	Generalizability (breadth)	Credibility (depth)
Value for policy	Succinct	Stories that resonate

The wavy line symbolizes that the distinctions between qualitative and quantitative research are sometimes blurry or permeable – a matter of degree – particularly in practice and through the language used in publications.

researcher might expose differences in experiences from distinct immigrant groups such that, for example, first-generation women immigrants have very different experiences from their male counterparts. The goal for the qualitative researcher may be to provide texture and details about those two realities (e.g., with interview quotations) that go beyond simply measuring gender as a variable (re: quantitative positivism). However, it is naïve to assume that all qualitative researchers subscribe to an ontology that values understanding these multiple realities. For example, critical theorists or feminists adopting a realist ontology may be much more comfortable than their humanist social-constructionist counterparts with the idea of assuming, or at least pursuing in their research, a single reality. Yet, unlike positivists, critical theorists and feminists, among others, are more focused on highlighting the reality that contradicts and undermines the status quo – the latter being considered the hegemonic reality of powerful elites. For these researchers, the goal is to highlight how the status quo reality is misleading and oppressive (e.g., blaming the victim for failing to access care) and that by acknowledging an alternative explanation/reality (e.g., social structures put up barriers to recent women immigrants, and we unwittingly or intentionally reproduce those social structures through our actions) we have a sound foundation for change.

Moving away from philosophical foundations toward qualitative *analytical strategies*, the differences between quantitative and qualitative research may better be characterized as a continuum. Although quantitative research emphasizes counting, it is problematic to assume qualitative researchers avoid it (Sandelowski, 2001). In the 1990s, feminist geographers were debating the idea of quantification as being traditionally aligned, it was assumed, with positivism and the very patriarchal and problematic status quo. The debate was based on the tacit assumption that feminists leaned heavily toward qualitative approaches. Most authors of a special issue of the *Professional Geographer* on the topic each concluded, though, that quantitative research indeed has a valuable role to play in feminist inquiry (Lawson, 1995; McLafferty, 1995; Moss, 1995), which was a strong signal that quantification was going to seep further into qualitative methodologies.

Meanwhile, outside of geography, Sandelowski (2001), though she likewise cautiously embraces the idea of qualitative researchers counting phenomena, warns that such quantification fosters ambiguities, which have spawned problematic practices among qualitative researchers. Like the feminist geographers, Sandelowski is very much against "anything goes" quantification. Instead, she details counting pitfalls in qualitative research (e.g., verbal counting, misleading counting, over-counting, acontextual counting); practices that step too far over the blurred line between quantitative and qualitative research, reminding us that there should remain limits to how qualitative researchers engage with quantification.

Other tensions in qualitative research

In the previous section of this chapter, I touched on some tensions that arise when we try to define "qualitative research," particularly with reference to quantitative/qualitative dualisms. So far, I have used the examples of epistemology, ontology, and counting (quantification). In this section, I will detail three others that, in one way or another, represent sets of decision-points for qualitative health researchers: (1) how to assess the quality of empirical qualitative research work, (2) the role of member checking for addressing rigor and (3) whether and how qualitative and quantitative methods should be mixed. I use the term *tension* to remind us that, when conducting empirical research, we must choose a path, and that each path is defined by our research decisions.

Fascinating, but the sample size is too small

Two key arguments against measuring the quality or rigor of qualitative research are that it tempts us to use concepts and strategies borrowed from positivistic quantitative research and it stifles interpretive creativity and flexibility in fieldwork (Sandelowski, 1993). Terms such as "rigor," "validity" and "reliability" have traditionally been the domain of statistically oriented quantitative researchers, and many of us have had our work unfairly judged against these criteria. For example, the title of this section signifies the type of comments many qualitative researchers have received in the peer-review process. Such comments make us wary of adopting generalizability, also known as external validity, as a criterion for rigor. For positivistic science, generalizability/external validity is based on the notion that if we carefully choose some people to study (i.e., a sample), we can draw conclusions about many more people (i.e., the population). This is achieved using large random samples, such that if the key characteristics of the sample and the population are the same (e.g., %women, %low income) we can reasonably assume that what we have learned about the sample applies to the entire population. Yet, large samples are generally unrealistic in qualitative work, given the focus on depth – several hours can be spent working with one participant doing in-depth interviews, re-interviews and perhaps member checking. Thus, qualitative practitioners emphasize the quality of the conceptual development (interpretation) (i.e., credibility – see Lincoln and Guba, 1985), not necessarily the number of interviewee transcripts consulted to develop the concept.

This brings us back to the issue of *counting*. Though counting the instances of some phenomenon may be useful in say, a grounded theory study where induction and deduction are both used to (re)shape a concept, counting is not necessary. A few articulate and insightful interviewees, or a single unique yet highly generative event (e.g., a pilot test of a community intervention) go a long way in qualitative interpretive research. Knowing whether the concept is more widely applicable may ultimately be a goal, but perhaps such goals are better accomplished through complementary quantitative methods (e.g., surveys). The qualitative researcher's job is to define concepts in such a way that they would likely inhere in contexts beyond the interviewees studied – a qualitative criterion for rigor called *transferability* (Lincoln and Guba, 1985).

Less tends to be said about reliability in qualitative research, a quantitative criterion for rigor, which is based on the idea of repeatability and firmly rooted in experimental science. That is, a set of procedures if carried out in the same conditions are meant to reproduce the same findings. In health geography, this seems to translate well for survey research – since the questionnaire is the same for every respondent, effectively satisfying that aspect of repeatability embedded in reliability.[3] The argument against such a criterion is ontological and fairly basic: the social world is different than the world studied in experimental labs. Thus, nobody can ever repeat your study – the procedures may be the same, but the conditions can never be identical – the social world is constantly in flux. Further, even with the same data, different researchers may glean different concepts (Golafshani, 2003), a point revisited in the section on member checking.

Repetition of this type is particularly problematic for interviewers and focus-group facilitators in the field if, in order to conform to the tenet of repeatability, we expect them to say and do things in exactly the same way in exactly the same order. Such a strategy violates a core feature of these methods – the researcher-as-instrument – whereby humans as socially aware are expected to react "on the fly" as social beings, rather than in a pre-scripted way as socially distant automatons. Nevertheless, many qualitative researchers will admit that something like reliability is important; aptly summarized by the familiar adage: *Those who cannot remember the past are doomed to repeat it.* As health geographers, we want to identify and avoid repeating practices that oppress or are otherwise identified as being threatening to well-being; likewise, we want to be able to replicate processes that are positive and restorative. Such repeating though is not of the procedural kind, as enshrined in the experimental sciences. In the social sciences, the conditions that reproduce social phenomena such as war, oppression, reduced access to health care, high uptake of health insurance, or positive impacts of healing landscapes, are socially structured. These structures cause us to analytically ponder, in our development of concepts, the necessary and sufficient conditions that would *reproduce* such processes and events.

Member checking

Moving from criteria for assessing rigor (generalizability, reliability) to strategies for enhancing rigor, one of the more controversial ways to guard against threats to rigor in qualitative research is member checking. This is a strategy that involves enlisting the participants (members) in your study to mull over (check) your interpretations (Doyle, 2007; Lincoln and Guba, 1985). Yet, member checking has raised a number of questions among qualitative practitioners: who should indeed check our interpretations – the participants or wider academia? What happens when your participants disagree with your interpretations? And, more pragmatically, what exactly should we check?

The member-checking literature tends to rather narrowly focus on returning transcripts to participants to check for transcription accuracy (Harper and Cole, 2012), rather than the relevance of the researcher's interpretations of all transcripts. Transcription checks generally only go as far as asking study participants to comment on the themes[4] attached to their own interview, but with little reference to the bigger picture. Given that there may be numerous codes that are not used in the final analysis and writing, one downside of transcript-based member checking is asking participants to do work on ideas that may not find their way into any publication. Those who do check the interpretations of the full set of study transcripts (not just individual transcripts for the relevant *member*) are faced with a different set of problems, in the sense that members of the study are asked to comment on interpretations that may not have been derived much from their own conversation(s) with the researchers (Turner and Cohen, 2008). Participants may feel they are being asked to comment on phenomena with which they have little experience, while, on the contrary, they may disagree with how something in their community has been characterized in the interpretation (Mason et al., 2016).

Borland (1991) provides a poignant example of such a disagreement. She and her grandmother eventually agree that concepts relevant to feminist theory could reasonably be used to understand her grandmother's interactions with a group of men at a horse race. However, her grandmother would not have initially chosen that frame to interpret the story. Conversations with granddaughter Borland persuaded the grandmother that a feminist interpretation was at least reasonable. Member checking of the main study findings may not always be so congenial, though (Bradshaw, 2001); while McConnell-Henry, Chapman and Francis (2011) suggest that member checking itself is counterproductive to interpretive research, due in part to interpretive sovereignty issues and the idea that participants are not necessarily meant to understand academic theory. Koelsch (2013) instead prefers to look on member checking as an opportunity to raise awareness and facilitate positive transformative actions among study participants – as may have been the case for Borland's grandmother. This moves away from mere representation as the goal of qualitative research; at the same time, it helps bridge a transactional divide between academia and non-academic publics.

Mixed methods

It may seem to geography students today that mixing research methods has always been common practice, but as recently as 1998, Philip wrote that it had "received little explicit attention in the geographical literature" (p. 276). At the time, Philip suggested that the way forward in human geography for mixing methods would be to leverage the similarities between the approaches – anti-dualistic thinking at work. For example, he highlights that both quantitative and qualitative researchers test and generate hypotheses. Qualitative researchers guided by grounded theory, for example, move back and forth between ideas (themes, codes, notions, hypotheses) and the data (e.g., interview transcripts) to generate refined theory both inductively and deductively. This has, unfortunately, fostered curious hybrid methods that may be unrecognizable to traditional practitioners. For example, instruments that combine an open-ended interview guide with closed-ended survey questions require the researcher to switch from having an open-ended conversation to interacting in the more stilted question-answer mode of the survey questionnaire. The danger of mixing in this manner is that we end up with results that do not provide sufficient depth to satisfy the discerning qualitative reviewer and not a large enough sample size to satisfy the quantitatively oriented reviewer (Baxter and Eyles, 1997).

Others have suggested that we should capitalize on the core differences between quantitative and qualitative methods as the basis for mixing them. For example, in the introductory article in a series on qualitative research in health geography in the *Professional Geographer*, Elliott (1999) proclaims in her article title that "the question shall determine the method." Elliott suggests that we should embrace the idea that quantitative and qualitative methods tend to operate in different modes answering different questions. For example, interviews may be used to answer so-called *why* questions (e.g., Why do recent immigrant women avoid regular physician visits?), while survey questionnaires can be used to address *what* and *how much* questions (e.g., What types of immigrants avoid regular physician visits nationally, and how many are there?). This type of method mixing generally assumes that the methods are kept separate, so it is more of a philosophically purist approach than Philip's approach implies. That is, Elliott is proposing that when we conduct interviews, focus groups or the like, we work in a strictly qualitative mode – for example, adopt a qualitative philosophy ontology and epistemology – whereas when we are conducting quantitative work (e.g., secondary data statistical analysis, surveys), we operate in a strictly quantitative mode – and that these happen at different times, not in the same interaction with study participants.

Conclusion

This chapter provides a concrete sense of the exponential growth of qualitative research in health geography, growth that nevertheless has not kept pace with that in geography generally. That said, health geographers have been fully engaged with various tensions involved in qualitative research. In the process of identifying how dualistic – "all or nothing" – language is problematic for defining "qualitative research," I nevertheless used qualitative-quantitative comparisons throughout the chapter, to highlight that qualitative research involves decisions that may simultaneously require thinking about more traditionally quantitative alternatives or even mixed-methods research. Fortunately, there are debates within the literature (e.g., special journal issues, books) to help practitioners sort through their own decisions, and this chapter highlights a few key ones: philosophical underpinnings, counting, criteria for rigor, and mixing methods.

Notes

1 Raw, meaning none of the finds is vetted to verify that the content is as expected. All indexes in Web of Science's "Core collection" were searched, e.g., Social Sciences Citation Index, Sciences Citation Index, Arts and Humanities Citation Index. Searches included the term "medical" with the Boolean "OR" with "health."
2 While qualitative researchers often refer to their research as *qualitative*, quantitative researchers do not tend to refer to their work as *quantitative* – i.e., not a reasonable comparator (only 68 articles).
3 For a more thorough treatment of various aspects of reliability and its qualitative analogs, see Baxter and Eyles (1997)
4 "Code," "theme" and "node" are synonymous in this chapter.

References

Andrews, G. J., Chen, S. and Myers, S. (2014). The "taking place" of health and wellbeing: towards non-representational theory. *Social Science & Medicine*, 108, pp. 210–222.

Asanin, J. and Wilson, K. (2008). "I spent nine years looking for a doctor": Exploring access to health care among immigrants in Mississauga, Ontario, Canada. *Social Science & Medicine*, 66(6), pp. 1271–1283.

Baxter, J. and Eyles, J. (1997). Evaluating qualitative research in social geography: establishing "rigour" in interview analysis. *Transactions of the Institute of British Geographers*, 22, pp. 505–525.

Borland, K. (1991). That's not what I said: interpretive conflict in oral narrative research. In: S. Gluck and D. Patai, eds., *Women's words: the feminist practice of oral history*. New York: Routledge, pp. 63–75.

Bradshaw, M. (2001). Contracts and member checks in qualitative research in human geography: reason for caution? *Area*, 33(2), pp. 202–211.

Coen, S. (2016). What can participant-generated drawing add to health geography's qualitative palette. In: N. E. Fenton and J. Baxter, eds., *Practicing qualitative methods in health geographies*. London: Routledge, pp. 131–152.

Crang, M. (2002). Qualitative methods: the new orthodoxy? *Progress in Human Geography*, 26(5), pp. 647–655.

Cutchin, M. P. (2007). The need for the "new health geography" in epidemiologic studies of environment and health. *Health & Place*, 13(3), pp. 725–742.

DeLyser, D., Herbert, S., Aitken, S., Crang, M. and McDowell, L. (eds.) (2010). *The SAGE handbook of qualitative geography*. Thousand Oaks, CA: Sage.

Doyle, S. (2007). Member checking with older women: a framework for negotiating meaning. *Health Care for Women International*, 28(10), pp. 888–908.

Dyck, I. (1999). Using qualitative methods in medical geography: deconstructive moments in a subdiscipline? *The Professional Geographer*, 51(2), pp. 243–253.

Elliott, S. J. (1999). And the question shall determine the method. *The Professional Geographer*, 51(2), pp. 240–243.

Fenton, N. E. and Baxter, J. (eds.). (2016). *Practicing qualitative methods in health geographies*. London: Routledge.

Golafshani, N. (2003). Understanding reliability and validity in qualitative research. *The Qualitative Report*, 8(4), pp. 597–606.

Hall, E. (2000). "Blood, brain and bones": taking the body seriously in the geography of health and impairment. *Area*, 32(1), pp. 21–29.

Harper, M. and Cole, P. (2012). Member checking: can benefits be gained similar to group therapy? *The Qualitative Report*, 17(2), pp. 510–517.

Hay, I. (2016). *Qualitative research methods in human geography*. 4th ed. Oxford: Oxford University Press.

Kearns, R. and Moon, G. (2002). From medical to health geography: novelty, place and theory after a decade of change. *Progress in Human Geography*, 26(5), pp. 605–625.

Koelsch, L. E. (2013). Reconceptualizing the member check interview. *International Journal of Qualitative Methods*, 12(1), pp. 168–179.

Lawson, V. (1995). The politics of difference: examining the quantitative/qualitative dualism in post-structuralist feminist research. *The Professional Geographer*, 47(4), pp. 449–457.

Lincoln, Y. S. and Guba, E. G. (1985). *Naturalistic inquiry*. Thousand Oaks, CA: Sage.

Longhurst, R. (1995). VIEWPOINT the body and geography. *Gender, Place & Culture*, 2(1), pp. 97–106.

Mason, S. A., Walker, C., Baxter, J. and Luginaah, I. (2016). Ethics and activism in environment and health research. In: N. E. Fenton and J. Baxter, eds., *Practicing qualitative methods in health geographies*. Abingdon: Routledge, pp. 54–72.

McConnell-Henry, T., Chapman, Y. and Francis, K. (2011). Member checking and Heideggerian phenomenology: a redundant component. *Nurse Researcher*, 18(2), pp. 28–37.

McLafferty, S. (1995). Counting for women. *Professional Geographer*, 47(4), pp. 436–441.

Moss, P. (1995). Embeddedness in practice, numbers in context: the politics of knowing and doing. *The Professional Geographer*, 47(4), pp. 442–449.

Philip, L. J. (1998). Combining quantitative and qualitative approaches to social research in human geography – an impossible mixture? *Environment and Planning A*, 30(2), pp. 261–276.

Plumwood, V. (1991). Nature, self, and gender: feminism, environmental philosophy, and the critique of rationalism. *Hypatia*, 6(1), pp. 3–27.

Sandelowski, M. (1993). Rigor or rigor mortis: the problem of rigor in qualitative research revisited. *Advances in Nursing Science*, 16(2), pp. 1–8.

Sandelowski, M. (2001). Real qualitative researchers do not count: the use of numbers in qualitative research. *Research in Nursing & Health*, 24(3), pp. 230–240.

Sheppard, E. (2001). Quantitative geography: representations, practices, and possibilities. *Environment and Planning D: Society and Space*, 19(5), pp. 535–554.

Turner, S. and Coen, S. E. (2008). Member checking in human geography: interpreting divergent understandings of performativity in a student space. *Area*, 40(2), pp. 184–193.

44

HEALTH GEOGRAPHIES OF ART, MUSIC AND SOUND

The remaking of self in place

Candice P. Boyd and Michelle Duffy

The need to consider and incorporate a place-sensitive *post-medical* perspective in health-geography research is a significant starting point for this chapter (Kearns, 1993), especially when considering the role of art and music in a geographical framing of health. As Kearns once urged, "little attention has been paid to the understanding of place as that experienced zone of *meaning* and *familiarity*" (Kearns, 1993, p. 140; emphasis in original). This idea is important, because what is expressed is a changed view as to how we might think about health. Rather than considering health as the study of disease, Kearns pointed to a shift toward health in terms of a socioecological framework, which leads to an examination of the sets of relations within a population and its social, cultural and physical environment (refer also to Andrews et al., 2012). As Kearns argued, "what occurs in a place (in terms of the relations between people and elements of their environment) has profound importance to health" (1993, p. 141). While Kearns' article generated much debate (e.g., Paul, 1994), his suggestion that health geography can be enriched through closer engagement with social and cultural geography opens up theoretical and methodological avenues of inquiry that can be readily taken up by therapeutic art practices.

Social scientists have turned to the arts and artistic practice as a means to not only capture the various ways in which the world is lived, but also share and communicate research more broadly (Barrett and Bolt, 2010). The creative arts offer significant opportunities to challenge traditional frameworks of knowledge production through re-presenting how we inhabit and engage with the world. Engagement with arts practice is more than simply *communicating*; rather, it offers a means to express our complex relations with human and non-human others. This can be liberating for those living with mental or physical ill health, yet it is also significant to achieving a better understanding of how community is constituted – that is, in and through the socio-spatial relations that generate the experience of well-being, resilience or vulnerability for its members.

Schwanen and Atkinson (2015) suggest that the history of geography's involvement in the study of health and well-being dates to the 1970s, when researchers became interested in the physical and social dimensions of geographic space as it is subjectively experienced. It was in the 1990s, however, with the introduction of the term *therapeutic landscapes*, that geographers began to theorize health and well-being as emerging out of practices, engagements, affective intensities, materialities and encounters that *take place* in everyday contexts (Atkinson, Lawson and Wiles, 2011). As such, health and well-being are no longer understood to be an outcome of intervention but understood as *becoming well*. Resolutely spatial and processual, this expanded conceptualization of health and well-being is particularly relevant to practices that have creativity at their core (Parr, 2012).

Research in cultural geography uses a range of modalities to tease apart the entangled, embodied and heterogeneous relationships that are fundamental to the constitution of place, identity and subjectivity. In this chapter, we consider how such modalities are shaped by and expressed through the geographies of art and music and their role in human health in two ways. In the first half of the chapter, we provide a brief overview of recent work on the geographies of therapeutic art practices and their role in promoting an individual sense of health and well-being. In the second half of the chapter, we consider the role of music-making in the creation of a collective sense of belonging to place. Through a concise summary of work in this field to date, we identify the historical trajectory of this field of research and recent contributions and identify prospects for the future.

Geographies of therapeutic art-making

Research on the geographies of therapeutic art-making is relatively new, and it is informed by developments in geographical thought that are attentive to creative process and encounter (Hawkins, 2014), the non-representational (Boyd, 2017), ethico-aesthetics (McCormack, 2014) and vernacular creativity (Edensor et al., 2010). Within each sphere is a commitment to critical exploration and methodological experimentation capable of providing some capture of processes that are otherwise fleeting, atmospheric and ineffable. In doing so, this research contributes to the broader remit of health geography to think *beyond evidence* to the sites and practices outside of health-care settings where health and well-being are spontaneously generated (Andrews, 2017; Doughty, 2013).

In her book *For Creative Geographies*, Harriet Hawkins (2014) argues that thinking about art as a series of creative encounters "oblig[es] us to be aware of the possibilities they present for experiencing and thinking the world differently" (p. 12). The artist does not simply make art but is *transformed* by it. In this way, art-making is regarded as a distributed set of creative relations that produces effects. These understandings shift the focus from what art *is* to what art can *do* – that is, from meaning to process. When we conceive of art as a process, rather than an object to be interpreted, we can appreciate the multitude of ways in which art brings about changes in subjectivities, both individual and collective (Hawkins, 2014).

In developing a theory of embodied aesthetics, Koch (2017) notes the lack of suitable models in psychology and cognitive science to adequately account for art therapies as a special source of healing. She instead turns to phenomenology and dynamic-systems theory to develop an understanding of art therapy that emphasizes its active-expression dimension. *Enactment*, she argues, is central to the healing process. Drawing on the work of Merleau-Ponty, she sees the constant action of our bodies with our environments as a dynamic source of therapeutic change, whereby the self is continually redefined through test-acting or *as-if* forms of experimentation. This opens up the subject to an aesthetic experience that cultivates an ethics of enchantment or *wondering with* the world. This experience is, however, deeply visceral and not always verbalizable, making it very difficult to assess.

In recent work, Boyd (2017) explored the non-representational geographies of therapeutic art-making using performative research methods. This involved participating in a range of therapeutic art practices – dance, painting, graffiti, poetic permaculture, and musical improvisation – while remaining open and attentive to the spatial, ecological and material qualities of therapeutic spaces. The purpose of the research, from a health-geography perspective, was to explore the notion of a geographically constituted self, hypothesized to continually emerge and reemerge from the singularity of affective encounters – a concept at the heart of Félix Guattari's ethico-aesthetic paradigm (Guattari, 1995). Adopting a performative approach to the research (rather than a qualitative or quantitative one) promoted a particular style of *thinking through practice*, as well as an opportunity to develop symbolic, presentational forms capable of translating some of the *felt* knowledges of therapeutic art-making into visual and audio forms. Working within a non-representational framework helps the artist-geographer remain open to the *eventhood of the moment*, which locates thought in place. Data took sensory forms, including photography, film, video and audio recordings. Performative insights, combined with sensory data collected in the field, were developed into a series of propositions relating to (1) the senses, (2) space and context, (3) materiality, the non-human and the ontic, (4) attunement and fielding, (5), affective knowledge translation, (6) change and transformation and (7) interstices of becoming.

Without rehearsing these propositions in detail, Boyd's (2017) research extends the work of other cultural geographers (e.g., McCormack, 2014) and contemporary philosophers (e.g., Manning, 2016) who understand therapeutic change and transformation as taking place in the event – not as psychological but as geographical (i.e., occurrences of space-time). In this sense, the self is geographically constituted; change and transformation cannot be engineered. What is cultivated in events of therapeutic art-making is an open and sensuous disposition, allowing *lines of flight* to occur more easily. Attention to affectivity and lines of flight effect an escape from tension and open up new ways of being in the world (Atkinson and Scott, 2015). Ontologically, creative practices *smooth* space-time, producing a liminal field enabling human counterparts to re-work their subjectivities unconsciously in relation to their environs (Manning, 2016). This is not something that can be judged easily in terms of measurable outcomes. It is a subjective experience of wellness.

Whereas geographies of therapeutic-art making reflect on the nature of becoming in relation to self, vernacular creativity refers to locally embedded, collective forms of arts and creative practice for the purposes of urban regeneration or civic boosterism (Edensor et al., 2010). Vernacular creativity invokes the concepts of *collective wellness* and *community well-being*. Perhaps more importantly, though, it foregrounds aspects of creativity that operate outside of the cultural economy – the un-hip, un-cool creative practices of everyday citizenship. As such, vernacular creativity highlights instances of hidden art, such as graffiti on the peri-urban fringe; or creative acts of resistance, such as guerilla gardening or creative destruction, as well as the more organized forms of community-based arts practice. It is to this collective understanding of the benefits of art and music to health that we now turn.

Geographies of music and sound in relation to well-being

Research in the broad area of sound and music in social and cultural geography covers a wide range of approaches that link to the concerns and concepts of health. The most recognizably connected are those studies in music and dance therapies, in which the benefits of participation are explored in terms of processes that may enhance socialization, social connectedness and well-being (Pascoe et al., 2005). The therapeutic value of music performance participation is a major research theme in geographies of health in terms of its social, emotional, physical and cognitive benefits (Packer and Ballantyne, 2011). This ranges from the use of music to increase workers' efficiency in repetitive work in industry (Jones, 2005), practices of listening to music that serves to create notions of home (Duffy and Waitt, 2013), the use of music to generate therapeutic spaces (Andrews, Kingsbury and Kearns, 2014; Little, 2013) and the impact of sound and music in producing or alleviating pain and stress (Andrews et al., 2011; Bell, 2016). In addition, a significant body of literature explores dance and its role in enhancing health and well-being (e.g., Atkinson and Scott, 2015). Nonetheless, certain music genres and practices can be associated with poor health (Andrews et al., 2011). For example, some music subcultural practices may be associated with drug use, violence and reckless behavior (Wilkinson, 2016), as well as other deleterious health impacts (Söderströma et al., 2016).

One approach that examines the fundamental dynamic between health and place (Andrews et al., 2011) is a range of geography literatures focused on defining *community* and its relationship to music in terms of "a lived experience happening within social and cultural contexts" (Ansdell, 2004, p. 67). For instance, Duffy and colleagues have explored music and health through a range of factors, such as the importance of musical culture to well-being and developing connections to place, particularly in culturally diverse communities (Duffy, 2005; Permezel and Duffy, 2007), as well as some of the ways in which music contributes to community resilience and social connectedness through activities such as singing or attending arts and music festivals (Waitt and Duffy, 2010).

Aligned more broadly with the ways in which place, community health and music intersect is the field of soundscape research, originating with Schafer's (1969) work that led to the establishment of the World Soundscape Project in the early 1970s (Sfu.ca, 2017). This approach arose out of concerns that changes in the sounds of the environment affected psychological and sociocultural well-being. Schafer's aim was to establish an acoustic urban design that provided noise abatement and hence lessened sound's negative impacts on the body. One branch of this research area, particularly in the fields of urban planning and architectural design, seeks to minimize the deleterious effects of urban and industrial sounds (Bijsterveld, 2008).

The soundscape tradition, which may be described in terms of geographies of well-being, also offers a framework through which to explore how we may feel connected or disconnected to place and community. Those working with soundscapes (and associated work conducted with sound walks – e.g., Gallagher, 2015) offer a means to explore how feeling, emotion and affect shape and communicate both who we are and how we interact, which then impact individual and subjective well-being. This body of work resonates with the development of emotional geographies (Davidson, Bondi and Smith, 2005), as well as renewed interest in embodiment (Longhurst, 2001) and arts-based research based on non-representational theories (Boyd, 2017).

Creative-arts practice bridges music, sound, place and health through drawing our attention to the ways in which we are intimately connected to our environment and communities. Sound and music permeate the entire body, not the ears alone (Duffy, 2009). In addition, sound elicits emotions and affects. Hence it is not simply that sound represents subjectivity and individual and community identity, thus creating notions of connection and belonging through representational and performative practices. Rather, it is that subjectivities are constituted within the very unfolding of the sonic event (Duffy, 2016). And, with relevance to health geographies, individuals are continually managing their emotions and affective relations in response to sound in their social, cultural and environmental contexts.

Conclusions and future directions

Recent research in social and cultural geography that informs contemporary health geography responds to calls within the field of arts in health for approaches and methods that go beyond evidence to situated under-standings of wellness (Parr, 2012; Stickley et al., 2017). However, research capable of producing in-depth accounts of the ways in which health and well-being are generated and performed in everyday contexts rarely meets positivist standards of rigor (Andrews, 2017). Researching *the creative* requires geographers to embrace performative methods of research and embodied, affective modes of knowledge production (Koch, 2017). The challenge for the future is to demonstrate the relevance of performative methods to fields whose sole concern is with the quantification of outcomes.

References

Andrews, G. J. (2017). Running hot: placing health in the life and course of the vital city'. *Social Science and Medicine*, 175, pp. 209–214.

Andrews, G. J., Evans, J., Dunn, J. and Masuda, J. (2012). Arguments in health geography: Oon sub-disciplinary progress, observation, translation. *Geography Compass*, 6, pp. 351–383.

Andrews, G., Kearns, R., Kingsbury, P. and Carr, E. (2011). Cool aid? Health, wellbeing and place in the work of Bono and U2. *Health & Place*, 17, pp. 185–194.

Andrews, G., Kingsbury, P. and Kearns, R. (2014). *Soundscapes of wellbeing in popular music.* Farnham: Ashgate.

Ansdell, G. (2004). Rethinking music and community: theoretical perspectives in support of community music therapy. In: M. Pavlicevic and G. Ansell, eds., *Community music therapy*. London: Jessica Kingsley, pp. 65–90.

Atkinson, S., Lawson, V. and Wiles, J. (2011). Care of the body: spaces of practice. *Social & Cultural Geography*, 12, pp. 563–572.

Atkinson, S. and Scott, K. (2015). Stable and destabilized states of subjective well-being: dance and movement as catalysts of transition. *Social & Cultural Geography*, 16, pp. 75–94.

Barrett, B. and Bolt, B. (2010). *Practice as research: approaches to creative arts enquiry.* London: I.B. Tauris.

Bell, S. (2016). The role of fluctuating soundscapes in shaping the emotional geographies of individuals living with Ménière's disease. *Social & Cultural Geography*, 18(6), pp. 831–850.

Bijsterveld, K. (2008). *Mechanical sounds: technology, culture, and public problems of noise in the twentieth century.* Cambridge: MIT Press.

Boyd, C. P. (2017). *Non-representational geographies of therapeutic art making: thinking through practice.* London: Palgrave Macmillan.

Davidson, J., Bondi, L. and Smith, M. (2005). *Emotional geographies.* London: Ashgate.

Doughty, K. (2013). Walking together: the embodied and mobile production of therapeutic landscape. *Health & Place*, 24, pp. 140–146.

Duffy, M. (2005). Performing identity within a multicultural framework. *Social and Cultural Geography*, 6, pp. 677–692.

Duffy, M. (2007). Negotiating cultural difference in local communities: the role of the body, dialogues and performative practices in local communities. *Geographical Research*, 45, pp. 358–375.

Duffy, M. (2009). Methods: sound and music. In: R. Kitchin and N. Thrift, eds., *International encyclopaedia of human geography.* London: Elsevier, pp. 230–235.

Duffy, M. (2016). The listening "I": children's emotional and affective representations of place. In: S. Gair and A. van Luyn, eds., *Sharing qualitative research: showing lived experiences and community narratives.* London: Routledge, pp. 96–109.

Duffy, M. and Waitt, G. (2013). Home sounds: experiential practices and performativities of hearing and listening. *Social & Cultural Geography*, 14, pp. 466–481.

Edensor, T., Leslie, D., Millington, S. and Rantisi, N. M. (2010). *Spaces of vernacular creativity: re-thinking the cultural economy.* Abingdon: Routledge.

Gallagher, M. (2015). Sounding ruins: reflections on the production of an audio drift. *Cultural Geographies*, 22, pp. 467–485.

Guattari, F. (1995). *Chaosmosis: an ethico-aesthetic paradigm.* Indianapolis: Indiana University Press.

Hawkins, H. (2014). *For creative geographies: geography, visual arts and the making of worlds.* New York: Routledge.

Jones, K. (2005). Music in factories: a twentieth-century technique for control of the productive self. *Social & Cultural Geography*, 6, pp. 723–744.

Kearns, R. (1993). Place and health: towards a reformed medical geography. *The Professional Geographer*, 45, pp. 67–72.

Koch, S. C. (2017). Arts and health: active factors and a theory framework of embodied aesthetics. *The Arts in Psychotherapy*, 54, pp. 85–91.

Little, L. (2013). Pampering, well-being and women's bodies in the therapeutic spaces of the spa. *Social & Cultural Geography*, 14, pp. 41–58.

Longhurst, P. (2001). *Bodies: exploring fluid boundaries.* London: Routledge.

Manning, E. (2016). *The minor gesture.* Durham, NC: Duke University Press.

McCormack, D. P. (2014). *Refrains for moving bodies: experience and experiment in affective spaces.* Durham, NC: Duke University Press.

Packer, J. and Ballantyne, J. (2011). The impact of music festival attendance on young people's psychological and social well-being. *Psychology of Music*, 39, pp. 164–181.

Parr, H. (2012). The arts and mental health: creativity and inclusion. In: T. Stickley, ed., *Qualitative research in arts and mental health: context, meaning, and evidence.* Monmouth: PCCS Books, pp. 1–21.

Pascoe, R., Leong, S., MacCallum, J., Mackinlay, E., Marsh, K., Smith, B., et al. (2005). *National review of school music education: augmenting the diminished.* Canberra: Australian Government.

Paul, B. (1994). Commentary on Kearns's "Place and health: toward a reformed medical geography." *The Professional Geographer*, 46, pp. 504–505.

Permezel, M. and Duffy, M. (2007). Negotiating cultural difference in local communities: the role of the body, dialogues and performative practices in local communities. *Geographical Research*, 45, pp. 358–375.

Schafer, M. (1969). *The new soundscape.* Scarborough: Berandol Music.

Schwanen, T. and Atkinson, S. (2015). Geographies of wellbeing: an introduction. *The Geographical Journal*, 181, pp. 98–101.

Sfu.ca. (2017). *World soundscape project.* [online] Available at: www.sfu.ca/~truax/wsp.html [Accessed 15 May 2017].

Söderströma, O., Empsonb, L., Codeluppia, Z., Söderströmc, D., Baumannb, P. and Conusb, P. (2016). Unpacking "the City": an experience-based approach to the role of urban living in psychosis. *Health & Place*, 42, pp. 104–110.

Stickley, T., Parr, H., Atkinson, S., Daykin, N., Clift, S., De Nora, T., Hacking, S., Camic PM., Joss, T., White, M. and Hogan, S. J. (2017). Arts, health & wellbeing: reflections on a national seminar series and building a UK research network. *Arts & Health*, 9, pp. 14–25.

Waitt, G., Duffy, M. (2010). Listening and tourism studies. *Annals of Tourism Research*, 37, pp. 457–477.

Wilkinson, S. (2016). Drinking in the dark: shedding light on young people's alcohol consumption experiences. *Social & Cultural Geography*, 18(6), pp. 839–757.

45

HEALTH GEOGRAPHY AND THE FUTURE OF DATA

Daniel Lewis

Health geography is concerned with the collection, analysis and presentation of data that measures or represents the health of individuals or populations in light of geographical factors. A guiding principle of health geography is that *places matter* for health (Kearns and Moon, 2002), the validity of which has been tested in according to a diversity of research practices (Macintyre, Ellaway and Cummins, 2002). In this chapter, the ongoing role of quantitative data in health geography is explored in light of developments sometimes termed the *data revolution* (Kitchin, 2014).

The chapter will first define what quantitative data are, describe how to distinguish between different types of quantitative data, and suggest some questions relevant to working with any dataset. The changing nature of data is then discussed, focusing on examples in health geography that illuminate *big* data, *open* and *volunteered* data and then data *infrastructures* and the value of *indexical* data. This should all be read as an introduction to thinking critically about data.

Health geographers are concerned with social justice (Kearns and Moon, 2002). For quantitative health geographers, evidence for social justice relies on quantitative data about people's personal and environmental circumstances. A solid appreciation of data is fundamental to making meaningful contributions to health geography and social justice.

Defining quantitative data: what are data?

Quantitative health geographers typically approach *data* as the building blocks of analysis. Data are collected, analyzed, and interpreted in the hope of producing meaningful knowledge about the world; this process is known as *enrichment*.

The word "data" comes from the Latin word *dare*, meaning *to give*, the idea being that facts about the world are given. However, Checkland and Holwell (2005) suggest that the real world is made up of an overwhelming mass of facts, or *data*, and so some practical selection must take place. They term data that has been selected *capta*, reflecting that most data are in fact *taken* for a specific purpose, from the Latin *capere*. Almost all data used in quantitative health geography are actually capta, reflecting both what it is feasible to measure and the decision of what to include in, and what to leave out of, our models of reality. For practical reasons, we tend to refer to this all as "data."

Quantitative data are numeric records and are usually seen as distinct from qualitative data, which are non-numeric and can take a range of forms: texts, images, audio and video recordings, material objects and so forth. The actual values that quantitative data take fall into two main groups: numerical and categorical, largely determined by what is actually being quantified. Most data are either *extensive* or *representative* (Kitchin, 2014); extensive data deals with quantities such as height, weight or distance, which relate to the physical reality of something being studied, while representative data encode non-physical or *latent* (Bollen, 2002) characteristics such as occupational, social class or quality of life. Table 45.1 further breaks down numerical and categorical data into sub-types of quantitative data.

In addition to the data types in Table 45.1, quantitative data are often typed based on how they are produced. Conventionally, *primary data* are collected for a specific study or purpose; a researcher either personally collects the data or sets the rules by which data are collected. *Secondary data* are primary data that have been made available to other researchers; these researchers reuse the data to answer their own questions. Sometimes a distinction is made between secondary data and *tertiary data* (Kitchin, 2014); census data and hospital-admissions data are generally released as tertiary data in which counts, rates, or summaries (e.g., means) are derived for aggregate units (such as census output zones, hospitals, clinics) to preserve confidentiality. Research with census secondary data or individualized hospital admissions may require visiting secure data labs.

When considering what quantitative data are, it can be tempting to see all data as just a mass of numbers; however, data require structure. Data about different phenomena are called *variables* (if the values of the data they hold vary; if not, they are called *constants*), which are gathered into datasets for specific research purposes. It is important to know what kind of data (numerical, categorical), each variable in a dataset holds and whether it is primary, secondary or tertiary data; however, we should also be aware of other aspects of the data. Table 45.2 gives an overview of some dimensions to take note of (identified by Kitchin, 2014) and the kinds of questions we could ask to get a better understanding of the *raw materials* of our research.

The answers to some of these questions will often be found in the *metadata*; that is, the data about the data. This will likely cover technical and spatial/temporal aspects and, depending on the type of data, may include references to ethical review. Other aspects, particularly the political/economic and philosophical dimensions of data, may not be easily identified, instead requiring independent critical thought.

There is much to consider, and much taken for granted, on the topic of data. Nonetheless, it is extremely important that researchers have a clear grasp of what data are, particularly in light of recent and ongoing changes to how data are produced.

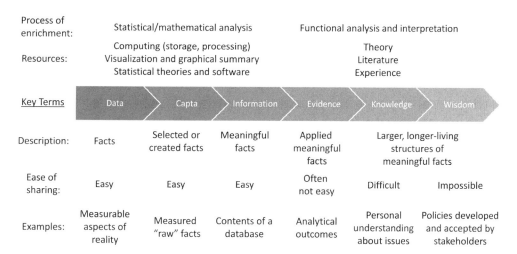

Figure 45.1 Enriching data: a hierarchy of information

Table 45.1 Types of quantitative data, their definitions and examples

Data type – sub-type	Definitions	Examples in health geography
Numerical	Continuous (decimal numbers) or discrete (whole numbers). The size of the value is meaningful, implying more or less of something.	The Townsend index is a measure of material deprivation based on census variables. Areas with high Townsend scores are more deprived relative to areas with low Townsend scores.
Interval	In addition to above, the distance between observations (the interval) is fixed and interpretable. Averages can be computed, but not ratios. Zero values (origins) are arbitrary.	Temperatures (Celsius/Fahrenheit) are interval data. A degree is a fixed unit, making differences between values as well as the size of a value meaningful. Temperature is associated with malaria transmission, affecting its geographic distribution and seasonality.
Ratio	As with interval data, but with true zero values, which indicate an absence of something. Taking ratios is also meaningful, so fractions can be computed (e.g., twice as much).	In addition to measures such as weight or distance, counts are also ratio data. The physician-to-population ratio is a simple measure of health-service access. The UK has 2.81 physicians per 1,000 people, more than the US, which has 2.55, and almost twice that of China at 1.49 (WHO, 2017).
Categorical	Data take on a fixed value indicating belonging to a group or category. Numerical data can be grouped according to certain schemes (e.g., quantiles – ordered, evenly sized groups). Categories are often representative, based on non-physical properties.	The Townsend index calculated for small areas can be divided into deciles (quantiles with 10 groups). Average life expectancy can be computed for each decile of deprivation and has been shown to decrease as deprivation increases in what is known as a social gradient.
Binary	Binary data express on-off, true-false, present-absent responses as data, or otherwise dichotomize (split into two categories) data.	In medical geography, whether or not someone has a specific disease at a given point in time is usually represented as a binary variable. A value of 1 indicates they are a case (i.e., they have the disease), while 0 indicates they are not a case.
Nominal	Nominal data is divided into named, discrete categories. There is no express order to the categories, and any given category is not intrinsically better or worse than any other.	Black, white, and Latino are examples of nominal categories expressing race or ethnicity. The welfare geography of who gets what, where, when and why (Smith, 1977) may consider disparities in access to green spaces by racial or ethnic group as evidence of environmental injustice.
Ordinal	Ordinal data are categorical data that can be sorted into an order. The position of a category relative to another is meaningful.	Disease severity is a classic example of ordinal data. In research on health-service utilization, a person's willingness to travel for surgery or care can be patterned according to the severity or stage of a disease. This can have implications for the provision of health care.

Table 45.2 Potential questions when dealing with data

Dimension	Relevant questions
Technical	What instruments were used to collect these data? ∘ How accurate are those instruments? ∘ How precise are those instruments? ∘ What are the uncertainties and/or biases in these data? ∘ Has the validity or reliability of these data been assessed? ∘ Should we be concerned about error? ∘ Were these data subject to any cleaning or post-processing? ∘ Should these data be cleaned or truncated?
Spatial/temporal	Are these data about particular places or times? ∘ Are they subject to particular aggregations: spatial zones or time periods? ∘ Are the data longitudinal? ∘ Are they about particular cohorts of people? ∘ Does the way these data are collected change over time or space? ∘ Have instruments or questions changed over time? ∘ How consistent are these data spatially and/or temporally? ∘ Can these data be compared across space and/or time?
Ethical	How were these data produced? ∘ Was any consent/permission obtained necessary and appropriate? ∘ Are there rules governing how these data are stored and/or shared? ∘ Are there any prohibitions, restrictions or guidance on how these data are to be used? ∘ Is it possible that generating or publishing these data will cause harm or distress? ∘ Are there any privacy or confidentiality issues that need to be addressed? ∘ Are these data anonymous or identifiable? ∘ Have disclosure controls been applied?
Political/ economic	What is the wider context within which these data were generated? ∘ Whom do these data represent? ∘ Are these data routinely collected? ∘ Is there anything controversial or unusual about these data, the way they were defined or the way they were generated? ∘ Whom are these data for? ∘ Are these data freely and openly available? ∘ Who benefits from these data? ∘ Who paid for the collection, creation and/or dissemination of these data, and why?
Philosophical	Why were these data created/collected? ∘ Do these data measure what they are supposed to, or claim to, measure? ∘ Can these data be taken at face value? ∘ How were these data originally conceived of? ∘ Are there omissions in what these data capture? ∘ Have the backgrounds/beliefs of the researchers producing these data influenced what is collected, and how?

Health geography and the data revolution

The way in which data are collected, stored, shared and analyzed is changing. Kitchin (2014) suggests that there are three main factors underpinning these changes: *big* data, *open* data, and data *infrastructures*.

Big data refers to the masses of data continually pulled from the world around us using sensors and information-communication technologies, documenting our actions, interactions and transactions. This data is given the prefix *big* because it differs from conventional *small* data, being high-volume (many observations/measurements), high-velocity (real-time or near real-time collection, continually updating) and high-variety (datasets come in many different formats – e.g., text files, PDFs, web data formats, SMS messages). *Volume*, *velocity* and *variety*, known as the *three Vs* (Sagiroglu and Sinanc, 2013) is one common way to think of big data, but many authors offer expanded or competing approaches (boyd and Crawford, 2012; Kitchin, 2014).

Open data are freely available to use, reuse and redistribute, provided you attribute the original source and make any changes or improvements to the data similarly open (Molloy, 2011). The intent is that the data be provided in a convenient form for use and that any monetary cost of data not exceed the reproduction cost.

Data infrastructure refers to the means to join everything together to promote the successful storage, sharing, linking, use and reuse of data. Government open data has been made possible by the digital data infrastructures that underpin online portals such as data.gov in the United States (www.data.gov/) and data.gov.uk in the United Kingdom (https://data.gov.uk/).

In the next three sections, examples of big data, open data and data infrastructures are given that have a health-geographic focus.

Big data: Google Flu Trends and electronic health records

Unlike familiar *small* data, which is captured for a specific research purpose, big data is often termed *exhaust* data (Kitchin, 2014). "Exhaust" implies that the data were not originally collected for the purpose for which they are being used but are a by-product of some other process, such as finding information on the internet or managing the treatment of patients.

The most well-known big-data experiment in health, Google Flu Trends, aimed to predict the rate of *influenza-like illness* (possible cases of influenza) by sifting through the tens of thousands of searches made on Google every second (Ginsberg et al., 2008). Using secondary data, researchers found the common search queries associated with physician visits for influenza-like illness, which they could then use to estimate present influenza activity. Conventional approaches to estimating influenza activity, such as those used by the US Centers for Disease Control and Prevention (CDC), rely on a number of sophisticated surveillance approaches, including using hospital admissions records, mortality reports, and clinical laboratory testing (CDC, 2016). However, estimates based on these data usually lag behind the actual rate of influenza activity; the CDC publishes estimated flu data with a one-to-two-week latency. Google Flu Trends was intended to *nowcast* influenza activity using big data, effectively predicting influenza outbreaks in near real-time (around a one-day latency) based on search activity. Google Flu Trends worked reasonably well, until suddenly in 2013 it did not, dramatically overestimating the flu activity that flu season. Lazer et al. (2014) discuss the reasons for Google Flu Trends' well-publicized failure of prediction, citing something they term *big data hubris* – the implicit assumption that big data could supplant conventional data rather than complement it.

The application of big-data approaches in health care has been described as inevitable (Murdoch and Detsky, 2013), partly due to the challenges and opportunities presented by electronic health records (EHRs). Nowadays, health-care providers can store all the data generated by hospital visits digitally as a diverse set of records including patient demographics; diagnosis, treatment and operations data; prescriptions; medical history; test results and scans; hospital administrative data, and so on. Data in EHRs are primarily aimed at improving the coordination of patient care by medical professionals. However, EHRs also offer researchers the chance to improve the health care of individuals and populations. EHRs allow for the creation of large observational datasets that integrate multiple data sources, as well as the potential for experimental datasets where patients have naturally received different treatments *as if at random* or for longitudinal datasets that track patients over time. Analytical uses for EHRs are still emerging, and the EHR is not yet a comprehensive resource in many countries. In the United Kingdom, for instance, Clarke et al. (2017) report limited or fragmented progress with EHRs, while Jha et al. (2009) have reported slow adoption among US hospitals.

Open data: health dashboards and volunteered health data

Typically, *small* data were expensive to produce, and, as a result, highly valuable. Often access to such data was tightly controlled. Pressure for transparency and accountability in government, as well as movements in the technology sector supporting open-source software and open collaboration, provided the conditions for a shift to open data. Open data provide opportunities for people to make informed decisions in their own lives as well as the tools for participating in and critiquing policy-making. Moreover, open data offers the possibility that communities of individuals could produce their own data and volunteer it for the reuse of others.

The diffusion of health data through *dashboards* is a visible product of the data revolution; dashboards are visual displays (statistical measures or graphical representations) that take often dynamic data and report key indicators or important summaries *at a glance*. Batty (2015) notes that monitoring human systems has long been a feature of modern medicine – for instance, the electrocardiogram (EKG) machine tracing the beating of a patient's heart is a kind of dashboard. Monitoring human organizations, such as cities, is more recent, requiring not only advances in computing, but also in many cases the advent of open data. Health and health-care dashboards offer the opportunity to transcend the EKG, curating multiscalar data relevant to individuals and their environments into effective summaries. Applications in personalized medicine (aka precision medicine) might involve integrating individual clinical data with smoking-behavior data from sensors in electronic cigarettes and open geospatial and residential neighborhood demographic data. Such an overview may support patients and doctors in choosing between different courses of action for preventative health care or, suitably anonymized, might help local policy-makers contextualize population health decision-making.

Volunteered data, or *crowdsourced* data, are an important part of the open-data picture. In general, crowd-sourcing means opening up a task for public contribution and collaboration (Ranard et al., 2014); by the same measure, crowdsourced data is data produced (knowingly or unknowingly) by members of the public. The idea that people choose to volunteer data is particularly prominent in geography (Elwood, Goodchild and Sui, 2012), where the term *volunteered geographic information* (VGI) covers the recent creation of web technologies that allow people to associate data with places, features or locations on Earth using a computer. Volunteered data has become increasingly important in humanitarian crises and disaster response, as in the case of the 2010 Haitian earthquake; Zook et al. (2010) describe how users of OpenStreetMap (a free and open-source web map) from around the world traced aerial photos to provide digital maps to aid workers on the ground. Ranard et al. (2014) also review a number of crowdsourcing initiatives that are rooted in conventional scientific inquiry.

Data infrastructure: indexical data for individual, longitudinal and geographical linkage

"Data infrastructures" refers to the hardware and software systems that make data available, shareable and reusable. They are present in the background of the examples discussed above, facilitating the use of big and open data. However, data infrastructures are also important for data linkage, a specific type of data – *indexical* data – underpinning this process. Linking previously independent data can enhance their usefulness for research. For instance, researchers may wish to link survey data on self-reported minor psychiatric morbidity and demographics with clinical admissions data on actual psychiatric episodes. Indexical data is crucial to achieving this aim. Often, a key outcome of data linkage is the integration of important *attributional* data; attributional data are measures such as those discussed in Table 45.1. Kitchin (2014) suggests a range of possible indexical data, developed in Table 45.3.

Indexical data offers the potential to transform conventional small data through data linkage and substantially improve the usefulness of those data. Indexical data can be used to track an individual over time, as well as to integrate data from diverse sources or devices and locate individuals in their geographical context. However, linkage can be extremely difficult – not only practically, but also legally and ethically. Indexical data may be subject to laws that govern whether, and under what conditions, data linkage can take place.

Consolidation: a critical appreciation of data

Krieger et al. (2015) argue that when people are killed by law enforcement it is a public-health as well as a criminal-justice issue and, further, that public responses such as protests or violence can have public-health consequences. Although killings of law-enforcement officials are well documented, similar data do not exist

Table 45.3 Types of indexical data

Indexical data	Description	Health-geography use case
Personal Name, date of birth, date of death, sex, ethnicity, place of birth/death.	Personal data is the basis for distinguishing between individuals.	Health-inequalities research makes extensive use of these data, often not indexically, but attributionally to define population subgroups of analytical interest.
Administrative Passport, social security, national identity card, national insurance, immigration documents, tax records, school/college/university enrollments, public health records.	Governments hold numerous unique identifiers covering both citizens and non-citizens to manage state services and systems.	Increasingly used by health researchers interested in integrating health data with other data. For instance, linking medical records to more general survey data using administrative identifiers.
Geographical Address, postal/zip code, district, polling area/constituency etc.	Addresses may allow linkage of individuals across datasets, with support from other identifying data (name, age, sex).	Commonly used by health geographers to link individuals or households to aggregate data about their spatial context or to measure proximity to resources (e.g., hospitals, green spaces).
Commercial Accounts and memberships, loyalty schemes/cards, financial services (bank accounts, credit cards etc.), social media/networking accounts.	Companies can build up substantial personal data based on transactions both online and offline. Accounts and memberships act as possible points of linkage for these data.	Limited current research in health geography, but considerable interest. Topics include food purchase through supermarket loyalty cards, ideally linked to attribute-rich survey data.
Electronic Email address, cellular phone, device ID (e.g., MAC (media access control) address, IP (internet protocol) address).	Electronic devices that connect ubiquitously to the internet broadcast unique identity numbers, which can be used to link activity on these devices to their owners.	Research in health geography very limited, but scope for applications in mobile health, and with the tech industry, such as the Apple health app, which tracks users' health.
Biomedical Fingerprint, DNA sequence, retina, voice, biomarkers.	Fingerprints and retinal scans are common access controls that uniquely identify individuals. Law enforcement and immigration commonly hold these data. DNA and other identifiable biological data are increasing a part of biomedical research data and could be used for linkage.	Health geography is currently more interested in the attributional value of these kind of data for population health than as a form of linkage.

on those people who were killed by the actions of law enforcement. Krieger et al. (2015) explain that these data must be gathered from multiple sources and combined. As attempts to obtain these data directly from law enforcement have hitherto proven unsuccessful, Krieger et al. (2015) suggest that deaths resulting from the actions of law enforcement should be made *notifiable* – a distinct cause of death, a *health condition*, in public-health mortality reporting.

What we think of as data are changing and, as a result, so are the conditions for health-geographic research. In the future, data will come from a diverse range of sources, in a diverse range of formats and at high speed. There will also be a lot more data, enabling health geographers to explore new questions and requiring new approaches to storing, analyzing and interpreting these *raw materials*. It is important, therefore, that we be clear on what these raw materials are: observations taken for a reason, conventionally for the direct purpose of learning something about the world, but increasingly as a by-product, an *exhaust*. Big data are being captured by machines, sensors, and proprietary algorithms whose primary intent is not research, and increasingly private interests are compiling and sharing data without disclosing their motivations for doing so.

Being critical about data and thinking carefully about what a given data source captures is as important as ever. Data are not strictly *neutral*; they reflect the context in which they were produced. Health geographers need a secure grasp of the changing nature of data if they are to continue to produce work that is timely and relevant.

References

Batty, M. (2015). A perspective on city dashboards. *Regional Studies, Regional Science*, 2(1), pp. 29–32.

Bollen, K. (2002). Latent variables in psychology and the social sciences. *Annual Review of Psychology*, 53, pp. 605–634.

boyd, d. and Crawford, K. (2012). Critical questions for big data: provocations for a cultural, technological, and scholarly phenomenon. *Information, Communication and Society*, 15(5), pp. 662–679.

Centers for Disease Control and Prevention (CDC). (2016). *Overview of influenza surveillance in the United States*. [online] Available at: www.cdc.gov/flu/weekly/overview.htm [Accessed 6 Aug. 2017].

Checkland, P. and Holwell, S. (2005). Data, capta, information and knowledge. In: M. Hinton, (ed.), *Introducing information management: the business approach*. London: Routledge, pp. 47–55.

Clarke, A., Watt, I., Sheard, L., Wright, J. and Adamson, J. (2017). Implementing electronic records in NHS secondary care organization in England: policy and progress since 1998. *British Medical Bulletin*, 121(1), pp. 95–106.

Elwood, S., Goodchild, M. and Sui, D. (2012). Researching volunteered geographic information: spatial data, geographic research, and new social practice. *Annals of the Association of American Geographers*, 102(3), pp. 571–590.

Ginsberg, J., Mohebbi, M., Patel, R., Brammer, L., Smolinski, M. and Brilliant, L. (2008). Detecting influenza epidemics using search engine query data. *Nature*, 457, pp. 1012–1014.

Jha, A., DesRoches, C., Campbell, E., Donelan, K., Rao, S., Ferris, T., Shields, A., Rosenbaum, S. and Blumenthal, D. (2009). Use of electronic health records in U.S. hospitals. *The New England Journal of Medicine*, 360, pp. 1628–1638.

Kearns, R. and Moon, G. (2002). From medical to health geography: novelty, place and theory after a decade of change. *Progress in Human Geography*, 26(5), pp. 605–625.

Kitchin, R. (2014). *The data revolution: big data, open data, data infrastructures and their consequences*. London: Sage.

Krieger, N., Chen, J., Waterman, P., Kiang, M. and Feldman, J. (2015). Police killings and police deaths are public health data and can be counted. *PLoS Medicine*, 12(12), e1001915.

Lazer, D., Kennedy, R., King, G. and Vespignani, A. (2014). The parable of Google Flu: traps in big data. *Science*, 343(6176), pp. 1203–1205.

Macintyre, S., Ellaway, A. and Cummins, S. (2002). Place effects on health: how can we conceptualise, operationalise and measure them? *Social Science & Medicine*, 55(1), pp. 125–139.

Molloy, J. (2011). The Open Knowledge Foundation: open data means better science. *PLoS Biology*, 9(12), e1001195.

Murdoch, T. and Detsky, A. (2013). The inevitable application of big data to health care. *Journal of the American Medical Association*, 309(13), pp. 1351–1352.

Ranard, B., Ha, Y., Meisel, Z., Asch, D., Hill, S., Becker, L., Seymour, A. and Merchant, R. (2014). Crowdsourcing – harnessing the masses to advance health and medicine, a systematic review. *Journal of General Internal Medicine*, 29(1), pp. 187–203.

Sagiroglu, S. and Sinanc, D. (2013). *Big data: a review*. International Conference on Collaboration Technologies and Systems (CTS), San Diego, CA, pp. 42–47.

Smith, D. (1977). *Human geography: a welfare approach*. London: Edward Arnold.

World Health Organization. (2017). *Global Health Conservatory (GHO) data*. [online]. Available at: www.who.int/gho/health_workforce/physicians_density/en/ [Accessed 1 Sept. 2017].

Zook, M., Graham, M., Shelton, T. and Gorman, S. (2010). Volunteered geographic information and crowdsourcing disaster relief: a case study of the Haitian earthquake. *World Medical and Health Policy*, 2(2), pp. 7–33.

46

HEALTH GEOGRAPHY AND THE BIG DATA REVOLUTION

Alec Davies and Mark A. Green

The era of big data is upon us. We have more data than ever before, greater computing power to process and analyze such data, and access to new forms of data opening up novel research avenues. The promise offered through big data has brought great potential for researchers to answer many long-standing questions and provide more accurate solutions to problems (Boyd and Crawford, 2012), although significant challenges remain (Khoury and Ioannidis, 2014). In this chapter, we outline the varying definitions of "big data," the applications across health and health geography and the challenges and opportunities associated with using big data.

Defining "big data"

The definition of "big data" is not universally agreed upon (Herland, Khoshgoftaar and Wald, 2014). "Big data" traditionally applies to data of size beyond the capabilities of commonly used software within a tolerable time frame, with size constantly increasing with computing capacity (Manovich, 2011). While size inherently matters (for example, data contained within the US health-care system exceeded 150 exabytes [or 150 billion gigabytes] in 2010, Andreu-Perez et al., 2015), a simpler means for thinking about what is considered as *big* might be data that cannot be opened, stored and processed in the memory of a desktop or laptop computer.

Kitchin (2014) identifies the three Vs of big data research. *Volume*, referring to the size of data, is often over-emphasized in big data (Birkin, Clarke and Clarke, 2017). Data size is making way to the capacity to search, aggregate and cross-reference large datasets as the defining feature of big data (Boyd and Crawford, 2012). *Velocity* refers to data that is generated (and received) quickly, often in or near to real-time. For example, Twitter data is generated frequently as individuals tweet throughout their days to provide a continuous stream of data. The high velocity of data received creates practical issues in the processing and analyzing of such data. Data are often received in a *variety* of heterogeneous types and forms, and often completely unstructured, thereby placing emphasis on manipulating such data (Andreu-Perez et al., 2015). *Variability* has also been touted as an additional *V*, in that what data is received may change in type or quality over time due to updated processes or practice.

Boyd and Crawford (2012) define "big data" as an interplay of technology, analysis and mythology. Technology enables computation, analysis allows interpretation and mythology is the belief that big data offer a higher form of intelligence. Such a holistic definition helps us move away from a technical definition to one that is more akin to the philosophy of *doing* big data research. In this vein, Andreu-Perez et al. (2015) recommend including *value* in any definition. There are many smaller datasets available that can help answer most research questions. Big data should complement or extend these analyses.

Alongside the growth in popularity of big data, we have also seen increasing usage of an interrelated phrase – *new forms of data*. The phrase refers to the increasing availability of data sources not traditionally used for research (e.g., internet search data, social media, loyalty-card records, mobile devices or wearable technology). Connelly et al. (2016) distinguish between *made* (collected for a specific research purpose) and *found* (which may be valuable for researchers but was collected for a non-research purpose – alternatively titled *repurposed*) data. They are *new* in the sense that these data have only recently started becoming available to researchers. For example, social-media organizations (e.g., Facebook, Twitter) did not exist 10–15 years ago, yet the data that they generate have increasingly been used by researchers to understand health-related behaviors (Khoury and Ioannidis, 2014). The *internet of things* (the interconnectedness of everyday objects through the inclusion of internet-based technologies) and *smart cities* (the interconnectedness of communication and technology used to manage a city) offer exciting directions for new forms of data to better understand urban dynamics.

These paradigms of *big data* and *new forms of data* have been enabled through the *datafication* of society. Data are generated in all aspects of our daily routines, whether it is purchasing food using our loyalty and credit cards at a supermarket, visiting a doctor or using public transport. Many of these data were being generated in the past at similar levels that would be considered big data today. The difference is that in the past many of these data were simply not being stored or analyzed. Costs and difficulties of creating, storing and analyzing data have caused limited scope and temporality in large datasets. As such, we have seen national datasets such as census compilation being confined to coarse temporal (e.g., every 10 years) and spatial (through aggregation to small geographical zones) resolutions (Kitchin, 2014). The continual improvements in computing power, alongside the development of cloud computing, have rendered these issues obsolete. The production of data through many actions of everyday life (and subsequent ability to exploit such data) have seen big data termed the *new oil* (Andreu-Perez et al., 2015).

Big data and health

If big data was around in 1854 when cholera swept through London, John Snow could have used Global Positioning System (GPS) data, mortality records and environmental data to identify the disease vectors and implement a solution within hours (Drexler, 2014; Khoury and Ioannidis, 2014).

Big data stand to improve health by providing insights into the causes and outcomes of disease, better drug targets for precision and personalized medicine, and enhanced disease prediction and prevention (Khoury and Ioannidis, 2014; Raghupathi and Raghupathi, 2014). Although traditional health-data sources are rarely the largest in terms of size, the potential offered through data linkage to newer forms of data (e.g., linking survey data to purchasing behaviors through loyalty-card records) could offer novel and exciting insights into health-related behaviors (Herland, Khoshgoftaar and Wald, 2014).

Evidence-based medicine has long adopted a practice of generating evidence from trials and experiments, using data to inform decisions. The evidence-based movement was founded on the belief that scientific inquiry is superior to expert opinion and testimonials, meaning the industry was ahead of many others in recognizing the value of data-guided decision-making (Murdoch and Detsky, 2013). Despite this long-standing support for evidence-informed decisions, health-care delivery has been slow to utilize the rich information routinely collected within its own data frameworks (Safran et al., 2007). There are significant opportunities for health-care researchers and policy-makers to learn from other data-driven revolutions that have been successful at applying large datasets (Kayyali, Knott and Van Kuiken, 2013; Weber and Kohane, 2014), such as astronomic telescopic information, retail-sales data or internet search engine data (Murdoch and Detsky, 2013).

That being said, big data have been used for some facets of health science. The management of pandemics such an influenza has seen opportunities for heterogeneous information from managed and unmanaged sources processed, mined and turned into decision actions to control outbreaks (Andreu-Perez et al., 2015). An example of such application is Google Flu Trends, which predicts outbreaks of flu from flu-related internet searches (Pentland, Reid and Heibeck, 2013; Herland, Khoshgoftaar and Wald, 2014). Similar studies have used Twitter data to estimate the prevalence of flu (Lampos, De Bie and Cristianini, 2010), as well as the likelihood of whence an individual developed a case of food poisoning (Sadilek et al., 2013). Cell-phone data were recently used in West Africa to track travel patterns and predict where Ebola might spread to after an outbreak in a town or village, allowing public-health officials to set up early preventative barriers for containing the spread of the disease (Wesolowski et al., 2014). Sensor information has accelerated in usage in recent years, bringing forward the notion of the internet of things, with wearables being at the forefront (Andreu-Perez et al., 2015). Such wearable technologies offer opportunities for improved prediction of health outcomes and are being combined with interventions to evaluate their success or monitor progress.

It is not the case, though, that big data will improve health applications overnight. In 2013, Google Flu Trends overestimated doctor visits for influenza-associated conditions by twice what validated data suggested it should be (Butler, 2013). In the past, the same service has also underestimated the H1N1 pandemic due to how people were searching for the condition (Cook et al., 2011). While these issues are partly due to the novelty of their algorithms (in Google Flu Trends' case, due to overfitting of data), issues that can be overcome, Lazer et al. (2014) acknowledge that *big data hubris* has seen many individuals overestimate the potential of big data. Big data should only supplement, rather than replace, traditional approaches for data collection and analysis. For these reasons, it is important to stress the *value* part of any big data application.

Applications in health geography

While the possibilities are wide-ranging for the applications of big data in health geography, we focus on a few examples to illustrate feasible opportunities within the field.

Developing new metrics

Much of the hype surrounding big data has been about the perceived endless analytical applications, but many of the actual applications have been somewhat basic in statistical and analytical terms. The most common application has been the aggregation and development of new indicators to aid the measurement of health phenomena. The opening up of big data (particularly new forms of data) creates opportunities for health geographers to better understand the contribution of environmental features on health. One example can be seen in Daras et al. (2017), who use consumer data on the location of retail outlets to calculate network distances for all postcodes (and also aggregated to small geographical zones) in Great Britain to their nearest amenities such as fast-food outlets, pubs or off-licenses. Developing national-level metrics from traditional and non-traditional sources and making these openly available (data are available here: http://maps.cdrc.ac.uk/) helps researchers and policy-makers better understand the role of neighborhood features on health and health-related behaviors without having to undertake major data manipulation.

Mapping services, such as Google Street View, have also been mined as a source of information about neighborhoods, such as aesthetics, location of food outlets and land use (Bethlehem et al., 2014). Fully automating the process has proved difficult but represents a useful research avenue, particularly with new developments in image processing and machine learning. Google has also incorporated air-quality measures onto their street-view cars to collect data on levels of nitric oxide, nitrogen dioxide and black carbon (Tuxen-Bethman, 2017), and the data can be requested freely from Google for research purposes. There is also growing interest in using remote-sensing data to develop desk-based audits of features of the physical and built environment (Charreire et al., 2016).

The increasing use of social media across the population has also offered a novel source of information on human relationships and social interactions. Twitter data is the most commonly used social-media data source, as unlike other platforms it makes a portion of its data openly available to researchers. The complexity of processing such data has been a limitation. Strategies employed have ranged from taking random samples of tweets, to selecting phrases, keywords or hashtags (e.g., Eichstaedt et al. 2015), to using machine learning to classify the content of posts (e.g., Nguyen et al., 2016). Many of these approaches can be combined with the geotagged information provided in tweets to explicitly incorporate the spatial dimension. For example, Nguyen et al. (2016) used machine-learning techniques to classify whether individuals were tweeting about fast food or high-calorie/energy-dense foods. Through linking these data to zip codes, they found that where there were higher rates of such tweets, there was a positive correlation to the number of fast-food outlets and the state-level prevalence of obesity. Similar studies have been undertaken using Twitter to measure geographical patterns in happiness (Gore, Diallo and Padilla, 2015), physical activity (Nguyen et al., 2016), diet (Nguyen et al., 2016; Widener and Li, 2014) and psychological distress (Eichstaedt et al. 2015).

Data linkage

One of the early promises of big data was the ability to connect multiple sources of information and combine their separate sources to create a comprehensive data resource. While there has been good progress in linking administrative sources (e.g., health records), there have been few examples using commercial data sources. This is partly a result of the difficulty of linking several datasets with no common identifier between them, but it is also due to the trepidation of data custodians who are wary of sharing or linking data to consumer sources that may reveal insights to competitors. Many organizations are understandably cautious, although initiatives such as the Consumer Data Research Centre (CDRC) have started a process of brokering conversations between organizations and researchers to help facilitate collaborations. Ethical issues around security, data ownership and privacy are important and require careful consideration before progress will be made. As boyd and Crawford (2012) argue, there is a fine line between accessibility and ethics. Some of these issues are being alleviated through examples of good practice, but progress remains slow.

One data source that has received much attention for data linkage has been the potential offered by loyalty-card records. Consumer loyalty cards were originally introduced as a marketing tool to provide customers with incentives to remain brand loyal (Mauri, 2003; Sharp and Sharp, 1997). Many organizations, particularly supermarkets, soon realized their potential for understanding consumer behaviors, helping them tailor experiences (e.g., through person-specific price coupons for items) and deciding which products to stock in stores. While they are an imperfect data source, since not all individuals have loyalty cards for the places where they shop (or some individuals have multiple cards), they provide objective data on behaviors that is important given the under-reporting of dietary behaviors common in traditional self-reported sources of data (Flegal, 1999).

A recent example demonstrating the potential of supermarket data can be seen in a study by Silver et al. (2017), who used point-of-sale data on purchasing behaviors in supermarkets to evaluate the impact of the introduction of a tax on sugar-sweetened beverages in the American city of Berkeley. Several countries and local governments have proposed introducing a similar tax as an approach to promote healthier diets (and therefore reduce the risk of developing conditions such as type 2 diabetes). The quality of evidence as to whether such a policy would be effective is limited and often relies on modeled estimates. Silver and colleagues' results suggested that one year after the introduction of the tax, sales of sugar-sweetened beverages declined by 9.6%, compared to an increase of 6.9% in supermarkets where the tax was not introduced. These findings were novel and important in understanding how successful a similar tax could be elsewhere. While the data were not loyalty-card records (loyalty-card data would have strengthened the quality of the study through providing repeated-measures data allowing for a better understanding of changes in behaviors post-intervention), they still demonstrate the usefulness and potential of consumer data.

Challenges of big data

While big data offer many possibilities for health geography, there are several important limitations. Ethics is important, and we have touched on security issues previously from the point of researchers and data custodians, but privacy concerns also matter to the public. The public may not have signed up to take part in external research knowingly, nor agreed for their data to be linked to other sources.

Datasets are considerably bigger, but not necessarily better. Moving away from thinking about the *big* in big data and incorporating boyd and Crawford's philosophy is more appropriate. We should not view a difference between *big* and *small* data; they are both data. Data should be judged on what they can add to a research question, as well as the quality that they offer (more eloquently put by the equation *garbage in equals garbage out*). Rather than a *big data revolution*, we should view future research within the *all-data revolution* paradigm whereby the importance is placed on innovative analytics to better understand the world (Lazer et al., 2014). In this regard, viewing the field as (geographic) data science seems a better fit, since it avoids many of these issues while preserving the approaches toward big data analytics that are important and sought after.

We need to continue placing emphasis on understanding what a sample offers, rather than using it purely because it ticks the *big data* box (a large amount of the studies using Twitter data fall into this category, with their primary interest in using *novel* data rather than necessarily improving on past approaches). This is important given that data are often not collected for research purposes, and, therefore, using repurposed data requires an understanding of the context and meaning of any data (boyd and Crawford, 2012) – which can be difficult to ascertain, as metadata is typically lacking. Classic research is hypothesis-driven, and data collection is therefore focused to answering a specific question, improving its *value*. Repurposed data move us away from this, and so data should not be taken at face value alone.

The issue of value in large datasets is also statistical: we need to understand how representative any data are. For example, Twitter data are inherently biased – Twitter is popular with younger populations, and only a small proportion (<1%) of tweets contain any geotagged information (and indeed those who geotag tweets are demographically different from those who do not: Sloan and Morgan, 2015). Understanding the appropriateness and validity of any data source is simply good statistical practice (hence our prior argument about the importance of discussing big data within a data-science or all-data context). Accuracy and precision are very different statistical issues, and the greater precision offered through larger sample sizes will not make up for a lack of accuracy.

Big data suffers gigantic noise, and big error can plague big data, where spurious correlations and ecological fallacies can multiply and mislead findings (Khoury and Ioannidis, 2014). Large sample sizes can distort many common statistical techniques and require us to approach analyses using newer methods (e.g., machine learning). This is one of the reasons why many studies opt for simple aggregation of data, since this is usually where the main patterns can be observed. Removing noise from the vast amounts of data also needs to be tackled for big data to be translated into gains in societal well-being (Khoury and Ioannidis, 2014).

Many big data analytics focus on leveraging large sample sizes to make better predictions; however, often in health geography we are more interested in causal inference. Health geographers have an important role to play here in ensuring that correct study design is applied to improve our (causal) understanding of the world. There are important and exciting opportunities available for natural experiments involving such data. For instance, one could envisage analyzing how changes in the environment are associated with changes in purchasing behavior. Although big data offer endless possibilities, only through carefully designed studies can we understand how best to utilize such data.

Conclusion

There has been a lot of interest around the opportunities big data pose, both across and within many disciplines. While many challenges remain, there is still a lot of promise, and we should not forget that such systems and technology are still fairly new. Indeed, many companies are only just realizing themselves what potential their data offers for improving their own insights! The opportunities opened up through the big-data revolution within health geography are due not only to new data avenues, but also to the analytical and modeling possibilities of larger datasets. Many of these possibilities have yet to be realized, as typical applications involve simple aggregations of data using non-complex methods. This should not be viewed as a criticism, though. Results may be initially unclear in large data, and only through the expertise of careful interpretation will we discover valuable insights (Van Dijck, 2014). As such, health geographers should construct their studies carefully to ensure that future developments are effective at helping us better understand the world.

References

Andreu-Perez, J., Poon, C., Merrifiedl, R., Wong, S. T. and Yang, G. Z. (2015). Big data for health. *IEEE Journal of Biomedical and Health Informatics*, 19(4), pp. 1193–1208.

Bethlehem, J., Mackenbach, J., Ben-Rebah, M., Compernolle, S. Glonti, K., Bárdos, H., Rutter, H. R., Charreire, H., Oppert, J.-M., Brug, J. Lakerveld, J. (2014). The SPOTLIGHT virtual audit tool: a valid and reliable tool to assess obesogenic characteristics of the built environment. *International Journal of Health Geographics*, 13(52), pp. 2–8.

Birkin, M., Clarke, G. and Clarke, M. (2017). *Retail location planning in an era of multi-channel growth*. London: Routledge.

boyd, d. and Crawford, K. (2012). Critical questions for big data. *Information, Communication & Society*, 15(5), pp. 662–679.

Butler, D. (2013). When Google got flu wrong. *Nature*, 494, pp. 155–156.

Charreire, H., Mackenbach, J., Ouasti, M., et al. (2016). Using remote sensing to define environmental characteristics related to physical activity and dietary behaviours: a systematic review (the SPOTLIGHT project). *Health & Place*, 25, pp. 1–9.

Connelly, R., Playford, C., Gayle, V. and Dibben, C. (2016). The role of administrative data in the big data revolution in social science research. *Social Science Research*, 59, pp. 1–12.

Cook, S., Conrad, C., Fowlkes, A. and Mohebbi, M. (2011). Assessing Google Flu Trends performance in the United States during the 2009 Influenza Virus A (H1N1) Pandemic. *PLoS One*, 6(8), pp. 1–8.

Daras, K., Davies, A., Green, M. and Singleton, A. (2017). Developing indicators for measuring health-related features of neighbourhoods. In: P. Longley, J. Cheshire and A. Singleton, eds., *Consumer data analytics*. London: UCL Press.

Drexler, M. (2014). Big data's big visionary. [online] *Harvard Public Health Magazine*. Available at: www.hsph.harvard.edu/magazine/magazine_article/big-datas-big-visionary/. [Accessed 20 June 2017].

Eichstaedt, J., Schwartz, H. A., Kern, M. L., Park, G., Labarthe, D. R., Merchant, R. M., Jha, S., Agrawal, M., Dziurzynski, L. A., Sap, M., Weeg, C., Larson, E. E., Ungar, L. H. and Seligman, M. E. (2015). Psychological language on Twitter predicts county-level heart disease mortality. *Psychological Science*, 26(2), pp. 159–169.

Flegal, K. (1999). Evaluating epidemiologic evidence of the effects of food and nutrient exposures. *American Journal of Clinical Nutrition*, 69, pp. 1339–1344.

Gore, R. J., Diallo, S. and Padilla, J. (2015). You are what you tweet: connecting the geographic variation in America's obesity rate to Twitter content. *PloS One*, 10(9), pp. 1–16.

Herland, M., Khoshgoftaar, T. and Wald, R. (2014). A review of data mining using big data in health informatics. *Journal of Big Data*, 1(1), pp. 1–35.

Kayyali, B., Knott, D. and Van Kuiken, S. (2013). The big-data revolution in US health care: accelerating value and innovation. *McKinsey & Company*, pp. 1–13.

Khoury, M. and Ioannidis, J. (2014). Big data meets public health. *Science*, 346(6213), pp. 1054–1055.

Kitchin, R. (2014). Big data, new epistemologies and paradigm shifts. *Big Data & Society*, 1(1), pp. 1–12.

Lampos, V., De Bie, T. and Cristianini, N. (2010). Flu detector – tracking epidemics on Twitter. In: J. L. Balcázar, F. Bonchi, A. Gionis and M. Sebag, eds., *Machine learning and knowledge discovery in databases*. ECML PKDD 2010. Lecture Notes in Computer Science, 6323. Berlin, Heidelberg: Springer, pp. 599–602.

Lazer, D., Kennedy, R., King, G. and Vespignani, A. (2014). The parable of Google Flu: traps in big data analysis. *Science*, 343(6176), pp. 1203–1205.

Manovich, L. (2011). Trending: the promises and the challenges of big social data. *Debates in the Digital Humanities*, 2, pp. 460–475.

Mauri, C. (2003). Card loyalty. A new emerging issue in grocery retailing. *Journal of Retailing and Consumer Services*, 10(1), pp. 13–25.

Murdoch, T. and Detsky, A. (2013). The inevitable application of big data to health care. *Journal of the American Medical Association*, 309(13), pp. 1351–1352.

Nguyen, Q., Li, D., Meng, H., Nsoesie, E., Li, F. and Wen, M. (2016). Building a national neighborhood dataset from geotagged Twitter data for indicators of happiness, diet, and physical activity. *JMIR Public Health and Surveillance*, 2(2), pp. 1–16.

Pentland, A., Reid, T. and Heibeck, T. (2013). *Big data and health: revolutionizing medicine and public health*. Report for the Big Data and Health Working Group. N.p.: World Innovation Summit for Health (WISH).

Raghupathi, W. and Raghupathi, V. (2014). Big data analytics in healthcare: promise and potential. *Health Information Science and Systems*, 2(1), pp. 1–10.

Sadilek, A., Brennan, S., Kautz, H. and Silenzio, V. (2013). nEmesis: which restaurants should you avoid today? *AAAI Publications, First AAAI Conference on Human Computation and Crowdsourcing*, pp. 138–146.

Safran, C., Bloomrosen, M., Hammond, W. E., et al. (2007). Toward a national framework for the secondary use of health data: an American medical informatics association white paper. *Journal of the American Medical Informatics Association*, 14(1), pp. 1–9.

Sharp, B. and Sharp, A. (1997). Loyalty programs and their impact on repeat-purchase loyalty patterns. *International Journal of Research in Marketing*, 14(5), pp. 473–486.

Silver, L. D., Ng, S., Ryan-Ibarra, S., et al. (2017). Changes in prices, sales, consumer spending, and beverage consumption one year after a tax on sugar-sweetened beverages in Berkeley, California, US: a before-and-after study. *PLoS Medicine*, 14(4), pp. 1–19.

Sloan, L. and Morgan, J. (2015). Who Tweets with their location? Understanding the relationship between demographic characteristics and the use of Geoservices and Geotagging on Twitter. *PLoS One*, 10(11), pp. 1–15.

Tuxen-Bethman, K. (2017). *Let's clear the air: mapping our environment for our health*. [online] The Keyword. Available at: https://blog.google/products/maps/lets-clear-air-mapping-our-environment-our-health/. [Accessed 20 June 2017].

Van Dijck, J. (2014). Datafication, dataism and dataveillance: Big Data between scientific paradigm and ideology. *Surveillance & Society*, 12(2), pp. 197–208.

Weber, G. and Kohane, I. (2014). Finding the missing link for big biomedical data. *Journal of the American Medical Association*, 311(24), pp. 2479–2480.

Wesolowski, A., Buckee, C., Bengtsson, L., et al. (2014). Commentary: containing the ebola outbreak – the potential and challenge of mobile network data. *PLoS Currents Outbreaks*, 29(1), pp. 1–20.

Widener, M. and Li, W. (2014). Using geolocated Twitter data to monitor the prevalence of healthy and unhealthy food references across the US. *Applied Geography*, 54, pp. 189–197.

47

NAVIGATING RESEARCH ETHICS IN HEALTH GEOGRAPHY

The case of big data

Nina J. Morris

Increasingly stringent governance of professional academic conduct in recent years has seen the review of proposed studies by institutional or national Research Ethics Committees (REC) become a standard feature of health-geography research (Boden, Epstein and Latimer, 2009; Burr and Reynolds, 2010; McCormack et al., 2012). This shift, stemming from a series of 20th century biomedical research scandals (e.g., medical experimentation in 1940s Nazi Germany, the 1932–1972 Tuskegee Institute Syphilis Study in the United States) and the ethical codes published in their wake (e.g., Nuremberg Code, Declaration of Helsinki, Belmont Report) (Dyer and Demeritt, 2009; Hoeyer, Dahlager and Lynöe, 2005; Schrag, 2011), has meant that academics' compliance with ethical standards is now a personal duty rather than simply a matter of "personal and professional integrity" (Dyer and Demeritt, 2009, p. 47). Recent interest in research ethics, however, has also been fueled by growing concerns over economic inequalities, international human rights, increasing public interest in scientific endeavors (Molyneux and Geissler, 2008) and broader social trends toward *accountability* (Boden, Epstein and Latimer, 2009). So, what began as an attempt to regulate medical research practice (Hammersley, 2009) has expanded to encompass "everything that might be considered 'human subjects' research, and [. . .] politically extended to almost all academic research" (Emmerich, 2013, p. 177). Health geography is no exception, and in this chapter I discuss the ethical implications of research with large datasets with particular reference to issues of consent and re-identification. These issues are important not only for researchers who undertake secondary data analysis, but also for those involved in primary data collection who intend to make their data available for reuse.

The rationale for ethical oversight is largely accepted (Klitzman, 2011; Librett and Perrone, 2010; Nind et al., 2012). RECs are credited with raising researchers' ethical awareness, promoting good-quality research with favorable risk-benefit ratios and reducing exploitative research (Molyneux and Geissler, 2008). Ethical codes of conduct are acknowledged to (a) promote the aims of research (knowledge, truth, avoidance of error) and the values that are essential to collaborative work (trust, accountability, mutual respect, fairness), (b) provide researchers with guidance and a consistent set of expectations, (c) ensure that researchers are accountable, build public support for research (trust in quality and integrity) and (d) promote other important values (social responsibility, human rights) (Madge, 2007; Metcalf and Crawford, 2016; Resnik, 2011).

Over the last decade, however, some have expressed concerns regarding the perceived bureaucratization of ethics through REC regulation (Hammersley, 2009; Pickersgill, 2012) and the applicability of biomedical-research governance procedures to social-science research including health geography (Bond, 2012; Emmerich, 2013; see Chattopadhyay and De Vries, 2008, for a discussion on non-Western contexts). For example,

RECs often require researchers to take precautions to ensure the safety of their research participants (Dingwall, 2008; Klitzman, 2011) without taking into consideration the varying power differentials that may be in operation depending on the status of the researcher/participant, the research topic and the methods being used (Burr and Reynolds, 2010). Research participants are not always vulnerable (Dyer and Demeritt, 2009), and those in powerful or elite positions may use ethical regulations (e.g., informed consent) to avoid close scrutiny (Boden, Epstein and Latimer, 2009). Conversely, those labeled "vulnerable" may feel that ethical standards such as anonymity disassociate them from the research (Metcalf and Crawford, 2016) or silence them entirely (Boden, Epstein and Latimer, 2009).

There is also frustration at the inability of standardized ethics-review procedures (often deemed to be rule-bound, formulaic and inflexible) to accommodate the often unpredictable (Hoeyer, Dahlager and Lynöe, 2005; Dyer and Demeritt, 2009; Hammersley, 2009; Burr and Reynolds, 2010), subjective and messy (Butz, 2008) nature of social-science research. A particular fear is that as REC protocols become ever more prescriptive, researchers will become more conservative in their choice of subject matter and methods (Bond, 2012; Librett and Perrone, 2010; Nind et al., 2012) or, worse, will be untruthful about the true nature of their research (Boden Epstein and Latimer, 2009; Hammersley, 2009) in order to get ethics approval. Likewise, there is a worry that if doing ethics becomes nothing more than a box-ticking exercise, researchers will lose their capacity for ethical reasoning (Dyer and Demeritt, 2009), thus limiting their ability to adapt their methods as circumstances change (McCormack et al., 2012), and that ethical debates will become entirely depoliticized (Blee and Currier, 2011; Molyneux and Geissler, 2008; Schrag, 2011).

A rapid rise in the application of big-data methods and secondary analysis (Zook et al., 2017), particularly in health geography, has brought with it a new series of ethical dilemmas (Metcalf and Crawford, 2016; Zook et al., 2017). Indeed, the "speed with which these new sources of data have emerged, as well as the increasingly imaginative ways that researchers are using them, risks running ahead of the development of an appropriate ethical framework for their use" (McKee, 2013, p. 299). For example, in the not-so-distant past, researchers were required to destroy their data on completion of a project as an absolute way of dealing with issues of confidentiality and data protection (Morrow, Boddy and Lamb, 2014). In recent years, however, opinion has shifted in favor of reusing quantitative and qualitative data – also known as secondary data collection or analysis – whenever possible. Using secondary data is said to be ethical in and of itself because it promotes research integrity by making data available to others for validation, thus acting as a safeguard against fraudulent research (Grinyer, 2009). Economically, reuse maximizes the value of any investment made in data collection (Morrow, Boddy and Lamb, 2014). Morally, it ensures that individuals from potentially vulnerable populations are not over-burdened by research. It ensures replicability of study findings and, as a result, greater transparency of research procedures and integrity of research work (Morrow, Boddy and Lamb, 2014). However, the reuse of data collected in either the course of one's own research or that of another (Grinyer, 2009) raises significant issues relating to consent, anonymity and representation (Blatt, 2012).

Of course, some argue that big-data research does not require ethical review because it does not involve direct intervention into the human body or because the data are public (Metcalf and Crawford, 2016) even if they originated from a breach or other illegal activity (Leetaru, 2016; Metcalf, Keller and boyd, 2016). Incorporating information derived from multiple sources (for health geographers, these might include hospital episode statistics, mortality and demographic data, land cover databases or social media), big data is characterized by its high volume, velocity, variety, exhaustive scope, resolution and indexicality, relationality, and flexibility (Kitchin, 2013), all of which can lead to an illusion of anonymity. Projects in the United States, for example, that make use of already existing, publicly available datasets are exempt from REC regulation because they are considered to "pose only minimal risks to the human subjects they document" (Metcalf and Crawford, 2016, p. 1). Others would argue that we should jettison consent altogether for research that uses anonymized personal data, because contributing our data is a social good and because requiring consent would limit researchers' potential to have an impact on long-term community health outcomes (Millett and

O'Leary, 2015). It has become rapidly clear, however, that the "scope, scale and complexity of many forms of big data creates a rich ecosystem in which human participants and their communities are deeply embedded and susceptible to harm" (Zook et al., 2017, p. 1).

Consent

Consent is a central pillar of ethical research. Previously, emphasis has been placed on the need for primary researchers to gain participants' voluntary informed consent prior to the data-collection phase; however, big-data research, involving the archiving and reuse of data, complicates this type of up-front contractual agreement (Morrow, Boddy and Lamb, 2014). Much attention has been paid to consent in relation to the ethics of *archiving* qualitative data, while relatively little attention has been paid to the ethical practicalities of *using* secondary data, be it qualitative or quantitative (Grinyer, 2009; Morrow, Boddy and Lamb, 2014), although it is generally agreed that context is crucial. If secondary data users understand the source of the datasets they are using, the rules and regulations with which they were gathered (Metcalf and Crawford, 2016; Morrow, Boddy and Lamb, 2014; Zook et al., 2017), they are much more likely to avoid misrepresenting or misinterpreting the data or acting contrary to the original consent agreements (Morrow, Boddy and Lamb, 2014). Exploring the use of Twitter as a method of surveillance for seasonal influenza, for example, Aslam et al. (2014) noted that, despite the many benefits of this technique, their inability to access demographic information such as age, gender and race made it difficult to determine precisely who was tweeting about the flu and to whom public-health efforts should be directed.

Of course, problems are more likely to arise when the data have been collected outside of research frameworks (Metcalf and Crawford, 2016). As our lives become increasingly digitized, we leave traces of personal data (e.g., records of commercial transactions, administrative data, medical test results, closed-circuit television images, internet usage records, photographs) everywhere we go, much of it "collected under mandatory terms of service rather than responsible research design overseen by university compliance officers" (Zook et al., 2017, p. 4). Social-media sites, for example, require new users to tick that they have read the terms and conditions, but many people will simply tick the box without reading the information, which is frequently written in legalese (Leetaru, 2016; Metcalf and Crawford, 2016). As a consequence, secondary data users cannot assume that informed consent for the data to be reused in research has been granted (Leetaru, 2016; Bishop, 2017). Opt-out consent is a possibility (Millett and O'Leary, 2015), but one with several drawbacks: it can be costly and time-consuming for large datasets, tracing participants can be impossible even when the owner grants permission to access them, it places a burden on participants and it can introduce bias (Bishop, 2017; Grinyer, 2009).

Likewise, it has become common for people to share highly detailed information about their daily lives online (McKee, 2013; Sui, 2011). Yet, this does not mean that they (as data creators) are making this information available on a public platform for anything other than their own private purposes (Zook et al., 2017). It is not easy to distinguish between public and private data on the internet (Bishop, 2017; Markham and Buchanan, 2012; Madge, 2007; McKee, 2013). For example, individuals:

> may operate in public spaces [e.g., Twitter, Instagram, discussion groups, forums] but maintain strong perceptions or expectations of privacy [or] they may acknowledge that the substance of their communication is public, but that the specific context in which it appears implies restrictions on how that information is – or ought to be – used by other parties.
>
> *(Markham and Buchanan, 2012, p. 6)*

The people who communicate in these public settings often do so believing that settings to be more private or, at least, fail to remember the longevity and accessibility of the information they post (Bishop, 2017;

Madge, 2007). Breaching privacy is the primary way in which big-data research can do the most harm (Zook et al., 2017). When using data derived from this type of source, an understanding of the research subjects' expectations is critical (Madge, 2007; cf. Doyle's (2013) work on the *Children's Health* section of Mumsnet Talk forum). One approach, suggested by McKee (2013, p. 299), in such instances is to apply a notion of reciprocity whereby the researcher asks himself or herself, *How would I feel if the roles were reversed?* "[W]hile the researcher might post information on his or her public profiles to be shared by friends or peers, this does not mean that they have consented for this information to be collated, analysed and published, in effect turning them into research subjects."

Re-identification

Datasets are said to have been de-identified if "elements that might immediately identify a person or organization have been removed or masked" (Bishop, 2017, p. 5). In recent years, as the volume of readily available personal data has increased (Gellman, 2007; Metcalf and Crawford, 2016) and our ability to connect disparate datasets through algorithmic analysis has improved (Metcalf, Keller and boyd, 2016), it has become increasingly easy for re-identification to occur. As Zook et al. (2017, p. 2) note, "even seemingly benign data can contain sensitive and private information" and, when used in conjunction with other datasets, has the potential to produce "unanticipated ethical questions and detrimental impacts." Indeed, there have been a number of cases in recent years where the profiling of individuals and groups (e.g., according to race, gender, class, propensity to certain diseases) in this way has led to severe discrimination (e.g., restricted access to treatments or services) (Zook et al., 2017). One method of combating re-identification is to release data only within safe settings; either secured physical sites, which researchers travel to in order to access confidential data (Richardson et al., 2015), or user- and/or function-restricted computer environments (Bell et al., 2006; Hartter et al., 2013) also known as geospatial virtual data enclaves (see Richardson et al., 2015). Researchers can analyze the data in, but not remove it from, these safe settings, and all outputs must be risk-assessed by a member of staff before release (Richardson et al., 2015).

For health geographers combining big health data – such as medical data or ambulance-dispatch records – with geospatial technologies, the ethical issues are even more complex (Blatt, 2012; Sui, 2011). GIS practitioners are now able to perform complex spatial analyses by conjoining databases that include not just people's individual attributes, but also their activities and details of their surrounding environment (Styblińska, n.d.). Datasets can range from large government or corporate databases (Richardson et al., 2015) down to an individual's geotagged tweets (Sui, 2011; Zook et al., 2017). Of course, there are a number of societal benefits that might accrue from "an improved understanding of how place influences an individual's space-time path, the activities they undertake, and ultimately their well-being" (Rainham et al., 2009, p. 671). However, Richardson et al. (2015, p. 102) note that "the potential of GIS to be far more invasive of personal privacy than many other information technologies has caused serious concern among privacy advocates, GIS researchers, and the public." Conjoining spatial information with another dataset can contradict confidentiality agreements made at the time the data were collected, in perhaps unforeseeable ways (Gellman, 2007; Kounadi and Leitner, 2014). Zook et al. (2017, p. 3) note that there are "numerous examples of researchers with good intentions and seemingly good methods failing to anonymize data sufficiently to prevent the later identification of specific individuals."

A number of masking methods – applied either before or after analysis – have been developed to protect locational privacy while allowing valid spatial analysis to be performed (Kounadi and Leitner, 2014). These include spatial, temporal or point aggregation, affine transformation, random perturbation, and flipping (Bell et al., 2006; Kounadi and Leitner, 2014; Richardson et al., 2015). Spatial and temporal aggregation have long been the most popular methods used to maintain the confidentiality of spatial health data used by health geographers (Bell et al., 2006); however, both have limitations (Kounadi and Leitner, 2014). They can limit the resolution of the data and thus their interpretability (e.g., the detection clusters, other spatial relationships), they increase the possibility of bias (under- or over-reporting the number of cases over time or in a subgroup) and they can greatly affect or prevent adjustments for bias, confounding and effect modification (Bell et al., 2006; Kounadi and Leitner, 2014).

Future ethics

Given that the ethics of big-data research are emergent, complex and often beyond the experience of many RECs, it is essential that health geographers seek advice from as many sources as possible (Zook et al., 2017). These sources might include fellow researchers/practitioners, relevant ethics guidelines, academic literature (within, and outside, one's discipline) and legal frameworks (Markham and Buchanan, 2012). As Zook et al. note (2017, p. 6), "understanding the different ways people discuss these challenges and processes provides an important check for researchers, especially if they come from disciplines not focused on human subject concerns." There can never be a universal checklist for ethical social-science research, especially where big data is concerned, given the range of data sources, research topics, and methodological approaches taken (Markham and Buchanan, 2012). Although formal ethics guidelines are a useful starting point, ethical decision-making will always be necessarily contextual, processual and negotiated. This requires health geographers to be sensitive to the specificities of the dataset(s) they are working with (Markham and Buchanan, 2012), as well as the governing regime of their local REC – or, as Heimer (2013, p. 371) puts it, the "gap between official ethics (ethics on the books) and ethics on the ground (ethics in action)." This is the case both for those who wish to archive their data and users of secondary data.

The most critical point for health geographers to remember when undertaking ethical big-data research, however, is that all big data relating to health are linked in some way to individual persons (Markham and Buchanan, 2012; Zook et al., 2017). As such, it is necessary to consider the ethical principles relating to research on human subjects, even if one disagrees with standardized pronouncements relating to the minimization of harm and/or consent. One should not assume that data cannot be re-identified (Zook et al., 2017), because identifiability is not a constant (Gellman, 2007). De-identification will always be an essential tool in minimizing risk, but only "as part of a broader approach to ensuring safe use of data" (Bishop, 2017, p. 5). This includes acknowledging that identifiability cannot be viewed in binary terms (and as such can never be eliminated) and that privacy is both contextual and situational (Zook et al., 2017).

References

Aslam, A. A., Tsou, M. H., Spitzberg, B. H., An, L., Gawron, J. M., Gupta, D. K., Peddecord, K. M., Nagel, A. C., Allen, C., Yang, J. A. and Lindsay, S. (2014). The reliability of tweets as a supplementary method of seasonal influenza surveillance. *Journal of Medical Internet Research*, 16(11), e250.

Bell, B. S., Hoskins, R. E., Williams Pickle, L. and Wartenberg, D. (2006). Current practices in spatial analysis of cancer data: mapping health statistics to inform policymakers and the public. *International Journal of Health Geographics*, 5(49), 14 pages.

Bishop, L. (2017). *Big data and data sharing: ethical issues*. University of Essex, Colchester. UK Data Archive. UK Data Service.

Blatt, A. J. (2012). Ethics and privacy issues in the use of GIS. *Journal of Map & Geography Libraries*, 8(1), pp. 80–84.

Blee, K. M. and Currier, A. (2011). Ethics beyond the IRB: an introductory essay. *Qualitative Sociology*, 34, pp. 401–413.

Boden, R., Epstein, D. and Latimer, J. (2009). Accounting for ethos or programmes for conduct? The brave new world of research ethics committees. *The Sociological Review*, 57(4), pp. 727–749.

Bond, T. (2012). Ethical imperialism or ethical mindfulness? Rethinking ethical review for social sciences. *Research Ethics*, 8(2), pp. 97–112.

Burr, J. and Reynolds, P. (2010). The wrong paradigm? Social research and the predicates of ethical scrutiny. *Research Ethics Review*, 6(4), pp. 128–133.

Butz, D. (2008). Sidelined by the guidelines: reflections on the limitations of standard informed consent procedures for the conduct of ethical research. *ACME: An International E-Journal for Critical Geographies*, 7(2), pp. 239–259.

Chattopadhyay, S. and DeVries, R. (2008). Bioethical concerns are global, bioethics is Western. *Eubios Journal of Asian and International Bioethics*, 18(4), pp. 106–109.

Dingwall, R. (2008). The ethical case against regulation in humanities and social science research. *Twenty-First Century Society: Journal of the Academy of Social Sciences*, 3(1), pp. 1–12.

Doyle, E. (2013). Seeking advice about children's health in an online parenting forum. *Medical Sociology Online*, 7(3), pp. 17–28.

Dyer, S. and Demeritt, D. (2009). Un-ethical review? Why it is wrong to apply the medical model of research governance to human geography. *Progress in Human Geography*, 33(1), pp. 46–64.

Emmerich, N. (2013). Between the accountable and the auditable: ethics and ethical governance in the social sciences. *Research Ethics*, 9(4), pp. 175–186.

Gellman, R. (2007). Appendix A: privacy for research data. In: M. P. Gutmann and P. C. Stern, eds., *Putting people on the map: protecting confidentiality with linked social-spatial data*. Washington, DC: National Academies Press, pp. 81–122.

Grinyer, A. (2009). The ethics of the secondary analysis and further use of qualitative data. *Social Research Update*, 56, Guildford: University of Surrey.

Hammersley, M. (2009). Against the ethicists: on the evils of ethical regulation. *International Journal of Social Research Methodology*, 12(3), pp. 211–225.

Hartter, J., Ryan, S. J., MacKenzie, C. A., Parker, J. N. and Strasser, C. A. (2013). Spatially explicit data: stewardship and ethical challenges in science. *PLoS Biology*, 11(9), e1001634.

Heimer, C. A. (2013). "Wicked" ethics: compliance work and the practice of ethics in HIV research. *Social Science & Medicine*, 98, pp. 371–378.

Hoeyer, K., Dahlager, L. and Lynöe, N. (2005). Conflicting notions of research ethics: the mutually challenging traditions of social scientists and medical researchers. *Social Science & Medicine*, 61, pp. 1741–1749.

Kitchin, R. (2013). Big data and human geography: opportunities, challenges and risks. *Dialogues in Human Geography*, 3(3), pp. 262–267.

Klitzman, R. (2011). The ethics police? IRBs' views concerning their power. *PLoS One*, 6(12), pp. 1–7.

Kounadi, O. and Leitner, M. (2014). Why does geoprivacy matter? The scientific publication of confidential data presented on maps. *Journal of Empirical Research on Human Research Ethics*, 9(4), pp. 34–45.

Leetaru, K. (2016). *Are research ethics obsolete in the era of big data?* [online] Forbes. [Accessed 17 June 2017] Available at: www.forbes.com/sites/kalevleetaru/2016/06/17/are-research-ethics-obsolete-in-the-era-of-big-data/#3794e0dd7aa3.

Librett, M. and Perrone, D. (2010). Apples and oranges: ethnography and the IRB. *Qualitative Research*, 10(6), pp. 729–747.

Madge, C. (2007). Developing a geographers' agenda for online research ethics. *Progress in Human Geography*, 31(5), pp. 654–674.

Markham, A. and Buchanan, E. (2012). *Ethical decision-making and internet research: recommendations from the AoIR Ethics Working Committee (Version 2.0)*. [pdf] Available at: https://aoir.org/reports/ethics2.pdf [Accessed 17 June 2017].

McCormack, D., Carr, T., McCloskey, R., Keeping-Burke, L., Furlong, K. E. and Doucet, S. Getting through ethics: the fit between research ethics board assessments and qualitative research. *Journal of Empirical Human Research Ethics*, 7(5), pp. 30–36.

McKee, R. (2013). Ethical issues in using social media for health and health care research. *Health Policy*, 110, pp. 298–301.

Metcalf, J. and Crawford, K. (2016). Where are human subjects in Big Data research? The emerging ethics divide. *Big Data and Society*, January–June, pp. 1–14.

Metcalf, J., Keller, E. F. and boyd, d. (2016). *Perspectives on Big Data, Ethics, and Society*. [pdf] A report for The Council for Big Data, Ethics and Society. Available at: http://bdes.datasociety.net/wp-content/uploads/2016/05/Perspectives-on-Big-Data.pdf [Accessed 1 June 2017].

Millett, S. and O'Leary, P. (2015). Revisiting consent for health information databanks. *Research Ethics*, 11(3), pp. 151–163.

Molyneux, S. and Geissler, P. W. (2008). Ethics and the ethnography of medical research in Africa. *Social Science & Medicine*, 67, pp. 685–695.

Morrow, V., Boddy, J. and Lamb, R. (2014). The ethics of secondary data analysis: Learning from the experience of sharing qualitative data from young people and their families in an international study of childhood poverty. *NOVELLA Working Paper: Narrative Research in Action*. London: Institute of Education.

Nind, M., Wiles, R., Bengry-Howell, A. and Crow, G. (2012). Methodological innovation and research ethics: forces in tension or forces in harmony? *Qualitative Research*, 13(6), pp. 650–667.

Pickersgill, M. (2012). The co-production of science, ethics, and emotion. *Science, Technology and Human Values*, 37(6), pp. 579–603.

Rainham, D., McDowell, I., Krewski, D. and Sawada, M. (2009). Conceptualizing the healthscape: contributions of time geography, location technologies and spatial ecology to place and health research. *Social Science & Medicine*, 70, pp. 668–676.

Resnik, D. B. (2011). *What is ethics in research and why is it important?* National Institute of Environmental Health Sciences. [online] Available at: www.niehs.nih.gov/research/resources/bioethics/whatis/ [Accessed 19 July 2014].

Richardson, D. B., Kwan, M-P., Alter, G. and McKendry, J. E. (2015). Replication of scientific research: addressing geoprivacy, confidentiality, and data sharing challenges in geospatial research. *Annals of GIS*, 21(2), pp. 101–110.

Schrag, Z. M. (2011). The case against ethics review in the social sciences. *Research Ethics*, 7(4), pp. 120–131.

Styblińska, M. (n.d.). *The GIS ethics: the code of ethics for GIS practitioners*. Katowice, Poland: University of Silesia.

Sui, D. Z. (2011). Legal and ethical issues of using geospatial technologies in society. In: T. L. Nyerges, H. Couclelis and R. McMaster, eds., *The SAGE handbook of GIS and society*. London: Sage, pp. 504–528.

Zook, M., Barocas, S., boyd, d., Crawford, K., Keller, E., Peña Gangadharan, S., Goodman, A., Hollander, R., Koenig, B. A., Metcalf, J., Narayanan, A., Nelson, A. and Pasquale, F. (2017). Ten simple rules for responsible big data research. *PLoS Computational Biology*, 13(3), e1005399.

48

SPATIAL MODELING'S PLACE IN HEALTH GEOGRAPHY

Trends, critiques and future directions

Sara McLafferty and Sandy Wong

Spatial modeling comprises a set of methods and approaches for measuring, analyzing and visualizing geographical relationships among people and the contexts in which they live, work and interact. For centuries, these methods have been central to health geographers' efforts to understand and improve population health and well-being and access to health services. The methods emphasize spatial relationships – those based on location, distance and proximity – as critical dimensions of health and health care. Looking back, the history of spatial modeling in health geography reveals both enthusiastic adoption of the methods and trenchant critiques of their validity and value. The give-and-take between embrace and critique led to the current pluralistic state in which spatial modeling stands alongside other methodologies as a well-used approach in health geography and a vibrant area of research endeavor. Spatial modeling has also achieved considerable success and impact as an arena for cross-fertilization with disciplines such as epidemiology, biostatistics and diverse health-related fields.

This chapter begins by discussing the historical and intellectual trajectory of spatial approaches in health and medical geography, focusing on key questions, debates and realignments. Later sections discuss current spatial methods and themes, along with ongoing critical directions and topics for future research.

Historical and intellectual trajectory

Much early work in medical geography relied on spatial association, the most basic form of spatial modeling. John Snow's map of cholera cases in the London epidemic of 1854 suggested clustering patterns and spatial associations with water pumps that offered clues about disease etiology. Jacques May's treatise on medical geography (May, 1950) provides many examples of using spatial association to shed light on geographical determinants of disease. In fact, his key insight – that disease agent, host and vector must coexist in space and time for disease transmission to occur – is based on spatial associational reasoning.

Building on these early works, the 1960s and 1970s saw the rapid adoption of quantitative methods of spatial analysis. Following the quantitative revolution that was sweeping geography at the time, many medical geographers embraced multivariate statistical methods in analyzing spatial associations between disease incidence and environmental factors. These methods were analogous to the visual and ad-hoc approaches adopted by earlier researchers, but their use of complex algorithms and hypothesis-testing procedures gave them an aura of scientific rigor and validity. Other researchers moved beyond spatial association. Peter Haggett and Andrew Cliff (in the United Kingdom), and Gerald Pyle (in the United States) developed dynamic models of disease diffusion that included both spatial and temporal elements. Research on health services adopted methods such as location-allocation modeling to identify optimal service locations and gravity modeling to analyze potential access to health services (Knox, 1978). These approaches advanced the field by explicitly modeling spatial relationships based on proximity, location and distance. Gesler (1986) offers a thoughtful summary of medical geographers' adoption of spatial-analytic methods in the 1960s and 1970s.

Debates and critiques

Important critiques of spatial analysis emerged in the 1980s and early 1990s as social-theory and political-economy perspectives gained prominence in medical geography and as emerging health concerns raised questions about traditional approaches. The HIV/AIDS pandemic revealed the complex politics of infectious disease, its uneven spread and the inadequate public-health response. Moreover, widening health inequalities in the 1980s and 1990s in many economically developed countries showed that people's health was driven by much more than environment and location.

Spatial modeling, and the quantitative methods on which it is based, were key targets in those critiques. Debates about the role of spatial modeling in medical geography focused on four main themes: First, spatial methods were criticized on epistemological grounds, for their inability to establish *causation* from evidence of spatial association. In discussing spatial analysis of disease, Mayer (1983) described the ambiguous relationship between spatial patterns of disease and the underlying processes that generate those patterns: spatial patterns reflect complex and multiple causal pathways that may not be evident in ecological associations. Mayer also raised a series of important conceptual challenges, including the lack of attention to the roles of emotions, experiences and social networks in disease etiology and varying cultural conceptions of disease. Second, spatial methods were criticized for their failure to shed light on the political-economic processes that shape relationships between places and health. According to Mohan (1989), many spatial studies adopted a position of "technical neutrality" (p. 169) devoid of consideration for the political and social forces that impact health and health-care inequalities. Spatial methods' failure to illuminate key health issues at that time, including health services privatization, HIV/AIDS and increasing disparities in population health, revealed the methods' inherent political-economic limitations.

The third area of critique focused on the inability of spatial models to reveal people's experiences, perceptions and agency (Dyck, 1999). People were often modeled as faceless and interchangeable entities. Characterized as "genderless and colourblind" (Pearson, 1989, p. 9), medical geography's spatial-analytic focus neglected key axes of social difference. Although this omission is not inherent to spatial methods, it represented the nature of most spatial-analytic research by medical geographers at the time. Pulling these themes together, the fourth critique centered on spatial methods' emphasis on space as opposed to place. By definition, spatial methods are concerned with space – location, proximity, distance and so on. Space is treated as a container in which events unfold, ignoring the recursive relationships between people and places. Kearns' (1993) described the spatial perspective as a *placeless* endeavor that fails to consider how people's health and well-being relates to their place-based experiences and active agency in place-making.

The emphasis on place in contrast to space significantly changed the course of medical geography and led to its newfound focus on health and place and the growing use of qualitative methods. Although qualitative methods were used earlier by researchers like Jenny Cornwell and John Eyles, and field-based research was always a mainstay in medical geography, qualitative methods gained great prominence in the 1990s, as evidenced by a special issue of the *Professional Geographer* devoted to qualitative methods in health geography (Dyck, 1999). The journal *Health & Place* was introduced to foster and represent this important shift from medical to health geography and from quantitative spatial analysis to qualitative and mixed methods.

Spatial modeling in the 1990s and 2000s

Despite these critiques, the spatial analysis and modeling tradition endured. Technological developments in computing and data storage greatly increased geographers' ability to analyze large geospatial datasets, and geographic information systems (GIS) for analyzing these datasets became ubiquitous. Medical/health geographers harnessed these developments to facilitate the kinds of spatial, locational and spatiotemporal analyses that were first proposed in the 1970s. Technological developments also enabled much more complex and computationally intensive types of spatial analysis. Multilevel modeling, in which health outcomes are modeled as functions of environmental and social factors at multiple spatial scales, gained a strong foothold among health geographers (Duncan, Jones and Moon, 1998). Health geographers developed and introduced important spatial epidemiological methods, such as methods for visualizing complex disease patterns and detecting disease clusters (Gatrell et al., 1996). At the same time, GIS emerged as a crucial tool for integrating, visualizing and analyzing large geospatial datasets on social and environmental factors and health outcomes (Cromley and McLafferty, 2002). A review by Rosenberg (1998) identifies the many developments in spatial-analytic perspectives in health geography in the 1990s.

The growth of spatial modeling in the 1990s and early 2000s was also closely tied to health geographers' involvement in multidisciplinary research teams. Health geographers' spatial-analytic skills were in demand as biomedical and public-health researchers began to explore socioecological and environmental influences on health. In Canada, the United Kingdom and the United States, health geographers spearheaded and participated in large, funded research projects addressing a wide range of health and health-care issues. In the 1990s in the United States, the Centers for Disease Control sponsored conferences on GIS and health, and agencies like the National Cancer Institute organized working groups and panels that addressed geospatial perspectives. These and other funded research endeavors contributed greatly to the field's expansion and innovation.

Thus, by the end of the millennium, a strong pluralistic model had emerged in health geography, in which both quantitative and qualitative research methods held prominence. Both sets of methods coexisted and thrived, each with a strong set of advocates. In addition, some prominent health geographers – Graham Moon, Susan Elliott, Mark Rosenberg, Sarah Curtis and many others – bridged both worlds. There was broad recognition that each set of methods could provide useful insights in particular research contexts. As Elliott (1999, p. 240) wrote: "And the question shall determine the method." Health geographers also advocated for mixed methods that combined the population-level insights derived from quantitative methods with the detailed process-based knowledge gained from qualitative methods. These mixed methods are widely used in contemporary health-geographic research.

Current trends

Several broad areas of research and application that leverage advances in web and mobile technologies and computing power are at the forefront of contemporary work on spatial modeling in health geography.

Space-time and longitudinal analysis

Researchers have long recognized that the processes impacting health and health care unfold in space and time, but the complexity and computing power needed to handle space-time data posed major barriers to spatiotemporal analysis. As these barriers have fallen, health geographers have developed and implemented novel methods for dynamic and space-time analysis. Some of these methods represent straightforward extensions of existing spatial methods, such as kernel density estimation, spatial autocorrelation analysis and spatial cluster detection, to include time as an added dimension. Clusters of health events in both space and time can be identified to provide clues about disease etiology and information for policy-making and intervention.

Cluster-detection methods that incorporate longitudinal data on residential histories, historical environmental exposures and disease latency periods have also been developed (Sabel et al., 2009).

Use of longitudinal methods is increasing in health geography. These methods represent the dynamic associations among variables over time, enabling analysis of complex relationships between people and places. Places change via economic, social and environmental processes, while people's behaviors, interactions and experiences both shape and are shaped by place change. Longitudinal methods that capture this dynamism have provided important evidence about the effects of migration on health and health inequalities. Longitudinal research in Scotland shows that healthy young people often move out of deprived areas to less-deprived, and generally healthier, areas (Norman, Boyle and Rees, 2005). These selective migration processes exacerbate health inequalities as flows of healthy out-migrants from deprived areas diminish health status in the places they leave and enhance it in the places they move to.

Efforts are underway to model and simulate the health-related behaviors of individuals over space and time. Geographical research using *agent-based models* is greatly expanding as the models are employed to simulate space-time patterns of disease spread based on data describing people's mobility patterns and spatial interactions, as well as characteristics of the social, built and natural environments (Mao, 2014). Results of these models are useful for predicting where and how quickly emerging disease outbreaks might spread. A related class of methods involves *spatial microsimulation*, a process of simulating population-level trends and patterns. Health geographers have implemented both dynamic and static versions of these models to provide small-area estimates of health-related behaviors and to analyze changes in health inequalities over time (Smith, Pearce and Harland, 2011).

Multilevel modeling

Multilevel modeling is increasingly used in health research, accelerated by computational advancements and software developments in the last three decades. By considering interactions between people and contexts, multilevel models avoid individualistic and ecological fallacies that limit other statistical methods (Owen, Harris and Jones, 2016). Health geographers were early adopters of multilevel modeling and have used it extensively to investigate social and contextual determinants of health (Duncan, Jones and Moon, 1998).

To account for multiple and overlapping spatial contexts that influence individual health outcomes, researchers have included two or more contextual effects in hierarchical models and used cross-classified models in which the contextual effects are not hierarchical. In one case of a four-level model, individuals were nested in households nested in communities nested in regions to examine the effect of income inequality on self-rated health (Subramanian et al., 2003). A study on childhood obesity used a cross-classified structure in which individuals were nested in both residential areas and schools (Townsend, Rutter. and Foster, 2012).

Multilevel models emphasize scalar relationships, but their treatment of space has tended to be naïve. *Spatial autocorrelation* can persist in multilevel models even after accounting for individual and contextual effects. Recent contributions tackle this issue by modeling spatial relationships among contextual units along with contextual effects (Arcaya et al., 2012). Results reveal clustering of health outcomes that emerges from spatial processes and interactions that cut across area boundaries. Other recent developments include *multiple membership models*, in which individuals are assigned to several neighborhood-level units simultaneously. These models acknowledge the important fact that people's health is affected by the many contexts they experience throughout their daily lives.

To model changes in contextual effects over time, there are emerging developments at the intersection of multilevel modeling and life-course epidemiology. A life-course perspective enriches our understandings on the relations between health and place by identifying critical time periods for and cumulative effects on health outcomes as people age. Næss and Leyland (2010) used a correlated cross-classified model, with data from four cohorts at four time points, to identify residential locations for people at particular life stages that most strongly affect their mortality risk.

Modeling of activities and exposures

To better understand contextual dimensions of health inequalities, opportunities, behaviors and risks, scholars are increasingly turning to spatial techniques to model individuals' health-related activities and exposures. This broad and diverse field increasingly relies on Global Positioning System (GPS)–based tracking devices and environmental sensors to track people's movements through space and time and record their exposures to changing environmental conditions.

Activity spaces are the physical spaces that people travel through and have access to over the course of daily activities. To move beyond the use of static residential units as proxies for context, activity spaces are utilized to capture multiple, non-residential and more accurate environmental influences on health outcomes over space and time (Kwan, 2012; Perchoux et al., 2013). Popular activity-space applications include the standard deviational ellipse, network buffer, kernel-density estimation and potential path area, which are typically used to identify the locations, sizes and features of activity spaces. Activity-space size has been found to be significantly correlated with diverse health-related issues, including health-care opportunities, dietary behaviors and risk of depression. Usually the larger the activity space, the better the health outcome and geographic access to health-promoting opportunities. Increasingly, researchers use GPS devices to record highly detailed space-time activity patterns associated with health outcomes and exposures (Zenk et al., 2011).

Using GPS-based activity tracking raises thorny methodological and conceptual issues. Detailed location data are often unavailable due to privacy and confidentiality restrictions. Tracking devices generate enormous streams of data, with varying, and occasionally unknown, levels of accuracy and precision, and complex algorithms are required to generate meaningful information. Most importantly, the fact that space-time activity spaces result from individual decision-making processes creates ambiguities in analyzing their role as health determinants. People actively choose spatial contexts, so an observed contextual effect on health may reflect preferences and choice rather than context, resulting in *selection bias* (Chaix et al., 2013). Although statistical methods exist for evaluating selection bias, their application to real-time tracking data that reflect mundane, everyday decisions is poorly understood.

Similar trends are underway in research on monitoring and measuring environmental exposures. Spatial patterns of air pollution have been estimated using universal kriging (Jerrett et al., 2001) and land-use regression models. In recent years, geographers have moved away from these traditional static models to more dynamic ones that are sensitive to space-time variations in individual mobility and air-pollution concentrations. Lu and Fang (2015) estimated personal exposure to both ambient and indoor air pollutants using space-time trajectories to identify danger zones where the health impact from air pollution was highest. Improvements in exposure assessment have been facilitated by technological advancements in personal environmental monitors and remote sensing (Steinle et al., 2015).

Spatial-accessibility modeling

Spatial-accessibility models estimate the local availability, supply and proximity of health-related services to populations in need. Although early applications of the models focused on health-care services, recent applications encompass parks and recreation facilities, food outlets, social services, risky spaces and other places that impact health and well-being. The simplest models include *container* measures that describe the ratio of services to population within fixed geographic zones and distance measures that determine Euclidean or network distance to the closest service facility; however, these have well-known limitations (Higgs, 2006). Also, widely used are gravity and potential models, which have been extended to incorporate factors, such as access to transportation, that restrict spatial access for vulnerable, low-income and elderly populations (Lovett et al., 2002).

Much recent research centers on *floating catchment area* (FCA) models, which determine geographically varying service-to-population ratios within overlapping spatial windows (Luo and Wang, 2003). The original two-step FCA (2SFCA) method involved determining the provider-to-population ratio within a catchment of each service provider and then summing the ratios within the catchment of each population zone. Researchers have made many enhancements to the 2SFCA, including incorporating distance decay effects, alternative transportation modes, variations in catchment sizes and differences in population health needs (Wang, 2012).

Researchers are also considering the role of time, as well as space, in service availability and access. Hours of operation for service providers and temporal variation in the types of services offered intersect with people's space-time constraints and daily activity patterns to restrict access (Widener and Shannon, 2014). Used in analyzing access to food stores, novel spatiotemporal accessibility measures integrate data on typical movement patterns throughout the workday with store opening and closing times (Widener et al., 2015).

These and other recent developments bring spatial-accessibility modeling closer to addressing the *uncertain geographic context problem* (Kwan, 2012) which argues that spatial contexts are imprecise, reflecting complex constraints, decisions and behaviors. Activity-tracking data map these contexts in real-time. However, with a few exceptions (Wang, 2007), these models pay insufficient attention to individual cultural, political-economic and psychosocial processes, the impact of which on access can be profound.

Future developments

Advances in spatial modeling go hand-in-hand with methodological and theoretical developments in health geography that offer exciting opportunities and challenges for future work. From a methodological perspective, a significant trend is the increasing adoption of *Bayesian methods* that are more robust and make less restrictive assumptions than widely used frequentist methods. Bayesian methods assume that the parameters of interest are random variables, rather than fixed quantities. By combining observed data with prior beliefs about the unknown values, we can estimate these parameter values and their distributional characteristics. Health geographers were among the earliest adopters of Bayesian methods in the discipline of geography, and use of the methods has expanded dramatically with improvements in software availability and access.

Another area at the forefront of spatial modeling is analysis of *spatial and social networks* and network spaces. Many spatial methods continue to rely on Cartesian conceptions of space that are based on Euclidean distance and geographical proximity. Yet people are increasingly embedded in far-flung social networks that do not map neatly onto Cartesian spaces. Such networks have critical effects on health behaviors and outcomes. Methods of social and spatial network analysis offer important tools for studying intersections between social and geographical networks, with implications for health outcomes (Kestens et al., 2017). Furthermore, such networks compel health geographers to think beyond traditional GIS-based models of space toward spaces that are socially defined.

The explosion of *big data* from social media, web searches, tracking systems, electronic medical records and other diverse sources also presents opportunities for spatial modeling. Much of this data is *volunteered geographic information*, contributed, either intentionally or unintentionally, by individuals and heterogeneous in content. For example, Twitter data have been used to map the US obesity epidemic (Ghosh and Guha, 2013). However, existing big-data applications emphasize geovisualization; few have taken the next step to model health determinants and processes. Moreover, we know little about the validity and reliability of volunteered geographic data and their embedded social and spatial biases.

Although spatial modeling has advanced significantly in recent decades, the debates and critiques of the 1990s continue to ring true. New models more accurately represent the dynamic and relational qualities of spatial contexts, but researchers rarely attempt to model how and why those contexts emerge and change (Pearce, 2015). Understanding contexts requires thinking about the political, economic, and decision-making processes that shape place environments at varying scales, a key point raised in earlier critiques. Researchers – for example, Larsen and Gilliland (2008) – are beginning to explore these issues by analyzing and modeling changes in food retail landscapes, but few models tie these contextual changes to people's health-related behaviors and experiences.

Critiques also highlighted the omission in spatial models of people's experiences and perceptions and of the recursive relationships between people and places. Although spatial modeling has inched closer to tackling these important issues by analyzing people's daily activities and exposures, a huge gap remains. Opportunities exist to humanize spatial modeling by incorporating cultural and psychosocial factors that mediate people's experiences in place environments. Techniques like *ecological momentary assessment* are being used to capture individuals' health-related emotions and behaviors and link them via GIS to specific place-time settings (Mason et al., 2016). Participatory methodologies such as *participatory GIS* are also attracting interest, as researchers share model results with community members and incorporate their input and feedback to gain new understandings (Beyer and Rushton, 2009).

Conclusion

Developments in spatial modeling, GIS and geospatial technologies have greatly advanced our ability to model spatial and spatiotemporal contexts and explore their impacts on health and access to health care. Future trends are likely to include increased reliance on real-time mobility and environmental data and further development of complex and dynamic models that depict health-related interactions between people and places. At the same time, the field will continue to grapple with issues raised in earlier critiques – issues of causation, selection bias, changing political-economic contexts and human decision-making and behavior – that underpin the models we develop and the insights we glean from them. Strengthening the ties between spatial modeling and qualitative research approaches, and integrating their perspectives and methods more fully, offers exciting possibilities for making spatial modeling a more place-based endeavor.

References

Arcaya, M., Brewster, M., Zigler, C. and Subramanian, S.V. (2012). Area variations in health: a spatial multilevel modeling approach. *Health & Place*, 18, pp. 824–831.

Beyer, K. and Rushton, G. (2009). Mapping cancer for community engagement. *Public Health Research, Practice & Policy*, 6, pp. 1–8.

Chaix, B., Méline, J., Duncan, S., Merrien, C., Karusisi, N., Perchoux, C., Lewin, A., Labadi, K. and Kestens, Y. (2013). GPS tracking in neighborhood and health studies: a step forward for environmental exposure assessment, a step backward for causal inference? *Health & Place*, 21, pp. 46–51.

Cromley, E. and McLafferty, S. (2002). *GIS and public health*. New York: Guilford Press.

Duncan, C., Jones, K. and Moon, G. (1998). Context, composition and heterogeneity: using multilevel models in health research. *Social Science & Medicine*, 46(1), pp. 97–117.

Dyck, I. (1999). Using qualitative methods in medical geography: deconstructive moments in a subdiscipline? *Professional Geographer*, 51(2), pp. 243–253.

Elliott, S. (1999). And the question shall determine the method. *Professional Geographer*, 51(2), pp. 240–243.

Gatrell, A. C., Bailey, T. C., Diggle, P. and Rowlingson, B. (1996). Spatial point pattern analysis and its application in geographical epidemiology. *Transactions of the Institute of British Geographers*, 21(1), pp. 256–274.

Gesler, W. (1986). The uses of spatial analysis in medical geography: a review. *Social Science & Medicine*, 23(10), pp. 963–973.

Ghosh, D. and Guha, R. (2013). What are we "tweeting" about obesity? Mapping tweets with Topic Modeling and Geographic Information System. *Cartography & Geographic Information Science*, 40(2), pp. 90–102.

Higgs, G. (2006). Measuring potential access to primary healthcare services. *Professional Geographer*, 58(3), pp. 294–306.

Jerrett, M., Burnett, R. T., Kanaroglou, P., Eyles, J., Finkelstein, N., Giovis, C. and Brook, J. R. (2001). A GIS–environmental justice analysis of particulate air pollution in Hamilton, Canada. *Environment & Planning A*, 33(6), pp. 955–973.

Kearns, R. (1993). Place and health: towards a reformed medical geography. *Professional Geographer*, 45, pp. 139–147.

Kestens, Y., Wasfi, R., Naud, A. and Chaix, B. (2017). "Contextualizing context": reconciling environmental exposures, social networks, and location preferences in health research. *Current Environmental Health Reports*, 4(1), pp. 51–60.

Knox, P. (1978). The intraurban ecology of primary care: patterns of accessibility and their policy implications. *Environment and Planning A*, 10(4), pp. 415–435.

Kwan, M. (2012). The uncertain geographic context problem. *Annals of the Association of American Geographers*, 102(5), pp. 958–968.

Larsen, K. and Gilliland, J. (2008). Mapping the evolution of "food deserts" in a Canadian city: supermarket accessibility in London, Ontario, 1961–2005. *International Journal of Health Geographics*, 18(7), pp. 1–16.

Lovett, A., Haynes, R., Sunnenberg, G. and Gale, S. (2002). Car travel time and accessibility by bus to general practitioner services: a study using patient registers and GIS. *Social Science & Medicine*, 55(1), pp. 97–111.

Lu, Y. and Fang, T. B. (2015). Examining personal air pollution exposure, intake, and health danger zone using time geography and 3D geovisualization. *ISPRS International Journal of Geo-Information*, 4(1), pp. 32–46.

Luo, W. and Wang, F. (2003). Measures of spatial accessibility to health care in a GIS environment: synthesis and a case study in the Chicago region. *Environment & Planning B*, 30, pp. 865–884.

Mao, L. (2014). Modeling triple-diffusions of infectious diseases, information, and preventive behaviors through a metropolitan social network: an agent-based simulation. *Applied Geography*, 50, pp. 31–39.

Mason, M., Mennis, J., Zaharakis, N. M. and Way, T. (2016). The dynamic role of urban neighborhood effects in a text-messaging adolescent smoking intervention. *Nicotine & Tobacco Research*, 18(5), pp. 1039–1045.

May, J. M. (1950). Medical geography: its methods and objectives. *Geographical Review*, 40(1), pp. 9–41.

Mayer, J. (1983). The role of spatial analysis and geographic data in the detection of disease causation. *Social Science & Medicine*, 17(16), pp. 1213–1221.

Mohan, J. (1989). Medical geography: competing diagnoses and prescriptions. *Antipode*, 21(2), pp. 166–177.

Næss, Ø. and Leyland, A. H. (2010). Analysing the effect of area of residence over the life course in multilevel epidemiology. *Scandinavian Journal of Public Health*, 38, pp. 119–126.

Norman, P., Boyle, P. and Rees, P. (2005). Selective migration, health and deprivation: a longitudinal analysis. *Social Science & Medicine*, 60(12), pp. 2755–2771.

Owen, G., Harris, R. and Jones, K. (2016). Under examination: multilevel models, geography and health research. *Progress in Human Geography*, 40(3), pp. 394–412.

Pearce, J. (2015). Invited commentary: history of place, life course, and health inequalities – historical geographic information systems and epidemiologic research. *American Journal of Epidemiology*, 181(1), pp. 26–29.

Pearson, M. (1989). Medical geography: genderless and colourblind. *Contemporary Issues in Geography and Education*, 3, pp. 9–17.

Perchoux, C., Chaix, B., Cummins, S. and Kestens, Y. (2013). Conceptualization and measurement of environmental exposure in epidemiology: accounting for activity space related to daily mobility. *Health & Place*, 21, pp. 86–93.

Rosenberg, M. W. (1998). Medical or health geography? Populations, peoples, and places. *International Journal of Population Geography*, 4, pp. 211–226.

Sabel, C., Boyle, P., Raab, G., Loytonen, M. and Maasilta, P. (2009). Modeling individual space-time exposure opportunities: a novel approach to unravelling the genetic or environment disease causation debate. *Spatial & Spatiotemporal Epidemiology*, 1(1), pp. 85–94.

Smith, D., Pearce, J. and Harland, K. (2011). Can a deterministic spatial microsimulation model provide reliable small-area estimates of health behaviours? An example of smoking prevalence in New Zealand. *Health & Place*, 17, pp. 618–624.

Steinle, S., Reis, S., Sabel, C. E., Semple, S., Twigg, M. M., Braban, C. F., Leeson, S. R., Heal, M. R., Harrison, D., Lin, C. and Wu, H. (2015). Personal exposure monitoring of $PM_{2.5}$ in indoor and outdoor microenvironments. *Science of the Total Environment*, 508, pp. 383–394.

Subramanian, S. V., Delgado, I., Jadue, L., Vega, J. and Kawachi, I. (2003). Income inequality and health: multilevel analysis of Chilean communities. *Journal of Epidemiology & Community Health*, 57(11), pp. 844–848.

Townsend, N., Rutter, H. and Foster, C. (2012). Age differences in the association of childhood obesity with area-level and school-level deprivation: cross-classified multilevel analysis of cross-sectional data. *International Journal of Obesity*, 36, pp. 45–52.

Wang, F. (2012). Measurement, optimization, and impact of health care accessibility: a methodological review. *Annals of the Association of American Geographers*, 102(5), pp. 1104–1112.

Wang, L. (2007). Immigration, ethnicity, and accessibility to culturally diverse family physicians. *Health & Place*, 13(3), pp. 656–671.

Widener, M. J., Farber, S., Neutens, T. and Horner, M. (2015). Spatiotemporal accessibility to supermarkets using public transit: an interaction potential approach in Cincinnati, Ohio. *Journal of Transport Geography*, 42, pp. 72–83.

Widener, M. J. and Shannon, J. (2014). When are food deserts? Integrating time into research on food accessibility. *Health & Place*, 30, pp. 1–3.

Zenk, S., Schulz, A. J., Matthews, S. A., Odoms-Young, A., Wilbur, J., Wegrzyn, L., Gibbs, K., Braunschweig, C. and Stokes, C. (2011). Activity space environment and dietary and physical activity behaviors: a pilot study. *Health & Place*, 17(5), pp. 1150–1161.

49

DIFFERENCE MATTERS

Approaches for acknowledging diversity in health geography research

Melissa Giesbrecht

Despite the vast diversity that characterizes our human population, research in health and social sciences tends to focus on singular dominant identities and population groups (men, women, children, etc.). It is true that categorizing people into single, distinct population groups is much easier for researchers, but according to Wilkinson, "[i]t has become increasingly apparent . . . that this way of doing research is rather limited in its ability to accurately represent the complexity of social life" (2003, p. 27). Particularly in the field of health, this approach to research has come under scrutiny from an increasing number of experts, as the scope of existing identities are not fully captured, and, as a result, meaningful and adequate solutions to health disparities are not realized (Hankivsky and Christoffersen, 2008). For example, if sex and gender were not adequately taken into account in earlier cardiovascular research, we would not now know that cardiovascular disease occurs ten years later in women than men, or that there are social differences in gender-related risk factors, such as smoking, depression, low income, obesity and lack of physical activity (Health Canada, 2010). Without acknowledging differences *between* groups of women, it would also not have been found that Indigenous women experience rates of diabetes five times higher than all other women in Canada (Health Canada, 2010). Such early research shows that some differences matter for how health is obtained and experienced. Meanwhile, these and other earlier studies have not fully embraced unpacking the true range of diversity that exists among people and populations – a challenge that now faces contemporary health researchers.

Health geographers are inherently interdisciplinary in nature, commonly collaborating with and drawing from work in fields such as sociology, anthropology, political science, disabilities studies, women's studies, public health and social gerontology, among many others. As such, health geographers are well-positioned to apply approaches and methods that are conducive to understanding the spatial aspects of everyday lives and experiences of diverse population groups in relation to health. The most general approach to exploring differentials within health research falls under the broad umbrella of what can be called diversity-based analyses (Giesbrecht and Crooks, 2016). This approach to analysis is not tied to particular theoretical paradigms or methods; it aims to simply explore and uncover differences between or within population groups. Here, "diversity" is used as a term to define all the ways that people are unique and different from others (Alberta Health Services, 2014). Broadly speaking, diversity-based analyses not only provide an understanding of how social and spatial patterns interact and impact one's health, but also emphasize issues of (in)equity by generating critical knowledge about how differences continually affect people's opportunities, choices, decisions and health outcomes. In this chapter, I present an overview of three existing approaches that health geographers can and do employ to acknowledge diversity among target population groups in their research. Namely, I introduce gender and sex based (+) analysis, health equity and the social determinants of health, and intersectionality. Each approach holds the potential to assist health geographers in untangling the complex ways in which differing axes of diversity inform the relationship between health and place.

Sex and gender–based analysis (+)

Originating from feminist health research, sex and gender–based analysis (SGBA) was one of the first diversity-based approaches to emerge and challenge the assumptions that males and females, women and men, are affected similarly by research, policies, programs, health-care delivery and health issues (Disabled Women's Network Ontario, 2014; Greaves, 2012; Health Canada, 2010). SGBA is a systematic approach to health research and policy formation that examines biological (sex-based) and sociocultural (gender-based) differences between women and men, boys and girls (Greaves, 2012; Johnson, Greaves and Repta, 2009; Health Canada, 2010). "Sex" refers to the biological characteristics, such as anatomy (e.g., body size and shape) and physiology (e.g., hormonal activity or functioning of organs), that distinguish males from females and vice versa (Health Canada, 2010), whereas "gender" refers to the array of socially constructed roles, attitudes, personality traits, behaviors, values and relative power and influence that society ascribes to two sexes on a differential basis (Health Canada, 2010). By considering both sex and gender, SGBA acts as a tool for understanding and accounting for biological and social processes of health, allowing for responses that embrace more equitable options (Disabled Women's Network Ontario, 2014).

Although SGBA has become the most common method for responding to health-based diversity, it has received an increasing amount of criticism from experts as it promotes binary conceptualizations and assumes that all men and women belong to one of these two sex-gender groupings (Greaves, 2012; Johnson, Greaves and Repta, 2009). Thus, recent applications of SGBA stress the use of a critical lens to consider sex and gender as *relational* issues that are context-specific, evolving across space and time (Greaves, 2012). Currently, sex and gender are recognized as concepts that "[are] tightly interrelated, exist on continua, and simultaneously interact iteratively with each other" (Greaves, 2012, p. 4). Recent applications of SGBA also emphasize that such approaches should always be undertaken in the context of *diversity*, whereby it is recognized that the differences between men and women are influenced by a variety of factors, including class, socioeconomic status, age, sexual orientation, gender identity, race, ethnicity, geographic location, education and physical and mental ability (Health Canada, 2010). In embracing diversity, SGBA has become renamed and is now known in some circles as SGBA(+), with the plus sign highlighting that the approach goes beyond sex and gender to include the examination of other intersecting identity factors (Status of Women Canada, 2016).

While some health geographers have begun unraveling gender and sex–based differences in health experiences, this has previously been done, most often, by drawing from feminist geography and focusing solely on the experiences of women (Dyck, 2010). While not all using feminist theory in their work, these researchers are concerned with exploring how gender (as expressed in what women do as mothers, daughters, and wives) differentiates them from the lives of men, which is ultimately reflected in their local experiences of place and their health, illness and disability (Dyck, 2010). For example, in her work examining the gendered implications of health-care reform in New Zealand, Wiles (2002) employed a fluid conceptualization of women's health, whereby the diverse social context of women's lives took precedent. More specifically, Wiles (2002) recognized that individual women had different, even contradictory, health-care issues and needs that will vary according to their differing roles, identities, contexts and access to resources. This conception of gender was applied to explore ideas about place and identity in relation to access to health-care services in New Zealand's restructured health-care system (Wiles, 2002). While little research focuses on the diverse experiences of men, Lewis (2016) has begun to address this gap by examining the process of immigration and resettlement, and post-migration urban encounters, that influence sexual health among gay and bisexual men. Health geographers have also become increasingly engaged with exploring, simultaneously, how health status and access to care varies across sociocultural, economic and political groups of men and women. For example, Setia et al. (2011) applied SGBA(+) to illuminate sex and gender differences, among other social factors, in the healthy-immigrant effect, a prominently geographic process. Setia et al. (2011) found that the country of origin played a role in the ethnic differential between women's and men's self-rated health statuses and had independent negative effects, particularly among women. There is great potential for health geographers to contribute further to SGBA(+) research, particularly by applying their specialized knowledge of space and place and how such dimensions shape the health experiences and outcomes of diverse men, women, boys and girls.

Health equity and the social determinants of health

Although sex and gender are important axes of diversity, they are not always the most significant factors shaping health and health outcomes. Rather, as mentioned in the previous sub-section, various other axes of diversity (e.g., class, socioeconomic status, age, sexual orientation, race, ethnicity, religion, geographic location, education, physical and mental ability) also play important roles. The social determinants of health (SDH) aim to capture these facets of diversity and are defined by the World Health Organization as those diverse "conditions in which people are born, grow, live, work and age" that shape one's health and health outcomes (World Health Organization, 2014, n.p.). Thus, the SDH aim to examine the multiple axes of diversity that can produce differentials in health experiences among diverse population groups. Furthermore, the SDH perspective recognizes that these determinants are dynamic, fluid, interrelated and continually evolving across space and time (Hankivsky and Christoffersen, 2008).

Since the emergence of the SDH approach, several models have been developed at national and international levels to guide health-policy creation and evaluation (Commission on Social Determinants of Health, 2008; Native Women's Association of Canada, 2007). While SDH models developed in different countries have some commonalities, they do differ with regard to which factors are included or excluded and in the ways in which these factors are interrelated. For example, the SDH identified by the Canadian Public Health Association (Canadian Public Health Association, 2017) include the following:

- Income and income distribution
- Education
- Unemployment and job security
- Employment and working conditions
- Early childhood development
- Food insecurity
- Housing
- Social exclusion
- Social safety network
- Health services
- Aboriginal status
- Gender
- Race
- Disability

Although debate exists regarding the need to develop such lists, thinking about SDH in this way facilitates recognition of the complex relationship between diversity and health.

SDH approaches emphasize that health is not an outcome of individual-level susceptibilities, choices, and behaviors, but rather is shaped by broad processes that occur at various scales, from the global to the local (Pauly et al., 2013; World Health Organization, 2014). This perspective highlights how unequal distributions of money, power and resources are thus largely responsible for differential health outcomes (World Health Organization, 2014). As such, SDH frameworks have become increasingly popular for use in research aiming to understand and address health inequities, which are the disparities in health outcomes that are avoidable, unfair and systemically related to social inequality and marginalization (Commission on Social Determinants of Health, 2008; World Health Organization, 2014). Not only are these differences in health an important social-justice issue, but acknowledgement of their importance has led to a growing understanding of the sensitivity of health to the social and physical environments in which we live our everyday lives (Pauly et al., 2013), something quite pertinent to the discipline of health geography.

With increasing recognition that social and physical environmental factors determine health outcomes, health geographers can and do make valuable contributions to the SDH area of diversity research. Dunn, Schaub and Ross (2007) argue, for example, that there has been an absence of geography in studies on the relationship between income inequality and population health. In their study, they use methods of spatial pattern visualization, outlier analysis, and comparative case-study analysis to investigate the role of geography in understanding the relationship between income inequality and health in Canada and the United States. Findings demonstrate that by applying a spatial approach, otherwise obscure patterns emerge, which opens the door for new questions regarding how unequal places may be less healthy than more egalitarian ones (Dunn, Schaub and Ross, 2007). Their findings highlight the role of differences in and between places and how they shape experiences of the SDH.

Numerous scalar and place-based inquiries into the relationship between health and socioeconomic status have been put forward by health geographers. For example, Wilson et al. (2009) deploy a highly geographic approach to SDH research and review the relative importance of the determinants for different health outcomes among those living in four neighborhoods in Hamilton, Ontario, Canada. Using neighborhood-level inquiry and applying both qualitative and quantitative methods, this study showed how varying health determinants impact neighborhoods with disparate socioeconomic status (e.g., income, education, housing tenure) and demographic diversity (e.g., percent married, visible minorities) (Wilson et al., 2009). Focusing on the health effects of land disposition, Richmond and Ross (2009) created a list of First Nation and Inuit SDH in Canada, which includes balance, life control, education, material resources, social resources and environmental/cultural connections. In further work, Richmond (2015) emphasizes that more health research should focus on understanding Indigenous linkages between the land, environmental dispossession, cultural identity and the SDH. Also focusing on Indigenous health, de Leeuw (2015) calls for researchers to move beyond prioritizing the *social* in the determinants of health, to acknowledge geography as a material and physical entity (e.g., place, earth, land, space, ecology, territory) that has profound impacts on Indigenous experiences of health and well-being. While these health geographers have begun contributing valuable research to the literature on SDH, there exists great potential to further employ SDH approaches in health geography research by examining further how geographic concepts, such as space and place, and various social and physical environmental factors together shape health experiences.

Intersectionality

It is widely acknowledged that social determinants play major roles in shaping health outcomes; however, to bring about change, the structural forces that create inequitable deprivations associated with the SDH must be addressed. Intersectionality has much to offer in this domain, particularly in providing a more precise identification of inequities, in developing equitable health intervention strategies and in ensuring research results are relevant within specific communities (Bauer, 2014; Hankivsky, 2012). Intersectionality emerged from black feminist scholars, most notably Crenshaw (1989), hooks (1990) and Collins (1990), to give a voice to black women's experiences in both feminist and anti-racist discourse, where analyses of the intersections of racism and sexism were largely absent. It is also rooted in Indigenous feminism, Third World feminism, and queer and postcolonial theory (Collins, 1990; Crenshaw, 1994; Hankivsky et al., 2011). Collins (1990) described how oppressions in society do not operate independently, but rather are experienced simultaneously, intersecting into complex patterns. Therefore, from an intersectional perspective, identity markers (e.g., age, culture, disability, ethnicity, gender, immigration status, race, sexual orientation, education level, social class, spirituality, indigeneity) each denote an axis of diversity that coalesces and intersects uniquely in each person's life, situating one at a particular location within the social hierarchy (Collins, 1990; Hankivksy et al., 2014). Similarly, oppressions related to these identity markers (e.g., classism, sexism, ageism, racism, heterosexism) also coalesce to simultaneously produce material and social dis/advantage (Hulko, 2009). Thus, intersectionality differs from the previously discussed approaches of SGBA and SDH, as these *additive* models tend to view various identities and forms of oppression as separate, rather than interlocking and interconnected and simultaneously experienced. Central tenets of intersectionality assert that human lives cannot be reduced to single categories, nor can the complexity of lived experiences be understood by prioritizing one single factor (e.g., gender) or a constellation of factors purposely selected in an additive manner (Bauer, 2014). Rather, intersectionality emphasizes that categories of differentiation are context-specific, fluid and dynamic and, in essence, inseparable and shaped by interacting and mutually constituting social processes and structures that, in turn, are shaped by power and influenced by both time and place (McGibbon and McPherson, 2011).

Although many geographers are committed to addressing issues of social justice, intersectionality has rarely been applied as an analytic lens. While it has been criticized for being overly relativistic and simply too difficult to operationalize (Bauer, 2014; Hankivsky, 2012), it has also been praised as having the potential to open new intellectual spaces for knowledge and research production, which can lead to theoretical and methodological innovation (Bauer, 2014; Bowleg, 2012; Valentine, 2007). Some geographers outside of health have applied intersectional approaches in their work, including Valentine (2007), who, through a case-study narrative technique, applied intersectionality to gain a better understanding of the ways that gender, sexuality, class, motherhood, disability and the cultural/linguistic identity of D/deaf interact with specific spatial contexts and biographical moments (Valentine, 2007). Using a quantitative approach, Veenstra (2011) also applied an intersectional lens to the Canadian Community Health Survey data in order to investigate the interconnections between race, gender, class, and sexuality and self-reported hypertension. Other examples from health geographers specifically include Holmes (2016), who applied an intersectional lens to explore experiences of violence, place and mental health in the lives of trans and gender-nonconforming people in Canada. Herron and Rosenberg (2016) explored how gender and age intersect throughout the life course to create challenges for those involved in care exchanges. Williams et al. (2016) also applied an intersectional approach to examine the diversity of experiences of Canadians providing care for a family member with multiple chronic conditions. With intersectionality only just emerging in the field of health geography, there is much room for researchers to contribute.

Future directions for health geographers

Currently, many health-related policies and programs are characterized by a one-size-fits-all approach, which attempts to achieve equality by treating a greatly heterogeneous population as all the same (Hankivsky et al., 2011). Those who advocate for diversity-based approaches to health research contend that if decision-makers continue to ignore the lived complexity of people's lives, the evidence base that is generated will only reify a range of inequities (Hankivsky et al., 2011). Thus, it is through recognizing and acknowledging diversity in research that health care, social supports, programs, and policies can be enhanced to address *who* needs *what* kinds of care and support, ultimately reducing inequitable health outcomes.

Whether one identifies as a quantitative or a qualitative researcher, or both, there is a great potential to apply diversity-based approaches simply by acknowledging that diverse lived experiences, or social and physical locations, have multiple impacts on one's health and health outcomes (Hankivsky, 2012). It is important for health geographers to consider that diversity will mean different things in different research contexts and that various methods and techniques can be used to explore diversity and, as such, there is no prescribed approach to their use in research. What matters the most is that time be taken to be reflexive throughout the research process and that existing assumptions regarding populations that are commonly conceptualized as homogenous be challenged. To embrace the potential of intersectional approaches in health geography, researchers can ask some of the following questions during the inception, data-collection and analytic phases: Whose voices or perspectives will be heard (or not heard) in this research? How are population groups differentially affected by particular issues? Do the research questions and/or analytic approaches reinforce dominant positioning and assumptions? Does this research recognize diverse knowledges and experiences? Where are the spaces and places within which diversity remains invisible (Crooks and Giesbrecht, 2016)?

References

Alberta Health Services. (2014). *Diversity and Alberta Health Services: Diversity definitions*. [online] Available at: www.calgaryhealthregion.ca/programs/diversity/diversity_resources/definitions/definitions_main.htm [Accessed 14 Feb. 2014].

Bauer, G. R. (2014). Incorporating intersectionality theory into population health research methodology: challenges and the potential to advance health equity. *Social Science & Medicine*, 110, pp. 10–17.

Bowleg, L. (2012). The problem with the phrase women and minorities: intersectionality-an important theoretical framework for public health. *American Journal of Public Health*, 102(7), pp. 1267–1273.

Canadian Public Health Association. (2017). *What are the social determinants of health*. [online] Available at: www.cpha.ca/en/programs/social-determinants/frontlinehealth/sdh.aspx [Accessed 14 Feb. 2017].

Collins, P. H. (1990). *Black feminist thought: knowledge, consciousness, and the politics of empowerment*. New York: Routledge.

Commission on Social Determinants of Health. (2008). *Closing the gap in a generation: health equity through action on the social determinants of health*. Geneva: World Health Organization.

Crenshaw, K. W. (1989). Demarginalizing the intersection of race and sex: a black feminist crique of antidiscrimination doctrine, feminist theory, and antiracist politics. *University of Chicago Legal Forum*, Article 8.

Crenshaw, K. W. (1994). Mapping the margins: intersectionality, identity politics, and violence against women of color. In: M. A. Fineman and R. Mykitiuk, eds., *The public nature of private violence*. New York: Routledge, pp. 93–120.

Crooks, V. A. and Giesbrecht, M. (2016). Conclusion: ways ahead in diversity-based health geography research. In: M. Giesbrecht and V. Crooks, eds., *Place, health, and diversity: learning from the Canadian experience*. New York and London: Routledge, pp. 238–242.

de Leeuw, S. (2015). Activating place: geography as a determinant of Indigenous peoples' health and well-being. In: M. Greenwood, S. De Leeuw, N. M. Lindsay and C. L. Reading, eds., *Determinants of Indigenous peoples' health in Canada: beyond the social*. Toronto: Canadian Scholars' Press, pp. 90–103.

Disabled Women's Network Ontario. (2014). *Gender-based analysis*. [online] Available at: http://dawn.thot.net/gender_analysis.html [Accessed 8 June 2014].

Dunn, J. R., Schaub, P. and Ross, N. A. (2007). Unpacking income inequality and population health: the peculiar absence of geography. *Canadian Journal of Public Health – Revue Canadienne De Sante Publique*, 98(S1), pp. S10–S17.

Dyck, I. (2010). Feminism and health geography: twin tracks or divergent agendas. *Gender, Place & Culture*, 10(4), pp. 361–368.

Giesbrecht, M. and Crooks, V. A. (eds.) (2016). *Place, health, and diversity: learning from the Canadian experience*. New York and London: Routledge.

Greaves, L. (2012). Why put gender and sex into health research? In: J. L. Oliffe and L. Greaves, eds., *Designing and conducting gender, sex, and health research*. Thousand Oaks, CA: Sage, pp. 3–14.

Hankivsky, O. (2012). Women's health, men's health, and gender and health: implications of intersectionality. *Social Science & Medicine*, 74(11), pp. 1712–1720.

Hankivsky, O. and Christoffersen, A. (2008). Intersectionality and the determinants of health: a Canadian perspective. *Critical Public Health*, 18(3), pp. 271–283.

Hankivsky, O., De Leeuw, S., Lee, J. A., Vissandjée, B. and Khanlou, N. (eds.) (2011). *Health inequities in Canada: intersectional frameworks and practices*. Vancouver: UBC Press.

Hankivksy, O., Grace, D., Hunting, G., Giesbrecht, M., Fridkin, A., Rudrum, S., Ferlatte, O. and Clark, N. (2014). An intersectionality-based policy analysis framework: critical reflections on a methodology for advancing equity. *International Journal for Equity in Health*, 13(1), pp. 50–78.

Health Canada. (2010). *Sex and gender-based analysis (SGBA)*. [online] Health Canada. Available at: www.hc-sc.gc.ca/hl-vs/gender-genre/analys/gender-sexes-eng.php [Accessed 14 Feb. 2017].

Herron, R. and Rosenberg, M. (2016). Aging, gender, and "triple jeopardy" through the life course. In: M. Giesbrecht and V. A. Crooks, eds., *Place, health, and diversity: learning from the Canadian experience*. London and New York: Routledge, pp. 200–219.

Holmes, C. (2016). Exploring the intersections between violence, place, and mental health in the lives of trans and gender nonconforming people in Canada. In: M. Giesbrecht and V. A. Crooks, eds., *Place, health, and diversity: learning from the Canadian experience*. London and New York: Routledge, pp. 53–75.

hooks, B. (1990). *Yearning: race, gender, and cultural politics*. New York: Routledge.

Hulko, W. (2009). The time- and context-contingent nature of intersectionality and interlocking oppressions. *Affilia*, 24(1), pp. 44–55.

Johnson, J. L., Greaves, L. and Repta, R. (2009). Better science with sex and gender: facilitating the use of a sex and gender-based analysis in health research. *International Journal for Equity in Health*, 8, pp. 1–11.

Lewis, N. M. (2016). Urban encounters and sexual health among gay and bisexual immigrant men: perspectives from the settlement and AIDS service sectors. *Geographical Review*, 106(2), pp. 235–256.

McGibbon, E. and McPherson, C. (2011). Applying intersectionality & complexity theory to address the social determinants of women's health. *Women's Health and Urban Life*, 10(1), pp. 59–86.

Native Women's Association of Canada. (2007). *Social determinants of health and Canada's Aboriginal women*. Ottawa: NWAC's submission to the World Health Organization's Commission on the Social Determinants of Health.

Pauly, B., Macdonald, M., Hancock, T., Martin, W. and Perkin, K. (2013). Reducing health inequities: the contribution of core public health services in BC. *BMC Public Health*, 13(1), pp. 1–11.

Richmond, C. (2015). The relatedness of people, land, and health: stories from Anishinabe elders. In: M. Greenwood, S. De Leeuw, N. M. Lindsay and C. L. Reading, eds., *Determinants of Indigenous peoples' health in Canada: beyond the social*. Toronto: Canadian Scholars' Press, pp. 47–63.

Richmond, C. A. and Ross, N. A. (2009). The determinants of First Nation and Inuit health: a critical population health approach. *Health & Place*, 15(2), pp. 403–411.

Setia, M. S., Lynch, J., Abrahamowicz, M., Tousignant, P. and Wuesnel-Vallee, A. (2011). Self-rated health in Canadian immigrants: analysis of the Longitudinal Survey of Immigrants to Canada. *Health & Place*, 17(2), pp. 658–670.

Status of Women Canada. (2016). *What is GBA+?* [online]. Status of Women Canada. Available at: www.swc-cfc.gc.ca/gba-acs/index-en.html [Accessed 14 Feb. 2017].

Valentine, G. (2007). Theorizing and researching intersectionality: a challenge for feminist geography. *Professional Geographer*, 59(1), pp. 10–21.

Veenstra, G. (2011). Race, gender, class, and sexual orientation: intersecting axes of inequality and self-rated health in Canada. *International Journal for Equity in Health*, 10(1), pp. 1–11.

Wiles, J. (2002). Helath care reform in New Zealand: the diversity of gender experience. *Health & Place*, 8(2), pp. 119–128.

Wilkinson, L. (2003). Advancing a perspective on the intersections of diversity: challenges for research and social policy. *Canadian Ethnic Studies Journal*, 35(3), pp. 3–23.

Williams, A., Sethi, B., Duggleby, W., Ploeg, J., Markle-Reid, M., Peacock, S. and Ghosh, S. (2016). A Canadian qualitative study exploring the diversity of the experience of family caregivers of older adults with multiple chronic conditions using a social location perspective. *International Journal for Equity in Health*, 15(40). doi: 10.1186/s12939-016-0328-6

Wilson, K., Eyles, J., Elliot, S. and Keller-Olaman, S. (2009). Health in Hamilton neighbourhoods: exploring the determinants of health at the local level. *Health & Place*, 15(1), pp. 374–382.

World Health Organization. (2014). *Social determinants of health*. [online] Available at: www.who.int/social_determinants/en/ [Accessed 9 June 2014].

50

PRACTICING HEALTH GEOGRAPHY IN PUBLIC HEALTH

A focus on population-health-intervention research

Mylene Riva and Sarah M. Mah

The affinity between health geography and public health is clear (Curtis, Riva and Rosenberg, 2009; Dummer, 2008). Addressing questions of space, place and scale as shaping health and well-being is crucial to public-health practice and to population-health research and intervention (Dummer, 2008) and constitutes the crux of health geography. Public health is "the organized efforts of society to keep people healthy and prevent injury, illness and premature death. It is a combination of programs, services and policies that protect, promote, and restore the health of all people" (Last, 2001, p. 145). In this chapter, we refer to "population health" as it is defined in Canada: as the science underpinning the practice of public health and understandings about how health is generated and distributed in populations (Dunn and Hayes, 1999; Hawe and Potvin, 2009). While the overlap between public and population health is extensive, population health is in the unique position to broaden our understandings of health, because the approach is interdisciplinary, is intersectoral and embraces complexity. Population health recognizes health as emerging from the interactions between individuals and social and political processes operating over the life course in particular settings.

There are parallels in the historical and intellectual trajectories of public health, population health and health-geography research. These parallels have ranged from improving living conditions, to controlling the spread of infectious diseases, to examining the role of environments in influencing lifestyle and non-communicable diseases, to a wider shift toward the social determinants of health and a renewed focus on place and health (Curtis, Riva and Rosenberg, 2009). More recently, efforts have been made to find and implement solutions to improve health and reduce health inequalities through population-health interventions (PHI). PHI are policies or programs that shift the distributions of health risks by addressing the underlying social, economic and environmental issues (Hawe and Potvin, 2009). These interventions can be deliberate efforts to improve health through actions across a variety of sectors; policies or programs designed and/or implemented outside of the health sector but that may lead to health improvement as a *side effect*; or natural phenomena such as earthquakes, flood, and economic recession that can be investigated for their impacts on health (Hawe, Di Ruggiero and Cohen, 2012). Increasingly, health geography is making significant advances in assessing the health impacts of such interventions (Harrington, McLafferty and Elliott, 2017; Learmonth and Curtis, 2013).

This chapter explores health geography's contribution to population-health-intervention research (PHIR). PHIR refers to "the use of scientific methods to produce knowledge about interventions that operate within or outside of the health sector and have the potential to impact health at a population level" (Hawe and Potvin, 2009, p. I-8). Using selected examples, we highlight some of the ways in which health geography contributes to PHIR in defining the context of an intervention and identifying the mechanisms through which an intervention can influence health. We also discuss the possibilities afforded by data linkage and geographic information systems (GIS) in conceptualizing and measuring the health effects of PHI targeting the built environment.

Defining "context" in PHIR

PHIs are complex. Upstream determinants of health almost always comprise multiple interacting actors and components. Simple linear relationships between exposures and outcomes are rare, and there can be long periods when no effects are observed and then, through micro-changes, effects emerge (Hawe, 2015). Context is key to understanding how and for whom an intervention creates change.

The context of an intervention is understood as encompassing several different features. The physical and social attributes of a setting are often accounted for, such as the school, the workplace, the neighborhood or the community. Context is also defined by the symbolic meaning ascribed to a setting, by the recursive interactions between actors and attributes of settings and by social networks and power relations between actors. Context also includes the political ecology of the setting, as well as its relational position within a broader system. A consideration of context enables us to examine the mechanisms through which a given intervention is hypothesized to influence health and to assess its effectiveness for improving population health and reducing health inequalities (Hawe, Shielland Riley, 2009).

Nonetheless, a recent review examining the representation of context in 21 empirical PHIR studies revealed that context was most often equated with the physical attributes of the intervention setting or as something that needed to be controlled for (Shoveller et al., 2016), as opposed to something that ought to be studied in its own right. Few of the studies reviewed provided a detailed description of the mechanisms through which the intervention was hypothesized to influence health or explored the interconnections between people, context and intervention. This is problematic because the success of an intervention at improving health and/or reducing health inequalities is closely connected to the context in which it is implemented (Hawe, 2009). Health geography, through its interest in perceptions and experiences of place, sense of place and everyday geographies, is well-positioned to contribute to our understanding and defining "context" in PHIR.

Housing and PHIR

Given that most people spend a significant time where they live, it makes sense to consider the home as the most proximal setting and a good candidate for health intervention. The Ottawa Charter for Health Promotion recognizes shelter as a fundamental condition and resource for health. In Western countries, most housing-intervention studies have assessed the effect of rehousing and housing improvement (e.g., improving energy efficiency) on physical and mental health outcomes (Thomson et al., 2013). Although the link between poor housing and poor health is known, the evidence of a *direct* effect of housing improvement on health outcomes ranges from strong to none or is conflicting (Thomson et al., 2013). This can be explained by numerous factors, including differences in the health outcomes selected for study, with more significant effects on physical health outcomes. Varying time to follow-up and the possibility of non-linear exposure effects on health outcomes are also challenges. Housing-intervention studies are often limited in defining the context of the intervention (i.e., the dwelling) by its physical attributes, whether these entail improvements such as warmth or structural integrity. Yet it is possible that housing interventions modify other contextual aspects that were unanticipated. For example, housing interventions may impact the interactions between the tenants and the house itself, or they may modify the meaning attached to the house. In turn, this may influence health outcomes, and especially mental-health outcomes, in ways that are different for men and women and different across age groups, family types and jurisdictions. The mechanisms through which housing interventions are thought to impact health are often not examined.

Importantly, there is an increased interest in the affective sense of home in PHIR – distinct from the functional necessity of being housed. Kearns and collaborators suggests that housing-related *psychosocial factors* might mediate the relationship between structural housing conditions and health (Kearns et al., 2011). These psychosocial factors refer to the ontological security the home conveys, the sense of being at home or of feeling in control of one's home (Padgett, 2007). In Scotland, a prospective controlled housing-intervention study showed improvements in mental health followed the relocation of low-income families to new housing units (Kearns et al., 2011). In the intervention group, rehousing and improved material conditions of the house did not have a direct effect on mental health. However, there were significant gains in psychosocial benefits, and these were associated with improved mental health. Psychosocial benefits included improved control of the domestic realm as well as perceived social status. In particular, significant positive effects were observed for families with dependent children, which suggests the intervention had a greater impact on certain subgroups. This serves to illustrate that failing to consider the potentially differential impacts of an intervention may obscure the scope of impacts observed in a group of heterogeneous individuals.

In a qualitative case study, Alaazi and collaborators used the therapeutic-landscapes framework to understand the sense of home among Indigenous homeless participants in the Canadian Housing First initiative, At Home/Chez Soi, in Winnipeg, Manitoba (Alaazi et al., 2015). The authors conceptualized sense of home as "a relational, social, and cultural construct that transcends the instrumental experience of being housed" (p. 30). This conceptualization extended beyond the physical attributes of the house to consider the symbolism and meaning of home. The authors observed that Indigenous participants reported that their home imparted a sense of place that reinforced connection to family, community and land. This representation of home was qualitatively different from non-Indigenous participants in the intervention, and it contrasted with the regimented approach of housing provision of At Home/Chez Soi, which precluded the support of culturally appropriate housing.

These studies illustrate how engaging with the sense, meaning and experience of place, as well as integrating cultural experiences of home, adds to our understanding of how the context of interventions may impact different groups of people. We need to integrate geographic inquiry into PHIR to understand how people relate to and interact with their environment, so that we might anticipate for whom and through which processes a given intervention may influence health outcomes.

Assessing the health impacts of large-scale interventions at the household level, but also at the neighborhood level, often requires hefty funding for studying their effects. Yet PHIs often occur in a time-sensitive period, which precludes large-scale primary data collection. The prominence of large-scale population surveys and ability to link these records to spatial datasets to characterize residential environment as well as to administrative data such as mortality, morbidity and hospitalizations offers unprecedented opportunities for conducting PHIR.

Embedding geography into population-health data: data linkage and GIS

Increasingly recognized by population-health researchers is the wealth of already existing data. However, the variables of interest are often located in different databases, collected by distinct agencies and sectors for purposes unrelated to research. Uniting individual-level records from disparate sources can be achieved using *data linkage*, which has been defined as "the bringing together of information from two records that are believed to relate to the same individual or family" (Black and Roos, 2005). Although statistics derived from administrative data (such as hospitalizations and cancer registries) are useful for understanding the broad patterning of diseases in populations and across geographies, they do not always tell us about the contributions of the social determinants of health or health-related behaviors (Sanmartin et al., 2016). In contrast, population-health surveys collect information about the health status and health behaviours of individuals but are typically cross-sectional and lack objective information on aspects such as health outcomes, utilization and medication. Combining health survey and administrative data through data linkage allows for robust investigation of the relationships between social determinants and health outcomes without the need for costly cohort studies (Sanmartin et al., 2016).

These linkage initiatives, population-based by design, are an efficient means of creating *information-rich* research environments (Roos, Menec and Currie, 2004) from previously disparate data sources – regardless of whether data was originally intended for population health. Importantly, however, the success and quality of record-linkage approaches (Harron, Goldstein and Dibben, 2016), rely on the ability to identify matching records from the same individual. Major work has been undertaken to maximize the use of existing administrative and survey datasets to answer population-health questions.

The integration of GIS and data-linkage approaches in PHIR is a particularly fertile avenue to account for context in shaping health outcomes and behaviors. GIS is a computer-based tool for collecting, analyzing and visualizing spatial data. Most forms of GIS consist of storage of spatial information via different layers. GIS inherently involves aspects of linkage between variables in different information layers. It has the potential to further advance linkage processes and data management, as well as enhance the population-based health research made possible by linkage. In the case of PHIR, embedding geospatial measures into routine population-data collection practices would enable richer analyses of health outcomes of an intervention. The next section reviews examples of GIS applications in studying the health effects of interventions targeting the built environment.

The built environment and PHIR

The built environment is an ideal candidate for population-wide intervention to improve health. Walkable environments, for instance, are thought to promote physical activity and fitness (Smith et al., 2008). A 2007 report from the World Health Organization on implementing population-based approaches for physical activity cited urban design friendliness and active transport as being key to reducing obesity and chronic conditions (World Health Organization, 2007). Increased attention, in both research and policy, has thus turned to the role of the built environment in improving health outcomes. While this geographic information is not often collected in administrative and health-survey data, geocoded items such as postal codes (e.g., zip codes) that are often present on these health records can provide the coordinates necessarily to derive measures of the built environment using other data sources, such as business registries, land-use maps and street network files. This approach of merging health data with environmental measures from geocoded information can serve as a means to evaluate the built environment's impact on health behaviors and outcomes across the life course (Villanueva et al., 2013). Environmental contaminants, neighborhood walkability, zoning for housing development and food outlets, and transit systems are some of the aspects of the built environment that have been studied as potential candidates for PHIR.

In the early stages of an intervention, applying GIS to map patterns of environmental factors, risks and prevalence is valuable for making a plan or defining and assessing the context of a planned intervention. GIS has been used to assess exposure to environmental health hazards such as lead (Akkus and Ozdenerol, 2014), which motivated lead-poisoning screening and prevention activities. In North Carolina, in the United States, GIS has played a key role in implementing geographically targeted community-intervention strategies (such as the North Carolina Childhood Lead Poisoning Prevention Program). Using a subset of data from a larger Geographic Health Information System, high-risk exposure areas across a variety of socioeconomic factors were mapped, identified and flagged as priority intervention areas (Miranda, Dolinoy and Overstreet, 2002). The models from this work have had substantive impact on screening children, monitoring progress on eliminating lead poisoning, and increasing community cohesion and advocacy on lead as a health hazard (Miranda et al., 2013). As illustrated, GIS plays an important role in the early-stage research for PHIs.

Some of the innovative uses of GIS in PHIR have been at the planning and implementation stages of the intervention. Several studies have combined GIS with community-based participatory approaches, forming partnerships with stakeholders, universities, organizations and individuals. A recent study combined GIS and neighborhood audits driven by youth for developing neighborhood-environment interventions related to health and physical activity (Topmiller et al., 2015). Engagement with youth in the project revealed safety concerns, and the study reported positive response from local government to improve their urban trails and sidewalks. Moreover, advances in communications technology and accessibility to this technology have given rise to Public Participatory GIS, in which members of the public or a certain community are able to directly contribute to the mapping of environmental health features. This application of GIS makes it possible for community members and stakeholders to participate in the design of interventions, and studies have begun to frame these initiatives as part of a holistic and democratic approach to designing PHIs.

PHIR can be a particularly effective approach for assessing both the implementation and impact of urban planning codes that mandate healthy environments. Western Australian State Government's Liveable Neighborhoods Guidelines is an operational policy consisting of urban design and assessment principles aimed at increasing safety and accessibility for walking, cycling and transit use (Western Australian Planning Commission, 2007). Introduced in 1998 as a voluntary alternative to the pre-existing design codes in Perth, these guidelines have been adopted and recognized as a unique natural experiment. The Residential Environments Project (RESIDE) monitored the implementation and effect of the policy on residential transport behaviors following new housing development (Giles-Corti et al., 2008). GIS-derived measures of design elements specified by the policy revealed that greater compliance with the urban design guidelines – particularly those related to community design (neighborhood structure, land use, access to amenities, public transit etc.), was associated with increased walking for transport (Hooper, Giles-Corti and Knuiman, 2014).

GIS also contributes to assessing the health impacts of large-scale PHIs, such as policies that promote active transport. For instance, the installation of major transit systems serve as natural experiments by which we can monitor population health changes. Public-transit use is associated with higher physical activity (Saelens et al., 2014) and might improve health outcomes by potentially preventing chronic conditions and reducing health inequalities (Turrell et al., 2013). An American study that utilized Global Positioning System tracking found that among study participants who lived close to a new light-rail system, new transit riders (which comprised about 10% of the sample) lost weight, while those who no longer took public transit gained weight (Brown et al., 2015). In assessing objectively measured accelerometer measures since the light-rail's installment, new transit riders were observed to have significantly higher physical activity than those who never used transit (Brown et al., 2015). Increasingly, the decision to evaluate substantive health changes before and after population-wide interventions is being made at the outset. One example is the Travel-Related Activity in Neighborhoods Study (TRAINS) (Durand et al., 2016), which will prospectively investigate the impact of a major expansion to the Houston light-rail system on physical activity and transit use, incorporating both macro-scale features (GIS-derived aspects such as street connectivity, land-use mix, and residential density) and micro-scale features (street-level characteristics such as transit access, pedestrian paths, and land use) of the built environment that are known to affect physical activity.

Conclusion

Interventions to improve population health aim to modify the characteristics of social and/or physical contexts, or the capacities of individuals in the population that influence the distribution of health risks. Trying to disentangle the intervention from its context is arbitrary, as they are mutually influencing one another (Potvin, 2017). The role that health geography can play in PHIR is unequivocal. The examples in this chapter were chosen to illustrate how health geography can be practiced in public health in the context of PHIR. Translating knowledge from geographical inquiry into place-based policy recommendations has the potential to contribute to the creation of settings that are supportive of health. To this end, both quantitative and qualitative methods used by geographers are necessary to understand the context of the intervention and the place-based mechanisms linking context, intervention and health outcomes.

References

Akkus, C. and Ozdenerol, E. (2014). Exploring childhood lead exposure through GIS: a review of the recent literature. *International Journal of Environmental Research and Public Health*, 11, pp. 6314–6334.

Alaazi, D., Masuda, J., Evans, J. and Distasio, J. (2015). Therapeutic landscapes of home: exploring Indigenous peoples' experiences of a Housing First intervention in Winnipeg. *Social Science & Medicine*, 147, pp. 30–37.

Black, C. and Roos, L. L. (2005). Linking and combining data to develop statistics for understanding the population's health. In: D. J. Friedman, E. L. Hunter and R. G. Parrish, eds., *Health statistics: shaping policy and practice to improve the population's health*. New York: Oxford University Press, pp. 214–240.

Brown, B. B., Werner, C. M., Tribby, C. P., Miller, H. J. and Smith, K. R. (2015). Transit use, physical activity, and body mass index changes: objective measures associated with complete street light-rail construction. *American Journal of Public Health*, 105, pp. 1468–1474.

Curtis, S., Riva, M. and Rosenberg, M. (2009). Health geography and public health. In: T. Brown, S. McLafferty and G. Moon, eds., *A companion to health and medical geography*. Chichester: Wiley-Blackwell, pp. 323–345.

Dummer, T. J. (2008). Health geography: supporting public health policy and planning. *Canadian Medical Association Journal*, 178, pp. 177–180.

Dunn, J. and Hayes, M. (1999). Toward a lexicon of population health. *Canadian Journal of Public Health*, 90(Supp 1), S7–S10.

Durand, C. P., Oluyomi, A. O., Gabriel, K. P., Salvo, D., Sener, I. N., Hoelscher, D. M., Knell, G., Tang, X., Porter, A. K., Robertson, M. C. and Kohl, H. W. (2016). The effect of light rail transit on physical activity: design and methods of the Travel-Related Activity in Neighborhoods Study. *Frontiers in Public Health*, 4(103).

Giles-Corti, B., Knuiman, M., Timperio, A., Van Niel, K., Pikora, T. J., Bull, F. C., Shilton, T. and Bulsara, M. (2008). Evaluation of the implementation of a state government community design policy aimed at increasing local walking: design issues and baseline results from RESIDE, Perth Western Australia. *Preventative Medicine*, 46, pp. 46–54.

Harrington, D., McLafferty, S. and Elliott, S. (2017). *Population health intervention research: geographical perspectives*, New York: Routledge.

Harron, K., Goldstein, H. and Dibben, C. (eds.) (2016). *Methodological developments in data linkage*. New York: Wiley.

Hawe, P. (2009). The social determinants of health: how can a radical agenda be mainstreamed? *Canadian Journal of Public Health*, 100(4), pp. 291–293.

Hawe, P. (2015). Lessons from complex interventions to improve health. *Annual Review of Public Health*, 36, pp. 307–323.

Hawe, P., Di Ruggiero, E. and Cohen, E. (2012). Frequently asked questions about population health intervention research. *Canadian Journal of Public Health*, 103, e468–471.

Hawe, P. and Potvin, L. (2009). What is population health intervention research? *Canadian Journal of Public Health*, 100(1), pp. I-8–I-14.

Hawe, P., Shiell, A. and Riley, T. (2009). Theorising interventions as events in systems. *American Journal of Community Psychology*, 43, pp. 267–276.

Hooper, P., Giles-Corti, B. and Knuiman, M. (2014). Evaluating the implementation and active living impacts of a state government planning policy designed to create walkable neighborhoods in Perth, Western Australia. *American Journal of Health Promotion*, 28, pp. S5–18.

Kearns, A., Whitley, E., Mason, P., Petticrew, M. and Hoy, C. (2011). Material and meaningful homes: mental health impacts and psychosocial benefits of rehousing to new dwellings. *International Journal of Public Health*, 56, pp. 597–607.

Last, J. M. (2001). *A dictionary of epidemiology*. Oxford: Oxford University Press.

Learmonth, A. and Curtis, S. (2013). Place shaping to create health and wellbeing using health impact assessment: health geography applied to develop evidence-based practice. *Health Place*, 24, pp. 20–22.

Miranda, M. L., Dolinoy, D. C. and Overstreet, M. A. (2002). Mapping for prevention: GIS models for directing childhood lead poisoning prevention programs. *Environmental Health Perspectives*, 110, pp. 947–953.

Miranda, M. L., Ferranti, J., Strauss, B., Neelon, B. and Califf, R. M. (2013). Geographic health information systems: a platform to support the "triple aim." *Health Affairs (Millwood)*, 32, pp. 1608–1615.

Padgett, D. (2007). There's no place like (a) home: ontological security among persons with serious mental illness in the United States. *Social Science & Medicine*, 64, pp. 1925–1936.

Potvin, L. (2017). Transforming local geographies to improve health. In: D. Harrington, S. McLafferty and S. Elliott, eds., *Population health intervention research: geographical perspectives*. New York: Routledge, pp. 163–169.

Roos, L. L., Menec, V. and Currie, R. J. (2004). Policy analysis in an information-rich environment. *Social Science & Medicine*, 58, pp. 2231–2241.

Saelens, B. E., Vernez Moudon, A., Kang, B., Hurvitz, P. M. and Zhou, C. (2014). Relation between higher physical activity and public transit use. *American Journal of Public Health*, 104, pp. 854–859.

Sanmartin, C., Decady, Y., Trudeau, R., Dasylva, A., Tjepkema, M., Finès, P., Burnett, R., Ross, N. and Manuel, D. (2016). Linking the Canadian Community Health Survey and the Canadian Mortality Database: an enhanced data source for the study of mortality. *Health Reports*, 27, pp. 10–18.

Shoveller, J., Viehbeck, S., Di Ruggiero, E., Greyson, D., Thomson, K. and Knight, R. (2016). A critical examination of representations of context within research on population health interventions. *Critical Public Health*, 26, pp. 487–500.

Smith, K. R., Brown, B. B., Yamada, I., Kowaleski-Jones, L., Zick, C. D. and Fan, J. X. (2008). Walkability and body mass index density, design, and new diversity measures. *American Journal of Preventative Medicine*, 35, pp. 237–244.

Thomson, H., Thomas, S., Sellstrom, E. and Petticrew, M. (2013). *Housing improvements for health and associated socioeconomic outcomes (Review)*. The Cochrane Collaboration, 2. Chichester: Wiley-Blackwell.

Topmiller, M., Jacquez, F., Vissman, A. T., Raleigh, K. and Miller-Francis, J. (2015). Partnering with youth to map their neighborhood environments: a multilayered GIS approach. *Family & Community Health*, 38, pp. 66–76.

Turrell, G., Haynes, M., Wilson, L. A. and Giles-Corti, B. (2013). Can the built environment reduce health inequalities? A study of neighbourhood socioeconomic disadvantage and walking for transport. *Health & Place*, 19, pp. 89–98.

Villanueva, K., Pereira, G., Knuiman, M., Bull, F., Wood, L., Christian, H., Foster, S., Boruff, B. J., Beesley, B., Hickey, S., Joyce, S., Nathan, A., Saarloos, D. and Giles-Corti, B. (2013). The impact of the built environment on health across the life course: design of a cross-sectional data linkage study. *BMJ Open*, 3(1).

Western Australian Planning Commission. (2007). *Liveable neighbourhoods: a Western Australian government sustainable cities initiative*. Perth: Western Australian Planning Commission.

World Health Organization. (2007). *A guide for population-based approaches to increasing levels of physical activity: implementation of the WHO global strategy on diet, physical activity and health*. Geneva: World Health Organization.

51

INTERVENTION RESEARCH FROM A PLACE-BASED PERSPECTIVE

Allison M. Williams

Although intervention design, implementation and evaluation are central to public-health research and practice, intervention research has and continues to be conducted by a wide range of health-focused disciplines, ranging from nursing to health geography. Whether at a population level (e.g., nation, province/state) or at a targeted population (e.g., a specific employee group in a small enterprise), intervention research is recognized as an applied form of research aimed at improving health and well-being through intentional activities – whether policies or programs, for example. As a scientific endeavor, intervention research involves the use of a range of research methods to produce knowledge about policy and program interventions that operate both within and outside of the health sector and have the potential to impact health (Canadian Institute for Health Information, 2014). As Hawe and Potvin (2009) note, health interventions act upon health to positively alter an established course, be it absolute health burden, inequality or inequity.

Using a rich and diverse set of methodologies, including a range of qualitative, quantitative, geographic information systems (GIS) and mixed-methods approaches in doing intervention research, health geographers pay close attention to how the characteristics of place and space – at varying scales of analysis – impact the environmental and social forces shaping people's health. Whether through intervention research design, through implementation or through a process of formative evaluation, health geographers are well-equipped to do effective intervention research, given their methodological toolbox and unique contextual lens. Given the social-justice emphasis of our discipline, a place-based-intervention research approach is commonly employed to determine whether and how interventions improve the social location and/or "health," broadly defined, of the vulnerable place, or place-based population, of concern. A recent collection of population-health-intervention research that uses a geographical lens (Harrington, McLafferty and Elliot, 2017) provides an illustration of the range and types of interventions that health geographers are involved in, including HIV, malaria and dengue prevention; natural experiments for preventing chronic disease and improving school health; food retail strategies to improve diet; and vacant lot redevelopment for improved community well-being. The actual interventions are similarly wide-ranging, and among the many examples are community-based initiatives such as walking groups to reduce the risk of developing various chronic diseases; the use of physical barriers such as bed nets; awareness campaigns facilitated through social media; and healthful school-nutrition policies.

Having used a geographical lens in all the various types of intervention research conducted in my career to date – whether via intervention implementation (Williams et al., 2017), evaluation (Giesbrecht, Crooks and Williams, 2010; Williams, 2010), or pre-intervention needs assessments (Williams, 1997), it is difficult to imagine not paying attention to the relevance of contextual factors – whether they be political, economic, cultural, demographic or other place-based characteristics that clearly have a bearing on the health and well-being of any particular place or population of concern. The health of individuals, communities or populations does not exist in isolation – it is influenced by a large range of contextual factors defined by space and place. Given the growing complexity of health issues that inevitably result from an increasingly diverse and aging population, context becomes all the more important to attend to in producing positive health and well-being. Thus, health geographers involved in applied health-intervention research recognize the role of space and place in the relationship between the numerous and often overlapping social determinants of health and individual, community and population health outcomes.

Intervention research

As noted, interventions can be wide-ranging and vary in scale. For example, individual interventions that target diet or physical activity operate to change individual behavior, whereas neighborhood interventions, such as block parents or vehicle-speed-reduction campaigns, have the ability to address social or health issues at a larger scale. Unlike descriptive research (Beyer et al., 2010), intervention research focuses on determining the impact of an intervention on the population for which it has been created, such as social-networking use on HIV-related risk/protective behaviors among homeless youth (Barman-Adhikari et al., 2016).

While intervention research often focuses on the design and development of an intervention to address a particular social or health concern (Fraser and Galinsky, 2010), it can also go one step further to answer broader questions about the effect of the intervention (Hawe and Potvin, 2009). Different types of interventions can be employed, depending on the target population, the size of the target population, desired outcomes and best practices. Hawe and Potvin (2009) note that intervention research related to population health is often diverse because studies may seek to answer different questions, such as the relevancy or impact of a particular intervention, instead of simply focusing on the effects achieved within the study population. Sallis, Bauman and Pratt (1998) provide a summary of various interventions aimed to encourage physical activity through changes to the environment and to policy, as well as suggesting a model for ongoing research that targets changes within those two contexts. This is an illustrative example of the variety and breadth of interventions that can be implemented within a single area of focus, such as physical activity.

Intervention research is truly one of the most practical approaches to research, often following a logical trajectory of pre-testing, intervention implementation and post-testing, to ultimately test whether a policy or program improves health in some capacity. Riley et al. (2015) concisely summarize how intervention research specific to population health differs from other fields, such as implementation science, evaluation and clinical epidemiology. The development of a set of competencies to guide the emerging field of population-health-intervention research was recently achieved in 2015. Through extensive consultation and literature reviews, Riley et al. (2015) identified 25 key competencies as preliminary principles of the field (p. 855). Among these principles, there is an emphasis on collaborating with interdisciplinary stakeholders and the "central role of contextual influences on interventions" (p. 855). Hawe and Potvin (2009) also outline a diverse set of skills that is needed by those engaged with intervention research. These competencies have yet to be adapted to provide a relevant set of guidelines specific to population-health interventions that employ a place-based approach and present a ready opportunity.

Furthermore, the application of mapping technology, such as GIS, continues to be employed to determine appropriate areas to for intervention implementation, whether defined by population characteristics (i.e., indices of vulnerability) or by disease rates. Examples of the use of GIS are many, including the identification of areas where there is a greater need for cancer-screening interventions for South Asians (Lofters, Gozdyra and Lobb, 2013) or diabetes-prevention programs (Gesler et al., 2004). Thus, health geographers' wide-ranging toolbox of techniques, from mixed-method approaches through to GIS, provide us with the skills and resources needed to best understand the contextual factors influencing health outcomes and to appropriately apply this knowledge to the interventions in order to positively impact health.

Employing a place-based approach to intervention research

Among the many substantive issues addressed in intervention research, a place-based approach has been used to address social and health inequities that affect a wide range of health concerns, including maternal and child health (Storey-Kuyl, Bekemeier and Conley, 2015); physical activity for seniors (Rosenberg et al., 2009); social support for informal caregivers (Duggleby et al., 2007); self-management of chronic disease (Lopez et al., 2017); targeted knowledge mobilization strategies for public health education (Giesbrecht et al., 2009; Williams et al., 2008); nutrition and physical activity in middle schools (Dzewaltowski, Estabrooks and Johnston, 2002); school-based health (Leatherdale, 2017; Vine, 2017); social isolation among parents in neighborhoods with high rates of poverty (Bess and Doykos, 2014); malaria control (Janko and Emch, 2017); racial discrimination among Indigenous and immigrant communities (Ferdinand, Paradies and Kelaher, 2013); dengue control (Dickin, 2017); and HIV/AIDS prevention (Lewis, 2017). Researchers involved in the field of population-health intervention are also actively engaged in the evaluation of large-scale place-based interventions in areas such as maternal and child health (Pies et al., 2016) and obesity levels in minority populations (Liao et al., 2016), to name a few. The diversity of health interventions using a place-based approach certainly highlights that consideration of place is increasingly important in efforts aimed to improve health.

With respect to place and its influence on both individual and population health, the unique aspects of living in either a rural or an urban community are important considerations for interventions designed to address complex health inequities. For example, research has compared the experiences between urban and rural family caregivers, noting that location can account for differences in access to and use of health-care or respite services and should be considered when designing or implementing targeted interventions (Brazil et al., 2013, 2014). This is highly important in order to utilize existing assets and to meet specific needs of those for whom interventions are being designed.

To illustrate, a Living with Hope program was designed for Canadian female caregivers living in rural Alberta and rural Saskatchewan as an intervention to improve levels of hope and overall quality of life (Duggleby et al., 2007). Recognizing that the caregiving experience is different for rural-dwelling caregivers compared to urban-dwelling caregivers, the researchers designed an intervention that could be self-administered by participants (Duggleby et al., 2007). This style of intervention was effective in significantly impacting health outcomes and found to be appropriate for rural caregivers, who are often unable to leave their home for lengthy periods of time while caring for their dependent(s) (Duggleby et al., 2013). This illustrates that the contextual factors of place operate in tailoring the design of health interventions, further promoting participation and successful health outcomes.

Nutrition and access to healthy foods is both a social and a health issue that is increasingly being addressed from a place-based perspective when developing interventions. These interventions directly relate to the obesity epidemic while also impacting broader health issues, such as diabetes and cardiovascular disease. Complex social issues, such as food insecurity (Minaker, Mah and Cook, 2017), can be isolated in neighborhoods characterized by low socioeconomic status and are often more prevalent in vulnerable populations that are geographically concentrated. Minaker, Mah and Cook (2017) suggest that effective programs are based on based on behavior-change theory and include a range of interventions in the area of healthy food retail, such as healthy corner stores, equipped with "display shelving, refrigerator units, store reorganization and novel process flows, to be able to provide perishable, nutritious foods (e.g., fruits and vegetables)" (p. 159) that are reasonably priced. Such interventions ideally require collaborative participation from community stakeholders in multiple sectors of concern; hence the development of a place-based partnership approach

(Burstein and Tolley, 2011; Riley et al., 2017). Understanding that social determinants typically vary across space and time and that opportunities for collaborative partnerships are often place-specific and based on any one community's assets, a context-specific place-based approach highlights the importance of a geographical perspective in conducting intervention research.

Tackling the obesity epidemic benefits not only from a collaborative framework among professionals in both the public and private sectors, but by using a place-based approach in the design of effective interventions. Liao et al. (2016) assessed a place-based intervention that was implemented to address obesity in 14 black communities in the United States. The place-based project was unique because the planners strategically targeted unique areas of improvement that were needed in each community, instead of focusing on changing individual-level behavior. Liao et al. (2016) noted that "[established] health initiatives were based on the unique historical and cultural context of the community" (p. 1447). Recognizing the central cultural and spiritual role that churches play in black communities, in that they "reach a large and consistent group," communities effectively established these places as spaces for health-promotion programming and asked respected clergy members to assist (Liao et al., 2016, p. 1447). Their efforts were successful due to the provision of "culturally appropriate health education" by largely respected local health workers and due to the "healthier and culturally preferable foods" offered when community members gathered at church or at programs (Liao et al., 2016, p. 1447). The authors who conducted the evaluation note that building capacity between community members was also crucial in effecting change at the community level (Liao et al., 2016). Here, a place-based approach offered a new lens through which to understand strategies to combat obesity (using the social-gathering spaces of a neighborhood) and, in so doing, influence obesity and communal spaces of support. Contextualization is increasingly important for the design of interventions so that policy-makers can effectively target appropriate and relevant factors that may play a role in health disparities in a particular place.

Limitations and future developments

Popular approaches to health-intervention research, which often focus on either elucidating the biologic mechanisms of disease or modifying individual health behaviors to improve well-being and prevent disease, may not always be sufficient to address the many health disparities influenced by space and place. Thus, health researchers, including health geographers, have looked for alternative interdisciplinary models that can integrate both social and biological factors in intervention design and uptake, recognizing the complexity of the systems within which health is often observed and researched.

Although employing a place-based approach has been successful in many cases, there still are certain limitations to the application of this perspective to health-intervention research, including those pertaining to measuring and evaluating the outcomes achieved through place-based approaches, given the many contextual factors at play (Bellefontaine and Wisener, 2011; Burstein and Tolley, 2011). Although rarely used in health geography, Shoveller et al. (2016) point to the established perception of randomized controlled trials (RCTs) as the gold standard for evaluating an intervention. Given that RCTs often aim to minimize the effects of context on the study results, they provide a stark contrast to intervention research that uses a health-geography lens, which works to understand the interplay between health and context. Shoveller et al. (2016) also found that descriptions of context within studies on population-health-intervention research were often vague, were broad in nature, and referred to different features, operating as a further limitation. Other studies have echoed the need to pay greater attention to scale, given that certain inequities are less evident when data is aggregated and differences in health between or among local communities is overlooked (Snyder et al., 2013). Given this, health geographers can continue to shed light on the contextual factors at play in both intervention practice and research.

Future developments in this field of research may seek to further understand the meaning that individuals derive from the places where they live, work and play and how these sentiments can affect health interventions (Eyles and Williams, 2008; Gallina and Williams, 2015; Williams et al., 2010). Nowell et al.'s (2006) findings underscore that "place-based initiatives would do well to attend to issues of place identity in structuring their interventions, as neighborhoods whose residents possess a poor or diffuse sense of place identity are likely to experience particular challenges in coming together for collective planning and problem solving" (p. 42). Consequently, place-based approaches should ideally involve a collaborative framework in which various players or stakeholders can cooperate to enhance the meaning and identity that residents draw from their local areas, which can be facilitated by health geographers and other researchers. This happens best when the importance of context is acknowledged. Another area of future development includes the application of a multimethod research approach to current randomized controlled trials in order to enhance the potential applicability of findings and resulting interventions while maintaining the strength of quantitative results (Hansen and Tjornhoj-Thomsen, 2016). As noted, taking a place-based lens to the key competencies that have been determined as preliminary principles of the field (Riley et al., 2015) is a ready direction for highlighting health geography in intervention research and practice. Lastly, the use of technology, such as social media, may play a larger role in population-health interventions and be an effective tool to reach target populations, as demonstrated by Barman-Adhikari et al. (2016).

The diversity of social and health concerns impacting health would benefit from using a place-based perspective in the design of intervention solutions. Many of these issues, such as food insecurity or obesity, are complex and often rooted in place-based social, cultural, political and environmental factors that need to be recognized when developing, implementing and evaluating interventions. There is tremendous potential for health geographers to continue contributing to and influencing intervention research and practice.

References

Barman-Adhikari, A., Rice, E., Bender, K., Lengnick-Hall, B., Yoshioka-Maxwell, A. and Rhoades, H. (2016). Social networking technology use and engagement in HIV-related risk and protective behaviours among homeless youth. *Journal of Health Communication*, 21, pp. 809–817.

Bellefontaine, T. and Wisener, R. (2011). *The evaluation of place-based approaches: Questions for further research.* [online] Government of Canada. Available at: www.horizons.gc.ca/eng/content/evaluation-place-based-approaches [Accessed 2 Feb. 2017].

Bess, K. and Doykos, B. (2014). Tied together: building relational well-being and reducing social isolation through place-based parent education. *Journal of Community Psychology*, 42(3), pp. 268–284.

Beyer, K. M. M., Comstock, S., Seagren, R. and Rushton, G. (2010). Explaining place-based colorectal cancer health disparities: evidence from a rural context. *Social Science & Medicine*, 72, pp. 373–382.

Brazil, K., Kaasalainen, S., Williams, A. and Dumont, S. (2014). A comparison of support needs between rural and urban family caregivers providing palliative care. *American Journal of Hospice & Palliative Medicine*, 31(1), pp. 13–19.

Brazil, K., Kaasalainen, S., Williams, A. and Rodriguez, C. (2013). Comparing the experiences of rural and urban family caregivers of the terminally ill. *Rural and Remote Health*, 13, pp. 1–9.

Burstein, M. and Tolley, E. (2011). *Exploring the effectiveness of place-based program evaluations.* [pdf] Policy Research Initiative. Available at: http://p2pcanada.ca/wp-content/uploads/2011/09/Place-based-Evaluations_Report_2011_FINAL.pdf. [Accessed 2 Feb. 2017].

Canadian Institute for Health Information. (2014). *Population health: intervention research for Canada (PHIRC).* [online] Available at: www.cihi.ca/en/factors-influencing-health/environmental/population-health-intervention-research-initiative-for [Accessed 1 June 2017].

Dickin, S. (2017). Making a place for health in vulnerability analysis: a case study on dengue in Malaysia and Brazil. In: D. W. Harrington, S. McLafferty, and S. J. Elliot, eds., *Population health intervention research: geographical perspectives.* New York: Routledge, pp. 78–90.

Duggleby, W., Williams, A., Holstlander, L., Cooper, D., Ghosh, S., Hallstrom, L. K., Thomas McLean, R. and Hampton, M. (2013). Evaluation of the living with hope program for rural women caregivers of persons with advanced cancer. *BMC Palliative Care*, 12(1), p. 36.

Duggleby, W., Wright, K., Williams, A., Degner, L., Cammer, A. and Holtslander, L. (2007). Developing a living with hope program for caregivers of family members with advanced cancer. *Journal of Palliative Care*, 23, pp. 24–31.

Dzewaltowski, D. A., Estabrooks, P. A. and Johnston, J. A. (2002). Healthy youth places promoting nutrition and physical activity. *Health Education Research*, 17(5), pp. 541–551.

Eyles, J. and Williams, A. (eds.) (2008). *Sense of place, health and quality of life*. London: Ashgate.

Fraser, M. W. and Galinsky, M. J. (2010). Steps in intervention research: designing and developing social programs. *Research on Social Work Practice*, 20(5), pp. 459–466.

Ferdinand, A., Paradies, Y. and Kelaher, M. (2013). Mental health impacts of racial discrimination in Victorian Aboriginal communities: The Localities Embracing and Accepting Diversity (LEAD) experiences of racism survey. Melbourne, Australia: The Lowitja Institute.

Gallina, M. and Williams, A. (2015). Variations in sense of place across immigrant status and gender in Hamilton, Ontario; Saskatoon, Saskatchewan; and, Charlottetown, Prince Edward Island, Canada. *Social Indicators Research*, 121(1), pp. 241–252.

Gesler, W. M., Hayes, M., Arcury, T. A., Skelly, A. H., Nash, S. and Soward, A. C. M. (2004). Use of mapping technology in health intervention research. *Nursing Outlook*, 52, pp. 142–146.

Giesbrecht, M., Crooks, V. A., Schuurman, N. and Williams, A. (2009). Spatially informed knowledge translation: informing potential users of Canada's compassionate care benefit. *Social Science & Medicine*, 69, pp. 411–419.

Giesbrecht, M., Crooks, V. and Williams, A. (2010). Scale as an explanatory concept: evaluating Canada's compassionate care benefit. *Area*, 42(4), pp. 457–467.

Hansen, H. P. and Tjornhoj-Thomsen, T. (2016). Meeting the challenges of intervention research in health science: an argument for a multimethod research approach. *The Patient*, 9, pp. 193–200.

Harrington, D. W., McLafferty, S. and Elliot, S. J. (2017). *Population health intervention research: geographical perspectives*. New York: Routledge.

Hawe, P. and Potvin, L. (2009). What is population health intervention research? *Canadian Journal of Public Health*, 100(1), pp. I-8–I-14.

Janko, M. and Emch, M. (2017). The geography of malaria control in the Democratic Republic of Congo. In: D. W. Harrington, S. McLafferty, and S. J. Elliot, eds., *Population health intervention research: geographical perspectives*. New York: Routledge, pp. 91–104.

Leatherdale, S. T. (2017). Shaping the direction of youth health with COMPASS: a research platform for evaluating natural experiments and generating practice-based evidence in school-based prevention. In: D. W. Harrington, S. McLafferty, and S. J. Elliot, eds., *Population health intervention research: geographical perspectives*. New York: Routledge, pp. 123–135.

Lewis, N. M. (2017). From cultural clashes to settlement stressors: a review of HIV prevention interventions for gay and bisexual immigrant men in North America. In: D. W. Harrington, S. McLafferty and S. J. Elliot, eds., *Population health intervention research: geographical perspectives*. New York: Routledge, pp. 65–77.

Liao, Y., Siegel, P. Z., Garraza, L. G., Xu, Y., Yin, S., Scardaville, M., Gebreselassie, T. and Stephens, R. L. (2016). Reduced prevalence of obesity in 14 disadvantaged black communities in the United States: a successful 4-year place-based participatory intervention. *American Journal of Public Health*, 106(8), pp. 1442–1448.

Lofters, A. K., Gozdyra, P. and Lobb, R. (2013). Using geographic methods to inform cancer screening interventions for South Asians in Ontario, Canada. *BMC Public Health*, 13, p. 395.

Lopez, P. M., Islam, N., Feinberg, A., Myers, C., Seidi, L., Drackett, E., Riley, L., Mata, A. et al. (2017). A place-based community health worker program: feasibility and early outcomes, New York City, 2015. *American Journal of Preventative Medicine*, 52(3S3), pp. S284–S289.

Minaker, L. M., Mah, C. L. and Cook, B. E. (2017). Food retail environments in Canada: evidence, framing and promising interventions to improve population diet. In: D. W. Harrington, S. McLafferty and S. J. Elliot, eds., *Population health intervention research: geographical perspectives*. New York: Routledge, pp. 150–162.

Nowell, B. L., Berkowitz, S. L., Deacon, Z. and Foster-Fishman, P. (2006). Revealing the cues within community places: stories of identity, history, and possibility. *American Journal of Community Psychology*, 37(1/2), pp. 29–46.

Pies, C., Barr, M., Strouse, C., Kotelchuck, M. and Best Babies Zone Initiative Team. (2016). Growing a best babies zone: lessons learned from the pilot phase of a multi-sector, place-based initiative to reduce infant mortality. *Maternal and Child Health Journal*, 20(5), pp. 968–973.

Riley, B., Harvey, J., Di Ruggiero, E. and Potvin, L. (2015). Building the field of population health intervention research: the development and use of an initial set of competencies. *Preventative Medicine Reports*, 2, pp. 854–857.

Riley, B. L., Robinson, K. L., Taylor, S. M. and Willis, C. D. (2017). Partnerships for population health improvement: geographic perspectives. In: D. W. Harrington, S. McLafferty, and S. J. Elliot, eds., *Population health intervention research: geographical perspectives*. New York: Routledge, pp. 10–25.

Rosenberg, D., Kerr, J., Sallis, J. F., Patrick, K., Moore, D. J. and King, A. (2009). Feasibility and outcomes of a multi-level place-based walking intervention for seniors: a pilot study. *Health & Place*, 15, pp. 173–179.

Sallis, J. F., Bauman, A. and Pratt, M. (1998). Environmental and policy interventions to promote physical activity. *American Journal of Preventative Medicine*, 15(4), pp. 379–397.

Shoveller, J., Viehbeck, S., Di Ruggiero, E., Greyson, D., Thomson, K. and Knight, R. (2016). A critical examination of representations of context within research on population health interventions. *Critical Public Health*, 26(5), pp. 487–500.

Snyder, R. E., Jaimes, G., Riley, L. W., Faerstein, E. and Coburn, J. (2013). A comparison of social and spatial determinants of health between formal and informal settlements in a large metropolitan setting in Brazil. *Journal of Urban Health: Bulletin of the New York Academy of Medicine*, 91(3), pp. 432–445.

Storey-Kuyl, M., Bekemeier, B. and Conley, E. (2015). Focusing "upstream" to address maternal and child health inequities: two local health departments in Washington State make the transition. *Maternal and Child Health Journal*, 19, pp. 2329–2335.

Vine, M. (2017). Exploring the implementation process of a school nutrition policy in Ontario, Canada: using a health geography lens. In: D. W. Harrington, S. McLafferty and S. J. Elliot, eds., *Population health intervention research: geographical perspectives*. New York: Routledge, pp. 135–149.

Williams, A. (1997). Canadian urban aboriginals: a focus on Aboriginal women in Toronto. *Canadian Journal of Native Studies*, 17(1), pp. 75–101.

Williams, A. (2010). Evaluating Canada's compassionate care benefit using a utilization focused evaluation framework: successful strategies and prerequisite conditions. *Evaluation & Program Planning*, 33(2), pp. 91–97.

Williams, A., Holden, B., Muhajarine, N., Waygood, K. and Spence, C. (2008). Knowledge translation strategies in a community-university partnership: examining local quality of life (QOL). *Social Indicators Research*, 85(1), pp. 111–125.

Williams, A., Kitchen, P., DeMiglio, L., Newbold, B., Eyles, J. and Streiner, J. (2010). Sense of place in Hamilton, Ontario, Canada: empirical results of a neighborhood-based survey. *Urban Geography*, 31(7), pp. 905–931.

Williams, A., Tompa, E., Lero, D., Fast, J., Yazdani, A. and Zeytinoglu, I. (2017). Evaluation of caregiver-friendly workplace policy (CFWPs) interventions on the health of full-time caregiver employees (CEs): implementation and cost-benefit analysis. *BMC Public Health*, 17(1), p. 728.

52

PRACTITIONER PERSPECTIVES

The case of nursing geographies

Elizabeth Peter and Joan Liaschenko

Health-care professionals have a long history of attending to concepts and theory related to health geography, often in more recent years explicitly drawing on theory in health geography. This chapter will focus on the history of nursing's contributions to and applications of health geography, which have uncovered the importance of place and space to the delivery and receipt of health-care services. We specifically focus on nursing because of nurses' unique perspectives and our own experience of using health geography in our scholarship. Nurses' insights are often the product of their lengthy observations of and experiences in various settings expressly as practitioners, along with their empirical studies of their patients and fellow colleagues. This work has added to the understanding of how communities, homes and hospitals, as complex social and cultural phenomena, are experienced by those who live, receive care and work in them.

We will describe three waves of inquiry that have characterized the intersection of nursing's scholarly endeavors with geography (Andrews, 2016; Kyle et al., 2016). In the first wave, the focus was on the environment, as identified by Nightingale (1859), and then later in the 20th century, environment became a meta-concept in nursing theory (Kyle et al., 2016). In the second wave, attention turned to nurses' work environments (Kyle et al., 2016). While the first and second waves do not explicitly make mention of health geography, they make reference to the importance of space and place in understanding disease and well-being and develop a number of related concepts that develop the notion of the work environment. We then will examine the third wave, in which scholars in nursing drew on human geography directly (Kyle et al., 2016). These studies have focused on interactions and specific kinds of interventions, often with a focus on ethics, power and decision-making, revealing the agency of places. Although space does not allow for a full portrayal of the breadth and depth of nurses' geographical work, we provide a snapshot of this scholarship, examining the interplay between place and nurse-patient relationships and between place and the outcomes of care.

History: the first wave

Nightingale (1859), arguably the founder of modern nursing, made multiple references to place and space in her work, especially in her book *Notes on Nursing: What It Is and What It Is Not* (Andrews, 2003). She made particular reference to hospitals and homes that she believed had a tremendous impact on health, describing the importance of clean air, pure water, light, warmth, cleanliness and efficient drainage. Not only did she believe that these environmental characteristics had an impact on decreasing the spread of infectious diseases, such as smallpox, scarlet fever and diphtheria, but also she believed that they could improve a patient's well-being and recovery (Nightingale, 1859). The importance of providing stimulation to patients through careful attention to the sick room by providing fresh flowers, attending to color and supplying familiar objects also revealed Nightingale's understanding of the significance of place.

On a macro-environmental scale, Nightingale (1859) identified the significance of spatial variations in morbidity and mortality. She recognized that the infant death rate in England was highly dependent on place of birth when she remarked that many more children under five years of age were dying in urban areas because of a lack of sanitary conditions. Commenting on her remarkable insights at that time, Andrews (2003) states, "Notably, spatial inequalities in health, and their reduction, were eventually to concern the sub-discipline of medical/health geography more than 100 years later, and to the present day" (p. 271). Nightingale's (1859) work closely followed John Snow's famous 1854 Broad Street map of cholera in London that illustrated that it was contaminated water, not air, that was the source of transmission (recognized as iconic in spatial public health and medical geography) (Shiode et al., 2015).

The environment as a meta-concept

A focused attention to the environment reemerged in late-20th-century American nursing scholarship. At the time, increasing attention was being given to identifying and explicating nursing as a unique discipline with distinct boundaries. A metaparadigm for nursing was developed to reflect the broad values, research methods and practice approaches of the discipline, which came to be represented by four central concepts: person, environment, health and nursing (Fawcett, 1989). "Environment" referred to the nursing-care recipient's "significant others and surroundings, as well as to the setting in which nursing actions occur" (Fawcett, 1989, p. 6). A number of theorists, such as Martha Rogers, Myra Levine, Jean Watson, Betty Neuman and Sister Callista Roy, developed their own unique understandings of what the concept of environment entailed in their theoretical frameworks. These theorists not only influenced the practice of nursing for a generation, but also were critical to debates surrounding the distinction of nursing as an academic discipline and an autonomous profession. While not engaging in the scholarship of health geography proper, these theorists also broadened Nightingale's work by more fully including the social and cultural aspects of care settings. Nevertheless, the lack of consensus and an inadequate conceptualization of space and place (Andrews and Moon, 2005) limited the continued use of their work, including their conceptualizations of the environment.

Nursing-work environments: the second wave

By the 1990s, increasing attention came to focus on nursing-work environments (Kyle et al., 2016) that centered not only on the patient, but also on nurses and their health and well-being. This burgeoning area of scholarship led to best-practice guidelines, such as the one created by the Registered Nurses' Association of Ontario (2008), who systematically reviewed a large body of evidence and further developed the concept of the work environment, articulating its complex and multidimensional character as including physical and policy components, professional factors and cognitive, social and cultural components. In his extensive review of geographical thinking in nursing, Andrews (2016) notes how the work-environment scholarship not only contributed to the understanding of the importance of a good layout and physical design, but also led to the recognition of the significance of empowerment, as well as psychological and moral health. The leadership within these environments is central to achieving these attributes, along with the development of high-performing professionals and the implementation of evidence-based practice (Andrews, 2016).

Moral environments

Attention to work environments has also been further refined to focus on their characteristics that affect the moral responsibilities of those who inhabit them. A variety of terms, including "moral environment" and "moral climate," have been used to portray the array of dimensions describing the complex interface between the moral agency of nurses, along with other health-care providers, and the moral particularities of the environment. Notable examples of this work (Peter, MacFarlane and O'Brien-Pallas, 2004; Vanderheide, Moss and Lee, 2013) draw on the ideas of *moral geography* and *moral habitability* by philosopher Margaret Urban Walker (1998), whose work conceives of moral environments, or *social-moral orders*, which can be critically scrutinized to determine whether they are morally habitable. While not making use of the geography literature, the work-environment and moral-environment developments in nursing benefited from incorporating knowledge from literature found in business, health services, sociology and philosophy to bring further depth and rigor to spatial concepts.

Geographies in nursing: the third wave

Although nursing scholarship has a lengthy history of interdisciplinarity, drawing not only on the basic sciences, but also on the social sciences and humanities, the inclusion of geographical scholarship has been a relative latecomer (Andrews, 2016). An explicit geographical approach in nursing has the benefit of drawing and building on the long and distinct history of geography, with its attention to the person-environment relationships that allows an analysis of the different experiences, negotiations and identities that patients, families, nurses and other health-care providers have of environment (Andrews and Moon, 2005).

Home care

The pioneering work of Liaschenko in the 1990s began the most comprehensive work in nursing to employ geographical scholarship. In a paper entitled "The Moral Geography of Home Care," Liaschenko (1994) introduced concepts from human geography to illuminate the moral concerns of nurses working in home care. She began "an exploration of the spatial organization of nursing in relation to the moral work of protecting patient agency" (p. 18). She argued that helping foster and protect a patient's agency under conditions of vulnerability was a central feature of the moral work of nursing. But importantly, how agency – both the patient's and the nurse's – could be enacted depended on where the care was delivered: the hospital or the home. While focusing on these two environments, she moved beyond the earlier discussions by Nightingale (1859) of light, ventilation, cleanliness and so forth by showing that places organized social space and, therefore, social relations, power, practices, resources and knowledge necessary to one's ability to act. Ultimately, places were understood as the site of the intersection of personal agency and institutional practices.

As in the work of Magnusson and Lützén (1999), Peter (2003), and Bender et al. (2011), the nurses in Liaschenko's (1994) study understood that while the hospital was their *turf*, the home was the patient's and the nurse was a guest. However, Liaschenko warned that the patient's agency might be impacted as the hand of institutionalized medicine moved more and more into the home. This concern was substantiated by Angus et al. (2005) and Seto-Nielsen et al. (2013), whose research of home-care clients, which used concepts from geography and sociology, demonstrated that the logics and practices of health care disrupted those of the home. Indeed, the monitoring and surveillance of patients that Liaschenko (1994) mentioned have only increased. Data can now be collected continually though surveillance devices and is transmitted in real time to clinicians and hospitals.

In related work, Peter (2002) explored the expression of moral agency as a function of place in her examination of historical documents that describe the experiences of American home-care (private-duty) nurses from 1900 to 1933. She recognized that moral agency is structured in part by place because of its capacity to influence physical activity, authority and meaning. Although this scholarship concerned the home, its implications are relevant to other settings. Because the taken-for-granted setting for contemporary health-care work is generally the hospital, which can be rather uniform in character, place can become invisible to its inhabitants. Studying the home in these ways can make the relevance of place recognizable and offer ways to reflect on all health-care settings.

Hospital care

Research in nursing focused on hospitals has also been advanced through the use of geography. For example, Cheek's (2004) work on older people shed light on acute-care settings and the need to foreground the uniqueness of older patients in these settings, and McKeever, O'Neill and Miller (2002) described the socio-spatial experience of women who cared for their infants in protective isolation in hospital. These mothers developed ways to manage this profound disconnection from their everyday lives by managing the space and themselves, managing their social relations and marking time. Similarly, in their study of children undergoing hospital-based hemodialysis, Zitzelsberger et al. (2014) described how the spatial arrangements resulting from being tethered to the machines constrained the children's agency, limited their privacy, and "magnified their isolation" (p. 116).

Other studies have focused on nurses in hospitals and how space and place have shaped their practice. For instance, Halford and Leonard (2003) compared nurses' and physicians' use of space, discovering that nurses' access of space was limited with less opportunity to roam and less private space available. In another study, conducted by Liaschenko, Peden-McAlpine and Andrews (2011), the purpose of which was to understand how nurses work with dying patients and their families in intensive care units (ICUs), space was central to the dying body and the dying process in three ways. First, when death was imminent, nurses felt a moral imperative to transform ICU space into a *sacred space*, a space more conducive to what they viewed as a *good death*, recognizing, like Gilmour (2006), that hospitals are hybrid spaces that can be transformed into caring and personal environments. Second, the body is mapped into medical spaces that reflect the specialized division of labor within medicine. The result is that while separate organ systems were cared for meticulously, the embodied whole that is the patient was sometimes lost. Third, space is directly tied to what kind of medical practices and procedures can be done where. The significance of this for dying patients was that patients' requests for *do not resuscitate* orders could change from place to place, depending on the procedures they were undergoing.

Long-term care and community settings

Nursing research focusing on the meaning of place in long-term and community settings has benefited greatly from geographical perspectives. Falk et al. (2013) examined the importance of attachment to place for older people to maintain a sense of identity and home when they move into residential care. Understanding aging as not only an embodied process, but also an emplaced process, this study has shed light on nursing interventions that facilitate processes that attend to the significance of space and place. Related work by Andrews and Peter (2006), which recognized autonomy as emplaced, identified nursing homes as potentially restraining because they could impede the capacity of residents to create meaningful lives.

The dynamic between identity and place has also been studied in community settings by nurses. For example, Vandemark (2007) explored the impact of displacement on people who were homeless. Without attachment to place, displaced people were challenged to find a sense of self-identity, belonging and self-efficacy. Similarly, Holmes, O'Byrne and Gastaldo (2007) studied how the architecture and design of gay bathhouses not only shaped sexual practices, but also contributed to the construction of positive gay identities. Although gay bathhouses have been delineated as sites of immorality, they note, bathhouses "provide the context for the creation of political beings" (p. 279) that lead to the development of gay pride and self-acceptance. Their findings reveal the normative force of places to structure social interactions and direct the development of public-health interventions to decrease HIV transmission and prevent other sexually transmitted diseases.

The spatiotemporal positioning of nurses and nurse-patient relationships

Liaschenko (1996, 2003) further discussed the theme of the place of nursing work in relation to the moral work of protecting patient agency. She examined the spatial connotations of the nurse-patient relationship in "Ethics and the Geography of the Nurse-Patient Relationship: Spatial Vulnerabilities and Gendered Space," noting that relationship *is* a spatial concept (Liaschenko, 1997). It is spatial in the sense of occurring *between* people and therefore across space. Relationship understood as this interpersonal bond is the way that nurses typically understand the nurse-patient relationship. But Liaschenko challenged nurses to think of the nurse-patient relationship as going beyond an interpersonal bond. For her, relationship is also the vehicle through which the work of nursing gets done, and work is always organized within a particular physical and social space. For this reason, interpersonal bonds are deeply affected by the politics of particular locales. Liaschenko advocated for an understanding of the nurse-patient relationship that encompassed the dimensionality of scale, from the intimate to the global and structural. While the former refers to the proximity of nurse and patient, the latter refers to their social positioning. Andrews and Peter (2006) described the social positioning of nurses best: "Nurses are located centrally in the circuitry of institutional power, constituting one of its most important and common cogwheels" (p. 5).

The examination of nurses' spatiotemporal positioning using geographical concepts has also yielded unique insights into nurses' work, their identity and their relationship to patients. For example, Malone (2003) demarcated three nested forms of proximity – physical, narrative and moral – which offered an understanding of nurses' opportunity to recognize patients' personal narratives and act on their behalf when nurses were physically near enough to patients to do so. Peter and Liaschenko (2004) further examined the physical proximity of nurses to patients through a moral and geographical lens in order to highlight the moral distress and moral ambiguity common in nursing practice. This work recognized that proximity is paradoxical in that it heightens nurses' moral responsiveness, yet the moral distress that it can create can lead nurses to want to distance themselves from patients. Both Malone (2003) and Peter and Liaschenko (2004) conclude that the current work environments of most nurses, with their corporate cultures, have a negative impact on nurses' positioning within health-care systems.

Limitations and future developments

One of the main limitations of this work in nursing has been a lack of a consistent and continued trajectory of scholarly development in the area of geography. This shortcoming may be limiting its potential to influence not only nursing, but the discipline of health geography as well. Another limitation is the tendency of nursing contributors to represent space and place as disconnected and static phenomena (Andrews, 2016). Geographical work on the impact of computerized information systems on nursing practice, however, may go a long way in addressing this concern because it does not conceptualize place as a static location. For example, Barnes and Rudge (2005) discuss the flow of information not only across "non-contiguous groups" (p. 307) but also across "professional values and allegiances" (p. 307) in order to deliver care. Sandelowski (2002) focuses on virtual environments in health care and the effect on embodiment, noting that the *fleshly bodies* of both patients and clinicians are disappearing. She argues that nurses need to think both "corporeally and informationally" (p. 67). This type of work may be helpful in reimagining future scholarship. In addition, geographic scholarship in nursing would benefit from a deeper understanding of human geography that has explicit relational thinking, such as non-representational theory, to offer new insights into various forms of nursing-driven inquiry (Andrews, 2016). Nevertheless, to date, nursing's relationship with geography has yielded a breadth of scholarship that has shed light and added depth into a wide array of nursing-related phenomena.

References

Andrews, G. J. (2003). Nightingale's geography. *Nursing Inquiry*, 10(4), pp. 270–274.

Andrews, G. J. (2016). Geographical thinking in nursing inquiry, part one: locations, contents, meanings. *Nursing Philosophy*, 17, pp. 262–281.

Andrews, G. J. and Moon, G. (2005). Space, place, and the evidence base: part II – rereading nursing environment through geographical research. *Worldviews on Evidence-Based Nursing*, 2(3), pp. 142–156.

Andrews, G. J. and Peter, E. (2006). Moral geographies of restraint in nursing homes. *Worldviews on Evidence-Based Nursing*, 3(1), pp. 2–7.

Angus, J., Kontos, P., Dyck, I., McKeever., P. and Poland, B. (2005). The personal significance of home: habitus and the experience of receiving long-term home care. *Sociology of Health and Illness*, 27(2), pp. 161–187.

Barnes, L. and Rudge, T. (2005). Virtual reality or real virtuality: the space of flows and nursing practice. *Nursing Inquiry*, 12(4), pp. 306–315.

Bender, A., Peter, E., Wynn, F., Andrews, G. and Pringle, D. (2011). Welcome intrusions: an interpretive phenomenological study of TB nurses' relational work. *International Journal of Nursing Studies*, 48(11), pp. 1409–1419.

Cheek, J. (2004). Older people and acute care: a matter of place? *Illness, Crisis and Loss*, 12(1), pp. 52–62.

Falk, H., Wijk, H., Persson, L. and Falk, K. (2013). A sense of home in residential care. *Scandinavian Journal of Caring*, 27(4), pp. 999–1009.

Fawcett, J. (1989). *Analysis and evaluation of conceptual models of nursing*. 2nd ed. Philadelphia: Davis.

Gilmour, J. A. (2006). Hybrid space: constituting the hospital as a home space for patients. *Nursing Inquiry*, 13(1), pp. 16–22.

Halford, S. and Leonard, P. (2003). Space and place in the construction and performance of gendered nursing identities. *Journal of Advanced Nursing*, 42(2), pp. 201–208.

Holmes, D., O'Byrne, P. and Gastaldo, D. (2007). Setting the space for sex: architecture, desire and health issues in gay bathhouses. *International Journal of Nursing Studies*, 44(2), pp. 273–284.

Kyle, R. G., Atherton, I. M., Kesby, M., Sothern, M. and Andrews, A. (2016). Transfusing our lifeblood: reframing research impact through inter-disciplinary collaboration between health geography and nurse education. *Social Science & Medicine*, 168, pp. 257–264.

Liaschenko, J. (1994). The moral geography of home care. *Advances in Nursing Science*, 17, pp. 16–25.

Liaschenko, J. (1996). A sense of place for patients: living and dying. *Home Care Provider*, 1, pp. 270–272.

Liaschenko, J. (1997). Ethics and the geography of the nurse-patient relationship: spatial vulnerabilities and gendered space. *Scholarly Inquiry for Nursing Practice*, 11, pp. 45–59.

Liaschenko, J. (2003). At home with illness: moral understandings and moral geographies. *Ethica*, 15(2), pp. 71–82.

Liaschenko, J., Peden-McAlpine, C. and Andrews, G. J. (2011). Institutional geographies in dying: nurses' actions and observations on dying spaces inside and outside intensive care units. *Health & Place*, 17(3), pp. 814–821.

Magnusson, A. and Lützén, K. (1999). Intrusion into patient privacy: a moral concern in the home care of persons with chronic mental illness. *Nursing Ethics*, 6(5), pp. 399–410.

Malone, R. (2003). Distal nursing. *Social Science & Medicine*, 56, pp. 2317–2326.

McKeever, P., O'Neill, S. and Miller, K. L. (2002). Managing space and marking time: mothering severely ill infants in hospital isolation. *Qualitative Health Research*, 12(8), pp. 1071–1083.

Nightingale, F. (1859). *Notes on nursing: what it is and what it is not*. London: Lippincott.

Peter, E. (2002). The history of nursing in the home: revealing the significance of place in the expression of moral agency. *Nursing Inquiry*, 9(2), pp. 65–72.

Peter, E. (2003). The home as a site of care: ethical implications for professional boundaries and privacy. *Ethica*, 15(2), pp. 57–69.

Peter, E. and Liaschenko, J. (2004). Paradoxes of proximity: a spatio-temporal analysis of moral distress and moral ambiguity. *Nursing Inquiry*, 11(4), pp. 218–225.

Peter, E., MacFarlane, A. V. and O'Brien-Pallas, L. L. (2004). Analysis of the moral habitability of the nursing work environment. *Journal of Advanced Nursing*, 47(4), pp. 356–364.

Registered Nurses' Association of Ontario. (2008). *Workplace health, safety and well-being of the nurse guideline*. Toronto: Registered Nurses' Association of Ontario.

Sandelowski, M. (2002). Visible humans, vanishing bodies, and virtual nursing: complications of life, presence, place, and identity. *Advances in Nursing Science*, 24(3), pp. 58–70.

Seto-Nielsen, L. S., Angus, J. E., Gastaldo, D., Howell, D. and Husain, A. (2013). Maintaining distance from a necessary intrusion: a postcolonial perspective on dying at home for Chinese immigrants in Toronto, Canada. *European Journal of Oncology Nursing*, 17(5), pp. 649–656.

Shiode, N., Shiode, S., Rod-Thatcher, E., Rana, S. and Vinten-Johansen, P. (2015). The morality rates and the space-time patterns of John Snow's cholera epidemic map. *International Journal of Health Geographics*, 14, p. 21.

Vandemark, L. M. (2007). Promoting the sense of self, place, and belonging in displaced persons: the example of homelessness. *Archives of Psychiatric Nursing*, 21(5), pp. 241–248.

Vanderheide, R., Moss, C. and Lee, S. (2013). Understanding moral habitability: a framework to enhance the quality of the clinical environment as a workplace. *Contemporary Nurse*, 45(10), pp. 101–113.

Walker, M. U. (1998). *Moral understandings: a feminist study in ethics*. New York: Routledge.

Zitzelsberger, H., McKeever, P., Peter, E., Chambon, A., Morgan, K. P. and Spalding, K. (2014). Doing "technological time" in a pediatric hemodialysis unit: an ethnography of children. *Health & Place*, 27, pp. 112–119.

INDEX

Note: Page numbers in *italic* indicate figures and in **bold** indicate tables on the corresponding pages.

Printed and bound by CPI Group (UK) Ltd, Croydon, CR0 4YY

30/10/2024

01781414-0003